T0342494

Mathematical Tools for Understanding Infectious Disease Dynamics

PRINCETON SERIES
IN THEORETICAL AND COMPUTATIONAL BIOLOGY

Series Editor, Simon A. Levin

Mathematical Tools for Understanding Infectious Disease Dynamics,
by Odo Diekmann, Hans Heesterbeek, and Tom Britton

The Calculus of Selfishness,
by Karl Sigmund

The Geographic Spread of Infectious Diseases: Models and Applications,
by Lisa Sattenspiel with contributions from Alun Lloyd

Theories of Population Variation in Genes and Genomes,
by Freddy Bugge Christiansen

Analysis of Evolutionary Processes,
by Fabio Dercole and Sergio Rinaldi

Mathematics in Population Biology,
by Horst R. Thieme

Individual-based Modeling and Ecology,
by Volker Grimm and Steven F. Railsback

Mathematical Tools for Understanding Infectious Disease Dynamics

Odo Diekmann, Hans Heesterbeek, and Tom Britton

PRINCETON UNIVERSITY PRESS

PRINCETON AND OXFORD

Published by Princeton University Press
41 William Street, Princeton, New Jersey 08540

In the United Kingdom: Princeton University Press
6 Oxford Street, Woodstock, Oxfordshire, OX20 1TW

Library of Congress Cataloging-in-Publication Data

 Diekmann, O.
    Mathematical tools for understanding infectious disease dynamics / Odo Diekmann,
Hans Heesterbeek, and Tom Britton.
       p. cm. – (Princeton series in theoretical and computational biology)
    Includes bibliographical references and index.
    ISBN 978-0-691-15539-5 (hardback)
    1. Epidemiology–Mathematical models–Congresses. 2. Epidemiology–Mathematical
models. 3. Communicable diseases–Mathematical models. I. Heesterbeek, Hans, 1960-
II. Britton, Tom. III. Title.
    RA652.2.M3D54 2013
    614.4–dc23

                                                               2012012058

British Library Cataloging-in-Publication Data is available

This book has been composed in LaTeX

The publisher would like to acknowledge the authors of this volume for providing the
camera-ready copy from which this book was printed.

Printed on acid-free paper. ∞

press.princeton.edu

Printed in the United States of America

10 9 8 7 6 5 4 3 2 1

I simply wish that, in a matter which so closely concerns the well-being of the human race, no decision shall be made without all knowledge which a little analysis and calculation can provide.

Daniel Bernoulli, 1760, on smallpox inoculation

As a matter of fact all epidemiology, concerned as it is with variation of disease from time to time or from place to place, *must* be considered mathematically (...) and the mathematical method of treatment is really nothing but the application of careful reasoning to the problems at hand.

Sir Ronald Ross, 1911, *The Prevention of Malaria*

We shall end by establishing a new science. But first let you and me unlock the door and then anybody can go in who likes.

Sir Ronald Ross in a letter to A.G. McKendrick, 1911

Contents

Preface

This book builds on two previous books on the same topic by the same set of authors (plus one). We feel it is important, right from the start, to make clear how the new book and the old books are related. Both the previous books appeared more than 10 years ago: *Mathematical Epidemiology of Infectious Diseases: model building, analysis and interpretation*, Diekmann and Heesterbeek, John Wiley & Sons, 2000; and: *Stochastic Epidemic Models and their Statistical Analysis*, H. Andersson and Britton Springer-Verlag, 2000. The first took a textbook approach to predominantly deterministic modeling — at least deterministic at the population level, but allowing for stochasticity at the level of individuals. The second had more a monograph-like approach to predominantly mathematical and statistical analysis of stochastic epidemic systems — concentrating on analysis, rather than model building. The present book is based on these two earlier volumes, and in fact makes both of them obsolete. It replaces them with a textbook in the spirit of 'Diekmann and Heesterbeek,' and the result is, in our (admittedly biased, but humble) opinion, more valuable than the sum of its parts. The new book integrates the deterministic and stochastic theory and approaches, rather than merely merging the old versions, treating both deterministic and stochastic modeling and analysis of infectious disease dynamics. New topics have been added, and for most topics already treated in one of the predecessors the text has been updated, or revised to improve exposition or integration.

We do not see our book as a mathematics monograph in the sense of instilling in the reader the beauty of the mathematical subject and prove theorems. The value of our book, in our view, is not in doing rigorous mathematics in 'theorem-proof style,' and also not in highlighting 'deep problems' from a mathematical point of view. The value of the book lies in showing how to be very precise in modeling phenomena in infectious disease dynamics, using mathematical reasoning and analysis. Mathematics is the tool, not the aim. We feel that for our aim the narrative style of doing mathematics is much more efficient in getting the message across. Our aim is to be very rigorous in the *modeling*. If we are being 'missionary' at all, it is in trying to get across what (often hidden) assumptions lie behind choices and concepts in modeling, what the consequences are of these choices, and how superficially different concepts are related. The book is about *translating* assumptions concerning biological (behavioral, immunological, demographical, medical) aspects into mathematics, about mathematical *analysis* of certain classes of equations aided by interpretation, about inference from data (measurements, observations), and finally about the *drawing of conclusions* where results from the mathematical and statistical analysis are translated back into biology. We try to offer insight into the relation between assumed mechanisms at the individual level and the resulting phenomena at the population level, both for 'small' and 'large' populations, and the grey area that lies in between.

Some books offer wisdom. They can be read at leisure in an armchair near a fireplace, provided one pauses every now and then for contemplation. This is not such a book. This book has a zillion exercises and begs to be read with pencil and paper at hand (or perhaps, in a more modern way, using a computer with a program for symbolic manipulation). Some of the exercises one may want to read simply to see what statements they concern. This reading is essential, since usually the exercises are an integral part of the exposition. For many exercises, however, mere reading is not enough: one actually needs to do them. Learning to translate, model, analyze and interpret involves training. Some exercises are ridiculously simple since we have tried not to omit arguments or to tire the reader with details that interrupt the exposition too much; where other writers would state 'one easily sees' or 'a simple argument shows,' etc., we have inserted an exercise. Other exercises, however, are difficult and elaborate. Many exercises point the reader to caveats, pitfalls, and to similarities and differences in concepts. We anticipate that our readers will feel at times frustrated or even irritated. We therefore provide complete elaborations of all exercises, even of the 'ridiculously simple ones,' as an integral part of the book. When a specific exercise seems beyond reach, we advise the reader to only glance at the elaboration as a kind of hint and then try again.

The authors are not sadists who like to pester their readers with exercises, even though it may sometimes feel like that. We are convinced that the reward is enormous. In literally working through this book the reader acquires modeling skills that are also valuable outside of epidemiology, certainly within population dynamics, but even beyond that. The reader receives training in mathematical argumentation, deterministic and stochastic modeling, analysis and inference.

Our hope is that the applied mathematicians learn to see i) the subtleties of model assumptions; ii) that continuous-time models not necessarily take the form of a system of ordinary differential equations; and iii) that often biological interpretation suggests how to proceed with the mathematical analysis.

Our hope is that the theoretical biologists and epidemiologists i) enlarge their tool kit considerably; ii) conclude that sometimes abstraction may actually make things simpler and more transparent; and iii) are inspired by the book to delve deeper into the mathematical tools used.

Our ideal reader feels attracted by these educational aims.

A BRIEF OUTLINE OF THE BOOK

This book is divided into four parts and 18 chapters. In Part I, we shall introduce the key questions, basic ideas, fundamental concepts and mathematical arguments in as simple a context as possible. This entails in particular that we treat all host individuals as identical with respect to behavior and physiology and that we deal with such concepts as thresholds, final sizes for epidemics, repeated outbreaks, the endemic state and population regulation, aimed both at small and large populations. In this simplest setting we also introduce methods to relate the simple models to data for inference.

When the host population is heterogeneous, we need more advanced mathematics, both for small and large populations. To describe the initial phase of epidemic spread, we can restrict attention to linear mathematics and a systematic approach is possible. The theory, with many examples, is presented in Part II. In addition we pay some (but not much) attention to nonlinear aspects in a general setting. We shall pay more attention to age structure and spatial structure in separate chapters, because these are particularly relevant for understanding of the population dynamics of many infective agents. To analyze nonlinear structured models one is often forced to make debatable simplifying assumptions. Even then, one needs to resort to tricks, for lack of a powerful general theory. We therefore do not forage deeply into nonlinear theory. For most of the examples in the book we have those infective agents in mind that are usually collectively called 'microparasites,' but in Chapter 11 we briefly touch upon some aspects where 'macroparasites' differ from 'microparasites' (and where they do not), and concentrate on the consequences that these differences and agreements have for the mathematical treatment of invasion. In the final chapter of Part II we pay attention to one of the fundamental and conceptually most difficult aspects of epidemic theory: the myriad ways in which one can model contacts between individuals (Chapter 12).

Part III consists of three chapters. Chapter 13 presents a selection of methods to estimate a value of the basic reproduction number R_0 from a variety of available data. Chapter 14 is a case study and shows in detail how to model the dynamics of a pathogen in a very small dynamic population (nosocomial infections in an Intensive Care Unit of a hospital) based on the type of data that will be routinely available. In Chapter 15 we briefly review computer-intensive statistical methods, that go beyond the methods of inference treated earlier in the book.

Part IV consists, as a consequence of our educational 'philosophy,' of complete elaborations for all exercises in three chapters, one for each Part of the book. These elaborations are detailed and sometimes lengthy, and in this way often serve as an extension of the main text. This makes the elaborations an integral part of the book.

It is good to point out that, as another consequence of our educational aim, this book is not an easy reference book in the sense that it can be used to quickly look up certain concepts and definitions or specific results. We have tried to help the reader somewhat in finding relevant information on specific topics of interest by providing a detailed index and by reiterating some concepts and notation in various places.

A final remark concerns our way of referring to the literature. The literature of epidemic theory is extensive and growing steadily. It would be very difficult, bordering on the impossible, to do justice to all valuable contributions to the literature. We have deliberately chosen to write a textbook and not a review of state-of-the-art epidemic theory. As a consequence we have two types of references: local specialist literature (mostly papers) and global general texts for further reading (books). The local literature is included in places where it is necessary for the exposition at that point and is given in footnotes. The global references are given in a short bibliography near the end of the book. They are ordered thematically and include background mathematical reading. To a large extent the choices of books we recommend for background are personal; in the current age we feel that a first impression of an unfamiliar mathematical concept or method can be obtained reliably from the internet, where notably Wikipedia is a good initial source.

The spirit of our view of modeling is captured by the following quote attributed to Picasso:

"Art is the lie that helps us to discover the truth."

Much, if not all, of our insight has been derived from reading the rich literature and notably from collaborations and discussions with many excellent scientists, colleagues and students over a period of many years. We cannot possibly list them all. We specifically want to mention (in no particular order): Hans Metz, Martin Bootsma, Mick Roberts, Håkan Andersson, Mirjam Kretzschmar, Karl Hadeler, Fred Brauer, Mart de Jong, Marc Bonten and Barbara Boldin. In addition, we want to thank the University of Utrecht, Stockholm University, the Netherlands Organization for Scientific Research (NWO) and the Swedish Research Council for financial support during the many years it took us to complete this book.

Part I

The bare bones: Basic issues in the simplest context

Chapter One

The epidemic in a closed population

1.1 THE QUESTIONS (AND THE UNDERLYING ASSUMPTIONS)

In general, populations of hosts show demographic turnover: old individuals disappear by death and new individuals appear by birth. Such a demographic process has its characteristic time scale (for humans on the order of 10 years). The time scale at which an infectious disease sweeps through a population is often much shorter (e.g., for influenza it is on the order of weeks). In such a case we choose to ignore the demographic turnover and consider the population as 'closed' (which also means that we do not pay any attention to emigration and immigration).

Consider such a closed population and assume that it is 'virgin' or 'naive,' in the sense that it is completely free from a certain infectious agent in which we are interested. Assume that, in one way or another, the infectious agent is introduced in at least one host. We may ask the following questions:

- Does this cause an epidemic?

- If so, at what rate does the number of infected hosts increase during the rise of the epidemic?

- What proportion of the population will ultimately have experienced infection?

Here we assume that we deal with *microparasites*, which are characterized by the fact that a single infection triggers an autonomous process in the host. We assume in addition that this process finally results in either death or lifelong immunity, so that no individual can be infected twice (this assumption is somewhat implicitly contained in the formulation of the third question).

In order to answer these questions, we first have to formulate assumptions about transmission. For many diseases transmission can take place when two hosts 'contact' each other, where the meaning of 'contact' depends on the context (think of 'mosquito biting man' for malaria, sexual contact for gonorrhea, traveling in the same bus for influenza, SARS, ...) and may, in fact, sometimes be a little bit vague (for fungal plant diseases transmitted through air transport of spores it is even far-fetched to think in terms of 'contact'). It is then helpful to follow a three-step procedure:

- Model the contact process.

- Model the mixing of susceptible and infective (i.e., infectious) individuals (which we shall refer to as 'susceptibles' and 'infectives,' respectively); that is, specify what fraction of the contacts of an infective are with a susceptible, given the population composition in terms of susceptibles and infectives.

- Specify the probability that a contact between an infective and a susceptible actually leads to transmission.

As an easy phenomenological approach to the first step we assume for the time being that individuals have *a certain expected number c of contacts per unit of time* with other individuals. So we postpone more mechanistic reasoning, and in particular a discussion of how c may relate to population size and/or density.

1.2 INITIAL GROWTH

1.2.1 Initial growth on a generation basis

During the initial phase of a potential epidemic, there are only a few infected individuals amidst a sea of susceptibles. So if we focus on an infected individual we may simply assume that all its contacts are with susceptibles. This settles the second step in the procedure sketched in Section 1.1.

For many diseases the probability that a contact between a susceptible and an infective actually leads to transmission depends on the time elapsed since the infective was itself infected. To be specific, let us assume that this probability equals

$$\begin{cases} 0 & \text{if} & \tau < T_1, \\ p & \text{if} & T_1 \leq \tau \leq T_2, \\ 0 & \text{if} & T_2 < \tau, \end{cases}$$

where τ denotes the *infection age* (i.e., the time since infection took place), $0 < p \leq 1$, and where we have assumed that there is a latency period (i.e., the period of time between becoming infected and becoming infectious) of length T_1 followed by an infectious period of length $T_2 - T_1$. (What happens at the end of the infectious period is unspecified at this point; it may be that the host dies or it may be that its immune system managed to 'defeat' the agent, with a then-immune host surviving the infection; we shall come back to this point later on.)

In order to distinguish between avalanche-like growth and almost-immediate extinction, we introduce the *basic reproduction number* (or basic reproduction ratio):

$$R_0 \quad := \quad \text{expected number of secondary cases per primary case}$$
$$\text{in a 'virgin' population.}$$

In other words, R_0 is the initial *growth rate* (more accurately: multiplication factor; note that R_0 is dimensionless) when we consider the population on a *generation basis* (with 'infecting another host' likened to 'begetting a child'). Consequently, R_0 has threshold value 1, in the sense that an epidemic will result from the introduction of the infectious agent when $R_0 > 1$, while the number of infecteds is expected to decline (on a generation basis) right after the introduction when $R_0 < 1$. The advantage of measuring growth on a generation basis is that for many models one has an explicit expression for R_0 in terms of the parameters. Indeed, from the assumptions above, we find

$$R_0 = pc(T_2 - T_1) \qquad (1.1)$$

where c is the contact rate introduced in Section 1.1.

We conclude that whether or not the introduction of an infectious agent leads to an epidemic explosion is determined by the value of the generation multiplication factor R_0 relative to the threshold value one. At least for simple sub-models for the contact process and infectivity, one can determine R_0 explicitly in terms of parameters of these sub-models.

Exercise 1.1 Female mosquitoes strive for a fixed number of blood meals per unit of time, in order to be able to lay eggs. Show that consequently the mean number of bites that *one* human receives per unit of time is proportional to $D_{\text{mosquito}}/D_{\text{human}}$, i.e., to the ratio of the two densities D.

Exercise 1.2 Consider one infected mosquito. Assume it stays infected for an expected period of time T_m during which it bites (different) people at a rate c. Assume that each bite results in successful transmission with probability p_m. How many people is this mosquito expected to infect?

Exercise 1.3 Consider one infected human. Assume it stays infected for an expected period of time T_h during which it is bitten by (different) mosquitoes at a rate k. Let each bite result in successful transmission with probability p_h. How many mosquitoes is this human expected to infect?

Exercise 1.4 Argue that for the above crude description of malaria transmission the quantity

$$c^2 T_m T_h p_m p_h \frac{D_{\text{mosquito}}}{D_{\text{human}}}$$

is a threshold parameter with threshold value 1. Spell out the meaning of 'threshold parameter' in some detail.

1.2.2 The influence of demographic stochasticity

Within the idealized deterministic description of infection transmission, we found in the preceding subsection that a newly introduced infectious agent starts to spread when $R_0 > 1$, while going extinct more or less immediately when $R_0 < 1$. However, for the deterministic description to be warranted, we need not only a large number of susceptibles but also a large number of infectives. Yet the very essence of the *introduction* of the agent is that it is present in only a few hosts. So we need to take account of demographic stochasticity, i.e., the chance fluctuations associated with the fact that individual hosts are discrete units, counted with integers and either infected or not (rather than fractionally). Only when the infectives form a small *fraction* of the large population, and not just a small number, does the deterministic description apply.

To refine the analysis, we need branching processes, as introduced below, but we keep counting on a generation basis. We postpone more precise formulations, and the development in real time, to Section 3.3. As infecting another host is very similar to producing offspring, we shall freely use the standard terminology of true reproduction, even though it does not apply literally to the context of disease spread that we consider here. We repeat that the number of susceptibles is assumed to be very large so that we can, in the initial phase of an epidemic, neglect the depletion of susceptibles by their conversion into infectives. So the finite population we are going to consider in the following is the sub-population of infected individuals.

Consider this finite population from a generation perspective and assume that individuals reproduce independently from each other, the number of offspring for each being taken from the same probability distribution $\{q_k\}_{k=0}^{\infty}$. This means that any individual begets k offspring with probability q_k and that $\sum_{k=0}^{\infty} q_k = 1$. We note right away that the expected number of offspring R_0 can be found from $\{q_k\}$ as

$$R_0 = \sum_{k=1}^{\infty} kq_k. \tag{1.2}$$

Exercise 1.5 To warm up, we consider the situation in which an individual has either two descendants, one descendant, or no offspring at all. In other words: $q_k = 0$ for $k \geq 3$. We exclude the uninteresting case $q_1 = 1$ (the results are not valid in that case).

i) Show that $R_0 > 1$ if and only if $q_2 > q_0$;

ii) Consider one individual. Let z be the probability that its line of descent will stop (sooner or later). Next consider two individuals. What is the probability that the line of descent of both of them will stop?

iii) Explain why z should satisfy the equation

$$z = q_0 + q_1 z + q_2 z^2.$$

iv) Rewrite this equation such that it becomes easy to see that either $z = 1$ or $z = \frac{q_0}{q_2}$.

v) Show that $z = 1$ if $R_0 \leq 1$. Does this surprise you?

vi) What do you expect z to be if $R_0 > 1$?

To deal with the general situation we introduce, as an auxiliary tool, the *(probability) generating function g* defined by

$$g(z) = \sum_{k=0}^{\infty} q_k z^k, \quad 0 \leq z \leq 1. \tag{1.3}$$

(Recall the convention $z^0 = 1$.)

Exercise 1.6 Check that

i) $g(0) = q_0$,

ii) $g(1) = 1$,

iii) $g'(1) = R_0$,

iv) $g'(z) > 0$,

v) $g''(z) > 0$.

Hint (for 1.6-iv and v): Realize that $q_k \geq 0$, with strict inequality for at least one value of k, which is implicitly contained in the interpretation. Which additional assumptions are needed to make the inequalities strict?

Now assume that $q_0 > 0$, which means that there is a positive probability that an individual will beget no offspring at all. Let us start the process with just one individual. Then clearly q_0 is also the probability that the population will be extinct after one step along the generation ladder.

Let z_n denote the probability that the population will be extinct after n steps along the generation ladder (so z_n equals the probability that the population will go extinct in at most n steps). Then our last observation translates into $z_1 = q_0$. We claim that the z_n can be computed recursively from the difference equation

$$z_n = g(z_{n-1}). \tag{1.4}$$

To substantiate this claim, we argue as follows. If in the first generation there are k offspring, then the lines descending from each of these should go extinct in $n-1$ generation-steps in order for the population to go extinct in or before the nth generation. For each separate line the probability is, by definition, z_{n-1}. By independence, the probability that all k-lines go extinct in $n-1$ steps is then simply $(z_{n-1})^k$. It remains to sum over all possible values of k with the appropriate weight q_k. Thus we find

$$z_n = q_0 + \sum_{k=1}^{\infty} q_k (z_{n-1})^k = g(z_{n-1})$$

as claimed.

Because g is increasing, the sequence z_n must be increasing and so, because it is also bounded by 1, it has a limit $z_\infty = \lim_{n\to\infty} z_n$. By definition z_∞ is the probability that the population started by the first individual will go extinct. We refer to this as the probability of a minor outbreak. When $z_\infty = 1$, the population goes extinct with probability 1. When $0 < z_\infty < 1$, there exists a complementary probability $1 - z_\infty$ that, as further arguments show (see books on branching processes, such as Haccou et al. 2005, Jagers 1975, Mode 1971 and Harris 1963), exponential growth sets in and the deterministic description applies. So we expect that $R_0 < 1$ implies $z_\infty = 1$, while for $R_0 > 1$ the inequality $0 < z_\infty < 1$ holds. That our expectation is correct follows most easily from a graphical consideration, see Figures 1.1 and 1.2.

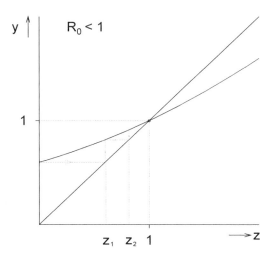

Figure 1.1: Schematic representation of the graphs of the functions $y = z$ and $y = g(z)$ for the case $R_0 < 1$, showing iteration toward the limit $z = 1$.

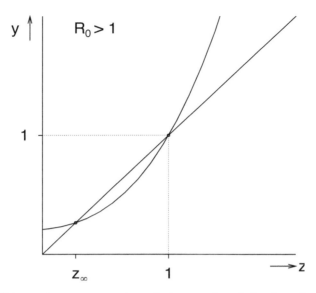

Figure 1.2: Schematic representation of the graphs of the functions $y = z$ and $y = g(z)$ for the case $R_0 > 1$, showing iteration toward the limit $z_\infty < 1$.

Exercise 1.7 Argue that z_∞ is the *smallest* root in $[0, 1]$ of the equation

$$z = g(z) \tag{1.5}$$

(and interpret this equation as a consistency condition that the probability to go extinct should satisfy. Hint: Recall the derivation of the difference equation (1.4)).

Exercise 1.8 Use Exercise 1.6-iii to show analytically that $z_\infty = 1$ for $R_0 < 1$ while $0 < z_\infty < 1$ for $R_0 > 1$. Hint: Also use Exercise 1.6-v.

Exercise 1.9 Determine z_∞ for the critical case $R_0 = 1$.

We conclude that even in the situation where the infectious agent has the potential of exponential growth, i.e., $R_0 > 1$, it still may go extinct due to an unlucky (for the agent) combination of events while numbers are low. The probability that such an extinction happens, when we start out with exactly one primary case, can be computed as a specific root of the equation $z = g(z)$.

But how do we derive the function g (or, equivalently, the probabilities q_k) from the kind of specification of a transmission model that we employed so far? (Recall Sections 1.1 and 1.2.1.)

When during a time interval of length $\triangle T$ contacts are made according to a Poisson process with rate c, the probability that k contacts are made equals

$$\frac{(c\triangle T)^k}{k!} e^{-c\triangle T}.$$

In other words, the number of contacts follows the Poisson distribution with mean parameter $c\triangle T$. When contacts lead to successful transmission with probability p, the number of 'successful' contacts is again Poisson-distributed with the parameter modified to $pc\triangle T$.

Exercise 1.10 Prove the last statement.

The generating function for the Poisson distribution with parameter λ equals

$$g(z) = \sum_{k=1}^{\infty} z^k \frac{\lambda^k e^{-\lambda}}{k!} = e^{\lambda(z-1)} \qquad (1.6)$$

where the first equality is the definition and the second equality follows from the Taylor expansion of the exponential function. Together with $R_0 = pc\triangle T = pc(T_2 - T_1)$ we arrive at the conclusion that for this particular sub-model z_∞ is, for $R_0 > 1$, the unique root in the interval $(0, 1)$ of the equation

$$z = e^{R_0(z-1)} \qquad (1.7)$$

(recall Figure 1.1). In order to avoid the wrong impression that this is a general result, we add that the minor outbreak probability z_∞ satisfies

$$z_\infty = \frac{1}{R_0} \qquad (1.8)$$

for another sub-model for infectivity, viz., the one where $\triangle T$ above is not a fixed quantity but a random variable following the exponential distribution. The following exercises provide the details underlying this assertion (see also Exercises 2.10 and 2.11).

We conclude that the probability z_∞ that the introduction of an infected host from outside does not lead to an epidemic can, for some simple sub-models, be either determined by a graphical construction or be expressed explicitly in terms of the parameters. Different sub-models that yield the same value for R_0 may lead to different values of z_∞.

Exercise 1.11 Convince yourself that when $\triangle T$ is exponentially distributed with parameter α,

$$\begin{aligned} g(z) &= \alpha \int_0^\infty e^{pc\triangle T(z-1)} \, e^{-\alpha\triangle T} \, d(\triangle T) \\ &= \frac{\alpha}{\alpha - pc(z - 1)}. \end{aligned}$$

Hint: Interchange summation with respect to k and integration with respect to $\triangle T$ in the definition of g. Note that the offspring distribution is geometric with success parameter $\alpha/(pc + \alpha)$.

Exercise 1.12 Compute that for this sub-model $R_0 = pc/\alpha$. Do this in two different ways:

i) using Exercise 1.6-iii;

ii) in a way involving the interpretation; more precisely, compute the expected length of the infectious period and multiply this with the expected number per unit of time of 'successful' contacts.

Exercise 1.13 Check that equation (1.8) is correct.

1.2.3   Initial growth in real time

By looking at initial growth on a generation basis, we now understand that a threshold phenomenon occurs governed by a parameter R_0, allowing for a clear biological interpretation, and that even for R_0 above threshold, there exists a positive probability z_∞ that introduction of one primary case does not lead to explosive exponential growth. For simple sub-models of contact and infectivity, we are able to derive explicit expressions (or simple equations) for R_0 and z_∞ in terms of the parameters of the sub-models.

The disadvantage of measuring growth on a generation basis is that, due to the fact that generations usually overlap each other in real time, it does not correspond to what we actually observe. Indeed, when we speak about exponential increase during the initial phase of an epidemic, we mean that

$$Ce^{rt}$$

for some $r > 0$ (and some constant $C > 0$), is the number of cases notified up to time t (where time is measured relative to some convenient, but otherwise arbitrary, starting point). The *incidence* $i(t)$ (i.e., the number of new cases arising per unit of time) will be proportional to the derivative, hence to $\exp(rt)$. (As an aside we note that the constant of proportionality may involve the probability that cases are actually reported — a quantity that is often less than one in the real world.)

Now let us return to the world of models. Is it possible to compute the exponential growth rate r, also called the intrinsic growth rate or the Malthusian parameter, in terms of the parameters of our sub-models for contact and infectivity?

New cases at time t result from contacts with individuals that were infected themselves before t and that are infectious at time t. In a deterministic description we pretend that the actual number of new cases equals the expected number of new cases. These arguments, our assumptions on contact and infectivity above, and elementary bookkeeping together lead to the equation

$$i(t) = pc \int_{T_1}^{T_2} i(t - \tau)\, d\tau \tag{1.9}$$

for the incidence *in the initial phase* of an epidemic.

Exercise 1.14 Give a detailed interpretation of the equality in (1.9), to check both the equation and your understanding of it.

Substituting the ansatz $i(t) = ke^{\lambda t}$, we find that λ should satisfy the so-called characteristic equation[1]

$$1 = pc \int_{T_1}^{T_2} e^{-\lambda \tau}\, d\tau. \tag{1.10}$$

The usual procedure in showing the existence of solutions for relations such as (1.10), with λ real, is to let $f(\lambda)$, say, be defined by the right-hand side of (1.10)

---

[1] An *ansatz* is an assumed relation that will be motivated, justified, or refuted by the consequences that one is going to derive from it. A *characteristic* equation is an equation in terms of a scalar quantity — usually complex and often denoted by λ — which constitutes a solvability condition (both necessary and sufficient) for a more complicated problem. For example, the equation $\det(M - \lambda I) = 0$ is a characteristic equation to the eigenvalue problem $Mv = \lambda v$, where M is a matrix and v a vector.

and to use monotonicity arguments as follows. For $\lambda = 0$ we obtain $f(0) = R_0$. Moreover, f is a decreasing function of λ, $\lim_{\lambda \to \infty} f(\lambda) = 0$, and $f(\lambda) \to +\infty$ for $\lambda \to -\infty$. See also Figures 1.3 and 1.4 for a qualitative picture.

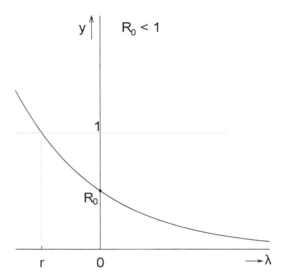

Figure 1.3: Schematic representation of the graph of a relevant decreasing function $y = f(\lambda)$ for the case $R_0 < 1$, showing intersection with the line $y = 1$ for a unique value r of $\lambda < 0$.

Exercise 1.15 Check these statements. Hint: Do *not* evaluate the integral, as this will make the proof much more cumbersome.

We conclude:

- There exists a unique real root $\lambda = r$; in other words, equation (1.10) tells us unambiguously what the exponential growth rate is for this model.

- We do not have an explicit formula for r (whereas we do for R_0).

- $r > 0$ if and only if $R_0 > 1$ and $r < 0$ if and only if $R_0 < 1$; this means that we have growth in real time if and only if we have growth on a generation basis. Also note that $r = 0$ if and only if $R_0 = 1$.

In general, a high value of R_0 does not necessarily imply a high value of r. To use demography as a metaphor: if people have many children, but only when they themselves are already quite old, population growth may still be slow. And in more mathematical language: the formula for R_0 depends only on the difference $T_2 - T_1$, whereas in the equation for r the magnitude of T_1 and T_2 matters a lot. In addition, R_0 is a dimensionless quantity, whereas r depends on the unit of time used.

Exercise 1.16 The aim of this exercise is to elaborate on the last observation. The task is to choose T_1^*, T_2^* and T_1^{**}, T_2^{**} such that $R_0^* > R_0^{**}$ but $r^* <$

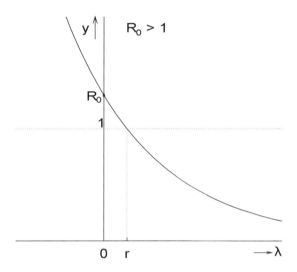

Figure 1.4: Schematic representation of the graph of a relevant decreasing function $y = f(\lambda)$ for the case $R_0 > 1$, showing intersection with the line $y = 1$ for a unique value r of $\lambda > 0$.

r^{**}, meaning that for the $*$ model growth on a generation basis is faster, while for the $**$ model growth in real time is faster. By 'to choose' we do not mean that you have to provide numerical values, but rather that you have to describe some procedure by which such T values could be obtained. (Incidentally, does this exercise shed any light on the one-child policy of the Chinese government?)

Exercise 1.17 If an epidemic has growth rate r in the initial phase, what is the doubling time?

The reason to consider the generation perspective is that the formula for R_0 is explicit, whereas for r we only have an equation. In Chapter 7 we will find that for more complicated situations, R_0 is also only characterized by some equation. That characterization, however, is still much more explicit than the corresponding characterization of r. The underlying reason is that for r the time course of infectivity matters, whereas for R_0 it does not.

Remark 1.18. We refrain here from a formulation and analysis of the real-time branching process that corresponds to our description of contact and infectivity, but postpone this to Section 3.3. Conditional on the infectious agent not going extinct, it would predict exponential growth with the rate r derived from the deterministic description in this section. But even when exponential growth occurs, there is an initial phase in which numbers are still low and stochastic effects manifest themselves. In particular one should think about the *duration* of this initial phase as a random variable. In other words, the continuous-time branching process predicts that either the agent goes extinct or exponential growth with rate r sets in after a random delay.[2] We return to 'duration' in Section 3.6, once we have introduced the relevant stochastic background in Chapter 3.

At this point we have more or less answered the first two of our three questions and it is time to turn to the third question concerning the final 'size' of the epidemic. It turns out that, in order to answer this question, we need to make additional model specifications. Before doing so we insert a brief interlude concerning terminology and notation, used in the rest of the book.

1.2.4   Interlude on terminology and notation

Although we have already used terminology and notation in the initial sections of this book, it is worthwhile pointing to several of the conventions we aim to follow. Trying to be precise does have the danger of stirring up more questions than are answered, for example when trying to define exactly what it means when a host individual is infected, so we will aim for workable conventions, rather than ironclad definitions of terms.

We will always consider a situation of a particular (usually unspecified) host species and a particular (usually unspecified) infectious agent that can infect individuals of that host species. A *susceptible* is an individual that is currently not infected, but is receptive to infection by the agent. In the plural form we will write either 'susceptible individuals' or 'susceptibles,' as has become customary. An *infected* is an individual that has been exposed to the infectious agent and in which the infectious agent can be detected by suitable tests (e.g., tests that detect genetic material of the agent in bodily fluids or tissues, so-called PCR tests). We will neglect the fact that such tests as a rule have a certain detection threshold, sensitivity and specificity that can lead to wrong conclusions as to infection status of an individual. These are some of the issues, hinted at above, that make defining 'infection' difficult, but they are outside the scope of our book. A group of infected individuals will often be referred to as 'infecteds' for convenience. An infected individual is not necessarily *infectious* (or synonymously, *infective*).

Upon successful infection the individual usually goes through a *latency* period where it is not able to pass on the infectious agent to other individuals. We will refer to these as 'latently infected individuals' or 'latent individuals.' Often such individuals are said to be 'exposed,' in the sense that they have come into contact with the infectious agent, but are not yet able to spread the agent themselves. We try to avoid this use as 'exposure' to an agent does not necessarily imply 'infection.' An infectious/infective individual is an infected that has the ability to infect, or at least expose, other individuals by shedding the infectious agent. Hence, an infected may be either exposed/latent (i.e., non-infectious), or infectious.

When the infectious agent can no longer be detected in an individual that was infected, we refer to the individual as having *recovered* from the infection, and assume it is (at least temporarily) immune to re-infection. These individuals can still be seropositive in that tests pick up antibodies to the infectious agent, as a sign that the individual has been infected at some time in the past.

We often view the various phases in the development of the infection within an individual as states that an individual can be in at a given point in time (see

---

[2]J.A.J. Metz: The epidemic in a closed population with all susceptibles equally vulnerable; some results for large susceptible populations and small initial infections. *Acta Biotheoretica*, **27** (1978), 75–123.

Chapter 6 for a more involved discussion). In model descriptions we will adhere to the convention of referring to the state with a capital letter in roman script: S for an individual in the susceptible state, E for a latently infected individual, I for an infectious individual, and R for a recovered individual. The same capital letter, but now in italic script (S, E, I, R) will then denote the size of the sub-populations in the various states. We will also refer to such sub-populations as classes, or compartments (in the context of compartmental models). All infected individuals combined (in the above case $E+I$) are usually referred to as the infection *prevalence*, the number of infected individuals present at a given point in time. Often, prevalence will refer only to a subset of infecteds, depending on the test used to assess the prevalence. We will not discuss methods to assess the prevalence from population samples and surveys.

The lower case letter s will be used in various places to denote the fraction, or proportion, of the susceptible sub-population, as part of the total population with size N, i.e., $s = S/N$. We will have only occasional use for the corresponding symbol for the fraction of infecteds, and will explicitly point this out where needed. We will use the letter i instead in another context, related to the growth of the sub-population of infected individuals, which we discuss next.

The quantity r has already been encountered in the previous section: the exponential growth rate of the infecteds in real time in the very early phase of an outbreak. The quantity $i(t)$ in equation (1.9) gives the *incidence* defined as the rate with which new infecteds arise, i.e., the number of new infected individuals per unit of time. This is also referred to as the population-level transmission rate. Related to this is the *force of infection*, denoted by Λ, and defined as the probability per unit of time for a susceptible to become infected: $i(t) = \Lambda(t)S(t)$.

For notation indicating the rate at which variables change with time, we take the freedom that the time derivative of $S(t)$, say, is sometimes denoted by dS/dt, sometimes by $\dot{S}$, and sometimes by S'. We let the circumstances (e.g., typographical complexity of an equation) dictate which form we use.

1.3 THE FINAL SIZE

1.3.1 The standard final-size equation

In a closed population and with infection leading to either immunity or death, the number of susceptibles can only decrease and so it must have a limit for time tending to infinity. Will this limit be zero? Or will some fraction of the population escape from ever getting infected? If so, what fraction (i.e., how does the fraction depend on the parameters)?

These questions were posed — and answered! — by Kermack and McKendrick in 1927.[3] We shall now first formulate and analyze the answers and only thereafter present the arguments to derive them, while scrutinizing the underlying assumptions.

Let $s = S/N$ denote the fraction of susceptibles in a total population of size N. Let $s(\infty)$ denote this fraction at the end of the outbreak (we will also write this occasionally as s_∞). The complementary quantity $1 - s(\infty)$ we shall call the

---

[3] W.O. Kermack and A.G. McKendrick: Contributions to the mathematical theory of epidemics, part I. *Proc. Roy. Soc. Lond. A*, **115** (1927), 700–721. Reprinted (with parts II and III) as *Bull. Math. Biol.*, **53** (1991), 33–55. We strongly encourage our readers to have a look at these papers, as their content is far richer than the standard reference they have become suggests.

final size of the epidemic, since it gives the fraction of the population that became infected sooner or later. The key result is that $s(\infty)$ is a root of the *final-size equation*

$$\ln s(\infty) = R_0 \left(s(\infty) - 1 \right) \tag{1.11}$$

also written as

$$s(\infty) = e^{-R_0(1-s(\infty))}.$$

We show how to derive this equation in Exercise 1.22. When $R_0 < 1$ the relevant root is $s(\infty) = 1$, meaning that the introduction of the agent does not lead to a major outbreak. When $R_0 > 1$ there exists a unique root between 0 and 1, which is the relevant one (the root $s(\infty) = 1$ still exists, but only has a meaning for stochastic models, see Chapter 3).

Exercise 1.19 Verify our statements about roots of equation (1.11). Hint: Derive inspiration from Figures 1.5 and 1.6.

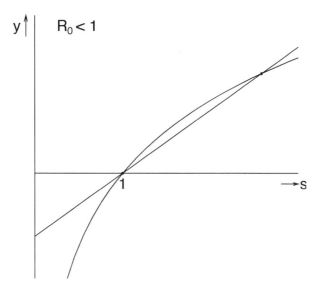

Figure 1.5: Schematic representation of the graphs of the functions $y = \ln(s)$ and $y = R_0(s - 1)$ for the case $R_0 < 1$, showing that the intersection in the interval $[0, 1]$ is at $s = 1$.

Exercise 1.20 Show that

  i) $s(\infty) \sim e^{-R_0}$ for $R_0 \to \infty$;

  ii) $s(\infty) \sim 1 - 2(R_0 - 1)$ for $R_0 \downarrow 1$.

We conclude that, within the framework of our assumptions, a certain fraction $s(\infty)$ escapes from ever getting the disease and that this fraction is completely determined by R_0 through equation (1.11). The larger R_0 is, the smaller the fraction that escapes. In fact, the fraction is negligibly small for large values of R_0.

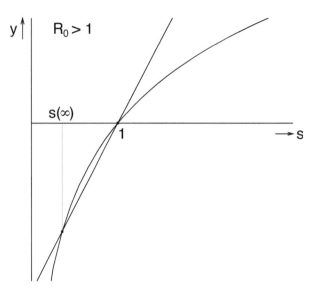

Figure 1.6: Schematic representation of the graphs of the functions $y = \ln(s)$ and $y = R_0(s - 1)$ for the case $R_0 > 1$, showing that the relevant intersection in the interval $[0, 1]$ is at a value < 1.

Below we shall derive the final-size equation by manipulating (integro)differential equations that correspond to a deterministic description. First, however, we present a quite different (less formal but much more direct) derivation in terms of a probabilistic consistency consideration. We advocate this method, since it generalizes much more easily to far more complicated situations (see Section 8.3, Exercise 8.25). A more refined presentation of the key arguments is given in Section 3.5.

An individual that is susceptible at time t_0 and experiences a *force of infection* $\Lambda(t)$ for $t > t_0$ will escape from being infected with probability $\mathcal{F}(t)$ defined by

$$\left.\begin{array}{l} \frac{d\mathcal{F}}{dt} = -\Lambda\mathcal{F} \\ \mathcal{F}(t_0) = 1 \end{array}\right\} \Rightarrow \mathcal{F}(t) = e^{-\int_{t_0}^{t} \Lambda(\tau)\,d\tau}.$$

Assume that the force of infection was zero prior to time t_0. For large populations we shall have $s(\infty) = \mathcal{F}(\infty)$, i.e., the fraction that remains susceptible equals the probability to remain susceptible. Let us call $\int_{t_0}^{\infty} \Lambda(\tau)\,d\tau$ the total cumulative force of infection. The fraction $z = 1 - s(\infty)$ that falls victim to the infection generates a total cumulative force of infection equal to

$$\frac{1}{N} pc \triangle T z N = R_0 z$$

(we have to divide by N since contacts are with probability $1/N$ with the susceptible individual that we consider). Hence

$$s(\infty) = \mathcal{F}(\infty) = e^{-R_0(1 - s(\infty))},$$

and by taking logarithms we arrive at (1.11). The fact that the equation is essentially identical to equation (1.7), which characterizes the probability of a minor outbreak, is not a coincidence:

Remark 1.21 It can be shown that $s(\infty)$ relates to the probability of extinction in a backward branching process. We will show this, by way of example, in Section 12.6.3. In Section 3.5, we look at random fluctuations around $s(\infty)$.

1.3.2   Derivation of the standard final-size equation (and reflection upon the underlying assumptions)

The Kermack-McKendrick model takes the form of an integral equation, see (1.16). A very special case can alternatively be formulated as a system of ODEs (Ordinary Differential Equations). We start this section with a rather long exercise that first introduces this ODE system and next derives (1.11) from it. In Section 3.4 we show how the system of ODEs can be obtained from a stochastic model by letting the population size go to infinity.

Exercise 1.22 Let S denote the size of the sub-population of susceptibles, I the size of the sub-population of infectives and R the size of the sub-population of removed individuals (meaning immune or dead or in quarantine, but neither susceptible nor infective). We postpone a discussion of the precise meaning of 'size,' in particular whether it refers to numbers or to spatial density. The *force of infection* is by definition the probability per unit of time for a susceptible to become infected. As an analog of the law of mass action from chemistry, we assume that the force of infection is proportional to I (see also Section 12.2). The constant of proportionality we shall call β (the transmission rate constant).

We assume that an infected individual becomes immediately infectious. In other words, upon infection, an individual labeled S turns into an individual labeled I.

We assume that infectives have a constant probability per unit of time α to become removed. Stated differently: the infectious period has an exponential distribution with parameter α, i.e., the probability to still be infectious τ units of time after infection is $e^{-\alpha\tau}$.

The system of differential equations

$$
\begin{aligned}
\frac{dS}{dt} &= -\beta SI, \\
\frac{dI}{dt} &= \beta SI - \alpha I, \\
\frac{dR}{dt} &= \alpha I
\end{aligned}
\tag{1.12}
$$

concisely summarizes our assumptions. (Incidentally, here and in the rest of this book we sometimes choose the layout of the right-hand side of a system of ODEs such that corresponding terms form a vertical row, in order to reflect the transition from one sub-population to another.) One speaks about the SIR model or the SIR system. (See Exercises 2.2 and 4.8 for an SEIR model, and Exercise 4.15 for an SIS model. These are also called *compartmental* models, with each letter referring to a 'compartment' in which an individual can reside, often (but usually not in the case of S) with an exponentially distributed sojourn time; for an exposition of theory and applications of compartmental models in general see e.g., the book by Jacquez 1998.) Note, however, that the last equation in (1.12) is redundant. There are two ways to arrive at this conclusion. The first is to note that the derivatives of S and I are given by expressions that do not involve R. The second is to note that $d/dt(S+I+R) = 0$, and hence $S+I+R = N$ is constant, where N denotes the total population size fixed by the initial condition.

i) Convince yourself that for this model $R_0 = \beta N/\alpha$ and $r = \beta N - \alpha$. Use these results to reformulate the threshold condition as one involving population size N.

ii) Show that $\frac{\alpha}{\beta} \ln S - S - I$ is a conserved quantity. Hint: Differentiate with respect to time and check that the result equals zero. To get an idea of how to derive this quantity, consider dI/dS and separate variables.

iii) Assume that $R_0 > 1$. If time runs from $-\infty$ (before the epidemic) to ∞ (after the epidemic), argue that

$$\frac{\alpha}{\beta} \ln S(\infty) - S(\infty) = \frac{\alpha}{\beta} \ln S(-\infty) - S(-\infty)$$

and that $S(-\infty) = N$. Derive (1.11) from these identities.

iv) Now let time run from 0 (the start of the epidemic) to ∞ (after the epidemic). Put $N = S(0) + I(0)$ (so $R(0) = 0$). Derive the following variant of (1.11) (recall that $s(t) = S(t)/N$):

$$\ln s(\infty) = R_0(s(\infty) - 1) + \ln s(0). \tag{1.13}$$

Analyze this equation graphically in the spirit of Exercise 1.19, paying attention to both cases $R_0 < 1$ and $R_0 > 1$. Deduce in particular what happens for $I(0) \downarrow 0$ or, equivalently, $s(0) \uparrow 1$.

v) Draw the phase portrait describing the reduced (S, I) system

$$\frac{dS}{dt} = -\beta SI,$$
$$\frac{dI}{dt} = \beta SI - \alpha I.$$

Hint: Use the conserved quantity (first integral) and restrict attention to the positive quadrant.

vi) For what value of S does the epidemic reach its peak, in the sense that I is maximal? Can you understand this? Hint: Recall the definition of R_0 and pretend that we can 'freeze' the dynamic variable S at any value we wish to consider.

vii) Verify that the root $s(\infty)$ of equation (1.11) becomes smaller when we increase N. (Observe first that R_0 increases as N increases.) This is essentially an overshoot phenomenon. Explain the mechanism underlying this 'overshoot phenomenon' by combining the results of v) and vi).

viii) Reformulate (1.12) in terms of *fractions* of the population, so using $s = S/N$ etc. What happens to the parameter β? Reflect upon the dimension of β and its relation to observable quantities, such as the c we introduced in Section 1.1.

After this lengthy exercise we take up our general discussion and set out to derive (1.11) while forming a clear idea of the underlying assumptions. So far we asserted that individual hosts have an expected number c of contacts per unit of time. Can we sustain this assumption when the disease makes victims? The answer presumably depends on what it really means to fall 'victim' to the disease.

When at the end of the infectious period hosts become immune and take part in the contact process as before, the epidemic does not interfere with the contact process and we can consider c as a fixed constant during the epidemic. This seems the easiest situation, so let us deal with it first.

If we focus on the contacts of an infected individual and contacts are 'at random,' the probability that the partner in the contact is susceptible equals $s = S/N$, the fraction susceptible of the total population of size N. The equation for the incidence then reads

$$i(t) = s(t)pc \int_{T_1}^{T_2} i(t - \tau) \, d\tau. \tag{1.14}$$

(Note that N does not depend on time and that we recover equation (1.9) for the incidence in the initial phase by putting $s(t)$ equal to 1.) Our assumption that the population is closed implies

$$i(t) = -\dot{S}(t) \tag{1.15}$$

i.e., the incidence equals the change in the susceptibles. After dividing both sides of (1.14) by N we can rewrite the equation as

$$\dot{s}(t) = s(t)c \int_0^{\infty} A(\tau) \, \dot{s}(t - \tau) \, d\tau, \tag{1.16}$$

where we have introduced the function A defined by

$$A(\tau) = \begin{cases} p & \text{if } T_1 \leq \tau \leq T_2, \\ 0 & \text{otherwise.} \end{cases} \tag{1.17}$$

Exercise 1.23 Derive (1.11) from (1.16). Hint: Divide by $s(t)$ and integrate from $t = -\infty$ to $t = +\infty$; use that $R_0 = c \int_0^{\infty} A(\tau) \, d\tau$ (see Exercise 1.29-i).

It is good to stress a fundamental point: In the following, we will work more and more with a general non-negative infectivity function $A = A(\tau)$ as a basic model ingredient. Such a function incorporates information about

- The degree to which an individual, that was infected τ units of time ago, is expected to take part in the contact process (e.g., an individual that has died does not take part at all);

- The probability of transmission, given a contact between such an individual and a susceptible individual.

At present, we still keep a multiplicative constant c, denoting the rate at which individuals are engaged in contacts, in the formulation to facilitate the comparison of various alternative assumptions concerning the contact process. Later on, when dealing with other issues, we shall absorb the c in the A, to simplify the notation. Of course, this has an impact on the interpretation of A and we encourage our readers to be aware of the difference.

For many situations it is reasonable to assume that c, and hence R_0, is proportional to population *density*, i.e., proportional to the number of individuals per unit area (but of course there are exceptions as well, e.g., when sexual contacts are concerned). The threshold for R_0 may be reformulated as a threshold for the density: Below a critical density, introduction of the disease is harmless; above the critical density, an outbreak will result (for instance, infectious diseases of humans became much more prominent when cities were formed; dense agricultural crops are very vulnerable for pests). Moreover, if densities are larger, the epidemic will affect a larger proportion of the population (this is an overshoot phenomenon: the expected number of successful transmissions will gradually decrease during the epidemic, but even when this number drops below one, there will be many more new cases due to the substantial part of the population that is infectious at that moment, cf. Exercise 1.22-v,-vii).

Next let us look at another extreme, where each individual is supposed to die at the end of the infectious period. Then total population size N depends on t. It is therefore quite conceivable that the expected number of contacts per unit of time per individual depends on t as well. Let us assume it is proportional to $N(t)$, with constant of proportionality θ. (The difference between θ here and β in Exercise 1.22 is that the latter incorporates a factor corresponding to the probability of transmission. Moreover, in Exercise 1.22 we consider an exponentially distributed infectious period. Here we have incorporated information about the probability of transmission in the function A.) Then we may write the equation for the incidence as

$$\dot{S}(t) = \frac{S(t)}{N(t)} \theta N(t) \int_0^\infty A(\tau) \dot{S}(t - \tau)\, d\tau. \qquad (1.18)$$

When $N(-\infty)$ denotes the original population size and $c = \theta N(-\infty)$ we obtain, after dividing by $N(-\infty)$, equation (1.16) again but now with

$$s(t) = \frac{S(t)}{N(-\infty)} \qquad (1.19)$$

(i.e., $s(t)$ measures susceptibles as a fraction of the original population size). So all our previous conclusions still apply, the only difference being that now $1 - s(\infty)$ is the fraction of the population that died, whereas before it was the fraction that became immune.

Exercise 1.24 Suppose that at the end of the infectious period an individual dies with probability $1 - f$, and otherwise becomes immune, while we still assume that $c = \theta N(t)$. Does the final size depend on f? The probability $1 - f$ is often referred to as the case fatality, usually defined as the proportion of the infected population that dies from the infection.

<u>We conclude</u> that the final-size equation (1.11) holds when the disease is harmless (in the sense that it interferes in no way with the contact process) or, if it is not harmless, (1.11) holds when contact intensity is proportional to population density.

1.3.3   The final size of epidemics within herds

The foregoing does not exhaust the catalog of assumptions that may be appropriate for certain specific real-life situations. For instance, phocine distemper virus is transmitted from seal to seal when they come to a sandbank to rest and sunbathe.

When group size diminishes (by death due to the disease), they occupy less area, while maintaining roughly the same contact intensity (in other words, the effective density remains constant).

More generally when animals live in herds, the number of contacts per unit of time per individual could very well be (almost) independent of herd size. If herd size diminishes due to a fatal infectious disease, the force of infection does not go down as quickly as it does in the model of the foregoing section, which assumed that the density, and hence the per capita number of contacts per unit of time, is proportional to population size. This has an effect on the final size. Exactly how much effect it has depends on the probability f to survive an infection. The final-size equation will now involve two parameters: R_0 and f. It will also involve two fractions: the fraction $s(\infty)$ that escapes from ever getting the infection and the fraction $n(\infty)$ that is still alive at the end of the epidemic.

Exercise 1.25 Argue that the interpretation of $n(\infty), s(\infty)$ and the constant f implies the relation

$$n(\infty) - s(\infty) = f(1 - s(\infty)). \tag{1.20}$$

A second relation between these quantities and R_0 reads

$$n(\infty) = s(\infty)^{\frac{1-f}{R_0}}, \tag{1.21}$$

which will be derived in Exercise 1.28 and beyond. There are various ways in which we can combine and/or rewrite these equations and each of these may be helpful for a certain purpose.

Exercise 1.26 i) Derive the equation

$$\ln s(\infty) = \frac{R_0}{1 - f} \ln(f + (1 - f)s(\infty)). \tag{1.22}$$

   ii) Analyze this equation graphically.

   iii) Show that in the limit $f \uparrow 1$ we find (1.11).

   iv) Show that $s(\infty)$ is a monotonically increasing function of f.

   v) Assume $R_0 > 1$. Show that $n(\infty) \downarrow 0$ when $f \downarrow 0$.

Exercise 1.27 When f and R_0 are unknown, but data about the situation after the epidemic allow us to make an estimate of $s(\infty)$ and $n(\infty)$, we can try to determine the parameters from the final-size equation. Show that

$$f = \frac{n(\infty) - s(\infty)}{1 - s(\infty)}, \qquad R_0 = \frac{1 - n(\infty)}{1 - s(\infty)} \frac{\ln s(\infty)}{\ln n(\infty)}. \tag{1.23}$$

Let us now discuss the difference between the situations described in this section and in Section 1.3.2. Death has two effects: a direct one and an indirect one. The direct one is simply that a fraction $1 - f$ of $1 - s(\infty)$ dies. The indirect one is that $s(\infty)$ itself decreases since, while immunes hinder contacts between infectives and susceptibles, dead individuals do not. The indirect effect makes the difference between the two models. When R_0 is big, $s(\infty)$ is very small ($\sim e^{-R_0}$), and the indirect effect is negligible. If, on the other hand, R_0 is only a little above the threshold value one, the indirect effect can be substantial.[4]

In the case of seals affected by phocine distemper virus the survival probability f may very well depend on physiological and immunological conditions, which, in turn, are determined by environmental aspects such as food availability and pollution.

We conclude that the precise form of the final-size equation depends on our assumptions concerning the contact process (and hence the force of infection), as affected by population size, and that this really matters when R_0 is only a little above threshold and the survival probability is low.

Exercise 1.28 i) Argue that the ODE system

$$
\begin{aligned}
\frac{dS}{dt} &= -\gamma \frac{SI}{N}, \\
\frac{dI}{dt} &= \gamma \frac{SI}{N} - \alpha I, \\
\frac{dN}{dt} &= -(1-f)\alpha I
\end{aligned}
$$

describes the spread of an infectious disease when the per capita number of contacts per unit of time is independent of the population size N. The parameter γ can be considered as the product of two more basic parameters. Which are these? The equations reflect a specific assumption about mortality as a result of infection. What assumption is this?

ii) Show that $R_0 = \gamma/\alpha$ (independent of population size!) and $r = \gamma - \alpha$.

iii) Show that, for any $-\infty \le \sigma \le t \le \infty$,

$$
\frac{S(t)}{S(\sigma)} = \left(\frac{N(t)}{N(\sigma)} \right)^{\frac{R_0}{1-f}}, \tag{1.24}
$$

and conclude that (1.21) is correct. Hint: Consider dS/dN.

iv) Derive that $N(\infty) - N(-\infty) = (1-f)(S(\infty) - S(-\infty))$ and next rewrite this identity in the form (1.20).

Hint: Integrate both $\frac{dN}{dt} = -(1-f)\alpha I$ and $\frac{dI}{dt} = -\frac{dS}{dt} - \alpha I$ from $-\infty$ to ∞.

v) Note from the ODE system that the force of infection equals $\gamma I/N$, whereas the force of mortality (i.e., the per capita death rate) $-\frac{1}{N}\frac{dN}{dt}$ equals $\alpha(1 - f)I/N$. Hence the two are proportional, with constant of proportionality $(1-f)/R_0$. Derive the identity (1.24) from this. Note that mathematically this amounts, in the end, to the same separation-of-variables method as in iii); this part of the exercise, however, is meant to prepare the reader for assumption (1.29).

We now set out to derive (1.20) and (1.21) for a more general class of models. This derivation may be skipped by readers who feel unhappy when abstract-looking formulas are manipulated.

---

[4]A. de Koeijer, O. Diekmann and P. Reijnders: Modelling the spread of phocine distemper virus among harbour seals. *Bull. Math. Biol.*, **60** (1998), 585–596.

Recalling the identity $i(t) = -\dot{S}(t)$ (see (1.15)), which expresses the incidence as the decrease in susceptibles, we start out from

$$\dot{S}(t) = c\frac{S(t)}{N(t)} \int_0^\infty \dot{S}(t-\tau)A(\tau)\,d\tau, \qquad (1.25)$$

where c is the number of contacts per unit of time, and where $A(\tau)$ describes the infectivity at infection age τ. The total population $N(t)$ at time t is composed of susceptibles and individuals who were infected τ units of time ago (at time $t - \tau$), and are still alive at t, for a range of values for τ. Symbolically,

$$N(t) = S(t) - \int_0^\infty \dot{S}(t-\tau)\mathcal{F}(\tau)\,d\tau, \qquad (1.26)$$

where $\mathcal{F}$ denotes the survival probability as a function of infection age τ. (We will generically denote 'survival' probabilities by $\mathcal{F}$, the exact interpretation of survival depending on the context. So the present $\mathcal{F}$ has nothing to do with $\mathcal{F}$ used in between Exercise 1.20 and Remark 1.21. In (1.26) $\mathcal{F}$ literally describes 'staying alive,' and should be considered as an additional model ingredient about which we make further assumptions below.) It is then reasonable to write

$$A(\tau) = a(\tau)\mathcal{F}(\tau), \qquad (1.27)$$

where $a(\tau)$ measures the output of infectious material at infection age τ (e.g., the amount of virus or bacteria shed), given that the individual is still alive.

It is unclear whether a final-size equation can be derived without further assumptions. What we shall do is add a relation between a and $\mathcal{F}$, which allows for a clear biological interpretation and which enables us to derive (1.20) and (1.21).

The hazard rate of death, $\mu(\tau)$, is by definition the probability per unit of time of dying at infection age τ, given that one survived until τ. Mathematically this translates into

$$\mathcal{F}'(\tau) = -\mu(\tau)\mathcal{F}(\tau). \qquad (1.28)$$

We now assume that μ is proportional to a, or, in other words, that the probability per unit of time of dying is proportional to the output rate of infectious material.

Exercise 1.29 Call the constant of proportionality q, i.e., assume that

$$\mu(\tau) = qa(\tau). \qquad (1.29)$$

The goal of this exercise is to express q in terms of c, f and R_0. The result reads

$$q = c\frac{1-f}{R_0}. \qquad (1.30)$$

i) Argue that

$$R_0 = c\int_0^\infty A(\tau)\,d\tau. \qquad (1.31)$$

ii) Derive that $f - 1 = -q\int_0^\infty A(\tau)\,d\tau$. Hint: Integrate, using (1.29) and (1.27), the identity (1.28) from 0 to $+\infty$.

Exercise 1.30 Our motivation for (1.29) did not derive from the interpretation, but rather from formula manipulation. We want to be able to integrate the identity obtained by dividing (1.25) by $S(t)$. So we want that

$$\frac{1}{N(t)} \int_0^\infty \dot{S}(t-\tau)A(\tau)\,d\tau \propto \frac{d}{dt} \ln N(t)$$

(where $\propto$ means 'is proportional to'). Show that this is indeed true if we assume (1.29). Hint: Write the integral term in (1.26) as

$$\int_{-\infty}^t \dot{S}(\tau)\mathcal{F}(t-\tau)\,d\tau.$$

A combination of the above two exercises leads to the conclusion that

$$\frac{N(t)}{N(\sigma)} = \left(\frac{S(t)}{S(\sigma)}\right)^{\frac{1-f}{R_0}} \qquad \text{for all } t, \sigma, \tag{1.32}$$

also for this more general class of models; and by letting $\sigma \to -\infty$, $t \to +\infty$, we thus arrive at (1.21).

Exercise 1.31 Derive, as a check on our model formulation, (1.20) by letting $t \to \infty$ in (1.26). Hint: See the hint in the previous exercise.

Exercise 1.32 Recapitulate the differences in the assumptions underlying the models of this and the foregoing section. Check that the difference does not really matter in the initial phase and that, in particular, R_0, r and the probability to go extinct are determined in exactly the same manner. Next imagine two widely separated seal colonies, one being twice as large as the other. Assume that the same virus is introduced in both. Compare the predictions made by the two models concerning R_0 and the final size. (If you wish to include r and the probability to go extinct in the comparison, go ahead.)

Exercise 1.33 With a sexually transmitted disease in mind, contemplate how the average number of sexual contacts per individual per unit of time depends on population size (see also Chapter 12).

1.3.4 The final size in a finite population

In a deterministic description of an epidemic in a closed population, as given above, we found a characterization of the *fraction* $1 - s(\infty)$ that is ultimately affected. In this section we will briefly indicate how that relates to the limit, for population size going to infinity, of the *distribution* of final size when we consider a finite population. We give a more thorough treatment in Chapter 3.

So imagine a population consisting of N individuals, one of whom is infected from an outside source. Once we specify a sub-model for contact and transmission, we can calculate the probability distribution for the number of its 'offspring' (cf. Section 1.2.2; but since we now consider a finite population of susceptible hosts, we will typically encounter binomial distributions instead of Poisson distributions). With a lot more effort, we can calculate the probability distribution for the sum of the first and second generation, see Exercise 3.4 below. At least in principle, we can extend this to any number of generations and so, by taking the limit, compute the distribution of the final size of the epidemic (note that there necessarily is a well-defined limit, because of monotonicity). There exist various sophisticated ways to compute the final-size distribution much more effectively than sketched above, see Section 3.5 (see also Ball 1995, Lefèvre and Picard 1995[5] and the references given there), but we emphasize that it is not at all an easy task. Here we shall first summarize some qualitative aspects of the result and then, by way of exercises, delineate a trick that leads to a convenient computational scheme for the simplest sub-model for infectivity.

When $R_0 < 1$, the final-size distribution is concentrated near zero. When we rescale by using k/N rather than k as the variable to describe the final size, and let $N \to \infty$, the distribution becomes more and more concentrated at zero, meaning that a negligible *fraction* is affected.

When $R_0 > 1$ the final-size distribution has a double peak. The first peak corresponds to so-called *minor outbreaks,* in which the infective agent goes extinct before affecting a substantial *fraction* of the population. The second peak corresponds to *major outbreaks,* in which approximately a fraction $1 - s(\infty)$ is infected. This distinction between minor and major becomes more and more prominent if we rescale to k/N and let $N \to \infty$. Then a fraction z_∞ of the total probability 1 becomes concentrated at zero, while a fraction $1 - z_\infty$ becomes concentrated at $1 - s(\infty)$ (with appropriate scaling the distribution around $1 - s(\infty)$ is accurately described by a normal distribution with standard deviation of order $1/\sqrt{N}$; see Section 3.5.3).

We conclude that the deterministic description has to be complemented by two observations deriving from the stochastic analysis:

- Even for $R_0 > 1$, introduction may lead to a minor outbreak only (cf. Section 1.2.2).

- $1 - s(\infty)$ is the *mean* size of major outbreaks.

Exercise 1.34 The setting for this exercise is as described in Exercise 1.22 (the Kermack-McKendrick ODE model). Note that this is identical to the setting of Exercise 1.28 under the parameter specification $f = 1$ (infection always leads to immunity, never to death) and $\gamma = \beta N$, with N = population size. Recall that $R_0 = \beta N/\alpha$. Let $P_{(n,m)}(t)$ denote the probability that at time t there are n susceptibles and m infectives, where $n, m \geq 0$ and $n + m \leq N$. It is helpful to introduce the convention that $P(n, m) = 0$ when the indices violate these constraints.

When the population is currently in state (n, m), two transitions are feasible: a susceptible is infected, in which case the new state is $(n - 1, m + 1)$; or an infective may lose its infectivity, in which case the new state is $(n, m - 1)$. The first event has probability per unit of time βnm and the second αm.

i) Argue that P should satisfy the ODE system

$$\frac{dP_{(n,m)}(t)}{dt} = -\beta nm P_{(n,m)}(t) + \beta(n + 1)(m - 1)P_{(n+1,m-1)}(t)$$
$$-\alpha m P_{(n,m)}(t) + \alpha(m + 1)P_{(n,m+1)}(t).$$

---

[5]F.G. Ball: Coupling methods in epidemic theory. In: Mollison (1995), pp. 34–52; C. Lefèvre and P. Picard: Collective epidemic processes: a general modelling approach to the final outcome of SIR epidemics. In: Mollison (1995), pp. 53–70.

Some jargon: we are dealing with a continuous-time Markov chain (see Section 4.6 and e.g., Taylor and Karlin 1984). Quite naturally, we have indexed the states by the combination (n, m). One can translate the current formulation to the standard formulation in terms of vectors and matrices by employing an appropriately defined (e.g., via lexicographic ordering) map $(n, m) \longmapsto i$. This helps to see the connection with the general theory, but otherwise it just complicates the bookkeeping. So we stick to the index (n, m).

ii) Describe in words the situation embodied in the initial condition

$$P_{(n,m)}(0) = \begin{cases} 1 & \text{for } n = N - 1, \ m = 1, \\ 0 & \text{otherwise.} \end{cases}$$

iii) The states $(n, 0)$ are *absorbing states*. First formulate the precise meaning of this statement. Next decide whether or not you agree. Are there any other absorbing states?

iv) Make plausible that $P_{(n,m)}(\infty) := \lim_{t\to\infty} P_{(n,m)}(t)$ exists and that this vector is concentrated on the set of absorbing states, i.e., $P_{(n,m)}(\infty) = 0$ for $m \neq 0$. Hint: Represent the set of states as a triangle of lattice points in the (S, I) plane. Indicate possible transitions by arrows.

v) Make the connection between $P_{(n,m)}(\infty)$ and the final-size distribution explicit.

vi) We proceed to derive an efficient computational procedure. First derive that the expected sojourn time in state (n, m) equals $(\beta nm + \alpha m)^{-1}$ provided $m > 0$. Next argue that the probability that the transition out of state (n, m) will lead to state $(n - 1, m + 1)$ equals

$$\frac{\beta nm}{\beta nm + \alpha m},$$

and, with the complimentary probability

$$\frac{\alpha m}{\beta nm + \alpha m},$$

the next state will be $(n, m - 1)$.

vii) Inspired by the generation point of view, we now base our bookkeeping on events (or, equivalently, jumps from one state to another). Let $Q_{(n,m)}(l)$ denote the probability that the lth event brought the population into state (n, m). (So now we are dealing with a discrete-time Markov chain; see e.g., Taylor and Karlin 1984.) Show that, provided $m > 1$,

$$Q_{(n,m)}(l + 1) = \frac{\beta(n + 1)}{\beta(n + 1) + \alpha} Q_{(n+1,m-1)}(l) + \frac{\alpha}{\beta n + \alpha} Q_{(n,m+1)}(l),$$

while, for $m = 0, 1$,

$$Q_{(n,m)}(l + 1) = \frac{\alpha}{\beta n + \alpha} Q_{(n,m+1)}(l)$$

and

$$Q_{(n,m)}(0) = \begin{cases} 1 & \text{for } n = N - 1, m = 1, \\ 0 & \text{otherwise.} \end{cases}$$

viii) Show how it is reflected in the recurrence relations for Q that $(n,0)$ is absorbing.

ix) Show that the number of events is bounded by $2N - 1$.

x) How can one compute the final-size distribution by using the recurrence relations for Q?

xi) To prepare for the next exercise, rewrite the recurrence relation of vii) in the form

$$Q_{(n,m)}(l+1) = \frac{R_0(n+1)}{R_0(n+1)+N}Q_{(n+1,m-1)}(l) + \frac{N}{R_0 n + N}Q_{(n,m+1)}(l)$$

for $m > 1$ and

$$Q_{(n,m)}(l+1) = \frac{N}{R_0 n + N}Q_{(n,m+1)}(l)$$

for $m = 0, 1$.

Exercise 1.35 Next consider the setting of Exercise 1.28 with $f < 1$. Denote the population state with n susceptibles, m infectives and k immune individuals by (n,m,k), with the constraints $n, m, k \geq 0$ and $n + m + k \leq N$. In the spirit of the last exercise, derive the recurrence relation

$$
\begin{aligned}
Q_{(n,m,k)}(l+1) &= \frac{R_0(n+1)}{R_0(n+1)+n+m+k}Q_{(n+1,m-1,k)}(l) \\
&+ (1-f)\frac{n+m+k+1}{R_0 n + n + m + k + 1}Q_{(n,m+1,k)}(l) \\
&+ f\frac{n+m+k}{R_0 n + n + m + k}Q_{(n,m+1,k-1)}(l)
\end{aligned}
$$

for $m > 1$, and

$$
\begin{aligned}
Q_{(n,m,k)}(l+1) &= (1-f)\frac{n+m+k+1}{R_0 n + n + m + k + 1}Q_{(n,m+1,k)}(l) \\
&+ f\frac{n+m+k}{R_0 n + n + m + k}Q_{(n,m+1,k-1)}(l)
\end{aligned}
$$

for $m = 0, 1$. What are the absorbing states? What is the maximum number of events? What is now the final-size distribution and how can we compute it?

1.3.5 What do we mean by population size?

In Chapter 3 we shall formulate stochastic models for finite populations of size N and also derive (see Section 3.4) the deterministic limit by letting N go to infinity. Yet in this chapter we analyzed deterministic models in which the total population size N figured as one of the parameters. What does this mean?

The deterministic limit is most easily formulated as well as interpreted in terms of *fractions*. Such fractions are expressed either as part of the initial population size at the moment of the introduction of the first infective individual (which is the population size throughout the epidemic if demography is assumed to be slow and death due to the disease is relatively rare) or, as we will see in Chapter 4, as part of the demographically steady population size (respectively the changing size of a growing population).

In reality, however, everything is finite and so the limit is a mental construction aiming at simplification. If we consider 1 million individuals as nearly infinitely many, then clearly we may also express the sub-populations of susceptibles and infectives in terms of millions and forget about the difference between an integer divided by 10^6 and a real number. We can put total population size at 1 and work with fractions.

At first, this seems to facilitate the comparison between epidemics in host populations of various sizes. But on second thought it just reveals that proportionality constants in quadratic transmission terms are not that easily interpreted, as indeed already emphasized in the last two subsections.

If one population is twice as large as another, what can we conclude about the per capita rate of having contact with other individuals? First of all, we should realize that the answer heavily depends on the character of 'contact.' For instance, for sexual contact it seems plausible that the per capita rate does NOT depend on population size at all. For 'contact' that consists of 'being near enough to inhale aerosols breathed out by the other individual,' things are far more subtle. Whether or not we consider Buda and Pest as two separate cities does, most likely, not matter at all. In contrast, the city of Shanghai and a rural part of China, inhabited by exactly as many people, are quite different even though the population sizes are the same. The point is not that in the large rural area spatial dependence (cf. Chapter 10) is of more importance (which indeed it is). The point is that the population *density* is so different and that this form of 'contact' is strongly influenced by the density.

We conclude:

- How we measure (sub)population size is a matter of bookkeeping. We have to be clear about our measuring stick and our conventions and be systematic and consistent, in particular when comparing results for different populations.

- The choice of the proportionality 'constant' in the quadratic transmission term has to be based on mechanistic considerations, often (but not always) involving population density. When comparing different populations, one should not only compare their sizes, but also the factors influencing the contact process underlying transmission.

1.4   THE EPIDEMIC IN A CLOSED POPULATION: SUMMARY

We have introduced several numbers to classify the 'infectiousness' of a disease, and we have indicated, by way of examples, how these might be calculated from sub-models for the contact process and the probability that transmission occurs, given a contact between a susceptible and an infective that was itself infected τ units of time earlier. These numbers give consistent information in the sense that

$$R_0 > 1 \Leftrightarrow r > 0 \Leftrightarrow s(\infty) < 1 \Leftrightarrow z_\infty < 1.$$

With other aspects of ordering we should be careful: a disease with a long latent period may have a big R_0 and yet a small r (Exercise 1.16). But the ordering of R_0 and $s(\infty)$ of two different non-lethal infections is always identical when considering uniformly mixing communities.

Exercise 1.36 Check that this statement is true. Do you understand why we added 'non-lethal'?

The tacit underlying assumption of deterministic models is that the number of individuals involved in the changes is large. The reward is the clear picture of a sharp threshold: $R_0 < 1 \Rightarrow$ no epidemic; $R_0 > 1 \Rightarrow$ epidemic affecting ultimately a fraction $1 - s(\infty)$ of the population.

In reality, populations are finite and thresholds manifest themselves at the individual level: a contact does or does not take place, and when it does, it either does or does not result in transmission. This blurs the picture: even when $R_0 > 1$, introduction of the agent may only lead to a minor outbreak, whereas for R_0 slightly less than 1 it may result in a substantial fraction of the population being affected. Rather than one precisely determined final size, we have to consider a probability distribution for final size.

The relationship between the intensity of contacts between individuals on the one hand and population size on the other, is a complicated issue that deserves a great deal of attention (much more than it usually gets). The question, 'what actually is population density?' is much harder than the seemingly obvious answer 'number divided by unit area' suggests. Behavioral patterns, as well as heterogeneity of the area (some parts may be attractive habitat, while other parts are only inhabited or visited for want of something better), create non-homogeneous distributions. Consequently, typical 'distance to' and 'contact with' nearest neighbors are the relevant quantities. (Recalling that Part I concentrates on the 'simplest context,' we stop here and refer to Chapter 10 for further remarks.) To compute the final size in the case of a potentially lethal infection, and to compare the final size of the same infection in different populations, one needs to address these issues. We have presented two consistent sets of assumptions leading to, respectively, (1.11) with R_0 proportional to population density, and (1.22) with R_0 independent of population size. Other meaningful variants may yet be uncovered. We do not claim completeness.

Exercise 1.37 The objective of this exercise is to illustrate how knowledge about R_0 comes in helpful when evaluating the chances that a particular control strategy will be effective. Suppose a perfect vaccine is available and we can keep a fraction v of the population vaccinated. Show that $v > 1 - 1/R_0$ leads to eradication of the infectious agent.

Remark: By 'perfect' we mean that a vaccinated individual is completely protected. Usually this is interpreted as 'showing no clinical symptoms after exposure.' In the present context the key feature is 'being not infectious at all after exposure.'

Exercise 1.38 If $R_0 < 2$, the probability z_∞, that introduction leads to a minor outbreak only, satisfies $z_\infty > 1/2$, if the duration of the infectious period is exponentially distributed, cf. (1.8). Reflect upon the possibility to draw conclusions from one or two field observations or experiments.

Exercise 1.39 Argue that the probability that a major outbreak occurs is

$$1 - (z_\infty)^k$$

when we are certain that in some early 'generation' there are exactly k cases.

Exercise 1.40 In animal husbandry, for example in dairy herds, it is a regular occurrence that several (replacement) animals enter the herd from outside. In contrast to the previous exercise it may be that the infection status is unknown or uncertain (because no reliable test exists for certain infectious agents). Assume the new animals all come from the same farm and that this farm has prevalence p for a certain infectious disease. Suppose n randomly selected animals from this farm enter a naive herd. Show that the probability of a major outbreak in that herd is given by

$$1 - (1 - p(1 - z_\infty))^n.$$

Exercise 1.41 Suppose $R_0 < 1$. Argue that the expected size (in terms of 'number of victims') of the epidemic equals $(1 - R_0)^{-1}$.

Exercise 1.42 We have repeatedly switched between a generation perspective and a real-time perspective. But so far we did not derive the deterministic final size in the generation framework. So let us do that now.

We start one generation 'step' with S_0 susceptibles and I_0 infectives, all of infection age zero. We put

$$\frac{dS(t)}{dt} = -cA(t)\frac{S(t)}{N}I_0, \quad S(0) = S_0,$$

where t is now both infection age (which we usually denote by τ) and time in between two generations and where $c, A(t)$ and N have the same meaning as before. We assume that the disease is non-lethal and that, consequently, population size N is constant. By integration, we find

$$S(t) = S_0 e^{-\frac{c}{N}\int_0^t A(\tau)\,d\tau\, I_0},$$

and in the limit $t \to \infty$

$$S(\infty) = S_0 e^{-R_0 \frac{I_0}{N}},$$

where we have used that

$$R_0 = c\int_0^\infty A(\tau)\,d\tau.$$

Clearly then there are

$$S_0(1 - e^{-R_0 \frac{I_0}{N}})$$

new cases created in this generation. Therefore the generation process is described by the recurrence relations

$$
\begin{aligned}
S_{k+1} &= S_k e^{-R_0 \frac{I_k}{N}}, \\
I_{k+1} &= S_k(1 - e^{-R_0 \frac{I_k}{N}}).
\end{aligned}
$$

i) Show that $I_k + S_k - \frac{N}{R_0}\ln S_k$ is a conserved quantity.

ii) Derive the final size equation.

iii) When you think about it, this is really strange. Since generations are not separated but overlap, each infective of a particular generation experiences a different 'environment,' in the sense of the time course of the susceptible fraction of the population that potentially can be infected. So the recurrence relations above, for non-overlapping generations, describe an altogether different process. Yet the final size coincides exactly with the final size of the real-time integral equations. Why?

The authors admit that this exercise is beyond their reach. But they know one argument, going back to Ludwig (1974), that makes the result less miraculous than it seems at first: for the final size it does not matter by whom you are infected, but only whether or not you are infected. To elaborate the argument, one has to think of a process in which infectives 'mark' other individuals, the mark indicating that the individual is now infected. For this process, the bookkeeping in generations and the bookkeeping in real time differ only in the order in which marks are made. So when at the end we count what fraction of the population carries a mark, we find that the outcome does not depend on our way of bookkeeping. See also the paper by Pellis et al. (2008), where the problem underlying this exercise is treated in a different manner.[6]

iv) In the literature one sometimes finds discrete-time epidemic models formulated as

$$
\begin{aligned}
S_{k+1} &= S_k - \beta I_k S_k, \\
I_{k+1} &= \beta I_k S_k.
\end{aligned}
$$

What is wrong with this formulation? How would that possibly show up in the behavior of the solution?

Exercise 1.43 Next consider an epidemic within a herd, as in Section 1.3.3. Show that the generation process is described by the recurrence relations

$$
\begin{aligned}
S_{k+1} &= S_k \left(1 - (1-f)\frac{I_k}{N_k} \right)^{R_0/(1-f)}, \\
I_{k+1} &= S_k - S_{k+1}, \\
N_{k+1} &= N_k - (1-f)I_k.
\end{aligned}
$$

Derive the final-size equations (1.20) and (1.21) from these.

Hint: Consider the ODE system

$$
\begin{aligned}
\frac{dS(t)}{dt} &= -ca(t)\frac{S(t)}{N(t)}I(t), \\
\frac{dI(t)}{dt} &= -\mu(t)I(t), \\
\frac{dN(t)}{dt} &= -\mu(t)I(t),
\end{aligned}
$$

and assume that $\mu(t) = qa(t)$ with $q = c\frac{1-f}{R_0}$ and

$$
f = e^{-\int_0^\infty \mu(\tau)\, d\tau}
$$

---

[6]L. Pellis, N.M. Ferguson and C. Fraser: The relationship between real-time and discrete-generation models of epidemic spread. *Math. Biosci.*, **216** (2008), 63–70.

as before. Show that $I(\infty) = fI_0$, $N(\infty) - N_0 = (f-1)I_0$ and $S(t)/S_0 = (N(t)/N_0)^{R_0/(1-f)}$. Deduce the recurrence relations from these identities. To derive (1.21), first show that

$$S_{k+1}/S_k = (N_{k+1}/N_k)^{R_0/(1-f)}$$

and then iterate. To derive (1.20), introduce as an auxiliary variable the number of immunes R_k, defined by $R_0 = 0$ (beware: here R_0 is the value of R in the zeroth generation and not the basic reproduction number) and $R_{k+1} = R_k + fI_k$. Show that $N_k = S_k + I_k + R_k$ and $R_{k+1} = f(N_0 - S_k)$. Take the limit $k \to \infty$.

Exercise 1.44 Check that the recurrence relations of Exercise 1.42 are recovered from those of Exercise 1.43 in the limit $f \uparrow 1$.

Chapter Two

Heterogeneity: The art of averaging

2.1 DIFFERENCES IN INFECTIVITY

We start with an example. Assume that the latency period and the infectious period of all individuals are the same, but that their infectivities during the infectious period may differ. With reference to the situation and notation introduced in Section 1.2.1, we say that T_1 and T_2 are fixed while p may differ from one individual to another. To describe the whole population, rather than each individual separately, we can specify the *distribution* of p-values (so we imagine that all individuals carry a label specifying the p-value they will have, should they happen to become infected).

During the very first period of the epidemic, the p-values of the few infected individuals, which are determined by chance, matter a lot. Once exponential growth takes off, all p-values are represented among the many infectives. Under the assumption that all individuals are equally susceptible (i.e., the p-label has no influence whatsoever on susceptibility), the occurrence of p-values among those actually infected is accurately described by the a priori given distribution. The force of infection is obtained by summing all contributions of the infectives. So effectively the force of infection is determined by the *mean value* of p. In other words, in our deterministic description (which ignores the demographic stochasticity of the very early stages) we can work with the *expected* infectivity, while ignoring the variance and other characteristics of the distribution. (We have already used this implicitly when working with c, the *expected* number of contacts per unit of time.)

This argument is not restricted to the special example. Given a stochastic submodel that determines infectivity at the individual level, we may invoke the law of large numbers to work with expected values at the population level when describing the dynamics of exponential growth and convergence to a final size (in the very late stages demographic stochasticity comes in again, but as this by definition concerns only few individuals, it hardly influences the final size in a large population).

So let us define

$$A(\tau) := \text{expected infectivity at time } \tau \text{ after infection took place,} \qquad (2.1)$$

having in mind that often the 'shape' of A will be as depicted in Figure 2.1.

The easiest interpretation of 'infectivity' is again the probability of transmission given a contact between a susceptible and an infective of disease age τ. But note that whenever death is a possibility, A necessarily incorporates the probability to survive to age τ (as a factor when the risk of dying is independent of the output of infectious material, but in a more complicated implicit manner otherwise). Moreover, there are contexts that ask for a slightly different interpretation, such as the rate of spore production in the case of fungal plant diseases. As attentive readers have undoubtedly noted, our notation in Chapter 1 anticipated this definition of $A(\tau)$.

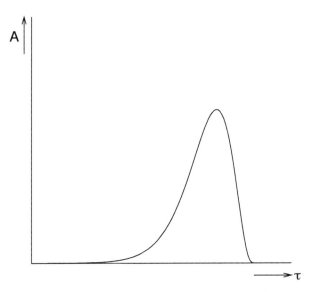

Figure 2.1: An example of a typical infectivity function $A(\tau)$ for an infection with a latency period, followed by an infectious period, and ultimately recovery.

We conclude that all results concerning R_0, r and $s(\infty)$ apply in this generality:

$$R_0 \;=\; c \int_0^\infty A(\tau)\, d\tau, \tag{2.2}$$

$$1 \;=\; c \int_0^\infty e^{-r\tau} A(\tau)\, d\tau,$$

and either (1.22) or $\ln s(\infty) = R_0(s(\infty) - 1)$.

To extend the argument from Exercise 1.16, we note that R_0 is determined only by the integral of A (i.e., only by the area under the graph of the infectivity function), whereas for r also the *shape* of the function A matters. It is possible that an infectious agent where the peak of infectiousness is attained at young infection age, shows faster real-time epidemic growth than an agent where most of the infectivity is shed later in the infectious period, but that both have the same value of R_0. This effect causes influenza virus to generate fast-growing outbreaks, even though the virus has a very low value of R_0 of 1.5–2.0.

Exercise 2.1 Show that for the Kermack-McKendrick ODE model of Exercise 1.22,

$$cA(\tau) = \beta N e^{-\alpha\tau}.$$

Exercise 2.2 To include a latency period and yet work with a system of ODEs, the following, so-called SEIR, system is used (here E denotes 'exposed and infected but not yet infectious'):

$$\frac{dS}{dt} = -\beta SI,$$

$$\frac{dE}{dt} = \beta SI - \theta E,$$

$$\frac{dI}{dt} = \theta E - \alpha I,$$

$$\frac{dR}{dt} = \alpha I.$$

i) Show that in this case

$$cA(\tau) = N\beta \frac{\theta}{\alpha - \theta}(e^{-\theta\tau} - e^{-\alpha\tau}).$$

Hint: In order to be in the 'infectious' state at time τ, one should have entered this state at some time $\tau_0 \in (0, \tau)$ and remained in it in the interval (τ_0, τ). Compute the probability for arbitrary τ and then integrate with respect to τ_0.

ii) Deduce that $R_0 = \beta N/\alpha$, independently of θ. Are you surprised?

iii) Demonstrate that in general, i.e., for arbitrary patterns of infectivity, the length of a latency period has no influence on R_0 whenever individuals do not die. What about r?

Exercise 2.3 So far we have pretended that infected individuals take part in the contact process in just the same way as healthy individuals do. Now imagine that the infectious period is from $\tau = T_1$ to $\tau = T_2$, but that individuals have, due to illness, a reduced contact rate in this period. Can you incorporate this in $A(\tau)$? How does this change the interpretation of $A(\tau)$?

(Since 'contacts' involve two individuals, we have to worry about the consistency of our assumptions in situations in which a non-negligible fraction of the population is infected. You are advised to postpone such justified worries until you reach Section 12.3, and concentrate here on the situation immediately after the introduction of the infectious agent.)

Exercise 2.4 i) Explain why a death rate μ (from 'natural,' i.e., infection-unrelated causes) is expressed as a factor $\exp(-\mu\tau)$ in $A(\tau)$.

ii) Deduce from this that a latency period of fixed length T_L is reflected in a factor $\exp(-\mu T_L)$ in R_0.

iii) Derive from i), or directly, that for the Kermack-McKendrick ODE model with death rate μ we have $R_0 = \beta N/(\alpha + \mu)$.

iv) To calculate R_0 for the SEIR system of Exercise 2.2 with death rate μ, we use a trick that we have encountered already in Exercise 1.34-vi. During the latency period there are competing 'risks,' viz., to die or to become infectious. If the latter occurs, we are again in the same situation as in iii) of the present exercise. Check that these arguments lead to the expression $R_0 = \frac{\theta}{\theta + \mu} \frac{\beta N}{\alpha + \mu}$.

v) If μ is small, death can be ignored. But what does 'small' mean? What are the quantities that one should compare μ with? Hint: Pay attention to the dimension.

Exercise 2.5 Consider a herpes infection, which, after a first infective period, may become latent, resurge with a certain probability per unit of time, etc. Let 1 denote the infective state and 2 the latent state. Let σ denote the rate of going from 1 to 2 and ν the rate of going from 2 to 1, while μ denotes the death rate. Assume that a newly infected individual is in state 1.

i) Give a priori arguments why $R_0 = \infty$ when $\mu = 0$ but $\nu > 0$.

ii) Show that $\int_0^\infty A(\tau)d\tau = h_1 \frac{\nu+\mu}{\mu(\sigma+\nu+\mu)}$, where h_1 measures the infectivity in state 1. Hint: First read on a bit, then return to the exercise.

Let us reflect upon some implicit underlying assumptions that correspond to the image we have of what is going on in a host after an infectious agent has invaded. We imagine that infection triggers an autonomous process within the infected individual. In particular, the invading organism reproduces within the host at such a rate that further infections with the same agent are irrelevant. Examples include measles, influenza, rabies and HIV, and fall under the general heading of *microparasites* (to distinguish them from another class, the *macroparasites*, such as worms, for which infection is not a unique event but rather a repeated process; we come back to these in Section 6.1.1 and Chapter 11). When taking the function $A(\tau)$ as our chief modeling ingredient, we give up on a precise biochemical-physical description of what is going on in the host and take instead the pragmatic view that all that matters for spread at the population level is the output of infectious material by the host. One increasingly important area not treated in this book is the interaction between the agent and the immune system of the host, and specifically the way in which this interaction shapes the infectious output A. In principle one could try to determine $A(\tau)$ experimentally, but for human infections this is in practice almost always impossible. For infections of animals one can often measure shedding of infectious material in controlled experiments where animals are inoculated and then sampled for a subsequent period. For plant pathogens a similar approach is possible, for example measuring spore production by a fungal pathogen. A first conclusion is that we should only have confidence in those inferences that are *robust*, i.e., independent of the details of A. The threshold $R_0 > 1$ is a clear example. A second is that we might try to work with parameter-scarce, coarse and quasi-mechanistic, sub-models of the autonomous process within the host, compute as many characteristics of A (such as R_0) from such a description as we need, and try to get insight and understanding by analyzing how the result depends on the parameters. (For this to work, the parameters should allow a quasi-mechanistic interpretation, of course.) If we are ambitious, we might try to fit the parameters on the basis of whatever field and experimental data are available.

An attractive class of sub-models, take the form of continuous-time Markov chains (see e.g., Taylor and Karlin 1984 for background). Here we assume that an infected individual can be in a finite number of states (which we shall call *d-states*, with 'd' denoting disease). The traditional SIR, SEIR, SIS, . . . , systems of ordinary differential equations are all based on such a description. Here we shall present a straightforward systematic procedure to compute R_0 for such models. The following should be seen as merely a prelude; we return to this theme in greater detail in Section 7.2.

Label the d-state by $i = 1, 2, \ldots, n$. Do not include 'dead' or 'removed' in this list, which correspond to the definite loss of infectivity. Let $\Theta = (\theta_1, \ldots, \theta_n)^\top$ (where $\top$ indicates taking the transpose) with $\sum_{i=1}^n \theta_i = 1$ denote the probability distribution for d-state at the moment immediately following infection. (Often one will arrange things such that $\theta_1 = 1, \theta_i = 0$, for $i \neq 1$, and let infected individuals progress through states $1, \ldots, n$ in consecutive order.) Let Σ denote the $n \times n$-matrix of transition probabilities per unit of time. By this we mean that σ_{ij}, $i \neq j$, is the probability per unit of time for an individual's state to change from j to i. In the absence of death/removal, the diagonal elements are defined by $\sigma_{jj} = -\sum_{i \neq j} \sigma_{ij}$ (columns therefore sum to zero). If states are passed through in their natural ordering $1, 2, \ldots, n$, we will have that $\sigma_{ij} = 0$ for all $i \notin \{j, j+1\}$ and so (when death/removal does not occur) $\sigma_{jj} = -\sigma_{j+1,j}$. We will refer to the matrix Σ as the transition matrix. Here and everywhere else in the text we adopt the convention for the elements of transition matrices that the second index gives the state of departure and that the first index gives the state of arrival (in contrast to common practice in probability theory). Note that transition probabilities are non-negative by definition; a negative element of Σ reflects that there is a positive probability of moving *out* of the departure state. Finally, note that when we include in Σ also changes that are interpreted as death or removal from the system, these give rise to additional negative terms on the diagonal of Σ.

If we denote by the vector $x(\tau)$ the probability to be in the various states at time τ, then

$$\frac{dx}{d\tau} = \Sigma x,$$
$$x(0) = \Theta,$$

(note that $x(\tau)$ may be defective, i.e., $\sum_{i=1}^{n} x_i(\tau) < 1$, due to the possibility of death/removal).

Let h denote the vector of infectivities associated with the various d-states. Then

$$A(\tau) = h \cdot x(\tau) = \sum_{i=1}^{n} h_i x_i(\tau) \tag{2.3}$$

where '$\cdot$' represents the inner product. (We are deliberately a bit vague about the interpretation of the 'infectivity' vector h. When infectivity is interpreted as 'probability of transmission, given a contact,' one should multiply $A(\tau)$ by a contact intensity parameter c and integrate with respect to τ from 0 to ∞ to obtain R_0. The alternative is to incorporate the contact intensity in h, in which case the integral of A itself equals R_0.) A key point is now to observe that in order to determine the integral of A, we do not need to determine $x(\tau)$. For if we integrate the differential equation from 0 to ∞, while noting that $x(+\infty) = 0$ by death/removal, we find

$$-\Theta = \Sigma \int_0^\infty x(\tau)\,d\tau \Rightarrow \int_0^\infty x(\tau)\,d\tau = -(\Sigma)^{-1}\Theta$$
$$\Rightarrow \int_0^\infty A(\tau)\,d\tau = -h \cdot (\Sigma)^{-1}\Theta. \tag{2.4}$$

So all we have to do is to invert the matrix Σ and apply the result to the vector Θ. We will return to the matrix Σ at length in Chapter 7, and also deal with the existence of Σ^{-1} there.

Exercise 2.6 Interpret $(-(\Sigma)^{-1})_{ij}$ as the expected total sojourn time in state i given that the system currently has state j.

Exercise 2.7 i) Do Exercise 2.5.

ii) Analyze how R_0 depends on the recrudescence parameter ν.

Exercise 2.8 Take $n = 3$. Let Σ be described by the scheme in Figure 2.2. Let $\theta_1 = 1$, $\theta_2 = \theta_3 = 0$, $h_1 = 0$, $h_2 = 1$, $h_3 = 2$. Calculate $\int_0^\infty A(\tau)\, d\tau$.

Exercise 2.9. In Section 1.3.3 we introduced the condition $\mu(\tau) = qa(\tau)$, see (1.29). Show that this translates into the condition $\mu = qh$ in the present context, where μ now denotes the vector of death rates on the diagonal of Σ. Go over the interpretation of this condition once more.

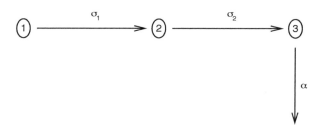

Figure 2.2: Schematic representation of a system where individuals can be in three different states, with transitions indicated by arrows.

In concluding this section, we analyze the influence of variability in infectivity on the probability of a minor outbreak. We return to this topic at length in Chapter 3. An abstract but very convenient approach is the following. Distinguish individuals from one another according to 'type,' and label types with a variable ξ taking values in a set Ω. Let the measure m on Ω describe the distribution of types (i.e., for any measurable subset ω of Ω, the number $m(\omega)$ equals the fraction of the population with type $\xi \in \omega$; in particular, $m(\Omega) = 1$). (An illuminating example, for those unfamiliar with the idea of a measure, may be when ξ denotes p, as discussed at the beginning of this section; then, with $\omega = [p_1, p_2)$, we have that $m(\omega)$ is the probability $\Pr\{p_1 \leq p < p_2\}$ that an arbitrary individual has p-value such that $p_1 \leq p < p_2$. For a very brief introduction to measures see Section 6.2.) Let $a(\tau, \xi)$ be the infectivity at infection age τ of individuals of type ξ. Then

$$A(\tau) = \int_\Omega a(\tau, \xi) m(d\xi), \tag{2.5}$$

while, reasoning as in Section 1.2, the generating function for the offspring distribution of the associated branching process is found to be

$$g(z) = \int_\Omega \exp\left(c(z-1) \int_0^\infty a(\sigma, \xi)\, d\sigma \right) m(d\xi). \tag{2.6}$$

Recall the assumptions: contacts are made according to a Poisson process with intensity c and transmission will occur with probability $a(\tau, \xi)$; so the probability that an infective of type ξ will have k offspring equals

$$\frac{(c \int_0^\infty a(\sigma, \xi)\, d\sigma)^k}{k!} e^{-c \int_0^\infty a(\sigma, \xi)\, d\sigma}$$

and this now has to be averaged over ξ, multiplied by z^k and summed over k to obtain g; by interchanging the integration with respect to ξ with the summation over k we arrive at (2.6).

Exercise 2.10 Verify that $g(z) = \exp(R_0(z-1))$ whenever infectivity is deterministic (i.e., there is but one type; formally Ω then consists of one point only). Conclude that the probability of a minor outbreak equals the (deterministic) complement $s(\infty)$ of the final size whenever infectivity is deterministic.

Exercise 2.11 i) Prove that in general $g(z) \geq \exp(R_0(z-1))$. Hint: Apply Jensen's inequality.

ii) Deduce from this that in general $z_\infty \geq s(\infty)$.

iii) <u>We conclude</u> that variability in infectivity has the tendency to increase the probability that introduction of the agent leads to a minor outbreak only. Do you agree? What exactly do we mean by such a statement? Do you find it intuitively plausible?

In Section 3.5.3 we shall show that the expected final size is determined by the expected infectivity, so by R_0. This was already made plausible by our 'derivation' of the final-size equation just preceding Remark 1.21. And in that remark we pointed to Section 12.6.3 for yet another way to provide intuitive understanding. <u>We conclude</u> that differences in infectivity, while they do complicate the characterization of the probability of a minor outbreak, do *not* complicate the characterization of the expected final size.

2.2   DIFFERENCES IN INFECTIVITY AND SUSCEPTIBILITY

Consider a population divided into two sub-populations, which we shall indicate by the labels 1 and 2. Although we are going to consider hypothetical numbers to illustrate a general phenomenon, it may be helpful to think of a sexually transmitted disease with sexual intercourse as a 'contact.' In such a context one then has to imagine a promiscuous population without any form of partnership. Suppose there are $N_1 = 10^5$ individuals of type 1, which have $c_1 = 100$ contacts/year, and $N_2 = 10^7$ individuals of type 2, which have $c_2 = 10$ contacts/year. Assume that a fraction $c_2 N_2/(c_1 N_1 + c_2 N_2)$ of the contacts of a 1-individual are with 2-individuals and a fraction $c_1 N_1/(c_1 N_1 + c_2 N_2)$ of the contacts of a 2-individual are with 1-individuals. Note that this means that individuals choose partners irrespective of type and that these rules satisfy the consistency condition that there are as many contacts between 1-individuals and 2-individuals as there are between 2-individuals and 1-individuals. Consider now an infective agent that yields an infectious period of exactly one year, during which transmission occurs with probability 1 for each contact with a susceptible. What is R_0?

The average value of c equals

$$\frac{c_1 N_1 + c_2 N_2}{N_1 + N_2} \approx 11,$$

and were new cases to be distributed over the types according to the relative sub-population sizes (i.e., as $N_1/(N_1 + N_2) : N_2/(N_1 + N_2)$) then this would, given our assumptions about infectivity, be equal to R_0. Actually, however, type i individuals account for a fraction $c_i N_i/(c_1 N_1 + c_2 N_2)$ of all $c_1 N_1 + c_2 N_2$ contacts and since transmission occurs during contact, the new cases are distributed over the types according to these relative numbers of contacts (i.e., as $c_1 N_1/(c_1 N_1 + c_2 N_2)$: $c_2 N_2/(c_1 N_1 + c_2 N_2)$), and therefore

$$R_0 = c_1 \frac{c_1 N_1}{c_1 N_1 + c_2 N_2} + c_2 \frac{c_2 N_2}{c_1 N_1 + c_2 N_2} = \frac{c_1^2 N_1 + c_2^2 N_2}{c_1 N_1 + c_2 N_2} \approx 18.$$

We conclude that the distinction of 'types' can have a profound influence on R_0 when it involves a correlation between infectivity and susceptibility (see also Section 12.4.1).

Exercise 2.12 Intuitively, one would think that whenever a group decreases its contact intensity (i.e., the value of the parameter c), the basic reproduction number would decrease. Yet this is not necessarily the case. What may happen is that when a less active group reduces its activity even further, a more active group is 'forced' to increase the fraction of 'within-group' contacts, in order to sustain its contact intensity, which then leads to an increase of R_0.

Substantiate this verbal argument by first computing the derivative of

$$R_0 = \frac{c_1^2 N_1 + c_2^2 N_2}{c_1 N_1 + c_2 N_2}$$

with respect to c_1, and then finding a condition for the result to be negative.

Exercise 2.13 The ratio of type 1 to type 2 individuals is 1:100. What is this ratio among new cases in the early phase of the epidemic? And what is this ratio among those responsible for transmission in the early phase of the epidemic?

In the present example there is correlation between the susceptibility and the infectivity of an individual, yet we can figure out directly how new cases will be distributed with respect to type. This allows us to compute the appropriate average directly. The key point is that the properties of the *two* individuals involved in the transmission event have *independent* influence, see Sections 7.4.1 and 8.4. A much more complicated situation arises when there is correlation between the infectivity of the 'donor' and the susceptibility of the 'receiver' involved in a transmission event. In that case it will be implicitly determined by the dynamics of the generation process how new cases will be distributed with respect to type. In Part II we shall present a systematic analysis based on the spectral theory of positive matrices and operators.

Exercise 2.14 The distribution of new cases, with respect to type, for sampling in real time, may or may not be identical to this distribution for sampling on a generation basis. Can you think of a condition that guarantees that they are identical?

When a population is subdivided into types, the monotonic relation between R_0 and $s(\infty)$ is lost. This is most easily seen by considering first the degenerate limiting case of two decoupled sub-populations: one small and one large. If we give the small population $R_0 = 1$ and the large population $R_0 = 2$, the final size, expressed as a fraction of the total population, will be close to the final size corresponding to $R_0 = 2$ for a homogeneous population. If we give the small population $R_0 = 2$

and the large population $R_0 = 1$, the final size, again expressed as a fraction of the total population, will be rather small. By introducing some coupling, we can arrange that in the first case the coupled population has R_0 slightly less than in the second case. On the other hand, the final size expressed as a fraction of the total population will hardly change by the coupling.

The conclusion is that a small *core group* (see also Section 12.4.2 and Exercise 8.37) may have a large impact on initial growth while, not surprisingly, it has little impact on the final size expressed as a fraction of the total population.

Even if we know that averaging introduces errors, we may still choose to do it. Tractable approximations are often to be preferred over scrupulous representations involving a multitude of badly known parameters. After all, modeling is the art of simplification without oversimplification.

So when building a model, a key question is which traits of individuals to take into account and which to neglect. Quite naturally, age and spatial position score highly when it comes to being included in a model.

The assumption that everybody is equally likely to be infected by a particular infectious individual is certainly not warranted when relative spatial position decides about the possibility of having contact. In such a case, $R_0 > 1$ is still necessary for spread, but not necessarily sufficient (see Section 10.7), and r is not always a good indicator of (initial) growth. In large spatial domains one finds wavelike spread with a characteristic velocity, the so-called *asymptotic speed of propagation*. We shall deal with this issue in Chapter 10.

Patterns of human social behavior and sexual activity correlate with age. The seriousness of many infectious diseases depends on age. Data on the distribution of the random variable 'age at infection' contain information about the prevailing force of infection in an endemic situation. These are three of the reasons to incorporate age as a variable characterizing individuals in models for the spread of contagious diseases, notably in human communities. We refer to Chapter 9 for elaboration.

Still other aspects of heterogeneity are the following. Space is neither homogeneous nor isotropic (winds, mountains!). Contact rates change during the year, for example as a result of the school system, or of the life history of an insect vector. Moreover, environmental conditions influence the transport of aerosols and the survival of a free virus. We shall pay some, but not much, attention to such aspects.

2.3   THE PITFALL OF OVERLOOKING DEPENDENCE

In Section 2.1 we argued that, when individuals differ only in infectivity, one can in a sense, simply work with the 'average' individual. The aim of this section is to expose a crucial, but hitherto hidden, assumption, viz., that the contact process is *uniform*. If, in contrast, the population is represented by a spatial lattice or a social network, with contact restricted to sets of neighbors (partners, acquaintances, ...), then *all* neighbors of a certain individual will experience the *actual* infectious pressure exercised by this *particular* individual, and not some kind of average. We shall deal with this kind of situation in Section 12.6, but we want to illustrate the key influence of dependence in experienced infectivity now, early on in the book, because this effect tends to be neglected in the literature.[1]

---

[1] Two examples are M.E.J. Newman: Spread of epidemic disease on networks. *Physical Review*, **E66** (2002), 016128; J.P. Aparicio and M. Pascual: Building epidemiological models from R_0: an

To set the scene, assume that an infectious individual transmits the infectious agent to a given susceptible neighbor with probability

$$1 - e^{-\delta T} \tag{2.7}$$

if the length of the infectious period of the infectious individual equals T.

We assume that, given T, transmission to different susceptible neighbors is independent and with equal probability. So if there are N susceptible neighbors, then the probability that, given T, i of these will become infected, is given by the binomial distribution

$$p_c(i; T) = \binom{N}{i} (1 - e^{-\delta T})^i (e^{-\delta T})^{N-i}. \tag{2.8}$$

By the well-known identity

$$\sum_{i=0}^{m} \binom{m}{i} i a^i (1-a)^{m-i} = am \tag{2.9}$$

we find that, given T, on average

$$(1 - e^{-\delta T})N \tag{2.10}$$

neighbors will become infected.

Next assume, for the sake of the exposition, that every individual has exactly N neighbors, but that the quantity T is allowed to differ from one individual to another, and is moreover considered as a random variable with a known distribution (by assumption). Even though it doesn't matter for the current exposition, it is natural to assume that the infectious periods of neighbors are not correlated.

The 'on average' that precedes (2.10) refers to the N copies of the chance process that determines whether, given an infection probability (2.7), transmission does or does not occur. But now we can introduce a second 'on average,' which refers to the chance process that determines T.

Let us, again for the sake of exposition, concentrate on the case that T is exponentially distributed with parameter γ. The average of transmission probability (2.7) is then given by

$$\int_0^\infty \gamma e^{-\gamma T} (1 - e^{-\delta T}) dT = 1 - \frac{\gamma}{\gamma + \delta} = \frac{\delta}{\gamma + \delta} \tag{2.11}$$

and the average of the average number (2.10) of secondary cases is given by

$$\frac{\delta}{\gamma + \delta} N. \tag{2.12}$$

So it is tempting to conjecture that, if we average the distribution (2.8) with respect to T, we obtain

$$\binom{N}{i} \left(\frac{\delta}{\gamma + \delta} \right)^i \left(\frac{\gamma}{\gamma + \delta} \right)^{N-i}, \tag{2.13}$$

but this is *not* true, simply because (2.8) is nonlinear in the quantity $e^{-\delta T}$ or, in more probabilistic terms, because the infection probabilities that different neighbors of the same infectious individual experience, are correlated.

implicit treatment of transmission in networks. *Proc. R. Soc. B*, **274** (2007), 505–512.

Exercise 2.15 Show that

$$\int_0^\infty \gamma e^{-\gamma T} \left(\begin{array}{c} N \\ i \end{array} \right) (1 - e^{-\gamma T})^i (e^{-\gamma T})^{N-i} dT = \frac{1}{N+1}.$$

Hint: Take the identity

$$\sum_{j=0}^i \frac{(-1)^j}{j!(i-j)!} \frac{1}{j+N-i+1} = \frac{(N-i)!}{(N+1)!}$$

for granted.

So when $\gamma = \delta$ the true distribution for the number of secondary cases is uniform, i.e., every number between 0 and N is equally probable, whereas the wrong description of (2.13) would suggest that it is binomial with parameter $1/2$. If you plot these distributions, you will see that the difference is rather striking.

Of course there are other pitfalls as well. When every individual has N neighbors, an infected individual has at most $N - 1$ susceptible neighbors, because the infected individual was infected by one of its neighbors, which hence cannot be susceptible (unless immunity is only temporary). However, if there is correlation, in the sense that two neighbors of any individual are more likely neighbors of each other than an arbitrary pair of individuals, the chance that, even at the start of an epidemic, the number of susceptible neighbors of an infected individual is less than $N - 1$, might be considerable. Populations represented by the points of a spatial lattice provide a good example, but the phenomenon also occurs in networks with more randomness. In such situations of locally finite populations with correlated contact structure, one cannot define R_0 in a meaningful way, because already in the initial phase the dependence (concerning epidemiological offspring production) in successive generations cannot be neglected. In particular, 'the expected number of cases produced by the first case' does not necessarily serve as a good predictor of the reproduction number in later generations and, hence, should not be called R_0 (and the correction of replacing N by $N - 1$ may not be sufficient). In Section 12.6.3, called 'A network with hardly any structure,' we shall eliminate the difficulty by making additional (debatable) assumptions, as is very often done in the network literature. The challenge to address analytically the difficulty imposed by correlated contact structure remains.

2.4   HETEROGENEITY: A PRELIMINARY CONCLUSION

Individuals of the host population show variation with respect to properties that are relevant for the transmission of the agent. These properties may affect susceptibility, infectivity, or both.

When every individual is equally susceptible, we can, in a deterministic setting (i.e., for large numbers of all categories of individuals involved), simply work with the *mean* infectivity, at least as long as we are only interested in R_0. When the susceptibility differs, but a certain form of independence holds between susceptibility and infectivity, we can still compute the appropriate averages directly. More precisely, this can be done when the properties of two individuals independently influence the probability of contact and transmission. This was demonstrated by way of example in Section 2.2 and will be covered in detail in Section 7.3.

How to handle the case of dependence cannot be explained at this stage. We simply refer to Chapter 7.

Chapter Three

Stochastic modeling: The impact of chance

In this chapter we introduce epidemic models that incorporate randomness at the population level, so-called demographic stochasticity. Even in the context of rather simple stochastic epidemic models it is quite complicated to characterize the distribution of the final size (in particular there is no closed-form solution). We therefore postpone this topic to Section 12.5.3 where we will treat a special situation concerning models for populations structured into households. Now we first focus on the derivation of approximations for large (but finite) population size. These approximations concern the initial phase of an epidemic, the main phase when the epidemic is full-blown and where most infections take place, as well as the final number infected.

In the present chapter we will encounter random variables following different *distributions*, some of which have already been mentioned in the previous chapters. We list some common distributions below and refer to, for example, Ross (1998), Balakrishnan and Nevzorov (2003) for more about properties of these and other distributions. We start with some discrete (integer-valued) distributions.

A non-negative integer-valued random variable X is said to follow the *Poisson distribution* with parameter $\lambda > 0$ if the probability function is given by: $P(X = k) = \lambda^k e^{-\lambda}/k!$, $k = 0, 1, 2, \ldots$. The mean of this random variable, defined by $E(X) = \sum_k kP(X = k)$, equals λ.

A non-negative integer-valued random variable Y is said to follow the *binomial distribution* with parameters n and p (where n is a positive integer and $0 < p < 1$) if the probability function is given by: $P(Y = k) = \binom{n}{k}p^k(1-p)^{n-k}$, $k = 0, 1, \ldots, n$. The mean of this random variable equals np.

A non-negative integer-valued random variable Z is said to follow the *geometric distribution* with parameter p if the probability function is given by: $P(Z = k) = p(1-p)^k$, $k = 0, 1, 2, \ldots$. The mean of this random variable equals $(1-p)/p$.

Another (main) class of random variables are those that may attain any value in some interval. Such random variables are called 'continuous.' The probability to exactly obtain any specific value is necessarily 0, so continuous random variables are most often described through their *density functions*. The density function $f_X(x)$ of a continuous random variable X is defined by $f_X(x) = \lim_{h\to 0} P[X \in (x, x+h)]/h$. We now give the density function for two common continuous random variables.

A continuous random variable X is said to follow the *exponential distribution* with parameter $\beta > 0$ (often referred to as the 'rate') if $f_X(x) = \beta e^{-\beta x}$ for $x \geq 0$ (and $f_X(x) = 0$ for $x < 0$). The mean of the exponential distribution, defined by $E(X) = \int x f_X(x)dx$, equals $1/\beta$.

A continuous random variable Y is said to follow the *normal (or Gaussian) distribution* with parameters μ and σ^2 if $f_Y(y) = (2\pi\sigma^2)^{-1/2}\exp\left((y-\mu)^2/2\sigma^2\right)$ (where $-\infty < \mu < \infty$ and $\sigma^2 > 0$) for $-\infty < y < \infty$. The mean of the normal distribution equals μ, and the variance, defined by $E((X - E(X))^2)$, equals $\int (y - \mu)^2 f_Y(y)dy = \sigma^2$. One often writes: $Y \sim \mathrm{N}[\mu, \sigma^2]$.

Finally, readers who have never come across the notation using 'O' and 'o' for asymptotic relations could consult http://en.wikipedia.org/wiki/Big_O_notation.

3.1   THE PROTOTYPE STOCHASTIC EPIDEMIC MODEL

In this section we describe and study a *stochastic* epidemic model for a finite homogeneous population using a similar description of the latency period, infectious period and contact processes as before, in line with standard terminology. We call the model the *prototype stochastic epidemic model*.

Recall that in the simple situation described in Section 1.2.1, an individual who becomes infected is first latently infected for a period T_1, then infectious for a period $\Delta T = T_2 - T_1$, and after that the individual recovers and becomes immune. The quantities T_1 and ΔT can be the same for all individuals or they may vary between individuals. In the latter case these durations are described by independent random variables having distributions F_L and F_I, respectively (these could of course be correlated for the same individual, but here we assume independence for simplicity). In other words, the latency periods of different individuals are independent and identically distributed, usually abbreviated as 'i.i.d.,' and the same holds for the infectious periods. Each pair of individuals is assumed to have contacts at random times according to a Poisson process with constant rate, and contacts between different pairs occur independently and have the same constant rate. This means that such contacts occur randomly in time, the number of contacts in disjoint intervals are independent, and the probability of a contact in a short interval of length h is proportional to h. As a consequence, an individual has contact with some random other individual in the community according to a Poisson process with a constant rate, being the sum of all (identical) individual rates. Let us denote this accumulated rate by c. Only contacts where one of the individuals is infectious and the other individual is susceptible may lead to disease transmission. Let p denote the probability that the susceptible individual becomes infected at such a contact — we then speak of a *successful* contact. Consequently, the susceptible remains susceptible with probability $1 - p$. Contacts between individuals in other states have no effect on the spread of the disease. The epidemic goes on until there are no latent or infectious individuals — then no one can get infected any more and the epidemic stops. Assume that initially there are N susceptible individuals and one recently infected individual (thought of as being infected from some source outside the population). To be precise, let $t = 0$ coincide with the start of this individual's latency period.

Exercise 3.1 It follows that the number of contacts an individual makes during a time interval of length t is Poisson distributed with parameter ct. Compute the probability that an individual has 4 contacts during one day ($t = 1$) if $c = 5$.

Exercise 3.2 An individual's over-all contact rate is c and the contact rate is the same between all pairs of individuals. There are in total $N + 1$ individuals, so for each individual there are N others to contact. What is the contact rate between a specific pair of individuals? What is the over-all population contact rate? Hint: How many pairs of individuals are there?

Because we have a finite community and the latency and infectious periods are assumed to be finite, the epidemic will stop within a finite time. If we let Y

denote the number of initially susceptible individuals who get infected during the course of the epidemic, i.e., Y is the final size in the stochastic system, then Y is a random variable taking values in $\{0, 1, \ldots, N\}$. The distribution of Y depends on the parameters c, p and on the distribution of the infectious period F_I (or on ΔT itself, if it is a constant). Somewhat surprisingly the distribution of Y does *not* depend on the latency period. The underlying reason for this is, as explained before, that the latency period only affects the time dynamics of the disease: to avoid being infected by a particular individual you only have to avoid being successfully contacted during its infectious period. The probability for this depends on the length of the infectious period but is independent of the length of the latency period; the latter only affects the timing of such a contact (cf. Exercise 2.2 where it was shown that the latency period had no effect on R_0 in the deterministic setting).

To compute the probability distribution of Y is actually quite difficult. The reason for this is two-fold. First, there are different infection chains by which a given set of k individuals can get infected, and the number of different routes grows very quickly with k and, even worse, the different routes have different probabilities. Secondly, unless the infectious period is constant, the events when two different individuals are successfully contacted by a given infectious individual are positively correlated and hence *dependent*: if the infectious period is random, then the information that a specific individual was successfully contacted should be incorporated in our expectations, which hence increases the probability that other individuals will be contacted successfully (cf. Exercise 3.3 below). It quickly becomes very complicated to compute $P(Y = k)$ when k and N grow. In fact, even with the aid of powerful computers it takes a prohibitive amount of time to compute these probabilities when k is in the hundreds. This illustrates the need for approximations. In the rest of this chapter we will hence study approximations relying on the population being large. In Section 12.5.3 we derive a recursive formula for the exact distribution, $P(Y = k)$, when considering an outbreak within a household (so for small N).

Exercise 3.3 Regard one infectious individual A and two susceptible individuals B and C, and neglect the rest of the population. Assume that A's infectious period has length $\Delta T = t$.

i) Show that the probability that B escapes infection from A is $e^{-cpt/N}$.

ii) This is also the probability that C escapes infection from A. All contact processes were defined to be independent, so what is the probability that both B and C avoid being infected by A?

iii) In other words, conditional on $\Delta T = t$, these two events are independent. However, if we don't condition on the infectious period, then the corresponding probabilities become $E(e^{-cp\Delta T/N})$ for B escaping infection from A (and the same for C escaping infection from A), and $E(e^{-cp\Delta T/N} \times e^{-cp\Delta T/N}) = E(e^{-2cp\Delta T/N})$ for both B and C escaping infection from A (expectation is with respect to the infectious period ΔT). Compute these two quantities for the case where ΔT has exponential distribution with parameter α, so for example $E(e^{-cp\Delta T/N}) = \int_0^\infty e^{-cpt/N} \alpha e^{-\alpha t} dt$.

iv) Show that the product of the probabilities of the two single events is not equal to the probability of the multiple event, i.e., that the two events are *dependent*. Is the probability of the multiple event larger or smaller than the square of the probability of the single event?

3.2 TWO SPECIAL CASES

Two special cases of the prototype stochastic epidemic model have received special attention in the literature, mainly because of mathematical tractability.

3.2.1 The Reed-Frost model

The (continuous-time version of the) Reed-Frost model, named after its inventors, is where there is no latency period ($T_1 \equiv 0$) and where ΔT is fixed and constant. This implies that successful contacts with different individuals are independent, as shown in the exercise above, and the probability to have a successful contact with a specific other individual is $1 - \pi$, where

$$\pi = e^{-pc\Delta T/N}.$$

This follows (cf. Exercise 3.3-i) because successful contacts occur at rate pc, and at rate pc/N with a specific other individual, and the expression for π is the Poisson probability of having no such contact during ΔT.

The discrete-time version of the Reed-Frost model is essentially the same as the continuous-time version but neglects real time and instead looks at *generations* of infected individuals. Each infectious individual in a given generation infects susceptibles independently with probability π and then recovers. As a consequence, a susceptible will remain available for infection in the next generation if and only if it escapes infection from all infectives in the present generation. So if there are I infectives, then this happens with probability π^I. If the parameters in the continuous time and discrete time version are calibrated by $\pi = e^{-pc\Delta T/N}$, the final size distributions coincide. The underlying reason for this is that, in both the discrete-time and the continuous-time version, contacts between different individuals are independent, and when calibrated the two versions have the same probability of infection (cf. Exercise 1.42).

Even for the Reed-Frost model, the distribution $Y \subset \{0, 1, \ldots, N\}$ of the final size is not easy to compute, as we now illustrate using the discrete time version. For Y to equal 0 all individuals must avoid being infected by the initially infected individual, this happens with probability $P(Y = 0) = \pi^N$. For Y to equal 1, one out of N must become infected by the initial infected and the rest must avoid infection from the initial *and* also from the newly infected. The probability hence equals $P(Y = 1) = \binom{N}{1}\pi^{N-1}(1 - \pi)^1\pi^{N-1}$. For Y to equal 2, the initial infected individual could either infect two new individuals and these fail to infect anyone else, or else the initial individual infects one who in turn infects one of the remaining and this individual fails to infect anyone else.

Exercise 3.4 Compute $P(Y = 2)$ and $P(Y = 3)$ (along the lines hinted at above).

3.2.2 The 'general' stochastic epidemic

The second model which has received special attention, is one with no latency period ($T_1 \equiv 0$) and an exponentially distributed infectious period ΔT (with parameter α and hence mean $1/\alpha$). This model goes under the somewhat unfortunate name *the 'general' stochastic epidemic*. No one really believes that an exponentially distributed infectious period mimics reality very well (see Figure 3.1), it is chosen for mathematical convenience. At time τ after being infected, an individual is expected to make successful contacts at a rate $cpe^{-\alpha\tau}$ (simply because it makes contacts at rate c, and these are successful with probability p if and only if the individual is still infectious, which is the case with probability $e^{-\alpha\tau}$). The graph of $\tau \mapsto cpe^{-\alpha\tau}$ has a similar form as the graph shown in Figure 3.1, which is clearly different from the form of infectivity curve generally believed to be most realistic (cf. Figure 2.1). Infectious periods being exponentially distributed would mean that there is no 'typical' duration for infectious periods — instead most are very short, a few are long, and fewer still are extremely long. However, compared to the Reed-Frost model it does capture the observed fact that the length of the infectious period varies between individuals.

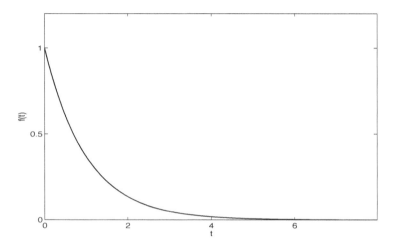

Figure 3.1: Plot of the exponential density function with parameter $\alpha = 1$.

The exponential distribution is characterized by the property of being memory-less or age-less. This means that the future 'life-length,' conditional upon having lived for some time a, is independent of a and distributed as the life-length at birth (cf. Exercise 3.15). The important consequence of this is that the epidemic model then possesses the *Markov property*. This means that the distribution of future events in the epidemic process, conditional on what has happened up until now, only depends on *how many* individuals are, at present, susceptible or infectious, respectively, and not on more detailed information about what has happened in the past (or, in the terminology to be introduced in Chapter 6, the *state* of the system is fully described by the current number of susceptibles and the current number of infectives). Stochastic processes possessing such a property are called Markov processes and many results have been established for this class of processes (e.g., Grimmet and Stirzaker 2001, Liggett 2010).

In the present chapter we focus on realizations of stochastic epidemic processes. For Markov processes one can alternatively adopt a Markov chain description. This involves the *probabilities* that at time t the population consists of certain numbers of susceptibles and infectives. On the basis of the model assumptions, one formulates differential equations for the rate of change of these probabilities and next uses the theory of differential equations to draw conclusions about how they change in the course of time, with special attention for the long time behavior (a great advantage being that the differential equations are *linear*). Attentive readers now realize that this is exactly what we did in Exercise 1.34.

As a consequence of the Markov property, the evolution of the 'general' stochastic epidemic can be described using so-called *intensity rates* (i.e., probabilities per unit of time) for the two types of event that can occur: infection and recovery.

Let $N_1(t)$ and $N_2(t)$ count the number of individuals that have been, respectively, infected and removed, by time t. So, $N_1(0) = 1$ and $N_2(0) = 0$. The numbers of susceptible, infectious and recovered individuals at t, $S(t)$, $I(t)$ and $R(t)$, can then be expressed using these two *counting processes* (also known as random jump processes): $S(t) = N + 1 - N_1(t)$, $I(t) = N_1(t) - N_2(t)$ and $R(t) = N_2(t)$. Due to the Markov property, the probability of an infection occurring at t depends only on the numbers of susceptible and infectious individuals just prior to t (denoted $t-$). Since each pair of individuals has contact at rate c/N, and only contacts between pairs where one is susceptible and the other is infectious can result in infection, with probability p, the probability that a new infection occurs in the small time interval $[t,\ t + h)$ equals

$$hS(t-)I(t-)\frac{cp}{N} + o(h),$$

where $S(t-)$ and $I(t-)$ denote the number of susceptibles and infectives, respectively, just before t. This is true without the assumption of exponentially distributed infectious period, but we need this assumption in order to specify the rate at which the other type of event, recovery, occurs. Because of the 'memory-less' property, each individual who is presently infectious recovers at the constant rate α independently of how long he/she has been infected. The probability that a recovery occurs in $[t,\ t + h)$ hence equals

$$h\alpha I(t-) + o(h).$$

These two so-called jump probabilities actually define the model. If the probabilities are divided by h, and h tends to 0, we obtain jump rates as limits. An alternative definition of the 'general' stochastic epidemic model is hence through the two counting processes $N_1(\cdot)$ and $N_2(\cdot)$ having stochastic jump rates $\lambda_1(t)$ and $\lambda_2(t)$ defined by:

$$\begin{aligned}
\lambda_1(t) &= cpI(t-)S(t-)/N, & (3.1) \\
\lambda_2(t) &= \alpha I(t-). & (3.2)
\end{aligned}$$

This description of the 'general' stochastic epidemic will be used later when making parameter inference from available data (Section 5.4), but it also helps to illustrate the resemblance with the deterministic epidemic. Indeed, the rate at which a counting process increases is closely related to the derivative in a deterministic setting. Since the number of susceptibles decreases whenever N_1 increases by 1, the jump rate for N_1 is equal to the negative derivative of the number of susceptibles in the deterministic model; see for example system (1.12) and note that β in the deterministic model corresponds to cp/N in the present model. Similarly, the number of infectious individuals increases when N_1 increases by 1 (an infection occurs) and decreases when N_2 increases by 1 (a recovery occurs). Therefore, the derivative of the number of infectives in the deterministic model agrees with the difference of the jump rates of N_1 and N_2 in the stochastic model.

3.3 INITIAL PHASE OF THE STOCHASTIC EPIDEMIC IN A LARGE POPULATION

We will now focus on what happens during the early stages of an epidemic in a large population. It turns out that the random number of infectious individuals is well approximated by the random number of living individuals of an associated continuous-time *branching process*. Since branching processes have been analyzed extensively, we can then obtain results for the epidemic by 'translation' from branching process theory. So, before saying more about the approximation of the epidemic with a branching process, we summarize the relevant results concerning continuous-time branching processes.

3.3.1 Continuous-time branching processes

Below we present a simple continuous-time branching process and some results for it, and introduce this at first using the context of demography, rather than epidemiology. A more detailed description as well as several extensions can be found in, for example, Jagers (1975) or Haccou et al. (2005).

A branching process aims at describing the evolution in time of some population. The main idea is that each individual acts independently of the others. In the simplest version, considered here, individuals are identical. If we only focus on the characteristic of being alive (versus dead), this means that an individual has a life-length that may be random but it should be independent of the rest of the population, and the distribution should be the same for all individuals. Let the life-length consist of two independent parts, the childhood having distribution F_1 and the adult part having distribution F_2. We adopt the convention that an individual has to go through both the childhood and the adult stage before dying. During the adult part of an individual's life, the individual gives birth to new individuals randomly in time at a constant rate (probability per unit time) γ. These new-born individuals behave in the same way, and independently. Suppose that, at time $t = 0$, we start with one new-born individual.

Let X be a random variable describing the adult life-length distribution, i.e., X has distribution F_2 with density function f_2. Further, let $E(X)$ denote the mean adult life-length. Then the expected number of times that an individual gives birth during its life equals $\gamma E(X)$: the birth rate multiplied by the average adult life-length. Depending on whether this number is smaller than, equal to, or larger than 1, the branching process is said to be *subcritical*, *critical* or *supercritical*, respectively. If $\gamma E(X) < 1$ (subcritical) then it will certainly die out (go extinct) since individuals will, on average, give birth to (so are replaced by) less than one individual. The critical case is more subtle, but it has been shown that in this situation as well the population will die out with probability 1 (but it may take a much longer time before extinction is a fact). Finally, in the supercritical situation, the branching process has positive probability to grow beyond all limits (of course the model is then no longer a good description of any real population, but this is not the issue here), however the opposite scenario of going extinct with finitely many people ever born is still possible. In other words, growing beyond all limits will not happen with certainty — the population can still die out (for example if the initial individual happens to die without giving birth at all). An interesting property in this situation is that either only few individuals will ever be born before the population dies out, or the population will become infinite. Indeed, the probability that the number Y_∞ of individuals ever alive exceeds n, but is finite, tends to 0 as n tends to infinity, i.e., $P(n \leq Y_\infty < \infty) \to 0$ as $n \to \infty$.

As already observed in Section 1.2.2 in the context of epidemiology, when successful contacts occur at rate pc (i.e., according to a Poisson process with rate pc) then the number of successful contacts during an interval of length x is Poisson distributed with mean pcx. So the present parameter γ corresponds to the product of parameters pc in the epidemic context. In the present demographical situation, conditional on that the adult life-length of an individual equals x, the number of offspring the individual produces is Poisson distributed with parameter γx: $P(k \text{ offspring}|X = x) = (\gamma x)^k e^{-\gamma x}/k!$, $k = 0, 1, \ldots$. To remove the condition, we have to take expectation with respect to X: $P(k \text{ offspring}) = E(P(k \text{ offspring}|X))$. The offspring distribution $\{p_k\}$, where $p_k = P(k \text{ offspring})$, therefore has what is called a mixed Poisson distribution:

$$p_k = E(P(k \text{ offspring}|X)) = E\left(\frac{(\gamma X)^k e^{-\gamma X}}{k!}\right) = \int_0^\infty \frac{(\gamma x)^k e^{-\gamma x}}{k!} f_2(x) dx \quad (3.3)$$

for $k = 0, 1, \ldots$.

Exercise 3.5 Compute the offspring distribution explicitly for i) the case where X is constant ($X \equiv x$) implying that there is no integral, and ii) the case where X is exponentially distributed with parameter α (so $f_2(x) = \alpha e^{-\alpha x}$, the exponential density function).

We now repeat, in different terminology and notation, the main results that were presented in Section 1.2.2. Let D denote a random variable having this (mixed Poisson) offspring distribution. Its (probability) generating function[1] is given by $E(\theta^D) = \sum_{k \geq 0} \theta^k p_k$ (cf. Section 1.2.2). The probability that the branching process dies out, the extinction probability, can be characterized using the probability generating function. Let z_∞ denote the probability that the branching process dies out. If we condition on the number of births D that the initial individual has, the branching processes initiated by these individuals are independent and for the original branching process to die out the branching processes of all of the offspring must die out. Each of them does so independently and with probability z_∞. The extinction probability z_∞ must hence satisfy the following equation:

$$z = \sum_{k=0}^\infty z^k P(D = k) = \sum_{k=0}^\infty z^k p_k = E(z^D). \quad (3.4)$$

It can in fact be shown that the extinction probability z_∞ is the *smallest* non-negative solution to the equation $z = E(z^D)$.

As discussed previously, the conditional offspring distribution, given that the length of the adult life equals x, is Poisson distributed with mean γx. The probability generating function of a Poisson random variable Z having mean m is $E(\theta^Z) = e^{-m(1-\theta)}$, cf. (1.6).

Exercise 3.6 Show this result, i.e., show that $\sum_{k \geq 0} \theta^k m^k e^{-m}/k! = e^{-m(1-\theta)}$.

---

[1]The word 'probability' is sometimes added in front to distinguish from other generating functions in probability theory, e.g., moment generating functions.

If we define ϕ_2 by

$$\phi_2(s) = E(e^{-sX}) = \int_0^\infty e^{-sx} f_2(x) dx$$

often referred to as the Laplace transform (of the adult life-length X in this case), it hence follows that

$$E(z^D) = E(E(z^D|X)) = E(e^{-\gamma X(1-z)}) = \phi_2(\gamma(1-\theta)) \tag{3.5}$$

where we first take expectations of D, given a value of x, and then take expectations with respect to x. All of this amounts to the observation that, when

$$p_k = \int_0^\infty \frac{(\gamma x)^k e^{-\gamma x}}{k!} f_2(x) dx$$

then

$$\sum_{k=0}^\infty z^k p_k = \int_0^\infty e^{-\gamma x(1-z)} f_2(x) dx.$$

The main result is thus that the extinction probability z_∞ is the smallest nonnegative solution to the equation

$$z = \phi_2(\gamma(1-z)), \quad \text{where} \quad \phi_2(\gamma(1-z)) = E(e^{-\gamma X(1-z)}). \tag{3.6}$$

Exercise 3.7 Write out the extinction probability equation and simplify as much as possible for i) the case where X is constant ($X \equiv x$), and ii) the case where X is exponentially distributed with parameter α (so $f_2(x) = \alpha e^{-\alpha x}$). Compute the solutions numerically for $\gamma = 2$ (in both cases) and $x = 1$ and $\alpha = 1$, respectively.

It is also possible to derive properties of the distribution of the total number of individuals that were ever born in the branching process (not counting the ancestor), the total progeny Y_∞. Given the offspring distribution $\{p_k\}$ (see (3.3) in our case), the distribution of the total progeny Y_∞ can be expressed in terms of sums of random variables having distribution $\{p_k\}$ as follows. Let $D_1, D_2, \ldots$ be independent random variables, all having distribution $\{p_k\}$. Then it is known (e.g., Jagers 1975) — meaning you are not required to produce a derivation, although of course you may — that the distribution of the total progeny Y_∞ is given by

$$P(Y_\infty = j) = P(D_1 + \cdots + D_{j+1} = j)/(j+1), \quad j = 0, 1, \ldots. \tag{3.7}$$

This distribution will be defective (meaning that $\sum_j P(Y_\infty = j) < 1$) if and only if $\gamma E(X) > 1$, i.e., the supercritical case. This is not obvious from the defining equation, but should still not come as a surprise: in the supercritical case we have positive probability to grow beyond all limits, so $P(Y_\infty = +\infty) > 0$. In general, there is no closed form expression for the total progeny. However, for some specific offspring distributions $\{p_k\}$ it is possible to compute $P(D_1 + \cdots + D_{j+1} = j)$, and hence also the total progeny distribution, explicitly.

Exercise 3.8 The distribution of the sum of j i.i.d. random variables is called the jth *convolution*. Compute the distribution of the jth convolution of i) a Poisson distribution, and ii) a geometric distribution.

Exercise 3.9 Use the results from Exercise 3.8 together with results from Exercise 3.5 to compute the distribution of the total progeny Y_∞ for the case when the adult life-length is i) constant and equal to x, and ii) exponentially distributed with parameter α.

If the branching process grows beyond all limits, i.e., if $Y_\infty = +\infty$, then the growth rate of the branching process, in the case where it grows beyond all limits (i.e., if $Y_\infty = \infty$), has also been analyzed. It is known that Y_t, the number of individuals born up to time t, grows like Ze^{rt} where Z is a random variable and r is the Malthusian parameter, identical in meaning to the exponential growth rate of infecteds in Section 1.2.3. Properties of Z and r are also known but we will only use that the number of individuals born grows exponentially. This will be used in Section 3.6 where we discuss the duration of the epidemic.

3.3.2 Approximation of the initial phase of the epidemic

Let us now return to the prototype stochastic epidemic and assume that the population size N (of initially susceptible individuals) is large. Then the number of latent and infectious individuals in the epidemic behaves very much like, respectively, the number of juvenile and adult individuals in the branching process defined in the previous section, with $F_1 = F_L$ and $F_2 = F_I$, and $\gamma = pc$, meaning that the birth rate is equal to the rate of successful contact. The reason for this similarity is that the latency period then agrees with the period of childhood, the infectious period agrees with the adult period and the rate at which an infective has successful contacts with other individuals pc agrees with the birth rate γ. The only difference between the two processes is when infectious contacts are with already infected individuals, since then no new infectious individual is created (or 'born'). However, during the early stages of the epidemic the vast majority of individuals will still be susceptible, so it is quite unlikely that an infectious contact will be with someone already infected (since we consider a well-mixed population). More precisely, the rate of successful contacts with susceptibles by an infective equals $pcS(t-)/N$, and during the early stages in a large community nearly all are still susceptible, i.e., $S(t-)/N \approx 1$.

The first infectious contact that ever occurs is necessarily (with probability 1) with a susceptible individual since all but the initially infective are susceptible. The second infectious contact has probability $1/N$ of being with someone already infected and probability $1 - 1/N$ of being with a susceptible — remember that an infectious individual has contact with all individuals with equal probability. The third infectious contact is with an already infected individual with probability $2/N$, with probability $1 - 2/N$ with a susceptible, and so forth. The probability that all the k first contacts are with susceptible individuals hence equals $1(1 - 1/N)(1 - 2/N) \cdot \ldots \cdot (1 - (k-1)/N)$. From this we see that the epidemic and branching process agree at least up until the kth contact/birth with a probability tending to 1 as $N \to \infty$. And this is true for any finite value of k.

Exercise 3.10 Replace k by k_N in the expression above and deduce that the same is true as long as $k_N = o(\sqrt{N})$, i.e., as long as $k_N/\sqrt{N} \to 0$ as $N \to \infty$. This is a variant of a phenomenon characterizing the 'celebrated' birthday problem, see e.g., Ross (2010, Chapter 2).

This means that, for N large, properties of the epidemic during the early stages may be understood in terms of the corresponding properties of the branching process. By 'early stages' we mean up until there are close to the order of $\sqrt{N}$ individuals infected. In the previous subsection it was stated that the number of individuals born grows exponentially with time (unless it dies out). As a consequence, the time until $\sqrt{N}$ individuals are born is of the order $\log(N)$ — at larger time scales the approximation breaks down and the epidemic no longer behaves like a branching process. Note that this statement means that, for any $\alpha > 0$, the probability that the final size exceeds αN converges to zero for $N \to \infty$, when $R_0 < 1$. From the point of view of the infectious agent, the situation improves when N increases. So intuitively, we expect that for any finite N the total number of cases will not be appreciable when $R_0 < 1$. The intricacies of chance make it close to impossible to make precise mathematical statements about finite populations, and therefore one turns to asymptotic results for large N.

Perhaps the most important feature of an epidemic/branching process is whether or not it takes off. We can approximate, for large N, the probability of a small outbreak by z, the smallest solution to (3.6) with X replaced by ΔT and γ by cp. We also know that only small outbreaks can occur, i.e., a minor outbreak happens with probability 1, if the expected number of offspring is smaller than or equal to 1. In the epidemic setting this number equals $cpE(\Delta T)$, the rate of successful contacts times the expected length of the infectious period. This quantity is hence central and we have encountered it before as the basic reproduction number and denoted it by R_0 (cf. Section 1.2). We hence recall that

$$R_0 = cpE(\Delta T). \tag{3.8}$$

and it follows that only small outbreaks occur if $R_0 \leq 1$, i.e., the probability that more than $\sqrt{N}$ individuals ever get infected tends to zero as $N \to \infty$. So when we consider the limit $N \to \infty$ (or use terminology like 'we approximate for N large'), we do this first of all to be able to use a branching process description of what happens shortly after the introduction of an infectious agent into a naive host population. The branching process description then yields the dichotomy: extinction or exponential growth. In the epidemic setting we translate this to: minor outbreak or major outbreak. The interpretation of these terms again involves the limit $N \to \infty$, as explained in Section 1.3.4.

It is only when R_0 exceeds 1 that there is positive probability that an outbreak affects a positive fraction of the population. The observation that qualitatively different things may happen depending on whether or not $R_0 > 1$ motivates us to call R_0 a threshold parameter with 1 as the threshold, as already discussed in Section 1.2. We repeat that, in the context of a stochastic model 'will take off' is replaced by 'has positive probability of taking off':

$$R_0 > 1 \iff P(\text{major outbreak}) > 0. \tag{3.9}$$

For the branching process we can compute the probability of a minor outbreak, and a major outbreak then occurs with the complimentary probability. The probability z_∞ of a minor outbreak is the smallest non-negative solution of

$$z = \phi_2(cp(1-z)), \tag{3.10}$$

where $\phi_2(\lambda) = E(e^{-\lambda \Delta T})$ is the Laplace transform of the infectious period ΔT.

As an example, consider the Reed-Frost model where the infectious period has a fixed (non-random) length ΔT. Then $\phi_2(cp(1-z)) = E(e^{-cp(1-z)\Delta T}) = e^{-cp\Delta T(1-z)} = e^{-R_0(1-z)}$ implying that the probability of a minor outbreak z_∞ is given by the smallest solution to

$$z = e^{-R_0(1-z)}, \tag{3.11}$$

(recall (1.7)). In Figure 3.2 we have plotted the graphs of $y = x$ and $y = e^{-R_0(1-x)}$ for $R_0 = 0.8$ (left) and $R_0 = 1.5$ (right).

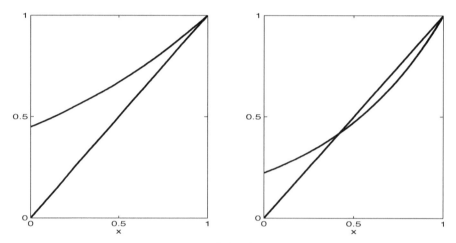

Figure 3.2: Plot of $y = x$ and $y = e^{-R_0(1-x)}$ vs x. The left figure is for $R_0 = 0.8$ and the right for $R_0 = 1.5$. The probability of a minor outbreak z_∞ for the Reed-Frost model corresponds to the smallest point of intersection of the two graphs. For $R_0 = 0.8$ this is $z_\infty = 1$ and for $R_0 = 1.5$ it is $z_\infty \approx 0.417$. We have seen a qualitative preview of this figure in Figures 1.1 and 1.2.

We have already computed the probability of a minor outbreak in a large community for the 'general' stochastic epidemic model. In Exercise 3.7 it was shown that the probability of a minor outbreak for the general epidemic equals $z_\infty = \min(1, \alpha/\gamma) = \min(1, 1/R_0)$. In Figure 3.3 we plot the two probabilities z_∞ as a function of R_0, the Reed-Frost case (lower curve) is obtained by numerical solution of (3.11). It is seen that the probability of a minor outbreak is smaller or, equivalently, the probability of a major outbreak is larger, for the case when the duration of the infectious period is constant (i.e., the same for all individuals). So, by introducing more randomness (random infectious period) one increases the risk for the branching process to go extinct or the epidemic not to take off, cf. Exercise 2.11-iii.

The distribution of Y_N, the number of individuals ultimately infected (the final size), in case of a small epidemic outbreak in a large population, may also be approximated by the distribution of the total progeny Y_∞ in the corresponding branching process (so in the supercritical case we condition on extinction for the branching process; note that the branching process approximation only breaks down when many people get infected). If the infectious period ΔT has distribution F_I and

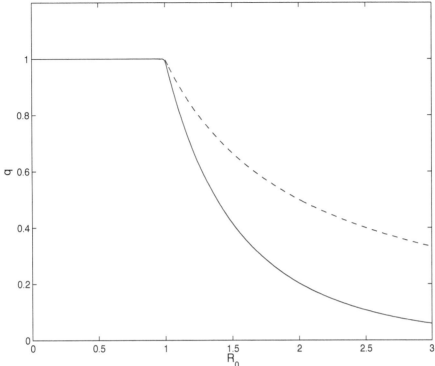

Figure 3.3: Plot of the probability of a minor outbreak z_∞ as a function of R_0, in a large community. The lower curve is for constant infectious period (Reed-Frost epidemic), and the upper for exponentially distributed infectious period ('general' stochastic epidemic).

density f_I, then, during the early stages of the epidemic, the number of successful contacts an individual has while infectious has the approximate distribution

$$p_k = P(k \text{ successful contacts}) \approx \int_0^\infty \frac{(\gamma t)^k e^{-\gamma t}}{k!} f_I(t)dt, \quad k = 0, 1, \ldots, \quad (3.12)$$

which corresponds to (3.3) in the branching process. The distribution of Y_N is given by (3.7) but with D_i having the distribution $\{p_k\}$ defined by (3.12).

Exercise 3.11 Compute the approximate final size distribution for i) the Reed-Frost model, and ii) the 'general' stochastic epidemic. Hint: Use Exercise 3.9.

Exercise 3.12 Numerically compute these probabilities for final size 0, 1, 2, and 3, both for the Reed-Frost and for the 'general' stochastic epidemic. Use $\Delta T = 1$ for the Reed-Frost, $\alpha = 1$ for the 'general' stochastic epidemic, and $cp = 2$ for both models, so that the two models are calibrated by having the same $R_0 = 2$. Compute also the accumulated probabilities (distribution functions): $F(k) = P(\leq k \text{ ultimately infected})$, $k = 1, \ldots, 4$, for the two models. If two distributions are compared and one of them has a smaller (or at least not larger) distribution function than the other for any argument k, then this distribution is said to be *stochastically larger* than the other. Determine which of the two models, if any, gives stochastically larger outbreaks than the other.

3.4   APPROXIMATION OF THE MAIN PART OF THE EPIDEMIC

In the previous section we focused on the initial stage of the epidemic by approximating the epidemic by a suitable branching process, the approximation relying on a large population. This approximation was seen to break down once the number of infected individuals is of the order $\sqrt{N}$. On the other hand, when that many have been infected, one would hope that some other large population approximation would start to hold. It can indeed be shown that the epidemic process, in terms of fractions rather than numbers, can then be approximated by the deterministic epidemic process defined in (1.12) (but note that here we shall adopt slightly different notation). In fact, one can do even better: it turns out that deviations of the scaled epidemic process from the deterministic model converge, when properly scaled, to a diffusion process (cf. Ethier and Kurtz 1986) as $N \to \infty$. The theory behind the proof is quite involved and we will only sketch the ideas.

But first an aside: In the present section we concentrate on *realizations* and show that for large population size the deterministic description provides a good approximation as soon as and as long as the fraction of infectives is bounded away from zero. In Section 1.3.4, in particular in Exercise 1.34, we introduced a Markov chain describing the *probabilities* to find the system in the various states that are feasible. One can also discuss the limit of population size going to infinity for that Markov chain description. With proper scaling one derives a linear first order partial differential equation. The so-called characteristics of that partial differential equation are exactly the solutions of the Kermack-McKendrick system of ordinary differential equations derived below. For details we refer to Diekmann and Heesterbeek (2000, Appendix A).

In the rest of this section we focus on the 'general' stochastic epidemic, having no latency period and with an exponentially distributed infectious period with mean $1/\alpha$. Similar results apply to the prototype stochastic epidemic model, but both formulation and analysis are easier when the epidemic process is Markovian. Since we have already dealt with the initial stages of the epidemic, we now assume that the number of infectives is large from the 'start,' i.e., when we set $t = 0$. More precisely, we assume that $S(0) = (1 - \epsilon)N$ and $I(0) = \epsilon N$ for some small $\epsilon > 0$ (of course there should be some recovered individuals as well, but when ϵ is small this is negligible). Recall from the 'general' stochastic epidemic model (Section 3.2.2) that, when the state is $(S(t-), I(t-))$ just before t, two events can occur at t: i) a susceptible becomes infected, which happens at rate $\lambda_1(t) = cpI(t-)S(t-)/N$, and ii) an infective recovers, which happens at rate $\lambda_2(t) = \alpha I(t-)$. In case of an infection, the number of susceptibles decreases by 1 and the number of infectives increases by 1, and when a recovery occurs, the number of infectives decreases by 1 and the number of susceptibles is unchanged. As a consequence, the stochastic epidemic process $(S(t), I(t)), t \geq 0$ has the following expected rates of change (jump rates):

$$\lambda_S(t) = -cp\frac{I(t-)S(t-)}{N},$$
$$\lambda_I(t) = cp\frac{I(t-)S(t-)}{N} - \alpha I(t-).$$

The state of the process changes by either $(-1, 1)$ or $(0, -1)$ at each jump. Since $S(0) = (1-\epsilon)N$ and $I(0) = \epsilon N$ are both of order N, the jump rates tend to infinity as $N \to \infty$. This means that when N is large, the epidemic process will make many jumps of size $(-1, 1)$ or $(0, -1)$ within a short time interval. If we scale the process and instead look at fractions: $\bar{S}(t) = S(t)/N$ and $\bar{I}(t) = I(t)/N$, we find the equations (divide both sides in the equations above by N):

$$\lambda_{\bar{S}}(t) = -cp\bar{I}(t-)\bar{S}(t-),$$
$$\lambda_{\bar{I}}(t) = cp\bar{I}(t-)\bar{S}(t-) - \alpha\bar{I}(t-).$$

There will be just as many jumps, but now the jumps are small: $(-1/N, 1/N)$ or $(0, -1/N)$. Our stochastic process hence makes very many, very small jumps — this is the set up where the law of large numbers comes in, allowing us to conclude that the process converges to $(s(t), i(t))$ satisfying the two-dimensional deterministic system of differential equations

$$s'(t) = -cpi(t)s(t),$$
$$i'(t) = cpi(t)s(t) - \alpha i(t), \tag{3.13}$$

with initial condition $(s(0), i(0)) = (1-\epsilon, \epsilon)$ (note that we use i here in the meaning 'the fraction of the population that is infected,' rather than denoting 'incidence' such as elsewhere in the book). Recall that we only aim to sketch the main ideas and leave out details. Note the resemblance between the jump rates of the scaled stochastic epidemic process $(\bar{S}(t), \bar{I}(t))$ and the differential equations for the deterministic function $(s(t), i(t))$. It can be proved that, as $N \to \infty$, $(\bar{S}(t), \bar{I}(t))$ converges uniformly on any finite interval $[0, T]$ to $(s(t), i(t))$; in symbols, for any $\delta > 0$,

$$P\left(\sup_{0 \le t \le T} |(\bar{S}(t), \bar{I}(t)) - (s(t), i(t))| > \delta\right) \to 0 \quad \text{as} \quad N \to \infty. \tag{3.14}$$

(The notation $sup_{0 \le t \le T}|(\bar{S}(t), \bar{I}(t)) - (s(t), i(t))|$ refers to the supremum, i.e., it is the largest value of the distance $|(\bar{S}(t), \bar{I}(t)) - (s(t), i(t))|$ over the interval $[0, T]$.) This type of result, stating that some random fraction becomes less and less random as some number tends to infinity, is called a law of large numbers. The assertion is illustrated in Figure 3.4. Three stochastic simulations were performed, for $N = 100$, $N = 1000$ and $N = 10,000$, respectively. All three simulations had $cp = 2$ and $\alpha = 1$ (so $R_0 = 2$) and were initiated with $\epsilon = 5\%$ initially infective; the same parameters and starting values were also used for the deterministic model, and the corresponding $i(t)$ is plotted in each figure.

It is seen that the simulated epidemic follows the deterministic curve with some random fluctuations around it. The law of large numbers is clearly visible: the deviations from the deterministic curve become smaller and smaller as N increases. Using theory for Markov population processes (Ethier and Kurtz 1986) it can be shown

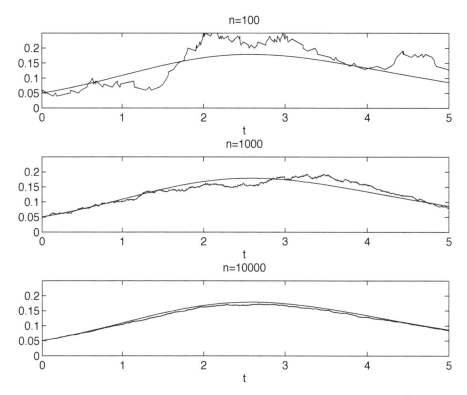

Figure 3.4: Plot of deterministic $i(t)$ and simulated stochastic process $\bar{I}(t)$ against t, for three different population sizes: $N = 100$, $N = 1000$ and $N = 10,000$. The simulations were based on the 'general' stochastic epidemic with parameters $cp = 2$ and $\alpha = 1$, so $R_0 = 2$.

that the fluctuations around the deterministic curve are asymptotically Gaussian. More precisely, the process $(\tilde{S}(t), \tilde{I}(t)) := (\sqrt{N}(\bar{S}(t) - s(t)), \sqrt{N}(\bar{I}(t) - i(t)))$ converges to a Gaussian process as $N \to \infty$. This type of result, stating that a whole process converges in distribution to a Gaussian process, is called *weak convergence* (cf. Section 4.7.2). It is harder to illustrate such a result using simulations. One possibility would be to simulate many trajectories and to try to show that, for each t, the different trajectories are normally distributed around the deterministic curve. Here we don't pursue this task and merely refer to Figure 3.4, hoping that the obliging reader sees that the fluctuations, over time, seem to fit a normal distribution quite well.

3.5 APPROXIMATION OF THE FINAL SIZE: THE SELLKE CONSTRUCTION

In the previous sections it was shown that the initial phase and the main phase of the epidemic behaved rather differently. So it might not come as a surprise that also the end phase is different (cf. Section 3.6). Perhaps more important than knowing how the epidemic process behaves over time is knowing how many eventually will get infected, i.e., knowing the final size. It turns out that this can be calculated without studying the time course of the epidemic process. We therefore now focus on the final size of the epidemic and give heuristic arguments for the derivation of its law of large numbers limit as well as arguments establishing the Gaussian nature of the deviation from this limit. This involves the so-called Sellke construction, which is elegant and hence worth presenting in its own right.

3.5.1 Law of large numbers limit: a heuristic argument

We start by giving an improved version of the derivation of the final size equation as presented at the end of Section 1.3.1. Consider a large population and a fixed $R_0 = cpE(\Delta T)$. Earlier, we deduced that the probability to avoid being infected by a specific other individual equals $E(e^{-cp\Delta T/N})$. Similarly, the probability to avoid being infected by each individual in a group of size k equals $E(e^{-cp(\Delta T_1 + \ldots + \Delta T_k)/N})$, where $\Delta T_1, \ldots, \Delta T_k$ are the (random but independent) lengths of the infectious periods of these individuals. When k is large, the law of large numbers implies that $(\Delta T_1 + \ldots + \Delta T_k)/N \approx kE(\Delta T)/N$, which is non-random, so the probability of escaping infection approximately equals

$$P(\text{escape infection from } k \text{ individuals}) \approx e^{-cpkE(\Delta T)/N} = e^{-R_0 k/N}.$$

Now, let Y denote the final number ultimately infected. Then this is the number of individuals one has to escape infection from in order to avoid getting infected during the entire epidemic outbreak. The final *fraction* infected is hence Y/N. Let $\sigma := 1 - Y/N$ denote the fraction escaping infection. The latter quantity is approximately the *probability* of not getting infected (we are neglecting that σ is random and also implicitly assume independence among individuals). Using this argument we obtain, by way of a consistency requirement, the following equation for the final fraction ultimately infected

$$
\begin{aligned}
\sigma &= 1 - \frac{Y}{N} = \text{fraction not infected} \\
&\approx P(\text{to escape infection}) \\
&\approx e^{-R_0 Y/N}
\end{aligned}
$$

which gives

$$\sigma = e^{-R_0(1-\sigma)}. \tag{3.15}$$

We have encountered this equation before as (1.7) and (3.11). Recall that the smallest solution to equation (3.11) was the (asymptotic) *probability* of a minor outbreak for the Reed-Frost model (beware that the corresponding probability for distributions of the infectious period other than a deterministic period is different). Now, the interpretation of the solution to (3.15) is that it yields the (asymptotic) fraction not getting infected, and the solution is valid not only for the Reed-Frost model, but for any member of the model family that we called the prototype stochastic epidemic model. There is always the solution $\sigma = 1$ corresponding to the entire population escaping infection — a minor outbreak. When $R_0 > 1$ there is a second non-trivial solution $0 < \sigma < 1$ of (3.15), and this will be the final fraction escaping infection in case of a major outbreak in a large community. Of course, the fraction escaping infection will not necessarily be exactly σ in a given realization. In Section 3.5.3 we will see that the deviation will not be larger than order $1/\sqrt{N}$ from σ.

<u>We conclude</u> that the distribution of the proportion escaping infection is concentrated around the two solutions of (3.15). If the proportion realized is close to $\sigma = 1$ we speak of a *minor outbreak*, and if the proportion is close to the solution in the interval $0 < \sigma < 1$ we speak of a *major outbreak*.

Exercise 3.13 Solve (3.15) numerically to determine the smallest solution for the values $R_0 =$ 1.1, 1.5, 2.0 and 3.0.

Exercise 3.14 The reproduction numbers for chicken-pox and polio are approximately 9 and 6, respectively (Anderson and May 1991, p. 70). Compute the approximate fraction getting infected in case of a major outbreak for these diseases when you assume that everyone is susceptible to the infection.

It is an interesting observation for the Reed-Frost model, that the proportion σ escaping infection in case of a major outbreak coincides with the probability z that the epidemic never takes off. We stress again that this is not the case in general. For example, Exercises 1.10–1.12 and 3.7 aimed at showing that the probability of the branching process going extinct when the life-length is exponentially distributed — corresponding to the 'general' stochastic epidemic never taking off — was equal to $z = 1/R_0$, which is clearly different from any solution to (3.15).

3.5.2   The Sellke construction

We now present a 'construction' of the prototype stochastic epidemic in the spirit of Sellke (1983),[2] who treated a slightly different epidemic model. This construction can be used to simulate an epidemic, and it also gives new insight into the epidemic process. The main idea is to keep track, as time evolves, of the accumulated 'infection pressure' generated by the infected individuals, and to let an individual become infected whenever this pressure exceeds this individual's 'resistance threshold.'

Label the individuals $0, 1, \ldots, N$, individual 0 being the initially infectious individual. Let $Q_1, \ldots, Q_N$ be exponential random variables with mean 1 (these will be interpreted as the 'resistance thresholds' of the initially susceptible). Further, let $(T_{1,0}, \Delta T_0), (T_{1,1}, \Delta T_1), \ldots, (T_{1,N}, \Delta T_N)$ be independent and identically distributed (i.i.d.) random vectors, and such that the random latency period $T_{1,i}$ and the random infectious period ΔT_i of an individual are pairwise independent with distributions F_L and F_I, respectively. Note however, that these variables will not be associated with individual i, i.e., with the correspondingly labeled individual. All random variables mentioned are defined to be mutually independent. These random variables, and no other random quantities, will now be used to construct the prototype stochastic epidemic.

Individual 0 has latency period $T_{1,0}$, followed by the infectious period ΔT_0 after which the individual is removed (recovered and immune). As time evolves, let $L(t)$ and $I(t)$ respectively denote the number of latent and infectious individuals, so $L(0) = 1$ and $I(0) = 0$. Define the *cumulative force of infection* (total infection pressure) up to time t by

$$\Lambda_c(t) = \frac{cp}{N} \int_0^t I(s)ds.$$

[2]T. Sellke: On the asymptotic distribution of the size of a stochastic epidemic. *J. Appl. Prob.*, **20** (1983), 390–394.

For $i = 1, \ldots, N$, individual i gets infected (and enters the latency state) when $\Lambda_c(t)$ reaches this individual's resistance threshold Q_i. The jth initially susceptible to become infected (which is very unlikely to be the individual labeled j) gets latency period $T_{1,j}$ followed by the infectious period ΔT_j. Note that $\Lambda_c(t)$ is random and depends on how many have been infected and also the timing of these individuals' infectious periods. The epidemic stops when there are no latent or infectious individuals left because then the cumulative force of infection cannot increase further, and consequently no further resistance thresholds will be reached.

An illustration of $\Lambda_c(t)$ is given in Figure 3.5, with the notation $Q_{(i)}$ yet to be defined.

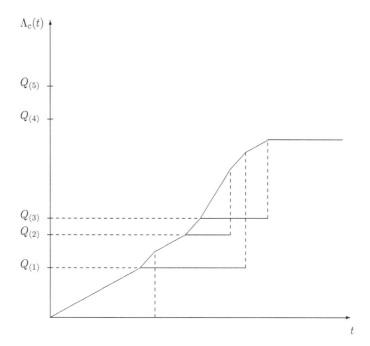

Figure 3.5: A realization of the cumulative force of infection $\Lambda_c(t)$ for the case of no latency period. The slope of $\Lambda_c(t)$ is proportional to the number of infectives $I(t)$. The infectious periods are marked by horizontal drawn lines. Note that in this example the infection pressure never reaches $Q_{(4)}$ so the epidemic stops and the final size is $Y = 3$.

In the illustration there is no latency period, so an infected individual becomes infectious immediately. Note that the first individual to become infected is the individual having smallest Q, the second to become infected is the individual having second smallest Q and so forth. The 'order statistics' of $Q_1, \ldots, Q_N$, are the variables listed in increasing order, and these are denoted by $Q_{(1)}, \ldots, Q_{(N)}$ in the figure. Whenever a new resistance threshold is reached, and a new infectious period is initiated (since there is no latency period), then the slope of $\Lambda_c(t)$ is increased by

cp/N, and whenever an infectious period ends, the slope decreases with the same amount.

We have now defined the Sellke construction of the prototype stochastic model. Let us check that this description really is equivalent with the definition of the model as given in Section 3.1. The latency and infectious periods have the correct distributions and these periods are, for different individuals, independent. It remains to compute the rate at which new individuals are infected in the Sellke construction and see that it agrees with the model definition. In the Sellke construction, new individuals are infected whenever the cumulative force of infection $\Lambda_c(t)$ reaches another resistance threshold. Given the number of infectives $I(t)$ we know from the definition of $\Lambda_c(t)$ that it increases linearly at rate $cpI(t)/N$. Further, given $S(t)$ we know that in total $S(t)$ resistance values are larger than $\Lambda_c(t)$ since these susceptibles have not yet been infected. The resistance values were defined to be exponentially distributed with mean 1, so because of the memory-less property of the exponential distribution $(P(Q > x + y | Q > x) = P(Q > y))$ each of the susceptibles becomes infected at rate $cpI(t-)/N$. The overall rate at which new individuals become infected is hence

$$\frac{cpI(t-)S(t-)}{N}, \tag{3.16}$$

as was shown already in the elaboration of Exercise 1.28.

In the model definition each infectious individual has contacts with other individuals at rate c. The individual to be contacted is chosen uniformly among all N individuals, so the contact rate with a specific individual is c/N and with all susceptible individuals hence $cS(t)/N$. However, each such contact only leads to transmission with probability p, so the rate at which an infective infects others is $cpS(t)/N$. Finally, if there are $I(t)$ infectious individuals the over-all rate at which new infections occur is hence this number multiplied by $I(t)$, which agrees with (3.16). This shows that the Sellke construction indeed is an equivalent description of the prototype stochastic epidemic.

Exercise 3.15 Show that $P(Q > x + y | Q > x) = P(Q > y)$ when Q is exponentially distributed. In words: show that the exponential distribution is memory-less.

3.5.3 Law of large numbers and central limit theorem for the final size

In this section we use the Sellke construction together with a so-called embedding argument due to Scalia-Tomba (1985)[3] to show that the outbreak size in case of a major outbreak satisfies a law of large numbers, i.e., that the fraction infected during a major outbreak has negligible randomness if the population is large. We also show that the randomness around this asymptotic limit is Gaussian. A heuristic argument for the law of large numbers was given in Section 3.5.1, we will now be more technical and use probability theory. Let

$$\Im(w) = \frac{pc}{N} \sum_{i=0}^{[w]-1} \Delta T_i, \quad 0 \leq w \leq N + 1,$$

---

[3]G. P. Scalia-Tomba: Asymptotic final size distribution for some chain-binomial models. *Adv. Appl. Prob.*, **17** (1985), 477–495.

where $[w]$ denotes the integer part of w. This implies that $\mathfrak{I}(w)$ is the total infection pressure caused by the first $[w]$ infected individuals — including the initial infective, which has label zero, hence $[w]-1$ is the upper summation limit — and hence that $\mathfrak{I}(w)$ increases only at the integers and is otherwise constant. Further, let

$$\mathfrak{Q}(v) = \sum_{i=1}^{n} \mathbf{1}_{\{Q_i \leq v\}},$$

so that $\mathfrak{Q}(v)$ counts the number of resistances among the n initial susceptibles that are smaller than or equal to v, a quantity called the 'empirical distribution.' The arguments w and v should not be interpreted as time, nor are they comparable. In the Sellke construction, an individual becomes infected when $\Lambda_c(t)$ reaches this individual's threshold. As soon as there are no infectious or latent individuals any more, $\Lambda_c(t)$ stays constant, implying that no further threshold can be reached. The final number infected Y can hence be written as

$$Y = \min\{k \geq 0; Q_{(k+1)} > \frac{pc}{N} \sum_{i=0}^{k} \Delta T_i\}, \qquad (3.17)$$

or, in words, the next ordered resistance is larger than the cumulative force of infection caused by all previously infected individuals. Using the notation just introduced this is equivalent to

$$Y = \min\{w \geq 0; \mathfrak{Q}(\mathfrak{I}(w+1)) = w\}. \qquad (3.18)$$

Exercise 3.16 Show that the two expressions for Y are equivalent. Hint: Show first that $Q_{(i)} \leq \mathfrak{I}(i)$ is equivalent to $\mathfrak{Q}(\mathfrak{I}(i)) > i - 1$ for each $i \geq 1$, and use this to show the result.

The probabilistic advantage of this characterization of the final size Y is that it is defined as the crossing point (the point where the two graphs intersect) of a straight line (w) and the graph obtained by the composition of two 'simple' and independent processes ($\mathfrak{Q}(\mathfrak{I}(w+1))$): $\mathfrak{I}(w+1)$ is the sum of i.i.d. random variables and $\mathfrak{Q}(v)$ is the empirical distribution of i.i.d. random variables. For both of these processes a law of large numbers and a central limit theorem is available.

We will now look at the processes $\mathfrak{I}(w)$, $\mathfrak{Q}(v)$ and their composition, assuming a large population N. We equip the processes with an index N and also define three related processes by

$$\bar{\mathfrak{I}}_N(w) = \mathfrak{I}_N(Nw), \qquad \bar{\mathfrak{Q}}_N(v) = \mathfrak{Q}_N(v)/N \qquad \bar{\mathfrak{C}}_N(w) = \bar{\mathfrak{Q}}_N(\bar{\mathfrak{I}}_N(w)).$$

Applying the law of large numbers for i.i.d. random variables we have the convergence

$$\bar{\mathfrak{I}}_N(w) \to cpE(\Delta T)w, \quad \text{and} \quad \bar{\mathfrak{Q}}_N(v) \to 1 - e^{-v},$$

in probability as $N \to \infty$ (meaning, for example for the first statement that, for any $\epsilon > 0$, $P(|\bar{\mathfrak{I}}_N(w) - cpE(\Delta T)w| > \epsilon) \to 0$ as $N \to \infty$). Further, since composition is a continuous operation it also follows that $\bar{\mathfrak{C}}_N(w) = \bar{\mathfrak{Q}}_N(\bar{\mathfrak{I}}_N(w)) \to 1 - e^{-cpE(\Delta T)w}$, in probability. Now returning to equation (3.18) for Y we have that $\mathfrak{Q}_N(\mathfrak{I}_N(w+1)) = \mathfrak{Q}_N(\bar{\mathfrak{I}}_N((w+1)/N)) = N\bar{\mathfrak{Q}}_N(\bar{\mathfrak{I}}_N((w+1)/N))$.

The definition of the final *fraction* ultimately infected Y_N/N is thus

$$
\begin{aligned}
\frac{Y_N}{N} &= \min\left\{\frac{w}{N} \geq 0;\ \mathcal{Q}(\mathcal{I}(w+1)) = s\right\} \\
&= \min\left\{\frac{w}{N} \geq 0;\ \bar{\mathcal{Q}}_N(\bar{\mathcal{I}}_N(\frac{w}{N} + \frac{1}{N})) = \frac{w}{N}\right\} \\
&= \min\left\{x \geq 0;\ \bar{\mathcal{Q}}_N(\bar{\mathcal{I}}_N(x + \frac{1}{N})) = x\right\}.
\end{aligned}
$$

Since $\bar{\mathcal{Q}}_N(\bar{\mathcal{I}}_N(x)) \to 1 - e^{-cpE(\Delta T)x}$ in probability (for any x), the same applies to $\bar{\mathcal{Q}}_N(\bar{\mathcal{I}}_N(x + \frac{1}{N}))$. So one expects that the final fraction of infected individuals should converge in probability to the smallest non-negative solution of the equation

$$
1 - e^{-cpE(\Delta T)x} = x.
$$

This is in fact not true, the reason being that there is one solution exactly at the 'boundary' $x = 0$ (corresponding to a minor outbreak). If $R_0 = cpE(\Delta T) > 1$, there is another strictly positive solution x^* corresponding to a major outbreak. Using more involved arguments one can show the following. If $R_0 \leq 1$, then Y_N/N converges to 0 in probability, but if $R_0 > 1$ then Y_N/N converges to a two-point distribution, the two points being the solutions 0 and x^* of the equation $1 - e^{-cpE(\Delta T)x} = x$. Note that the equation is identical to (3.15) with $x = 1 - \sigma$ (the proportion getting infected, x, is one minus the fraction escaping infection, $1 - \sigma$).

The convergence to 0 in both cases actually follows from results presented in Section 3.3.2. There it was shown that, in case of a minor outbreak, the final size distribution agrees with the corresponding distribution of the total progeny of a branching process in case the latter never took off. In this case the *number* of infected individuals was bounded (in probability) implying that the *proportion* infected converges to 0. The *probability* (z_∞) that Y_N/N converges to 0, corresponding to the branching process going extinct, was shown to be the smallest solution to $\theta = E(e^{-cp\Delta T(1-\theta)})$.

In case there is a major outbreak, a central limit theorem can in fact be derived (i.e., it can be shown that the randomness around the deterministic limit has a normal (Gaussian) distribution). The underlying reason for this is that for the two simple processes $\bar{\mathcal{I}}_N(w)$ and $\bar{\mathcal{Q}}_N(v)$ a central limit theorem holds, because they are constructed from i.i.d. random variables. For any v and w, the two quantities

$$
\sqrt{N}(\bar{\mathcal{I}}_N(w) - cpE(\Delta T)w) \qquad \text{and} \qquad \sqrt{N}(\bar{\mathcal{Q}}_N(v) - (1 - e^{-v}))
$$

both converge to normally distributed random variables as $N \to \infty$. In fact, viewed as processes (in w and v, respectively) they converge to Gaussian processes, and so does their composition. This can be used to formally prove that the outbreak size Y_N in case of a major outbreak (assuming $R_0 = cpE(\Delta T) > 1$) is asymptotically normally distributed with mean size Nx^* and standard deviation of order $\sqrt{N}$, where x^* is the positive solution to the equation $1 - e^{-R_0 x} = x$. More precisely, the asymptotic variance of Y_N in case of a major outbreak can be shown to equal

$$
V(Y_N) = N\frac{x^*(1 - x^*)}{(1 - (1 - x^*)R_0)^2}\left(1 + r^2(1 - x^*)R_0\right), \qquad (3.19)
$$

where $r^2 = V(\Delta T)/E(\Delta T)^2$ is the squared coefficient of variation of the infectious period (not to be confused with the Malthusian parameter, which we also denote by r). We conclude that the final size Y_N will either be small (minor outbreak), or else it will be normally distributed with mean Nx^* and variance given by equation (3.19). The latter case, a major outbreak, is possible only if $R_0 = cpE(\Delta T) > 1$.

Exercise 3.17 The basic reproduction number for influenza is usually[4] considered to be approximately $R_0 = 1.6$. Assume that the coefficient of variation of the infectious period equals $r = 0.3$. Compute the mean and the standard deviation for the number infected in an influenza outbreak in a community consisting of 10,000 susceptible individuals.

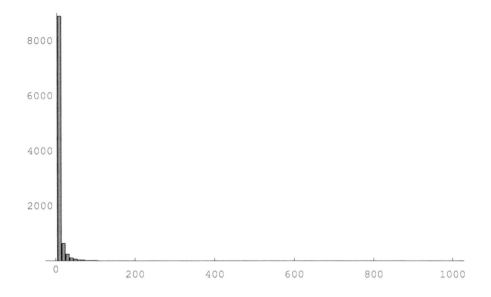

Figure 3.6: Empirical distribution of outbreak sizes in 10,000 simulated epidemics in a population of size 1000. $R_0 = 0.8$.

An illustration of the result is given in Figures 3.6 and 3.7. In the two figures the Reed-Frost epidemic has been simulated in a population of size $N = 1000$. Both figures show histograms of the final outbreak sizes from 10,000 simulations. In Figure 3.6, $R_0 = cpE(\Delta T)$ was set to 0.8 and in Figure 3.7, $R_0 = cpE(\Delta T) = 1.5$.

These figures actually illustrate most of the results mentioned in this section. We see that when $R_0 < 1$ there are only small outbreaks but when $R_0 > 1$ a fraction of the outbreaks are small and the remaining outbreaks have sizes around Nx^* (when $R_0 = cpE(\Delta T) = 1.5$ the positive solution to $1 - e^{-cpE(\Delta T)x} = x$ is $x^* \approx 0.583$, so $Nx^* \approx 583$). The agreement with the normal distribution in case of a major outbreak is also clearly visible. The probability of a minor outbreak (corresponding to the branching process going extinct) is the smallest solution to the equation $\theta = E(e^{-cpE(\Delta T)(1-\theta)})$, which in case of the Reed-Frost model having constant infectious period and $R_0 = 1.5$ becomes $\theta = e^{-1.5(1-\theta)}$. The solution to

[4]e.g., S. Cauchemez, A.J. Valleron, P.Y. Boelle, A. Flahault and N.M. Ferguson: Estimating the impact of school closure on influenza transmission from sentinel data. *Nature*, **452** (2008), 750–754.

this is $1 - x^* \approx 0.417$. This also agrees quite well with the simulations: the number of simulations resulting in outbreak sizes smaller than 50 (say) was 4083 (out of 10,000). The only thing that is not evident from the figures, but was observed in the simulations, is that the distribution of outbreak sizes in case of a minor outbreak are very similar to the distribution of the total progeny of the corresponding branching process.

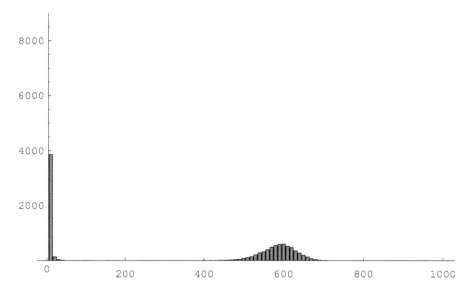

Figure 3.7: Empirical distribution of outbreak sizes in 10,000 simulated epidemics in a population of size 1000. $R_0 = 1.5$.

3.5.4 The final size in a partially vaccinated community

Suppose now that a vaccine giving 100% immunity is available, and that a fraction v of the community is vaccinated prior to the outbreak. How does this alter the size of the epidemic? We present in this sub-section a slightly more refined elaboration for Exercise 1.37. From a modeling point of view, the only difference from the situation without vaccination is that now the number of initially susceptible individuals is changed from N to $N(1-v)$, since $N(1-v)$ are now vaccinated and immune. As a consequence all results remain true with this change, for example the law of large numbers, the branching approximation and so forth.

One important effect of vaccination is that we get a new reproduction number, denoted R_v, where the subscript indicates the dependence on v. Its interpretation is that R_v is the expected number of individuals an infected individual infects during the early stages of an outbreak when a fraction v are initially vaccinated. Since $R_0 = cpE(\Delta T)$ (cf. equation 3.8) this number will be reduced by the factor $1 - v$ (the fraction not vaccinated). As a consequence we have

$$R_v = (1 - v)R_0 = (1 - v)cpE(\Delta T). \tag{3.20}$$

For the same reason as before (but now for R_v instead of R_0), only minor outbreaks can occur if $R_v \leq 1$. In terms of the fraction vaccinated this is equivalent to $v \geq 1 - 1/R_0$, i.e.,

$$v \geq 1 - \frac{1}{R_0} \iff R_v \leq 1 \iff P(\text{major outbreak}) = 0. \qquad (3.21)$$

The quantity $v_c = 1 - 1/R_0$ is called the *critical vaccination coverage*. Its interpretation is that if at least the fraction v_c is vaccinated, then the whole community is protected from major outbreaks. The community is then said to have *herd immunity*.

3.6 THE DURATION OF THE EPIDEMIC

Several results have also been obtained concerning the duration of an epidemic in a large community. Because such analyses are quite technical, we will only provide a brief description of what is known and refer to for example Barbour (1975)[5] for details.

Of course, the duration will depend on whether the outbreak is major or minor. In the latter case, the duration will be of order 1 as $N \to \infty$. The more interesting case is when $R_0 > 1$ and there is a major outbreak. In Section 3.3 it was shown that the initial phase of the epidemic looked very much like a branching process, and it was also shown there that the number of individuals born up to t grew like Ze^{rt}, where Z is a random variable and r is the so-called Malthusian parameter. The branching process approximation breaks down when approximately $\sqrt{N}$ individuals have been infected. This will hence take a time of the order $\log(N)$. In fact, for a small fixed $\epsilon > 0$ the time until ϵN individuals are born is of the same order (but with a different constant in front).

Once many individuals have been infected, things start to happen more quickly. The epidemic then behaves much like the corresponding deterministic model. Starting with any positive fraction ϵ of initial infectives, the main part of the epidemic, where the number of infectives first increases and then decays down to some small fraction δ, takes place in a time interval with length of order 1, just as for the deterministic model, i.e., not growing with the community size N.

The final part of the epidemic, when the number of infectious individuals has decreased to some small fraction δ, can in fact also be approximated by a branching process. The difference is that this branching process is started with many individuals (δN) and is sub-critical. The time for such a branching process to die out is also of order $\log(N)$, but with a different constant.

To summarize, the time-duration T_N of a major outbreak in a population of size N equals

$$T_N = c_1 \log(N) + c_2 + c_3 \log(N) + X, \qquad (3.22)$$

where c_1, c_2 and c_3 are constants and X is a random variable, all depending on the model parameters. As explained, the first term comes from the initial phase, the second from the main part of the epidemic (during which nearly all infections take place) and the third from the end phase of the epidemic. The random variable X is the sum of two independent components (X_i and X_e, say), one stemming from the initial phase and the other from the end phase.

In Figure 3.8 the result is illustrated by simulations of the 'general' stochastic epidemic model, for three different population sizes: $N = 1000$, $N = 10,000$ and

---

[5]A.D. Barbour: The duration of the closed stochastic epidemic. *Biometrika*, **62** (1975), 477–482.

$N = 100,000$. In each case $R_0 = 2$, so we are above threshold. In each of the three cases, the epidemic was initiated with one initial infective and the rest susceptible, and the figure shows 'typical' realizations of epidemics that take off (so if a minor outbreak occurred, a new simulation was performed).

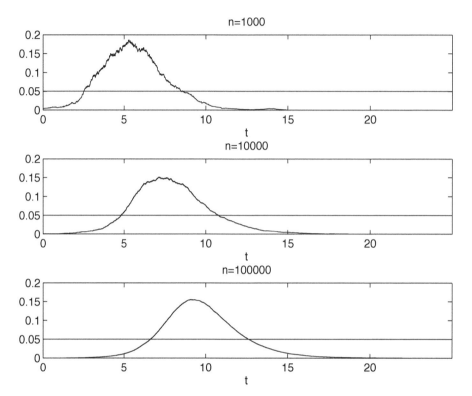

Figure 3.8: Plot of $\bar{I}(t)$ for a 'typical' simulation that takes off, for $N = 1000$, $N = 10,000$ and $N = 100,000$. In all three cases the simulations are from the 'general' stochastic epidemic with $cp = 2$ and $\alpha = 1$, so $R_0 = 2$.

From the figure one can see that the duration of the epidemic increases with population size. The durations were 15.08, 19.88 and 27.69, respectively. Let us, somewhat arbitrarily, define the initial phase as the time from the start until 5% are infective. The main phase can then start at this point and continue over the peak to the point where the proportion of infectives has dropped to 5%. The end phase then runs from this time until the epidemic has terminated. These parts are visible in the figure as the time before crossing the horizontal line 0.05 upwards, the time between the upward and the downward crossing, and the time between the downward crossing and the end of the epidemic. With these definitions, the initial phases lasted 2.53, 4.81 and 6.63 time units for $N = 1000$, $N = 10,000$ and $N = 100,000$, respectively. These durations hence increase in time, albeit rather slowly — note that the population size increases ten-fold in each step. To see that the duration increases as $\log(N)$ is not possible from these few simulations, in particular since there is also the random component X_i that is different for the three (independently generated) simulations.

The end phases show a similar pattern: the durations were 6.58, 9.07 and 15.02, respectively. Hence, for the end phase the durations are also increasing in N and larger than the initial phases; the latter observation is explained by the fact that the jump rate *out of* the I state in the rate of change of $\bar{I}(t)$ at the end is of smaller magnitude than the jump rate *into* the I state during the initial phase. Even though during the end phase the tendency of the process is to have jumps out of the state at a higher rate, it may well be, due to randomness, that the process continues to make increasing steps.

Finally, the middle phase (which is very similar to simulations presented in Figure 3.4) is hardly affected by the population size. These durations were 5.97, 6.00 and 6.04, respectively. The corresponding duration for the deterministic model (3.13) was 6.024; this was obtained by solving the differential equations numerically (the deterministic model was initiated with 0.1% initial infectives, and the time between hitting the 0.05-line for the first and second time was measured).

As mentioned previously it requires quite a technical argument to derive these results rigorously. The hardest part is to 'glue' the different pieces together. Our intention here has been to make the results plausible and to illustrate that the initial and end phase grow (rather slowly!) with N, whereas the duration of the main phase is hardly affected by the community size.

3.7   STOCHASTIC MODELING: SUMMARY

In the present chapter we have defined a stochastic counterpart to the homogeneous deterministic epidemic model introduced in Chapter 1. The model considers a homogeneous community of individuals that *mix* uniformly, meaning that there is no social structure in the community. We use the word 'mix' in the sense of engaging in a type of contact that may possibly lead to transmission; what mixing is will therefore depend on characteristics of the infectious agent and the host. The randomness in the model stems from the latency and infectious periods being random (i.e., typically different for different individuals), and also from the contact process: infectious contacts of infected individuals occur randomly in time and with randomly selected individuals in a finite population. We highlighted two special cases, called the 'general' epidemic and the Reed-Frost epidemic in the literature.

Our main focus was on deriving results for large population size N (the case with small N is postponed until Section 12.5.3 where the distribution of an outbreak within a small household is derived). The basic reproduction number R_0 was derived, equation (3.8), and it was shown that a minor outbreak is the only possibility if $R_0 \leq 1$, whereas a major outbreak can occur if $R_0 > 1$. Using a branching process approximation (relying on N being large) the *probability* z_∞ of a minor outbreak was characterized: it is given by the smallest solution to (3.10). In case of a minor outbreak the distribution of the final size Y_N has distribution given by (3.7), with D_i having the distribution $\{p_k\}$ defined by (3.12). If $R_0 \leq 1$ this is a proper distribution, i.e., summing to unity, whereas if $R_0 > 1$ the distribution sums to z_∞. The remaining probability corresponds to a major outbreak. In this case Y_N will lie close to $N(1 - \sigma)$, where σ was defined as the smallest solution to (3.15). It can in fact be shown that in case of a major outbreak Y_N is normally distributed:

$$Y_N \sim \mathrm{N}\left[N(1-\sigma),\ N\frac{x^*(1-x^*)}{(1-(1-x^*)R_0)^2}\left(1+r^2(1-x^*)R_0\right)\right], \qquad (3.23)$$

where $r^2 = V(\Delta T)/E(\Delta T)^2$. As a consequence, the final *proportion* infected is concentrated around 0 (a minor outbreak) and $1-\sigma$ (a major outbreak) with Gaussian 'noise' of order $1/\sqrt{N}$ in case of a major outbreak (this was illustrated with simulations in Figure 3.7).

It was also shown that, assuming the epidemic takes off, the whole epidemic process, scaled by N, behaves approximately like the corresponding deterministic model with normally distributed deviations of order $1/\sqrt{N}$. Finally, we have indicated why the duration of a major epidemic outbreak is proportional to $\log(N)$, the initial and final phase being of this order, whereas the main phase, where almost all infections occur, lasts for a period of time that does not grow with N.

A relevant question is what new insights are gained by studying a more complicated stochastic model above those obtained in the deterministic setting of Chapters 1 and 2. Perhaps the most important difference is that when $R_0 > 1$ a major outbreak is possible, but not certain, as it is in the deterministic setting. In the stochastic counterpart there is a probability z_∞ that the outbreak remains minor, and in this case the actual distribution of the random final size can be derived. In case of a major outbreak, the final proportion infected is close to $x^* = 1 - \sigma$. In the deterministic model it is exactly this, but in the stochastic setting the distribution is actually normally distributed with this as the mean and an explicit expression (3.19) for the variance (to be divided by N^2 when considering proportions). In Section 5.4 we will make use of this randomness to obtain standard errors for parameter estimates, something that is not possible if we restrict our attention to a deterministic setting. Furthermore, the variance term depends explicitly on the distribution of the infectious period through its mean but also through its variance: the more random the length of the infectious period, the larger the variance we obtain for the final size.

Chapter Four

Dynamics at the demographic time scale

4.1 REPEATED OUTBREAKS VERSUS PERSISTENCE

If population turnover is slow relative to the transmission of infection, we reach almost the final size of the epidemic in a closed population before the gradual inflow of new susceptibles has any effect. When an agent has struck a virgin (or naive) population, the susceptible fraction of the population is then of the order of e^{-R_0} (for R_0 large, see Exercise 1.20), and it will therefore take a long time before susceptibles will constitute a substantial fraction of the population again. During this period there are so few infectives that demographic stochasticity will lead to extinction of the infective agent. When, after a possibly long time, the population of susceptibles is above threshold again, re-introduction of the agent from outside leads to another epidemic. Thus we expect to see recurrent outbreaks with fade-out and irregular periods in between.

Data about measles and influenza on Iceland show exactly this pattern (see Cliff and Haggett 1988). (In some cases it was even possible to trace the ship that carried the infected sailor who triggered a specific measles epidemic!) But data about measles in New York show a different pattern: large outbreaks every two years with low but non-zero incidence in the years between outbreaks.

There are many ways in which Iceland and New York differ. Two relevant ways seem to be 1) the isolation from the 'outside world' and 2) the population size. How do such factors influence the probability that a virus goes extinct after a large outbreak? Is there a *critical community size* for virus persistence? (Note that Ne^{-R_0} may still be reasonably large if total population size is large; a low density over a large domain may yield an appreciable number.) Or is geographical expanse the key point? If local epidemics are out of phase then the proneness to global extinction may be much smaller, cf. metapopulation models in ecology (Gilpin and Hanski 1991, Hanski and Gilpin 1997). Note, in addition, that whether or not we find a 'fade-out followed by re-introduction' in the data, may depend on the somewhat arbitrary geographical division of the public health administration. When trying to analyze the wealth of available measles data (Grenfell and Harwood 1997, Grenfell et al. 1995),[1] one is forced to include other complicating factors, such as age structure and seasonality (both the weather and the school system).

The question, will the agent go extinct after the first outbreak? cannot be answered within the context of a deterministic description. So we would like to be able to switch back to a stochastic description at the end of the epidemic outbreak. While it is well known how to calculate the probability of extinction from a branching process in a constant environment (where we can work within a generational

---

[1] B.T. Grenfell and J. Harwood: (Meta)population dynamics of infectious diseases. *TREE*, **12** (1997), 395–399; B.T. Grenfell, B.M. Bolker and A. Kleczkowski: Seasonality, demography and the dynamics of measles in developed countries. In: Mollison (1995), pp. 248–268.

perspective; see Sections 1.2.2 and 3.3), it seems difficult to do so when environmental quality (from the point of view of the agent, i.e., the presence of susceptibles!) is improving linearly at a certain rate. We are not aware of any work in this direction. In fact we know only of a few papers[2] in which the relevant probability is calculated for a stochastic version of the Kermack-McKendrick ODE model of Exercise 1.22. The calculation is based on approximate solutions of the Fokker-Planck equation and constitutes an ingenious piece of work. It is to be hoped that this will trigger more work in this direction, concentrating on other models and different methods such that in the end a robust picture emerges.

In stochastic models, in which the number of individuals cannot have unlimited growth, the endemic state can only be *quasi-stationary*, which means that it may exist for a long period of time. Ultimately, however, a rare combination of chance events will drive the infective agent to extinction, which is an 'absorbing' state from which return is only possible by the deus ex machina of re-introduction from outside. The expected time until extinction is an important quantity, since it tells us on what time scale the quasi-stationary state is a reasonable description. See Nåsell (1995) and Grasman and van Herwaarden (1999) for efficient methods for the computation of this quantity.[3] We will treat such methods in Section 4.7.

In the next section we will at first simply forget about fade-out and concentrate on the dynamical behavior near the endemic steady state, in which the inflow of new susceptibles is balanced by the incidence (and by death). We take the rate at which newborn susceptibles are added to the population as a given constant (and we ignore immunity from maternal antibodies). This means that, on the time scale considered, demography can influence infection dynamics, but not vice versa. In the subsequent Section 4.3, we use this framework to analyze and discuss various issues related to vaccination.

In Section 4.4 we shall investigate the influence of disease on demography: can the infective agent regulate the host population? There we consider a population that, on the time scale considered, grows exponentially in the absence of the agent, and we ask to what extent the growth rate of the host is affected by the agent.

Thus we are naturally led into evolutionary questions: how prudent should an infectious agent be? Section 4.5 is intended as an invitation to use the approach of Adaptive Dynamics for answering such questions. We illustrate this approach by applying it to a very caricatural nested model that, simplifications notwithstanding, exhibits some of the subtleties associated with the evolution of virulence.

Intensive care units (ICUs) are characterized by small numbers of individuals present at any given time, but also by a high turnover. In Section 4.6 we kill two birds with one stone: we introduce continuous-time Markov chains and show that they provide a very convenient framework to describe the spread of bacteria

---

[2]O.A. van Herwaarden: Stochastic epidemics: the probability of extinction of an infectious disease at the end of a major outbreak. *J. Math. Biol.*, **35** (1997), 793–813; H. Andersson and T. Britton: Stochastic epidemics in dynamic populations: quasi-stationarity and extinction. *J. Math. Biol.* (2000), **41**, 559–580; B. Meerson and P.V. Sasarov: WKB theory of epidemics fade-out in stochastic populations. *Physical Review E*, **80** (2009), 041130; and in the master thesis of I. Taxidis: Epidemic dynamics, disease extinction and critical community size in homogeneously mixed and network populations. Utrecht University, 2007 (available from the author OD on request).

[3]I. Nåsell: The threshold concept in stochastic epidemic and endemic models. In: Mollison (1995), pp. 71–83; J. Grasman and O.A. van Herwaarden: *Asymptotic Methods for the Fokker-Planck Equation and the Exit Problem in Applications*. Springer-Verlag, Berlin, 1999.

resistant to antibiotics in ICUs, a topic that we then return to in detail in Chapter 14.

Section 4.7 is devoted to the various ways in which one can, in the context of models, estimate the expected time to extinction and, in addition, get some insight in the distribution of this quantity. In a short heuristic section (Section 4.7.5) we shall then comment on the concept of critical community size and its relation to the ratio of the time scales of demography and transmission of infection, respectively.

We end this introduction by noting a paradox for transmission situations in which the standard final-size equation (1.11) applies: for the infective agent, a virgin host population is both a best and a worst case. It is best since it gives the highest initial growth rate. It is worst since the probability of extinction after the outbreak is highest. Indeed, as Figure 1.6 and Exercise 1.20 show, $s(\infty)$ is a decreasing function of R_0 and when only a fraction p of the host population is susceptible, one has to replace R_0 by pR_0 in equation (1.11). We encourage the reader to do some computations to ascertain the quantitative effect, which is quite substantial.

4.2   FLUCTUATIONS AROUND THE ENDEMIC STEADY STATE

To describe demographic turnover in the absence of any infection, we use the caricature

$$\frac{dN}{dt} = B - \mu N, \tag{4.1}$$

where N denotes population size, B the population birth rate and μ the per capita death rate. So life expectancy is μ^{-1} and the population stabilizes at the size

$$\overline{N} := B/\mu \tag{4.2}$$

at which inflow and outflow match (where the symbol ':=' signifies 'is defined by').

To model the spread of the agent, we use the Kermack-McKendrick ODE model of Exercise 1.22. In combination with demographic turnover as in (4.1), this yields

$$\begin{aligned} \frac{dS}{dt} &= B - \beta SI - \mu S, \\ \frac{dI}{dt} &= \beta SI - \mu I - \alpha I, \end{aligned} \tag{4.3}$$

to which we could add an equation for the removed (which we think of as immunes, such as for relatively innocent children's diseases, in order not to have to model how contact intensity changes with population size):

$$\frac{dR}{dt} = -\mu R + \alpha I.$$

But because system (4.3) is a closed system (i.e., R does not appear on the right-hand side), we can disregard R when doing our analysis.

Note that when we start with

$$S + I + R = \bar{N} = \frac{B}{\mu}$$

this relation will hold for all time. Also note that with $S + I + R = N$ we recover (4.1), and hence $S+I+R$ will converge to $\bar{N}$ for $t \to \infty$. By focusing our attention on

(4.3), while forgetting about R, we avoid having to deal with these issues, they are relegated to the background. Whenever the infection has no impact on demography — so whenever neither death nor reproduction is influenced — we are in this kind of situation, and we can study infection dynamics while assuming a demographic steady state for the host population. To improve readability, we shall continue to indicate total population size by N, even though $\bar{N}$ would be more precise.

Exercise 4.1 Show that the 'virgin' (or 'infection-free') state $(S, I) = (N, 0)$ with $N = B/\mu$ is stable if and only if $R_0 < 1$ where

$$R_0 = \frac{\beta N}{\alpha + \mu}. \tag{4.4}$$

Exercise 4.2 Show that in an endemic steady state $(S, I) = (\overline{S}, \overline{I})$ with $\overline{I} > 0$ we have

$$\frac{\overline{S}}{N} = \frac{1}{R_0}. \tag{4.5}$$

Does this surprise you? (If so, return to Exercise 1.22-vi and consider that the same argument now applies to minima of I at which S is increasing; in more mathematical terms, verify that $S = N/R_0$ is an isocline (also called nullcline by many authors) and draw a picture of how orbits may proceed through the (S, I) plane.) Reflect upon the possibilities of estimating R_0 from endemic steady-state data.

Exercise 4.3 Show that in an endemic steady state

$$\overline{I} = \frac{\mu}{\beta}(R_0 - 1) \tag{4.6}$$

and that consequently such a state exists if and only if $R_0 > 1$. (Recall the requirement $\overline{I} > 0$!). Note that the equivalent formula

$$\frac{\overline{I}}{N} = \frac{(\alpha + \mu)^{-1}}{\mu^{-1}} \left(1 - \frac{\overline{S}}{N}\right)$$

expresses the relative steady-state prevalence in terms of measurable quantities: the life expectancy μ^{-1}, the expected length of the infectious period $(\alpha + \mu)^{-1}$, and the steady-state fraction of susceptibles $\overline{S}/N$.

Existence of a steady state does not guarantee that the balance between inflow of new susceptibles and the combined effect of infection and death is actually exact at every instant. There has to be a balance, but it may be over a longer time interval. Fluctuations around the steady state are not necessarily damped. In order to find out what happens in the present model, we linearize around the steady state. The linearized system has solutions that depend on time through a factor $\exp(\lambda t)$ for special values of λ. When λ is real, this gives information about growth or decay rates. When λ is complex, the real part determines the growth or decay rate, whereas the imaginary part determines the frequency of the oscillations that accompany the growth or decay. The principle of linearized stability guarantees that, provided the real parts of the λ's are non-zero, the information about solutions of the linearized system carries over to solutions of the nonlinear system, as long as these stay in a small neighborhood of the steady state (see e.g., Hirsch and Smale

1974), i.e., there is local asymptotic stability. The last proviso matters only if we find instability, i.e., growing exponential terms; in that case we can only conclude that there are solutions that leave a given neighborhood of the steady state and not that the distance to the steady state keeps increasing exponentially, since quadratic and higher-order terms matter much more further from the steady state.

The linearized system is fully characterized by a matrix M and the λ's are precisely the *eigenvalues* of M, which can be found by solving the *characteristic equation* $\det(\lambda I - M) = 0$, which is a polynomial in λ of degree n, where n is the dimension of the system. In the two-dimensional case the characteristic equation reads

$$\lambda^2 - T\lambda + D = 0 \tag{4.7}$$

where T is the *trace*, i.e., the sum of the diagonal elements $T = m_{11} + m_{22}$, and D the *determinant*, i.e., $D = m_{11}m_{22} - m_{12}m_{21}$, of the matrix $M = (m_{ij})_{1 \leq i,j \leq 2}$. It follows at once from the explicit formula

$$\lambda = \frac{T \pm \sqrt{T^2 - 4D}}{2}$$

that

$$T < 0 \text{ and } D > 0 \tag{4.8}$$

is the additional condition for linearized stability (i.e., decaying exponential terms in the solution) and that

$$T^2 < 4D \tag{4.9}$$

is the condition for an oscillatory approach to the steady state.

Exercise 4.4 Show that the matrix

$$\begin{pmatrix} -\beta\bar{I} - \mu & -\beta\bar{S} \\ \beta\bar{I} & 0 \end{pmatrix}$$

corresponds to the linearization of system (4.3) around the endemic steady state. Deduce that the endemic steady state is stable.

Exercise 4.5 Show that the characteristic equation can be written as

$$\left(\frac{\lambda}{\mu}\right)^2 + R_0\frac{\lambda}{\mu} + \left(1 + \frac{\alpha}{\mu}\right)(R_0 - 1) = 0,$$

where each term is dimensionless. Consider the situation where the life expectancy μ^{-1} is much bigger than the expected duration α^{-1} of the infectious period. Show that, unless R_0 is only slightly above 1, the model predicts damped oscillations around the steady state, with relaxation time $2/\mu R_0$ (this means, by definition, that in a time interval of length $2/\mu R_0$ the amplitude diminishes by a factor e^{-1}) and approximate frequency $\sqrt{\mu\alpha(R_0 - 1)}$. The approximate period is $2\pi/\sqrt{\mu\alpha(R_0 - 1)}$. Use once more that $\mu \ll \alpha$, to deduce that the relaxation time is much longer than the period and that, consequently, we can expect to see many oscillations before the steady state is reached.

Exercise 4.6 In steady state the force of infection (recall that this is the probability per unit of time for a susceptible to become infected) is a constant, say Λ. Show that the *mean age at infection* $\bar{a}$ is given by

$$\bar{a} = \frac{1}{\mu + \Lambda} = \frac{1}{\mu R_0}. \tag{4.10}$$

Hint: The safe way is to write out the probability density function for exit from the susceptible state, while conditioning on not dying. A short cut is obtained by arguing that $\bar{a} = $ expected sojourn time in the susceptible state (without any condition), since exits to 'death' and to 'infectious' occur throughout in the same fixed proportion (determined by μ and Λ). Finally, observe that $\Lambda = \beta \bar{I} = \mu(R_0 - 1)$.

Combining the results of the last two exercises, we see that we can give a rather complete description in terms of three *observable* quantities, all with the dimension of time: the life expectancy $L = 1/\mu$, the expected duration of the infectious period $1/\alpha$ and the mean age at infection $\bar{a}$. When $1/\mu \gg 1/\alpha$, the relaxation time for approach to the stable endemic state equals $2\bar{a}$, while the period of oscillation equals $2\pi\sqrt{\bar{a}/\alpha}$ (which involves the geometric mean of the two time scales $\bar{a}$ and $1/\alpha$). Here we assume that R_0 is big enough for the difference between μR_0 and $\mu(R_0 - 1)$ to be negligible.

As an important side-remark we note that (4.10) can be rewritten in the form

$$R_0 = \frac{L}{\bar{a}},$$

which indicates a possibility to estimate R_0 from data (a second possibility for endemic infections; cf. Exercise 4.2). See also Exercise 13.6 and 13.8-ii.

Data about measles from many towns and regions show *sustained*, rather than damped, oscillations. So we are naturally led to the question: what is missing in the present model? Various possibilities present themselves. Stochastic effects may enhance the deterministic fluctuations (Bartlett 1960). Weather conditions may influence the probability of transmission and make β periodic (see Kuznetsov and Piccardi 1994)[4] and the references given there). Age structure may necessitate the use of a more complex model, including seasonal effects of the school system.[5]

We conclude this section with several exercises dealing with generalizations, variations on the theme, different aspects, etc. The first intends to demonstrate quantitatively that our estimate of the period is in fact not bad at all.

Exercise 4.7 For measles in pre-vaccination western Europe and the United States one estimates $\bar{a}$ as somewhere between 4 and 5 years, while $1/\alpha$ is approximately 12 days. Compute $2\pi\sqrt{\bar{a}/\alpha}$ and compare the result with the observed period of two years.

Our next two exercises are intended to test the robustness of our conclusions by investigating the influence of minor modifications to the basic model.

---

[4]Y. Kuznetsov and C. Piccardi: Bifurcation analysis of periodic SEIR and SIR epidemic models. *J. Math. Biol.*, **32** (1994), 109–121.

[5]D. Schenzle: An age-structured model of pre- and post-vaccination measles transmission. *IMA J. Math. Appl. Med. Biol.*, **1** (1984), 169–191.

Exercise 4.8 Consider the SEIR system (cf. Exercise 2.2) with demography described by

$$
\begin{aligned}
\frac{dS}{dt} &= B - \beta SI - \mu S, \\
\frac{dE}{dt} &= \beta SI - \mu E - \theta E, \\
\frac{dI}{dt} &= -\mu I + \theta E - \alpha I.
\end{aligned}
\tag{4.11}
$$

i) Recall that now $R_0 = \frac{\theta}{\theta+\mu} \frac{\beta N}{\alpha+\mu}$ (Exercise 2.4-iv).

ii) Show that in the endemic steady state $\overline{S}/N = 1/R_0$, $\overline{I} = \mu(R_0 - 1)/\beta$ and $\overline{E} = \frac{\alpha+\mu}{\theta}\overline{I}$.

iii) The linearized system is now described by the 3×3 matrix

$$
\begin{pmatrix}
-(\beta\overline{I} + \mu) & 0 & -\beta\overline{S} \\
\beta\overline{I} & -(\mu + \theta) & \beta\overline{S} \\
0 & \theta & -(\mu + \alpha)
\end{pmatrix}
$$

and the eigenvalues are the roots of the characteristic equation

$$
\lambda^3 + (\mu R_0 + 2\mu + \alpha + \theta)\lambda^2 + \mu R_0(\alpha + 2\mu + \theta)\lambda + \mu(R_0 - 1)(\alpha + \mu)(\theta + \mu) = 0.
$$

Algebraically inclined readers are invited to check these statements, while others are asked to believe them.

iv) When both α and θ are large relative to both μ and μR_0, roots of the characteristic equation should lie close to roots of the simplified equation

$$
\lambda^3 + (\alpha + \theta)\lambda^2 + \mu R_0(\alpha + \theta)\lambda + \mu(R_0 - 1)\alpha\theta = 0,
$$

which we can rewrite as

$$
\lambda^3 + (\alpha + \theta)\left(\lambda^2 + \mu R_0\lambda + \mu(R_0 - 1)\frac{\alpha\theta}{\alpha + \theta}\right) = 0.
$$

In Anderson and May (1991, Appendix C, p. 668), it is concluded that this cubic equation has one root $\lambda \approx -(\alpha + \theta)$ (corresponding to perturbations that decay rapidly) and two other roots given approximately by the roots of the quadratic equation between braces. Thus one finds that now the period of the oscillations is given by $2\pi\sqrt{\overline{a}\frac{\alpha+\theta}{\alpha\theta}}$, or, in other words, that $1/\alpha$ has to be replaced by $(\alpha + \theta)/\alpha\theta = 1/\alpha + 1/\theta$, which is still the expected duration of 'infection,' in the sense of the period between being infected and becoming immune.

In order to sustain this claim by formal asymptotics, one has to consider the limit $\mu \to 0$. Readers who like to do asymptotic calculations are invited to derive these approximations for the roots.

Exercise 4.9 Let us now consider a model in which expected infectivity is described by a general integral kernel A. We still take

$$
\frac{dS}{dt} = B - \mu S - \Lambda S,
$$

with $\Lambda(t)$ the force of infection at time t, but we assume that

$$\Lambda(t) = \frac{c}{N} \int_0^\infty A(\tau)\Lambda(t-\tau)S(t-\tau)\,d\tau$$

(to understand this expression, note that $\Lambda(t-\tau)S(t-\tau)$ is the incidence τ units of time before the current time t, so that the individuals infected then, now have infection age τ; recall that $A(\tau)$ is the expected infectivity at infection age τ and that the possibility of death is incorporated when calculating the expectation).

i) Recall that $R_0 = c \int_0^\infty A(\tau)\,d\tau$ (see Section 2.1).

ii) Use the equation for Λ to deduce that in an endemic steady state necessarily $\overline{S}/N = 1/R_0$.

iii) Use the variation-of-constants formula (see any book on ODEs, for instance Hale (1969), Arnold (2006), or the first paragraph of the elaboration) to rewrite the differential equation for S as the integral equation

$$S(t) = B \int_{-\infty}^t e^{-\mu(t-\tau)} \exp\left(-\int_\tau^t \Lambda(\sigma)\,d\sigma\right) d\tau$$

or, equivalently,

$$S(t) = B \int_0^\infty e^{-\mu\sigma} \exp\left(-\int_{t-\sigma}^t \Lambda(s)\,ds\right) d\sigma.$$

iv) Show that in an endemic steady state $\overline{\Lambda} = \mu(R_0 - 1)$ and compare this expression with the expression (4.6) for $\overline{I}$.

v) Write $S(t) = \overline{S} + x(t)$ and $\Lambda(t) = \overline{\Lambda} + y(t)$ and derive the linearized system of integral equations

$$
\begin{aligned}
x(t) &= -\frac{N}{R_0} \int_0^\infty e^{-\mu R_0 \sigma} y(t-\sigma)\,d\sigma, \\
y(t) &= \frac{c}{R_0} \int_0^\infty A(\tau)y(t-\tau)\,d\tau + \frac{c}{N}\mu(R_0-1)\int_0^\infty A(\tau)x(t-\tau)\,d\tau.
\end{aligned}
$$

Verify that this system has solutions of the form

$$\begin{pmatrix} x(t) \\ y(t) \end{pmatrix} = e^{\lambda t} \begin{pmatrix} x_0 \\ y_0 \end{pmatrix}$$

if and only if λ is a root of the characteristic equation

$$1 = c\frac{\lambda + \mu}{R_0(\mu R_0 + \lambda)}\overline{A}(\lambda),$$

where

$$\overline{A}(\lambda) := \int_0^\infty e^{-\lambda\tau} A(\tau)\,d\tau$$

or, in words, $\overline{A}$ is the Laplace transform of A.

vi) Consider the characteristic equation for $R_0 = 1$. Show that $\lambda = 0$ is a root and that all other roots lie in the left half-plane, i.e., have negative real part. Hint: Use the non-negativity of A.

vii) Use the fact that the roots depend continuously on parameters to deduce that for R_0 slightly bigger than 1, all roots lie in the left half-plane.

viii) Convince yourself that roots can only enter the right half-plane by crossing the imaginary axis. In other words, make plausible that roots cannot enter the right half plane at infinity.

ix) Show that for $R_0 > 1$ the characteristic equation cannot have a root on the imaginary axis. Hint: Take the modulus of both sides of the characteristic equation and use that $c|\overline{A}(i\omega)| \leq R_0$ and $|(i\omega+\mu)/(i\omega+\mu R_0)| < 1$ for $R_0 > 1$.

x) Accepting that the principle of linearized stability holds for these systems of integral equations and that the growth or decay of solutions of the linear system is completely determined by the position of the roots relative to the imaginary axis, conclude that the endemic steady state is locally asymptotically stable for *every* non-negative and integrable kernel A. See Diekmann et al. (2007)[6] for justification of the assumptions about linearized stability and about the exponential decay of all solutions of the linearized problem when all roots of the characteristic equation lie in the left half-plane. We conclude that the model predicts *damped* oscillations and that therefore other mechanisms are responsible for the observed sustained oscillations.

Exercise 4.10 Returning to the basic model (4.3), show that the endemic steady state is in fact *globally* asymptotically stable.

Hint: Consider the Lyapunov function

$$V(S, I) = S - \overline{S} \ln S + I - \overline{I} \ln I.$$

Verify that

$$\frac{dV}{dt} = \frac{\partial V}{\partial S}\frac{dS}{dt} + \frac{\partial V}{\partial I}\frac{dI}{dt} = -\frac{\mu N}{S\frac{N}{R_0}}\left(S - \frac{N}{R_0}\right)^2.$$

Check that on the line $S = N/R_0$, at which $dV/dt = 0$, the maximal invariant subset is precisely the endemic steady state. Conclude that all orbits converge to this point. See Hale (1969, Chapter X) for background information.

Exercise 4.11 The so-called *microcosm principle* (cf. Mollison (1995)[7]) asserts that for a quite general population process in steady state, the fraction π_j of the population in state j is proportional to the mean time τ_j an individual spends in that state. Hence $\pi_j = \tau_j/L$, where L denotes life expectancy, so $L = 1/\mu$. Apply this principle to both the basic model described by (4.3) and the SEIR model of Exercise 4.8, and verify the correctness.

Exercise 4.12 Check that for the basic model the identity

$$\frac{\text{incidence}}{\text{prevalence}} = \frac{1}{\text{infectious period}}$$

holds for the endemic steady state. Reflect upon the possibility to estimate the infectious period from data at the population level.

---

[6]O. Diekmann, Ph. Getto and M. Gyllenberg: Stability and bifurcation analysis of Volterra functional equations in the light of suns and stars. *SIAM Journal of Mathematical Analysis*, **39** (2007), 1023–1069.

[7]D. Mollison: The structure of epidemic models. In: Mollison (1995), pp. 17–33.

Exercise 4.13 If a disease is lethal with high probability, would a model with *constant* inflow B of new susceptibles make any sense?

Exercise 4.14 Formulate and analyze a model in which births are all concentrated in a very short period once a year, which we describe somewhat caricatural as an event taking place at the integer values of time t. In a sense the model is hybrid, in that it combines continuous-time and discrete-time features. To investigate the analog of the endemic steady state, we have to adopt a stroboscopic way of monitoring the population; that is, we look only at the values of S, I, etc. immediately after (or, if you prefer, before) each birth pulse.

Exercise 4.15 Formulate and analyze a simple SIS model for a closed population. For a disease like gonorrhea, in which one returns to the class of susceptibles immediately after treatment, one can assume that those leaving the I-class return to the S-class. For other diseases an SIRS formulation in which immune individuals have a certain fixed probability per unit of time of losing immunity and re-enter the S-class may be more appropriate.

Exercise 4.16 Consider a disease for which *vertical transmission* (i.e., transmission from mother to fetus) cannot be ignored. How would you modify the basic model (4.3)? Analyze the modified system. (Among other things, calculate R_0.)

Exercise 4.17 Consider a host population in which infectious agent 1 is in steady state. Now let infectious agent 2 enter the population. Assume cross-immunity (i.e., hosts that have ever been infected by agent 1 cannot become infected by agent 2, and vice versa). Show that the number of individuals infected by agent 2 will increase if and only if $R_0^2 > R_0^1$ and conclude that, given our assumptions, natural selection will tend to increase the R_0 of the agent.

In general, one expects a trade-off between infectiousness (a component of β) and the length of the infectious period (as determined by α). A very virulent agent incurs a high death rate of the host or at least a strong reaction of the immune system. To model this, we consider β as a 'free' parameter under natural selection, while α is constrained as depicted in Figure 4.1.

Devise a graphical procedure to find the 'uninvadable' value of β, i.e., the value of β at which R_0 is maximal. See if you agree with the following statement: intermediate virulence is favored by natural selection, since the price for high infectiousness is a very short infectious period.

Exercise 4.18 Greenwood and co-workers (1936)[8] carried out experiments with mice infected with *Pasteurella muris*. Some mice die from the clinical effects of this infection and some recover to become temporarily immune. The parameter B was experimentally tuned to 6, 3, 2, 1 or 1/2 mice being added every day (the last value meaning, of course, one mouse every two days). Every day, except on Sunday, the mice were transferred to sterilized and cleaned cages while dead mice were removed. When transferring the mice, the number of cages was adjusted so as to keep the number of mice per unit area

---

[8]M. Greenwood, A.T. Bradford Hill, W.W.C. Topley and J. Wilson: *Experimental Epidemiology*. MRC Special Report Series **209**, HMSO, London, 1936.

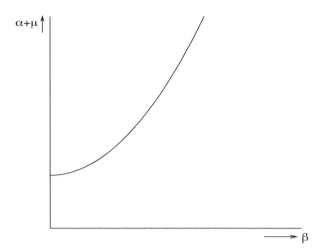

Figure 4.1: Schematic graph of trade-off between host mortality, both natural mortality (rate μ) and mortality due to the infectious agent (rate α), and infectiousness (quantified by β).

constant. All cages were connected so as to form one large cage. Reproduction did not take place under the circumstances. When B was held fixed at a certain value, the total number N of mice settled to a rather constant level after a while. The value of N was monitored as a function of B.

In this exercise we want to address two related questions: how should we model this situation and what information can we deduce from the experimental data?

i) Would you model the incidence term by βSI or by $\gamma SI/N$? To compare the implications of each, we shall analyze both. To make the calculations relatively easy, we shall ignore the possible loss of immunity.

ii) Show that for the system

$$\frac{dS}{dt} = B - \beta SI - \mu S,$$
$$\frac{dI}{dt} = \beta SI - \mu I - \alpha I,$$
$$\frac{dR}{dt} = -\mu R + f\alpha I,$$

we have

$$\overline{N} = \overline{S} + \overline{I} + \overline{R} = \frac{1 + \frac{f\alpha}{\mu}}{\mu + \alpha} B + (1 - f)\frac{\alpha}{\beta},$$

which is, as a function of B, a straight line with positive intercept of the N axis.

(Digest the following statement as a test for your understanding: for $B < \mu(\alpha+\mu)/\beta$ we have $R_0 < 1$ and we expect to find experimentally that N settles at the infection-free value B/μ. So in that range the previous expression refers to a mathematical extrapolation, not to actual measurements.)

iii) Show that for the system

$$
\begin{aligned}
\frac{dS}{dt} &= B - \gamma \frac{SI}{N} - \mu S, \\
\frac{dI}{dt} &= \gamma \frac{SI}{N} - \mu I - \alpha I, \\
\frac{dR}{dt} &= -\mu R + f \alpha I,
\end{aligned}
$$

with $N = S + I + R$, we have in steady state that

$$
\overline{N} = \frac{1 + \frac{f\alpha}{\mu}}{(1 - \frac{1-f}{\gamma}\alpha)(\mu + \alpha)} B,
$$

which is a straight line through the origin.

iv) Conclude that in principle the data could be used to falsify one of the alternatives, by fitting a straight line and finding whether or not it passes through the origin. Unfortunately this test was inconclusive. See de Jong et al. (1995).[9] There, a slightly different variant is analyzed that, in particular, includes return to the susceptible class by loss of immunity. In that reference one can also find other data that point more clearly to the system of iii) as the appropriate description.

4.3 VACCINATION

From a medical or economic point of view, it is the ability to prevent or suppress clinical manifestation of infection in individuals that determines the efficacy of a vaccine. From an epidemiological point of view, it is the reduction in susceptibility and infectivity that counts, as these matter for the dynamics at the population level. Within the context of the basic model, introduced in Section 4.2 and described by (4.3), we now briefly consider the effect of removing a fraction v of the susceptible population by vaccination. First let us assume that we have a perfect vaccine for the infectious agent that we consider. In Exercise 1.37 and Section 3.5.4 we already used this word, but there we mentioned only one aspect, i.e., the vaccine induces 100% reduction of infectiousness upon exposure to the wild type (i.e., the type or strain of the infectious agent present in the population, as opposed to the, often weakened, or killed, types or strains, or even fragments, of the same agent utilized in the vaccine). With 'perfect' we mean in addition that the vaccine works in every individual, and gives protection that lasts forever.

We note first that one has to make several choices when implementing vaccination, even in a simple unstructured model system. This refers in particular to the choice of the sub-population that is targeted and to the timing of the vaccination. An obvious option is to vaccinate only newborn individuals (i.e., vaccinate at birth), continuously; a second option is to vaccinate randomly and continuously; a

---

[9]M.C.M. de Jong, O. Diekmann and J.A.P. Heesterbeek: How does transmission of infection depend on population size? In: Mollison (1995), pp. 84–94.

third option is to vaccinate only at certain times (usually at fixed intervals) and not in between. The latter option is usually referred to as pulse vaccination.

We concentrate in this section on calculating the amount of vaccination effort needed to prevent the wild type from spreading when introduced into the partially vaccinated population, rather than into a fully susceptible population. Let us denote the reproduction number for invasion into the vaccinated population by R_v. As a rule, the condition $R_v < 1$ guarantees that the infectious agent goes extinct if it is currently present in the population (the exception occurs in systems with backward bifurcation, a situation that we ignore). In other words, the condition guarantees elimination of the agent (local reduction to zero prevalence). This is not as strict as eradication, where the agent is eliminated globally. Not vaccinating, i.e., choosing $v = 0$, leads to the basic reproduction number R_0.

Exercise 4.19 i) Suppose we vaccinate a fraction v of all newborns. How do the differential equations in (4.3) change? Give an expression for R_v. In Exercise 1.37 we saw that from this expression one concludes that the minimal fraction that needs to be vaccinated in order to prevent spread of the wild type is $1 - 1/R_0$.

ii) Suppose we now vaccinate randomly and continuously so that each individual has a probability v per unit of time to be vaccinated. Assume that the vaccine has no effect when the individual is already infected by the wild type or has recovered from natural infection and is permanently immune. How do the differential equations in system (4.3) change in this case? Note that there are two options for the vaccinated individuals, depending on whether vaccination gives permanent protection (see Exercise 4.22 for an illustration). Give an expression for R_v and give the minimal fraction to be vaccinated in order to prevent spread of the wild type.

Suppose that the vaccination coverage is not sufficient to eliminate the wild type from the population. Let us elaborate, in that situation, the effects of continuous vaccination at birth in terms of some of the quantities introduced in the previous section: force of infection, endemic fraction susceptible, average age at infection, relaxation time and period of oscillations near the stable endemic state. We will see in the next exercise that even in the context of the simple model one can get insight into likely population consequences of vaccination.

Exercise 4.20 i) How is $\overline{S}/N$ affected by continuous vaccination at birth? Are you surprised?

ii) How are the force of infection Λ and the average age at infection $\overline{a}$ (given by (4.10)) affected? Comment on the repercussions that this effect on $\overline{a}$ has for diseases such as rubella that are rather innocent, except when contracted during pregnancy (when there is a chance of leading to the serious complication of CRS, congenital rubella syndrome). This effect has been observed, for example, in Greece, see Exercise 9.18.

iii) What happens to the relaxation time and the period of oscillations around the endemic steady state? Hint: How does the characteristic equation from Exercise 4.5 change in the presence of vaccination?

Exercise 4.21 We now consider pulse vaccination where random susceptible individuals are vaccinated, but where vaccination effort is not continuous but

rather concentrated in campaigns, where nobody receives a vaccination in
between. In particular, we are interested in seeing whether there are any ad-
vantages compared to continuous vaccination at birth. Concerning the pulse
vaccination we assume that a fraction v of the population of susceptibles gets
vaccinated per pulse. Note that this presupposes that the public health system is capable of knowing which individuals are susceptible and which are
not.

When we have a constant population birth rate B and a constant per capita
death rate μ, the population density stabilizes at the level

$$\bar{N} = \frac{B}{\mu}.$$

When vaccination is periodic, say with period T, the susceptible density may
nevertheless fluctuate with that period.

If S is a given positive T-periodic function then the zero steady state of the
linear differential equation

$$\frac{dI}{dt} = (\beta S - \mu - \alpha)I$$

is determined by whether the dominant Floquet multiplier (see Section 7.9,
and Hirsch and Smale 1974)

$$e^{\beta \int_0^T S(t)\,dt - \mu T - \alpha T}$$

is greater or less than one. Hence it is determined by the sign of

$$\beta \int_0^T S(t)\,dt - \mu T - \alpha T.$$

To achieve stability of the infection-free steady state, we should have that

$$\frac{1}{T} \int_0^T S(t)\,dt < \frac{\mu + \alpha}{\beta} = \frac{N}{R_0}$$

i.e., the average value of S over the period should be below the inverse of
the basic reproduction number. Define the vaccination effort as the average
number of vaccinations per unit of time. For pulse vaccination at time intervals of length T, how does the effort required for elimination depend on T?
How does this compare to the effort required for elimination by vaccinating
successfully a fraction v of all newborns (cf. Exercises 1.37 and 4.19)?

In reality there is no vaccine that is perfect according to the definition given
above. There are various reasons for this. For example the wild type has evolved
and has become genetically far enough removed from the vaccine strain that 100%
protection is no longer guaranteed (the vaccines against tuberculosis and pertussis
are examples). Another possibility is that the vaccine gives sufficient protection,
but that this situation only lasts for a limited period (which can still be many
years) after which the acquired immunity needs a booster. Indeed, some vaccines
that were once thought to give life-long protection appear not to do so (smallpox is

an example). Finally, it is always a possibility that the vaccine does not take hold in a fraction of the individuals it is given to.

How does 'the degree of imperfection' influence the vaccination levels required for effective prevention? Several authors have addressed this issue. For the exercise below we have used the ideas from McLean and Blower (1993).[10] They consider the three ways in which vaccines can fail to be perfect in the sense described above, and call these 'degree' (for the level of protection, the reduction in susceptibility), 'duration,' and 'take' (for the probability to be successful when administered). Let us denote the degree of protection by ψ, i.e., ψ is the fraction by which the probability of infection is reduced upon exposure. Assume that protection wanes with rate w, and assume finally that the probability that the vaccine takes in a given individual is π. Note that the choice $\pi = 1$, $\psi = 1$, and $w = 0$ corresponds to a perfect vaccine. We regard how the minimal fraction v to vaccinate depends on π, ψ and w. We do so for the case where we vaccinate continuously at birth.

Exercise 4.22 Give, for a vaccine that is imperfect in the above-mentioned aspects, the differential equations analogous to those of system (4.3). Give an expression for R_v and show that spread of the wild type is prevented when

$$v > \frac{1}{\vartheta}(1 - \frac{1}{R_0})$$

where

$$\vartheta = \frac{\pi\psi\mu}{\mu + w}.$$

We will return to vaccination in Section 9.6 after we have introduced models for populations with age structure. Incorporating age structure allows one to compute effects of more realistic vaccination schemes, where as a rule specific age groups are targeted.

Note that we have concentrated on effects of vaccination on susceptibility. There can also be effects on infectivity, in the sense that vaccinated individuals can still be infected by the wild type present in the population, but are less infectious to others, or for a shorter period of time, for example because of reduced replication of the wild type infectious agent in the vaccinated host. We briefly touch upon the difference between effects on susceptibility and infectivity in Section 8.1.2.

4.4   REGULATION OF HOST POPULATIONS

From the point of view of the infectious agent, the host population constitutes a renewable resource. Depending on the time scale considered and on the, perhaps implicit, incorporation of other factors, we can consider the birth term of the host population as zero, a constant or a per capita constant. The first is adopted when we want to model and understand an epidemic outbreak, the second when we want to study the dynamical balance of supply and 'consumption' of susceptibles and the third when we want to investigate the possible long-term influence of infectious diseases on population growth. So our own objectives matter. In this section we consider a population that grows exponentially in the absence of the infectious agent and we address the regulation problem, the third option above, and ask a number

---

[10]A.R. McLean and S.M. Blower: Imperfect vaccines and herd immunity to HIV. *Proc. R. Soc. B*, **253** (1993), 9–13.

of questions. Under what conditions on the parameters does the infectious agent: go extinct, grow but at a slower rate than the host, grow at the same rate as the host (and if so, how much is that rate reduced relative to the infection-free growth), or induce a steady host state (i.e., stop population growth), turn exponential host growth into exponential decline and eventual extinction?

But what does host population growth mean? In any case it means that numbers grow. But does it also entail that the density increases? Or is simultaneously the occupied area increased such that the density remains roughly constant (as in an expanding city). The key issue is of course whether or not the number of contacts per unit of time per individual increases, and if it increases, in what manner?

We shall start to consider the situation in which the per capita contact rate does not change. That is, we imagine a population that expands while growing so as to keep the density constant. The system

$$
\begin{aligned}
\frac{dS}{dt} &= bS + bR - \mu S - \gamma \frac{SI}{N}, \\
\frac{dI}{dt} &= -\mu I + \gamma \frac{SI}{N} - \alpha I, \\
\frac{dR}{dt} &= -\mu R + f\alpha I
\end{aligned}
\tag{4.12}
$$

with $N := S + I + R$, incorporates in addition the assumption that S and R individuals have a per capita birth rate b, while I individuals do not reproduce successfully. Recall from Exercise 1.24 and Section 1.3.3 that $1 - f$ is the probability that an individual dies at the end of the infectious period. Note that the system is first-order homogeneous, i.e., if (S, I, R) is a solution then so is (kS, kI, kR) for any constant k. This makes exponential solutions feasible even though the system is nonlinear.

Quite in general it is worth the effort to change to relative quantities, i.e., fractions, when studying regulation problems. Therefore we define

$$
y = \frac{I}{N}, \quad z = \frac{R}{N}.
\tag{4.13}
$$

Exercise 4.23 Verify that (4.12) decouples into the two-dimensional system

$$
\begin{aligned}
y' &= y\{\gamma(1 - y - z) - \alpha + \alpha(1 - f)y + b(y - 1)\}, \\
z' &= y(f\alpha + (1 - f)\alpha z) - bz(1 - y),
\end{aligned}
\tag{4.14}
$$

with a scalar equation for N appended,

$$
N' = \{b - \mu - (b + \alpha(1 - f))y\}N.
$$

We assume that $b > \mu$, so in the infection-free situation N grows exponentially with rate $b - \mu$. Before going into the analysis, we summarize the conclusions by plotting in Figure 4.2 the rate of growth of both host and infectious agent, as a function of the parameter γ which measures the infectivity.

At $\gamma = \gamma_0 = \alpha + \mu$ we have $R_0 = 1$. For $\gamma_0 < \gamma < \gamma_1$ the infected sub-population y grows exponentially with rate $\gamma - \alpha - \mu$, but N grows with the larger rate $b - \mu$, and the 'dilution' effect is that the proportion y of hosts infected still remains zero. At $\gamma = \gamma_1$ the two rates become equal, i.e., $\gamma_1 - \alpha - \mu = b - \mu$, so $\gamma_1 = b + \alpha$.

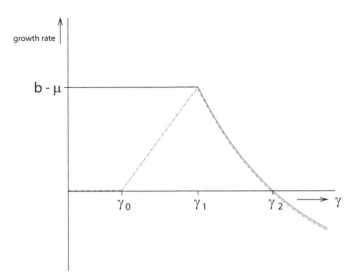

Figure 4.2: Qualitative graphs of the rate of growth of the host population N (solid line) and the 'infectious agent,' i.e., the class of infected hosts y (dash-dotted line), from system (4.14), as a function of the infectivity parameter γ.

An alternative way to characterize γ_1 has the advantage that it carries over to more complicated situations. The idea is to compute the expected number of secondary cases, while discounting for the host population growth by using the proportion of the population as the 'currency,' rather than absolute numbers. In more detail, consider a population of size $N(t) = N_0 e^{rt}$. Consider one individual infected at time zero. It constitutes a fraction N_0^{-1} of the population. Any secondary case produced by this individual after a time interval of length τ constitutes a fraction $N_0^{-1} e^{-r\tau}$ of the population. We want to determine whether or not the *fraction* of infecteds will grow. So we compute the ratio of fractions, which is $e^{-r\tau}$, and consider this as the appropriate weighing factor for the secondary case. Finally, integrate the product of the rate at which secondary cases are produced and this weighing factor with respect to τ, to obtain a number with threshold value one (for *relative* growth). Thus we define

$$R_{0,\text{relative}} = \gamma \int_0^\infty e^{-(\alpha+\mu)\tau} e^{-(b-\mu)\tau} d\tau = \frac{\gamma}{b+\alpha}.$$

(Note that the first factor of the integrand is the probability to be still alive and infectious at infection age τ, while the second derives from the fact that the host population has meanwhile increased by a factor $\exp(b-\mu)\tau$.) At $\gamma = \gamma_1$ the $R_{0,\text{relative}}$ passes the threshold value one. For $\gamma > \gamma_1$ the relative prevalence y is positive; the total host and infected host populations grow at the same reduced rate $b - \mu - (b + \alpha(1-f))\overline{y}$. Whether or not a further increase of γ leads to a reduction of the growth rate to below zero is determined by another quantity admitting a biological interpretation. The expected number of offspring produced by an individual that is infected right at birth[11] equals

$$\frac{\alpha}{\alpha + \mu} f \frac{b}{\mu}.$$

If this quantity is greater than one, the host population (and therefore the sub-population of infected hosts) will increase exponentially no matter how prevalent the infection is. If the quantity is less than one, increasing γ will inevitably lead to a situation in which the population will decline. Figure 4.2 is for the latter case.

The mathematical arguments underpinning these conclusions derive from a phase-plane analysis of the (y, z) system. For $R_{0,\text{relative}} < 1$, all solutions in the positive quadrant converge to the origin. For $R_{0,\text{relative}} > 1$, there exists a locally stable non-trivial steady state and the coordinate $\overline{y}$ of this steady state increases monotonically with the parameter γ, with limit

$$\overline{y}_\infty = \frac{\alpha + 2b - \sqrt{\alpha^2 + 4b\alpha f}}{2(b + (1 - f)\alpha)}$$

for $\gamma \to \infty$. So the population growth rate reduces from $b - \mu$ to the value $1/2\sqrt{\alpha^2 + 4b\alpha f} - \mu - \alpha/2$, which is negative if and only if $b\alpha f/(\alpha + \mu)\mu < 1$. The critical value γ_c at which the growth rate changes sign is given explicitly by

$$(\alpha + \mu)(b + \alpha(1 - f))(\alpha + \mu - b\alpha f \mu^{-1})^{-1}.$$

Exercise 4.24 Perform the phase-plane analysis and check the calculations.

The conclusion that the host is possibly driven to extinction for high values of γ hinges upon the homogeneous expression for the incidence. For low values of N this form is in fact debatable (for instance, for certain species of whale, the number of animals in parts of their habitat is so low that finding a mate becomes a problem). The next exercises discuss an alternative formulation.

Exercise 4.25 Consider the system

$$\frac{dS}{dt} = b_1 S + b_2 I - \mu S - \frac{\gamma S I}{K + N},$$

$$\frac{dI}{dt} = -\mu I + \frac{\gamma S I}{K + N} - \alpha I,$$

where now $N = S + I$. To keep things simple, we have eliminated the removed class (think of a lethal disease). Check that the behavior as a function of γ shows exactly the same pattern as before, but that now a (locally stable) steady state exists for $\gamma > \gamma_c$. Derive the condition for the existence of γ_c and check its interpretation. The transmission term here is only so-called asymptotically (for large population sizes) homogeneous. This means that for large N the influence of the fixed factor K in the denominator becomes negligible.

Exercise 4.26 Repeat the analysis for the system of the preceding exercise, but with the incidence term replaced by

$$\gamma C(N) \frac{S I}{N}.$$

---

[11]V. Andreasen: Disease regulation of age-structured host populations. *Theor. Pop. Biol.*, **36** (1989), 214–239.

What properties should the function $C(N)$ have? What is its interpretation? We will return to this incidence in Section 12.2.

Exercise 4.27 Consider the system

$$\frac{dS}{dt} = b_1 S + b_2 I - \mu S - \beta SI,$$
$$\frac{dI}{dt} = -\mu I + \beta SI - \alpha I.$$

i) Show that a steady state exists provided that $b_2/(\mu + \alpha) < 1$ (a condition with an interpretation that, we hope, has by now become familiar). Show that this steady state is at least locally stable.

ii) What happens when the condition does not hold?

Hint: Prove first that $S(t) \geq b_2/\beta$ for large t, next that $I(t) \to \infty$ for $t \to \infty$, and subsequently that $S(t) \to b_2/\beta$ for $t \to \infty$. Conclude that the agent reduces the host growth rate to $b_2 - \mu - \alpha$ and that, essentially, every individual spends its whole life in the class of infecteds.

We conclude that the parameter space is divided into regions according to three threshold criteria: one for the absolute growth of the agent, one for relative growth and one for stopping host growth. We have emphasized the biological interpretation of the various criteria, since these are robust, while explicit expressions are usually out of the question for more complicated models.

4.5 TOOLS FOR EVOLUTIONARY CONTEMPLATION

4.5.1 Some evolutionary considerations about virulence

This section touches upon only the tip of an enormous iceberg of largely open questions. For example, can we understand how evolution by natural selection has shaped the various ways in which infective agents exploit their hosts (and travel from one host to the next)?

The adage 'optimal adaptation to the environment' presupposes that the environment is somehow given and fixed. Yet if ever it is clear that individuals interact by feedback to components of the environment (which are therefore called 'environmental interaction variables'), it is when we consider parasitism (think of both the availability of susceptible hosts, the provoked immune response within one host and the death of a host). The key issue is *invasion potential*: is a new type, the mutant or rival, able to increase in number in the environment as set by the current type, the resident? Compare Exercise 4.17. For background reading on the theory of *adaptive dynamics* (or, as it is called when genetics is properly dealt with, *evolutionary dynamics*) we mostly refer to the literature,[12] and give one detailed

---

[12]O. Diekmann, F.B. Christiansen and R. Law (guest eds.) (1996): *Evolutionary Dynamics*, J. Math. Biol., **34**, issue 5/6, 483–688; F. Dercole and S. Rinaldi: *Analysis of Evolutionary Processes: the Adaptive Dynamics Approach and its Applications.* Princeton University Press, 2008; U. Dieckmann and J.A.J. Metz (eds.): *Elements of Adaptive Dynamics.* Cambridge University Press, Cambridge, to appear; O. Diekmann: A beginner's guide to adaptive dynamics. *Banach Center Publications*, **63** (2004), 47–86; U. Dieckmann: *Adaptive dynamics in context*, to appear, Cambridge University Press; a good reference list can be found at the url: http://mathstat.helsinki.fi/~kisdi.

example in Section 4.5.2. An inspiring study of the evolution of infectious disease can be found in the book by P.W. Ewald[13] (see also van Baalen and Sabelis 1995).[14] The general conclusion that emerges is that infectious agents are 'forced' to deal more carefully with their hosts when transmission is more difficult. So the virulence of the infectious agent depends on the host contact process and structure, on which transmission is superimposed. For a review of the evolutionary dynamics of virulence see Pugliese (2002).[15]

A metapopulation is a population consisting of many local populations (or colonies), linked by migration. When reproduction is both local (within the colony) and global (the founding of a new colony), it is often the second process that matters for evolution by natural selection, in particular when colonies consist of individuals with a high degree of genetic correlation (and perhaps even of individuals that are genetically identical). The problem that the organism has to face is to tune the reproduction within the colony in such a way that reproduction at the colony level cannot be improved (i.e., no mutant competitor can do any better). When a rival mutant cannot invade an existing colony, this is essentially the problem of optimal exploitation of a renewable resource. But it becomes a more subtle game when a patch of habitat consisting of a renewable resource has to be shared with the rival, once that rival has arrived at the patch. Indeed, in such a situation the economically attractive strategy of prudent exploitation of the resource (the so-called 'milker strategy') may be invaded because a more aggressive exploitation strategy (the so-called 'killer strategy') will secure a larger part of the resource. The dilemma is whether one should gain a lot at a slow pace or quickly take what can be gotten right away? The answer certainly depends on the likelihood of arrival of potential competitors.

Microparasites form a metapopulation. Indeed, from their point of view, the world consists of inhabitable patches, viz., hosts. Virulence is strongly coupled to reproduction (usually leading to genetically identical copies) within the host, while transmission corresponds to reproduction at the patch level. In our basic models, we ignore *superinfection* (i.e., the arrival, within the host, of yet another microparasite from outside the host) since reproduction within the host is fast and enormous. However, as we have just argued, for evolutionary considerations superinfection is crucial.[16]

A feature specific to the evolution of infective agents is the interaction with the immune system of the host. The host is not just a (renewable) resource, it also actively defends itself to prevent or stop its exploitation. Acquired immunity assures that a host cannot be colonized repeatedly. When evaluating the fitness (in the sense of Metz et al. 1992)[17] of a mutant rival, the issue of cross-immunity plays a crucial role. We think that the related aspects of superinfection and cross-immunity make the evolution of virulence such a complicated process to describe and analyze. (Note the paradox that the fact that for many current microparasites

---

[13] P.W. Ewald: *Evolution of Infectious Disease*. Oxford University Press, Oxford, 1994.

[14] M. van Baalen and M.W. Sabelis: The scope for virulence management — a comment on Ewald's view on the evolution of virulence. *Trends Microbiol.*, **3** (1995), 414–415.

[15] A. Pugliese: Evolutionary dynamics of virulence. In: U. Dieckmann, J.A.J. Metz, M.W. Sabelis and K. Sigmund (eds.): *Adaptive Dynamics of Infectious Diseases: in Pursuit of Virulence Management*. Cambridge University Press, 2002.

[16] M. van Baalen and M.W. Sabelis: The dynamics of multiple infection and the evolution of virulence. *Am. Nat.*, **146** (1995), 881–910.

[17] J.A.J. Metz, R.M. Nisbet and S.A.H. Geritz: How should we define 'fitness' for general ecological scenarios? *TREE*, **7** (1992), 198–202.

superinfection is irrelevant may be due to the key role of superinfection in shaping these microparasites by natural selection.)

A special evolutionary puzzle is posed by HIV. Within any particular host, the agent evolves by mutation and selective force exerted by the immune system, as well as, nowadays, cocktails of drugs. An understanding of the adaptive dynamics of the quasi-species (i.e., coexisting collections of individuals with only marginally different traits that are almost faithfully propagated from generation to generation) of HIV within the microcosm of one host is emerging. However, whether (and, if so, how) there is a concerted influence on adaptive dynamics at the metapopulation level remains to be seen/investigated.

This subsection is relatively short. We have tried to sketch some key features. We are aware that we have missed out on others (e.g., coevolution of the host and infectious agent or the battle between the host's immune system and the infectious agent). The next subsection, in contrast, is relatively long. At times the reader may wonder where it leads. We advise to at least skim the subsection, and to return to it for a closer reading while consulting the adaptive dynamics literature, if one is intrigued.

4.5.2   A hopefully illuminating example

We now present an analysis of a particular class of nested models. By 'nested' we mean that a sub-model for within-host dynamics serves as a building block for a transmission model at population level (between-host dynamics, see for example Mideo et al. 2008).[18] Our aim here is to illustrate some of the issues discussed previously in a more concrete context. In addition, our aim is to introduce the methodology of adaptive dynamics. As a first building block we take the following model for within-host dynamics, incorporating competition for target cells. Consider a virus and denote the cells that it infects by T, for 'target cells.' We denote the infected target cells by T^* and the free virus by V, and assume that

$$\begin{aligned}
\frac{dT}{dt} &= \lambda - kVT - \delta T, \\
\frac{dT^*}{dt} &= kVT - (\mu(p) + \delta)T^*, \\
\frac{dV}{dt} &= pT^* - kVT - cV.
\end{aligned} \qquad (4.15)$$

So target cells are produced at a constant rate λ and die at a constant per capita rate δ. Free virus dies at a per capita rate c. The rate at which target cells come into contact with free virus, and hence can become infected, is assumed to be given by the law of mass action, with rate constant k. We incorporate the reduction of free virus due to virus particles entering target cells. This may have a negligible effect at later stages (and is, for this reason, often ignored), but it is an essential component of the invasion process. Infected target cells are assumed to produce free virus at rate p. This is costly for the cell, in the sense that infected target cells have a death rate that is increased by μ, which depends on p. One can either think of continuous production of virus by infected cells (p virus particles per unit of time for on average $1/(\mu(p)+\delta)$ time units), or of cell death (called 'lysis' in this

---

[18]N. Mideo, A. Alizon and T. Day: Linking within- and between-host dynamics in the evolutionary epidemiology of infectious diseases. *TREE*, **23** (2008), 511–517.

context), in which a cell produces $p/(\mu(p)+\delta)$ virus particles at once, after a length of time that we assume to be exponentially distributed with mean $(\mu(p) + \delta)^{-1}$. We denote this *burst size* by B_0, so

$$B_0 = B_0(p) = \frac{p}{\mu(p) + \delta}. \tag{4.16}$$

One easily verifies that (4.15) has a non-trivial equilibrium

$$\begin{aligned}
\hat{T} &= \frac{c}{k(B_0 - 1)} \\
\hat{T}^* &= \frac{B_0}{p}\left(\lambda - \frac{c\delta}{k(B_0 - 1)}\right) \\
\hat{V} &= \frac{\lambda}{c}(B_0 - 1) - \frac{\delta}{k}
\end{aligned} \tag{4.17}$$

provided that

$$\tilde{R}_0 = \frac{k\lambda}{k\lambda + \delta c}B_0 > 1. \tag{4.18}$$

(Note that the first factor is the probability that a free virus particle enters a target cell, when the target cell population is at its 'virgin' size λ/δ.)

Imagine such a steady situation.[19] We call the virus the 'resident' and the corresponding value of p the 'trait' (or 'trait value') of the resident. Next imagine that, either from outside the host or by mutation inside the host, a slightly different virus emerges, which we call the 'invader.' The only difference between resident and invader is in the value of p, and so we shall talk about the 'trait of the invader.'

Two basic questions arise:

- Will the invader's population start growing?

- If yes, will it outcompete the resident population and become the new 'ruler' of the within-host world, i.e., will it become the new resident?

Before we can answer these questions we have to specify one more rule of the game: if a target cell is infected by virus with an arbitrary trait, it is protected from further infections from the same infectious agent (so we assume complete cross-immunity at the cellular level).

To answer the first question, we introduce in complete analogy with (4.18) the quantity

$$\tilde{R}_0 = \tilde{R}_0(\hat{T}, q) = \frac{k\hat{T}}{k\hat{T} + c}\left(\frac{q}{\mu(q) + \delta}\right) \tag{4.19}$$

(we will return to this in Exercise 7.14), where q denotes the trait of the invader and $\hat{T}$ denotes the 'environmental condition set by the resident' (in other words, $\hat{T}$ is given by the first formula in (4.17), with p the trait of the resident). It helps to incorporate the dependence on p into the notation and to write $\hat{T}(p)$ and $B_0(p)$. We can then rewrite (4.19) as

$$R_0(\hat{T}(p), q) = \frac{B_0(q)}{B_0(p)}. \tag{4.20}$$

---

[19]In P. de Leenheer and H.L. Smith: Virus dynamics: a global analysis. *SIAM J. Appl. Math.*, **63** (2003), 1313–1327, it is proved that the steady state is locally asymptotically stable for all relevant parameter values, while global stability is guaranteed if $\delta(\delta + \mu) > k\lambda$.

If we ignore the intricacies of demographic stochasticity we can now formulate an answer to the first question: Yes, whenever $B_0(q) > B_0(p)$; no, whenever $B_0(p) > B_0(q)$. (The effect of demographic stochasticity is that 'yes' should be read as 'possibly.')

We next turn to the second of the basic questions. For the system under consideration the answer should be 'yes,' but we do not know a published paper in which this is elaborated in detail (the situation is reminiscent of competition in the chemostat, for which there is a large body of literature dealing with this question, see the book of Smith and Waltman 1995). There is, however, a general result stating that the answer is 'yes' if p and q differ very little, so if we assume that mutations are necessarily small.[20]

The final conclusion is that, provided $B_0(q) > B_0(p)$, a *trait substitution* occurs, in which the resident with trait p is replaced by a new resident with trait q.

We add another crucial assumption, that of *time scale separation*: the ecological time scale at which q replaces p is short relative to the time scale of mutations. The point is that we then can repeat the game and hence, at the long evolutionary time scale set by the rate of successful mutations, picture evolution as a *trait substitution sequence*. Necessarily the burst size increases along this sequence, so we might say that evolution in this case maximizes burst size (or, for that matter, $\tilde{R}_0$). The more mechanistic formulation, however, is that evolution minimizes $\hat{T}$. Quite generally this is called the *pessimization principle* and it applies whenever interaction is via a one-dimensional environmental variable.

We turn to the between-host transmission model, which we describe by

$$\frac{dS}{dt} = b - \delta_h S - \beta SI,$$
$$\frac{dI}{dt} = -\delta_h I + \beta SI - \alpha I \quad (4.21)$$

(we replace, compared to (4.3), B by b and μ by δ_h ('h' for host) to avoid notational confusion with B_0 introduced in (4.16), and $\mu(p)$ in (4.15)). The idea is that both α and β are not, as we usually assume, constants, but depend on, and are determined by, the within-host situation. To avoid having to consider within-host transients, we assume that the within-host dynamics are fast relative to the time scale of transmission and of host demographic turnover. The upshot is that we can assume that α and β are determined by the steady state levels given in (4.17).

If we disregard super-infection, the evolutionary story is identical to the one for the within-host model: evolution will maximize

$$R_0 = \frac{\beta}{\alpha + \delta_h} \frac{b}{\delta_h} \quad (4.22)$$

or, equivalently, minimize

$$\hat{S} = \frac{\alpha + \delta_h}{\beta}. \quad (4.23)$$

The key observation in Gilchrist and Coombs (2006)[21] is that the evolutionary forces trying, respectively, to minimize $\hat{S}$ and to minimize $\hat{T}$, may work in opposite

---

[20]See S.A.H. Geritz (2005), Resident-invader dynamics and the coexistence of similar strategies. *J. Math. Biol.*, **50**, 67–82; as well as the book on adaptive dynamics by F. Dercole and S. Rinaldi: *Analysis of Evolutionary Processes: the Adaptive Dynamics Approach and its Applications*. Princeton University Press, 2008.

[21]M.A. Gilchrist and D. Coombs: Evolution of virulence: interdependence, constraints and selection using nested models. *Theor. Pop. Biol.*, **69** (2006), 145–153.

directions. Boldin and Diekmann (2008)[22] have shown that one can investigate the possibilities for a balance between these opposing forces by allowing for superinfection at the host level. To illustrate these issues we now focus on the relatively simple situation in which β is constant, i.e., independent of the within-host state. As it seems very natural to assume that α increases if $\hat{T}$ decreases, it is now completely obvious that the two forces do indeed act in opposing directions, whenever α is determined by $\hat{T}$ (and not, for instance, by $\hat{T}^*$).

How do we incorporate the possibility of super-infection? First we have to say what we mean by 'super-infection.' Here we understand this term as follows: if a host infected by virus with trait p is infected once more, but now with virus with trait q, and the invader 'q' outcompetes (within that host) 'p' (in an interval of time that can be considered as having zero length at the time scale of transmission and host demographic turnover), then we say that super-infection (of that host) has occurred.

Given that a host infected by virus with trait p is infected once more, now with virus of trait q, what is the probability that the invader outcompetes the resident? According to our analysis of the within-host model, it is zero if $B_0(q) < B_0(p)$. The drastic deterministic point of view suggests to declare it to be one if $B_0(q) > B_0(p)$. In the paper by Boldin and Diekmann it is shown that, in this so-called 'jump' case, the within-host evolutionary forces overrule the between-host evolutionary forces and consequently evolution will maximize burst size. The more subtle stochastic point of view suggests to give the 'outcompete-probability' a name, say $\phi(p, q)$, and to derive an expression for ϕ from a branching process description of the multiplication of virus with trait q in the environmental conditions (assumed to be constant) as set by the resident with trait p. More precisely, ϕ should be the probability of non-extinction of the super-critical branching process, given the parameter q and the number n of virus particles with which the process starts (so, unfortunately, ϕ depends on the *dose* about which we know very little, if anything at all). But there is even more indefiniteness: as we already mentioned in between equations (4.15) and (4.16), one can either think of continuous production of virus by infected cells, or of cell death, and the corresponding branching processes are different!

Here we shall choose the view that an infected cell lives for an exponentially distributed amount of time, with parameter $\mu(q) + \delta$, and produces during this period free virus (i.e., virus not in a host cell) according to a Poisson process with intensity q. As we know from Exercise 1.11, the corresponding generating function is

$$\frac{1}{1 - \frac{q}{\mu(q)+\delta}(z - 1)},$$

which can be rewritten as

$$\frac{1}{1 - B_0(q)(z - 1)}.$$

But if we start from a free virus, it will with probability

$$\frac{c}{k\hat{T}(p) + c}$$

[22]B. Boldin and O. Diekmann: Super-infections can induce evolutionary stable coexistence of pathogens. *J. Math. Biol.*, **56** (2008), 635–672.

fail to infect a target cell, so the overall generating function is

$$g(z) = \frac{c}{k\hat{T}(p) + c} + \frac{k\hat{T}(p)}{k\hat{T}(p) + c}\left(\frac{1}{1 - B_0(q)(z-1)}\right), \tag{4.24}$$

which we can also write as

$$g(z) = 1 - \frac{1}{B_0(p)} + \frac{1}{B_0(p)}\frac{1}{1 - B_0(q)(z-1)}. \tag{4.25}$$

The equation $g(z) = z$ has, apart from the root $z = 1$, the root

$$z = 1 - \frac{1}{B_0(p)} + \frac{1}{B_0(q)}, \tag{4.26}$$

which is meaningful if and only if

$$B_0(q) > B_0(p).$$

Now recall from Exercise 1.7 and the exposition in Chapter 3 that we can interpret this root as the probability that the offspring recursively originating from a single free virus with trait q in a host already 'occupied' by virus with trait p, goes extinct, despite its potential for exponential growth when still small (relative to the virus population with trait p). So if we inoculate such a host with n free virus particles of trait q, the corresponding probability to fail is given by

$$\phi(p,q) = \begin{cases} 1 - (1 - \frac{1}{B_0(p)} + \frac{1}{B_0(q)})^n, & B_0(q) > B_0(p); \\ 0, & \text{otherwise.} \end{cases} \tag{4.27}$$

(Note that, as to be expected, the limit $n \to \infty$ brings us back to the 'jump'-case.)

Imagine a steady situation at the host population level with a resident virus of trait p. Imagine that a mutation leads to an invader virus of trait q. We postulate that the initial dynamics of the population I_q of q-infected hosts are, when we ignore demographic stochasticity, described by the linear equation

$$\frac{dI_q}{dt} = -\delta_h I_q + \beta\hat{S}(p)I_q - \alpha_q I_q + \beta\hat{I}(p)\phi(p,q)I_q - \beta I_q\phi(q,p)\hat{I}(p),$$

which we write as

$$\frac{dI_q}{dt} = s_p(q)I_q \tag{4.28}$$

with the so-called *invasion exponent* $s_p(q)$ given by

$$s_p(q) = -\delta_h + \beta\hat{S}(p) - \alpha_q + \beta\hat{I}(p)(\phi(p,q) - \phi(q,p)). \tag{4.29}$$

Note that, of course, $s_p(p) = 0$. Using (4.23) and

$$\hat{I}(p) = \frac{b}{\alpha_p + \delta_h} - \frac{\delta_h}{\beta} \tag{4.30}$$

we rewrite (4.29) as

$$s_p(q) = \alpha_p - \alpha_q + \left(\frac{\beta b}{\alpha_p + \delta_h} - \delta_h\right)(\phi(p,q) - \phi(q,p)) \tag{4.31}$$

In view of (4.27), $s_p(q)$ depends on α_p, $B_0(p)$ and α_q, $B_0(q)$, but not directly on either p or q. This makes it tempting to bypass p and q, in order to expose the essential features without being distracted by messy computational details. To do so, we make another assumption, viz., that $p \mapsto \alpha_p$ is invertible. In a rather implicit way this yields a function f such that

$$f(\alpha) = \frac{1}{B_0(p(\alpha))}. \tag{4.32}$$

Note that, from a mechanistic point of view, α should increase if B_0 does increase and that, therefore, it is reasonable to assume that f is a decreasing function of α.

So, rather than studying (4.31), we shall study

$$s_{\alpha_1}(\alpha_2) = \alpha_1 - \alpha_2 + (\frac{\beta b}{\alpha_1 + \delta_h} - \delta_h) H(f(\alpha_1) - f(\alpha_2)) \tag{4.33}$$

where the function H is defined by

$$H(x) = \begin{cases} 1 - (1-x)^n, & \text{if } x > 0; \\ 0, & \text{if } x = 0; \\ -1 + (1+x)^n, & \text{if } x < 0. \end{cases} \tag{4.34}$$

Note that H is differentiable at $x = 0$ with

$$H'(0) = n \tag{4.35}$$

but that the second derivative of H is discontinuous at $x = 0$, with limit from the left

$$H''(0-) = n(n-1) \tag{4.36}$$

and limit from the right

$$H''(0+) = -n(n-1). \tag{4.37}$$

In order to determine whether evolution has a tendency to increase or to decrease α, we compute the so-called *selection gradient*

$$\frac{\partial}{\partial \alpha_2} s_{\alpha_1}(\alpha_2)|_{\alpha_1 = \alpha_2 = \alpha} = -1 - (\frac{\beta b}{\alpha + \delta_h} - \delta_h) H'(0) f'(\alpha). \tag{4.38}$$

When we recall (4.30), we realize that necessarily α is restricted to satisfy

$$\frac{\beta b}{\alpha + \delta_h} - \delta_h > 0 \tag{4.39}$$

(this is equivalent to $R_0 > 1$, recall (4.22)) and so the two terms in (4.38) have opposite sign, reflecting that for transmission to susceptibles it is advantageous to have a low α, while to avoid losing a host to a competitor by super-infection it is advantageous to have a high α. Clearly at the upper boundary of the feasible α-region, characterized by the condition $R_0 = 1$, the selection gradient is negative. Provided $-f'(\alpha)$ is sufficiently large for small values of α, the selection gradient is positive for small α. So there has to exist at least one *singular point*, i.e., a value of the trait α at which the selection gradient vanishes. Under further conditions on $\alpha \mapsto f'(\alpha)$ there will be a unique singular point. Note that in such a point the selection gradient changes sign from positive to the left to negative to the right, and that hence a *trait substitution sequence* will bring us towards the singular point (in questionable, yet standard, terminology this is expressed by saying that the singular point is *convergence stable*).

There exists a classification of singular points on the basis of second-order derivatives.[23] In the present situation, however, we have to deal with some lack of smoothness, as demonstrated by (4.36)–(4.37).

In any case, one of the revelations of the classification is that a singular point may be convergence stable, yet able to be invaded (so not an evolutionarily stable strategy). What happens is that near such a singular point the population turns *dimorphic* (meaning that two types of resident coexist), due to *mutual invasion potential*. This motivates us to look for coexistence of two sub-populations (characterized by different α's) in the present model. We speak of mutual invasion potential if simultaneously $s_{\alpha_1}(\alpha_2) > 0$ and $s_{\alpha_2}(\alpha_1) > 0$. In view of (4.33) and the fact that $H(-x) = -H(x)$ we can rewrite these conditions in the form

$$(\frac{\beta b}{\alpha_2 + \delta_h} - \delta_h)H(f(\alpha_1) - f(\alpha_2)) < \alpha_2 - \alpha_1 < (\frac{\beta b}{\alpha_1 + \delta_h} - \delta_h)H(f(\alpha_1) - f(\alpha_2)). \tag{4.40}$$

The dynamics of a dimorphic population are generated by

$$\begin{aligned}
\frac{dS}{dt} &= b - \beta S(I_1 + I_2) - \delta_h S, \\
\frac{dI_1}{dt} &= \beta S I_1 - (\alpha_1 + \delta_h)I_1 + \beta I_1 I_2 H(f(\alpha_2) - f(\alpha_1)), \\
\frac{dI_2}{dt} &= \beta S I_2 - (\alpha_2 + \delta_h)I_2 + \beta I_1 I_2 H(f(\alpha_1) - f(\alpha_2))
\end{aligned} \tag{4.41}$$

and so the coexistence steady state equations are

$$\begin{aligned}
b - \beta S(I_1 + I_2) - \delta_h S &= 0, \\
\beta S - \alpha_1 - \delta_h + \beta I_2 H(f(\alpha_2) - f(\alpha_1)) &= 0, \\
\beta S - \alpha_2 - \delta_h + \beta I_1 H(f(\alpha_1) - f(\alpha_2)) &= 0.
\end{aligned} \tag{4.42}$$

If we subtract the last equation of (4.42) from the middle one and use once more that $H(-x) = -H(x)$ we find

$$I_1 + I_2 = \frac{\alpha_1 - \alpha_2}{\beta H(f(\alpha_2) - f(\alpha_1))}, \tag{4.43}$$

which upon substitution in the first equation of (4.42), yields

$$S = \frac{b}{\delta_h + \beta(I_1 + I_2)} = \frac{bH(f(\alpha_2) - f(\alpha_1))}{\alpha_1 - \alpha_2 + \delta_h H(f(\alpha_2) - f(\alpha_1))}, \tag{4.44}$$

which upon substitution in the last two equations of (4.42), yields explicit expressions for both I_1 and I_2. Note, however, that whereas the fact that $\alpha_1 - \alpha_2$ and $g(f(\alpha_2) - f(\alpha_1))$ have the same sign guarantees that the right-hand sides of (4.43)

---

[23]See S.A.H. Geritz, J.A.J. Metz, E. Kisdi and G. Meszena: The dynamics of adaptation and evolutionary branching *Phys. Rev. Letters*, **78** (1997), 2024–2027; and for expositions F. Dercole and S. Rinaldi: *Analysis of Evolutionary Processes: the Adaptive Dynamics Approach and its Applications.* Princeton University Press, 2008; O. Diekmann: A beginner's guide to adaptive dynamics. *Banach Center Publications*, **63** (2004), 47–86.

and (4.44) are positive, the explicit expressions for I_1 and I_2 do not necessarily yield a positive number. If, for definiteness, we concentrate on the case $\alpha_2 > \alpha_1$, then the requirement that both I_1 and I_2 are positive, translates into the condition

$$\frac{\alpha_1 + \delta_h}{\beta} < S < \frac{\alpha_2 + \delta_h}{\beta}.$$

Using the explicit expression (4.44) and some straightforward formula manipulation we find that this, in turn, translates into exactly the condition (4.40). We conclude that *steady state coexistence is possible if and only if there is mutual invasion potential*!

Note that it is as yet unclear whether (4.40) holds in a non-empty region of the (α_1, α_2)-plane. Keeping in mind that we often restrict our attention to small mutations, we first look for such a region near the diagonal. If we consider the choice $\alpha_1 = \alpha, \alpha_2 = \alpha + \epsilon$, with ϵ small, and expand both the left and the right-hand side of (4.40) with respect to ϵ, we find that both sides are up to first order given by

$$(\frac{\beta b}{\alpha + \delta_h} - \delta_h)H'(0)(-f'(\alpha))\epsilon.$$

So to be able to satisfy both inequalities in (4.40) we find that necessarily α needs to be a singular point! Next, one has to compute second-order terms to find a condition that guarantees that (4.40) holds in a non-empty region. Unfortunately, such computations tend to become rather messy.

The diagonal $\alpha_1 = \alpha_2 = \alpha$ is a so-called *neutrality curve*, by which we mean that $s_\alpha(\alpha) = 0$. At the singular point a second neutrality curve intersects the diagonal. The classification of singular points involves the slope of this second neutrality curve. In the present situation the lack of smoothness of H has as an effect that the second neutrality curve has a 'corner' at the singular point. The condition for mutual invasion potential involves the relationship between the curve and its reflection in the diagonal. We now show that in the simple case of (4.33) one can use the reflection without having to compute the local shape of the curve. Indeed from (4.33) it follows that

$$s_{\alpha_1}(\alpha_2) + s_{\alpha_2}(\alpha_1) = H(f(\alpha_1) - f(\alpha_2))[\frac{\beta b}{\alpha_1 + \delta_h} - \frac{\beta b}{\alpha_2 + \delta_h}]$$

and whenever $\alpha_1 \neq \alpha_2$ the right-hand side is positive! So if $s_{\alpha_1}(\alpha_2) = 0$ then $s_{\alpha_2}(\alpha_1) > 0$. If we now move from the point on the neutrality curve to neighboring points at the 'positive' side, then by continuity, the invasion exponent in the reflected point is positive as well. We conclude that the region of mutual invasion potential is non-empty.

Computations become much easier if one replaces (4.27) by the phenomenological

$$\phi(p, q) = \psi(q - p)$$

with $\psi(x) = 0$ for $x \leq 0$, $\psi'(x) > 0$ for $x > 0$ and $\psi(\infty) = 1$, see Boldin and Diekmann (2008; Appendix B), cited in footnote 22.

In conclusion we summarize the main points. Infectious agents reproduce at two levels: within an individual host and by way of transmission to another host. As a consequence, they are exposed to opposing evolutionary forces: to win within-host competition they should be relentless, but to let the host serve as the vehicle bringing them to the next host they should be prudent. One can imagine that this either leads to compromise or to diversification. How to find out? (i.e., how to determine the outcome of a long process of selection and mutation within an artificial universe characterized in terms of plausible assumptions?) Adaptive Dynamics offers both a conceptual framework and a tool kit, but at the price of certain simplifying assumptions. The aim of this section has been to provide a gentle introduction by way of a somewhat caricatural example.

4.6   MARKOV CHAINS: MODELS OF INFECTION IN THE ICU

In the rest of this chapter we shall deal with stochastic models for finite populations with demographic turnover. We begin with a rather specific example: the intensive care unit (ICU) in a hospital. The key characteristics of the ICU are low numbers (so the limit $N \to \infty$ is irrelevant) but high turnover (most patients stay only a couple of days). By concentrating on long time intervals, we may still obtain a large total population and do statistics. This point of view leads in a very natural way to a Markov chain description: we consider the probabilities of the various compositions of the ICU in terms of infected and susceptible individuals, and how these probabilities change in time (rather than describing the actual realizations of the stochastic process). A great advantage of the Markov chain description is that it leads to *linear* mathematical problems for which we have many systematic mathematical tools. So this section serves simultaneously to introduce some elementary/basic ideas, concepts and methods from the general area of Markov chains, and to consider a very specific example that is relevant in its own right, for reasons we explain below.

Infections that are acquired in the hospital are called nosocomial infections. Prominent examples concern infection by antibiotic-resistant bacteria like VRE (vancomycin-resistant enterococci), MRSA (methicillin-resistant *Staphylococcus aureus*), and bacteria producing ESBLs (extended-spectrum beta-lactamases). The incidence of such infections is increasing at an alarming rate and somewhat pessimistically one could argue that they jeopardize the health care system. Indeed, the prospect of bacteria developing resistance against *all* available antibiotics is rather terrifying.

The reason why such bacteria circulate much more in hospitals than in the population at large is that within hospitals one finds vulnerable hosts (a weakened constitution entails a weakened immune system) who, moreover, are very often treated with antibiotics in order to eliminate or avoid infections. It is thought that there might be a cost to resistance for bacteria, so that they are competitively inferior to the wild type in healthy hosts who are not treated with antibiotics, but the precise mechanisms of competition are not known (in other words, the idea of 'cost to resistance' is based on circumstantial evidence only).

Of all the wards in a hospital, the ICU harbors the weakest patients. In addition to this, the type of treatment in the ICU, in particular ventilator-assisted breathing, helps bacteria to cross barriers that are (as yet) insurmountable under normal conditions. Accordingly the incidence of nosocomial infections is highest in the ICU.

A typical ICU has on the order of 10 beds. There is a relatively rapid turnover of patients by way of admission and discharge (either to another ward or as euphemism for death). We therefore need a *stochastic* description (because of the small numbers) of an *open* population (because of the 'demographic' turnover). The aim of this section is to introduce such a description.

This description takes the form of a Markov chain. We start with the continuous-time case, as it is easier from a modeling point of view. The discrete-time variant is considered in Exercise 4.32. In this section our aim is to introduce the main ideas, concepts and rules by way of very simple toy examples. In Chapter 14 we return to the topic of antibiotic resistant bacteria that spread in ICUs. There we will be more ambitious in the sense that we shall make inferences concerning transmission parameters from hospital admission and discharge data.

4.6.1 The simplest example

Consider, as the simplest case, an ICU consisting of just two beds that are occupied at all times (i.e., we simply assume that a discharge is followed immediately by an admission). We assume that a patient is either susceptible or infectious. In other words, a patient is assumed to be either colonized or not colonized, and the transition is instantaneous. We ignore therefore subtle differences between being colonized and suffering from an infection in either an organ or in the blood, while also ignoring that it takes time for the colonizing bacterial population to grow to an appreciable size.

Let us call the state of the ICU in which both patients are susceptible '0,' the state in which one is susceptible while the other is colonized '1,' and the state in which both are colonized '2.' Note that these numbers carry a meaning; also note that we consider the beds as indistinguishable (exchangeable) in the sense that we assume it is irrelevant which of the two beds is occupied by a colonized patient when the state is '1.'

If we want to conform to the standard way of bookkeeping, then we should conceive of time as a discrete variable measured in days. But, as illustrated by the details of Exercise 4.32, discrete-time models are in some ways more complicated than continuous-time models, simply because of the possibility of multiple events during a single time step. Therefore we choose to work in continuous time.

The model that we aspire to make should describe how states change in the course of time. As we cannot be certain about the changes, this description will be probabilistic. The model that we will formulate belongs to the category of *queuing models*.[24]

Even though we are merely interested in the prevalence of colonization, we are forced to make assumptions concerning the distribution of the length-of-stay (LOS) of patients in the ICU, for the simple reason that the state may change as a result of the discharge of a colonized patient, followed by the admission of a susceptible patient. The assumption we make is that every patient (whether colonized or not) has a probability $1/\delta$ per unit of time of being discharged. The LOS then has an exponential distribution with mean δ, which admittedly may not be very realistic, but serves the purpose for the simple model.

Strictly speaking, the parameter δ corresponds to the average LOS of *colonized* patients and the LOS of uncolonized patients is irrelevant (as an uncolonized patient who is discharged is instantaneously replaced by a newly admitted uncolonized patient). We have chosen the more restrictive assumption above, in view of the

---

[24] See P. Trapman and M.C.J. Bootsma: A useful relationship between epidemiology and queueing theory: The distribution of the number of infectives at the moment of the first detection. *Math. Biosci.*, **219** (2009), 15–22, and the references given there.

need to estimate δ from data concerning LOS, see Section 5.2. In Section 5.3 we shall consider the situation in which both the uncolonized and the colonized patients are assumed to have an exponentially distributed LOS, but with possibly different parameters.

We assume that newly admitted patients are with certainty susceptible. Hence, if the ICU is in state 1, there is a probability per unit of time $1/\delta$ that it changes to state 0. Similarly, if the ICU is in state 2, there is a probability per unit of time $2/\delta$ to go to state 1, because each one of the two colonized patients may be discharged, independently. In continuous time it is impossible to go directly from state 2 to state 0, because always one of the two colonized patients is discharged first (in the continuous-time framework 'events' never happen simultaneously; of course the ICU may sojourn in state 1 for a very brief time only and then move on to state 0).

We assume that detectable colonization can arise in a few different ways. The first is by transmission. If the ICU is in state 1, a susceptible patient can acquire colonization by way of bacteria being carried from the already colonized patient to the susceptible patient, for instance via the hands of a health-care worker. We assume that this happens with (unknown!) probability per unit of time β. Another way in which detectable colonization can arise, in case of antibiotic resistant bacteria, is that the susceptible patient already carried these bacteria, but at an undetectably low level. Once treatment with the antibiotic has eliminated the wild-type competitor, the resistant bacteria find themselves in an environment in which they can grow to a high, and detectable, population level. A third scenario for colonization is that the resistant form arises in the susceptible host by way of a mutation of the wild type, followed by growth of the mutant population due to a selective advantage in a host under antibiotic treatment. Because the mathematical descriptions of the second and third scenario are identical, we choose not to distinguish between the two. This mathematical description amounts to postulating a per-susceptible, per unit of time probability α of turning into a colonized patient. At the ICU level this gives a rate, i.e., a probability per unit of time 2α at which state 0 changes to state 1, and a rate $\alpha + \beta$ at which state 1 changes into state 2.

When we finally assume that a colonized patient remains colonized for at least as long as the patient stays in the ICU, then our description of the state transitions is complete and we have arrived at the scheme depicted in Figure 4.3.

By taking samples from patients and culturing these, one may obtain partial, or even complete, information about the state of the ICU at various points in time. But in general we are somewhat in the dark as to the state. We acknowledge this uncertainty by introducing

$$p_i(t) := \text{probability that the ICU is in state } i \text{ at time } t. \qquad (4.45)$$

The scheme depicted in Figure 4.3 immediately translates into the system of differential equations:

$$
\begin{aligned}
\frac{dp_0}{dt} &= -2\alpha p_0 + \frac{1}{\delta}p_1, \\
\frac{dp_1}{dt} &= 2\alpha p_0 - (\frac{1}{\delta} + \alpha + \beta)p_1 + \frac{2}{\delta}p_2, \qquad (4.46) \\
\frac{dp_2}{dt} &= (\alpha + \beta)p_1 - \frac{2}{\delta}p_2.
\end{aligned}
$$

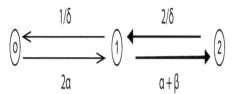

Figure 4.3: Schematic representation of the states and possible transitions for the ICU system; see text for details.

Note that, as required by the interpretation,

$$\frac{d}{dt}(p_0 + p_1 + p_2) = 0$$

so that

$$p_0 + p_1 + p_2 = 1 \qquad (4.47)$$

provided that this equality holds at the initial time, say $t = 0$. If we know for certain that at $t = 0$ both patients are susceptible, we should adopt the initial condition

$$p_0(0) = 1, \qquad p_1(0) = p_2(0) = 0.$$

What kind of information can we extract from the description (4.46)? The parameter δ can be estimated relatively easily from the data concerning the actual LOS of patients, see Section 5.2; note that it also has to be checked whether or not the data support our assumption of an exponentially distributed LOS. The parameters α and β, however, are essentially unknown. Yet, whether or not a particular control measure (such as a more strict hygiene protocol, or a more strictly adhered-to protocol, concerning hand washing for health-care workers) is effective depends on whether colonization is mainly due to transmission or to opportunistic growth of already present (or newly mutated) resistant bacteria. Hence we like to have information about α and β. Chapter 5 is devoted entirely to inference, i.e., to drawing conclusions from data by way of a model, but here we take some initial steps, mainly to illustrate some aspects of continuous-time Markov chains, such as the one corresponding to (4.46).

In terms of the three-vector p we can write (4.46) in the form

$$\frac{dp}{dt} = Ap \qquad (4.48)$$

where the 3x3-matrix A is given explicitly by

$$\begin{pmatrix} -2\alpha & \frac{1}{\delta} & 0 \\ 2\alpha & -(\frac{1}{\delta} + \alpha + \beta) & \frac{2}{\delta} \\ 0 & \alpha + \beta & -\frac{2}{\delta} \end{pmatrix}.$$

This is a positive-off-diagonal matrix with columns that sum to zero (the fact underlying the 'conservation of total probability' equality (4.47)). As explained in the books listed under the heading 'Non-negative matrices and operators' in our Bibliography (and in Sections 7.1 and 8.2), this entails that the matrix A has 0 as an eigenvalue and that the other two eigenvalues have (as a rule strictly) negative real part. The first of these two properties can easily be verified directly: the vector

$$\bar{p} = \frac{1}{1 + 2\alpha\delta(1 + \frac{\delta}{2}(\alpha + \beta))} \begin{pmatrix} 1 \\ 2\alpha\delta \\ 2\alpha\delta\frac{\delta}{2}(\alpha + \beta) \end{pmatrix} \tag{4.49}$$

is such that $A\bar{p} = 0$. Together with (4.47) the two properties imply that the solution $e^{tA}p(0)$ of (4.48) converges for $t \to \infty$ to $\bar{p}$ for any positive initial condition $p(0)$ satisfying (4.47). We therefore call $\bar{p}$ the stationary distribution of the Markov chain. The meaning is that, when we consider a very long time period, the ICU will, to a good approximation, spend a fraction of the time $\overline{p_i}$ in state i. Equivalently, if one's latest information about the state of the ICU dates from a long time ago, then the probability to find the ICU in state i equals $\overline{p_i}$. Note that the state keeps on jumping back and forth between 0, 1 and 2, and that $\bar{p}$ gives a probabilistic description of the occupation fractions. If one has information about these fractions from actual data, one can estimate the parameters α and β by using the formula (4.49). In Chapter 5 we will introduce a somewhat more sophisticated version of this idea.

When is a time period very long? What is our standard for estimating the length of a time interval? The time scale of convergence to $\bar{p}$ is determined by the non-zero eigenvalue with the largest real part. If we denote this eigenvalue by λ, then a time interval of length τ is long when $\exp(\operatorname{Re}\lambda\tau)$ is small (recall that necessarily $\operatorname{Re}\lambda < 0$). And, finally, it is the user's choice how small is small enough.

If we scale time (i.e., if we choose another unit of time, for instance milliseconds instead of centuries), the parameters α, β and $1/\delta$ are all multiplied by the same constant. This manifests itself in the fact that $\bar{p}$ only depends on the dimensionless quantities $\alpha\delta$ and $\beta\delta$. So one can only estimate α and β from a data-based estimate of $\bar{p}$ if one has an independent way of estimating δ.

Exercise 4.28 Show that one can estimate the *ratio* $\alpha/(\alpha + \beta)$ from an observed $\bar{p}$.

In the (somewhat academic) special case $\beta = 0$ (corresponding to perfect hygiene), the two beds behave independently and the number of infections in stationarity is binomially distributed. We elaborate a little more on the (equally academic) special case $\alpha = 0$, where no return is possible from the desirable state 0. Accordingly, 0 is called an absorbing state when $\alpha = 0$. The stationary distribution $\bar{p}$ degenerates into $(1,0,0)^{\top}$, reflecting that with certainty the state will become 0 at some time (and then remain 0 forever). The most relevant information then is how long it will take before state 0 is reached. Clearly this depends on whether the current state is 1 or 2. Let's assume it is 1. There is a very simple efficient method, called first-step analysis, to calculate the expected time T to absorption (this method will feature extensively in Section 7.8). Recalling Exercise 1.34-vi we first note that the next state will be 0 with probability $(1/\delta)/(1/\delta + \beta) = 1/(1 + \beta\delta)$ and it will be 2 with the complementary probability $\beta\delta/(1 + \beta\delta)$. From state 2 we will with certainty return to state 1, after which the expected time to absorption is again T (because we are back to exactly the same position from which we started; this is the so-called Markov property). The expected duration of a sojourn in state 1 is $(1/\delta+\beta)^{-1}$ and the expected duration of a sojourn in state 2 is $\delta/2$. Combining these pieces of information we deduce the identity

$$
\begin{aligned}
T &= \frac{1}{(1+\beta\delta)}\frac{\delta}{(1+\beta\delta)} + \frac{\beta\delta}{1+\beta\delta}\left(\frac{\delta}{1+\beta\delta} + \frac{\delta}{2} + T\right) \\
&= \frac{\delta}{1+\beta\delta} + \frac{\beta\delta^2}{2(1+\beta\delta)} + \frac{\beta\delta}{1+\beta\delta}T.
\end{aligned}
$$

Solving this equality for T leads to

$$T = \delta(1 + \frac{\beta\delta}{2}). \tag{4.50}$$

Exercise 4.29 To train yourself in this kind of argument, re-derive (4.50) in a slightly different way as follows. Let s denote the expected number of visits to state 1 before absorption in state 0 (the starting visit included). Derive an equation for s and solve it. Next express T in terms of s and show that (4.50) is obtained.

Exercise 4.30 Assume the ICU has three beds. Formulate the analog of the scheme of Figure 4.3 and the matrix A given following (4.48). Include into your deliberation how the force of infection acting on susceptible patients can depend on the number of colonized patients in the ICU. Check that total probability is conserved (cf. (4.47)).

Exercise 4.31 For an ICU with N beds, specify the rate at which the transition $i \to i+1$ occurs (here, as before, i denotes the number of colonized patients).

Exercise 4.32 The administration of a hospital works on a day-by-day basis, and hence a discrete-time model with time unit one day seems more appropriate. Develop the analog of Figure 4.3 for such a setting.

In conclusion, we introduce the notion of *quasi-stationary distribution* and illustrate how one can make use of it (not so much because of the relevance in the present context, but because it will be used in the next subsection, while being easier to explain in the current simple setting).

We return to system (4.48) with $\alpha = 0$. So we know that ultimately we will end with an ICU without any colonized patients. If we start from a situation in which there is at least one colonized patient, we may *condition* that this 'extinction' has not yet happened. Technically we do this by dividing the quantities p_1 and p_2 by $1-p_0$, the probability that absorption in state 0 has not yet occurred. Remembering that $p_0 + p_1 + p_2 = 1$, we introduce

$$q_1 = \frac{p_1}{p_1 + p_2}, \qquad q_2 = \frac{p_2}{p_1 + p_2}$$

and compute, using (4.48) and also $q_1 + q_2 = 1$, that

$$\frac{dq_1}{dt} = \frac{2}{\delta} - (\frac{3}{\delta} + \beta)q_1 + \frac{1}{\delta}q_1{}^2$$

from which we conclude that q_1 converges for $t \to \infty$ to $\bar{q}$, the unique root in $(0,1)$ of the equation

$$q^2 - (3 + \beta\delta)q + 2 = 0.$$

The distribution

$$\begin{pmatrix} \bar{q} \\ 1 - \bar{q} \end{pmatrix}$$

is called the quasi-stationary distribution. It describes, in this application, what fraction of the time the ICU will have either one or two colonized patients, *given* that it has at least one.

While the ICU has one colonized patient, extinction (by way of the colonized patient being discharged) occurs at rate $1/\delta$. While the ICU has two colonized patients, extinction cannot happen. So starting from the quasi-stationary distribution, the time till extinction is exponentially distributed with parameter $\bar{q}/\delta$, i.e., expected duration $\delta/\bar{q}$.

Consistence of course requires that, with T defined by (4.50),

$$\bar{q}T + (1 - \bar{q})(T + \frac{\delta}{2}) = \frac{\delta}{\bar{q}}.$$

We leave it to the reader to verify that this is just another way of writing the defining quadratic equation for $\bar{q}$.

4.7   TIME TO EXTINCTION AND CRITICAL COMMUNITY SIZE

We now return to stochastic models for large populations, and consider infectious agents that may become endemic. We consider the same type of models as described in Section 4.2, i.e., where new individuals are born into the community, individuals can get infected, recover and die, and where immunity is life-long. In the present section our focus is on the time it takes for such an infectious agent to go extinct, and how this is affected by the initial condition as well as by the model parameters in general and, particularly, the community size. Since extinction is necessarily preceded by a period of low number of infecteds, we use a stochastic epidemic model for this exploration.

4.7.1   The stochastic epidemic model with demography

We choose the following requirements and conventions for the model: the community size should fluctuate around a value N (the 'average' population size), individuals live for an exponentially distributed amount of time (the average life-length we denote by $1/\mu$), an individual makes pairwise contacts with other individuals with rate γ (such contacts are assumed to lead to infection if one individual is infectious and the other susceptible), and the average length of the infectious period should equal $1/\alpha$. To keep the model as simple as possible, we consider a homogeneously mixing community in which all individuals are identical with respect to susceptibility and infectivity. Finally we assume that the (random) length of an infectious period follows the exponential distribution. This last assumption helps to achieve that it is enough to keep track of the present number of susceptible and infectious individuals, i.e., to achieve that the epidemic process is Markovian (the future is independent of the past given the present state, in terms of numbers, of the epidemic).

We now define more precisely the stochastic epidemic model with demography that we are going to study. This is the stochastic counterpart of system (4.3).

The components of the population dynamics are defined as follows: (susceptible) individuals are 'born' at the constant rate μN, and each individual lives an exponentially distributed time with mean $1/\mu$. The number of (susceptible) individuals is then a birth-and-death process $S(t)$ with constant birth rate μN and death rate $\mu S(t)$. Within this dynamic community with demographic structure an infectious disease spreads as follows. An individual who gets infected immediately becomes infectious and remains so for an exponentially distributed period with mean $1/\alpha$. During the infectious period such an individual transmits the infection to others at rate γ, where, as before, the transmission rate constant $\gamma = cp$ can be factorized into the contact rate c and the probability p of disease transmission given contact and given that the other individual is susceptible. The close contact rate with a *specific* other individual is γ/N since we are assuming homogeneous mixing. Finally, we assume that the death rate is unaffected by the disease, so both susceptible and infectious individuals die at the rate μ. The infectious disease is introduced by one individual being externally infected and it persists until the first point in time that there is no infectious individual present in the community. After that the rate of infection is 0, which implies that no one can become infected. We say that the epidemic has gone extinct, and the time until this happens is called the extinction time.

A few comments about the model are worth making. The reason for having the constant birth rate μN rather than a birth rate depending on the actual population size (e.g., $\mu S(t)$) is simply that we want the equilibrium value N to be stable — otherwise there would be no equilibrium and the whole community would die out sooner or later. It would be most natural to define the close contact rate with a specific other individual, due to the homogeneous mixing assumption, to be γ divided by the actual population size. Here we are cheating slightly by dividing by the *average* population size N. This approximation allows us to neglect the number of removed individuals, thus reducing the dimension of the process from three to two. Finally, an increased death rate for infectious individuals could be included in the recovery rate α — a person who dies also stops being infectious.

Let $(S(t), I(t))$ denote the number of susceptibles and infectives at time t, respectively (when needed we write $(S_N(t), I_N(t))$ to emphasize the dependence on N). The process $\{(S(t), I(t)); t \geq 0\}$ is a two-dimensional Markov jump process, each component being integer valued. If at time t the process is in state $(S(t), I(t)) = (n, m)$, the process can make four different jumps:

$$
\begin{array}{llll}
\text{to} & (n+1, m) & \text{at rate} & \mu N, \\
\text{to} & (n-1, m) & \text{at rate} & \mu n, \\
\text{to} & (n-1, m+1) & \text{at rate} & \gamma nm/N, \\
\text{to} & (n, m-1) & \text{at rate} & (\alpha + \mu)m,
\end{array}
$$

as a result of, respectively, birth, death of a susceptible, transmission of infection, and recovery or death of the infective. Note that the number of removed individuals does not enter in any of the expressions, implying that there is no need to keep track of them. We scale by N and denote the corresponding proportions by $(\bar{S}(t), \bar{I}(t)) = (S(t)/N, I(t)/N)$.

An infectious individual stops being infectious either by recovery (rate α) or death (rate μ) so the average duration of the period an individual is infectious is $1/(\alpha + \mu)$. During the initial phase of an outbreak in a large community nearly all individuals are susceptible, so an infectious individual infects others at rate γ while infectious. During the initial phase of an outbreak the expected number of infections caused by one infectious individual is hence

$$R_0 = \frac{\gamma}{\alpha + \mu}. \tag{4.51}$$

As a numerical illustration, using years as the time-unit, the expected life-length is $1/\mu = 75$ years (so $\mu = 1/75$) and an average length of the infectious period of one week, $1/\alpha = 1/52$ (so $\alpha = 52$). A typical value of the basic reproduction number for childhood diseases in human populations is $R_0 \approx 10$, implying that $\gamma = R_0(\mu + \alpha) \approx 500$. Later we will approximate $\mu + \alpha \approx \alpha$ (so $R_0 \approx \gamma/\alpha$). In the numerical example this is clearly valid: $1/75 + 52 = 52.01 \approx 52$.

As in the case of the prototype epidemic model for a closed and fixed population, an outbreak may or may not occur (by outbreak we now mean a 'major outbreak' as opposed to a minor outbreak, where the number ever infected is of smaller order than N). If $R_0 \leq 1$ there cannot be an outbreak, but if $R_0 > 1$ there is positive probability that an outbreak occurs. The question that we focus on in the present section is: what will happen if the epidemic takes off. For this reason we assume from now on that $R_0 > 1$. The scaled process $(\bar{S}(t), \bar{I}(t))$ will, when N is large, make small jumps (of size $1/N$) but with high rates (the rates given above). This process can then, as for the 'general' stochastic epidemic model, be approximated by a deterministic process, similar to the one defined in Section 3.4:

$$s'(t) = \mu(1 - s(t)) - \gamma s(t)i(t), \tag{4.52}$$
$$i'(t) = \gamma s(t)i(t) - (\mu + \alpha)i(t). \tag{4.53}$$

These rates are obtained from the previous ones, for s and i separately, by dividing all expressions by N, replacing $\bar{S}$ by s and $\bar{I}$ by i, and adding, subtracting or leaving out each of the four terms depending on how it affects the state change ('add' if the state is increased, 'subtract' if it is decreased, and 'leave out' if the state is unchanged).

This deterministic process has two equilibrium points: $(s, i) = (1, 0)$ (infection-free equilibrium) and $(s, i) = (\hat{s}, \hat{i})$, where

$$\hat{s} = \frac{1}{R_0} \quad \text{and} \quad \hat{i} = \frac{\mu}{\mu + \alpha}\left(1 - \frac{1}{R_0}\right) = \varepsilon\left(1 - \frac{1}{R_0}\right), \tag{4.54}$$

where

$$\varepsilon := \mu/(\mu + \alpha) \approx \mu/\alpha \tag{4.55}$$

is the ratio between the average length of the infectious period and the average life length, hence typically a small number ($\varepsilon \approx 1/3750$ in our numerical example). Still assuming $R_0 > 1$, the infection-free state is unstable (a small perturbation making $i(t)$ positive initiates an outbreak) whereas the endemic state is stable.

Exercise 4.33 Compute the endemic level (i.e., give the proportions of the population that are infectious and susceptible, respectively) for the numerical values given, i.e., $\alpha = 52$, $\mu = 1/75$ and $R_0 = 10$.

We end this section with a short digression on the form of the distribution of the time to extinction. Consider once again the situation that an infectious agent is introduced into a population of size N. Assume that $R_0 > 1$ and that a major outbreak results. We start a clock when the prevalence reaches, say, the level $N/100$ for the first time. We stop the clock when the agent has become extinct. The time that the clock shows when we stop it is a stochastic variable. What can we say about its distribution?

For large N the time course of the variables s and i is, to a good approximation, described by the solution of the ODE system (4.52)–(4.53). Yet, in order to capture the extinction event we have to switch to a stochastic description when the number of infecteds becomes small. As one cannot 'undo' the limit $N \to \infty$, underlying the deterministic ODE description, we are confronted with an 'unsolvable' problem.

In such situations, a way around is to argue as follows. The deterministic dynamics lead, by way of damped oscillations, to the endemic steady state. There exists a stochastic counterpart to that steady state: the quasi-stationary distribution. As we shall see in Section 4.7.2, the time to extinction is exponentially distributed with an exponent that one can characterize, if one starts in the quasi-stationary distribution. It is now plausible that the time on the clock has a distribution with an exponential tail and, moreover, that it is possible to derive approximations for the exponent for large values of N.

Especially when R_0 is substantially larger than one and when, in addition, demographic turnover is slow relative to infection transmission, one expects that the deterministic dynamics embodied in (4.52)–(4.53) will lead to very low values for i at the end of what might be called the 'first outbreak.'

This creates a bottleneck for the infectious agent. In the distribution for the time to extinction this manifests as a peak for the density. Although it is hard to give a precise mathematical definition, the mass in the peak corresponds to what we call 'the probability to go extinct after the first outbreak.' As we already emphasized in Section 4.1, it is a major challenge to complement and extend the work reported there in footnote 2 on the characterization and computation of this probability.

The deterministic damped oscillations around the endemic steady state will yield further, but far less pronounced, peaks in the density. Our conjecture is therefore that the density has a first substantial peak, a few much smaller peaks that we can neglect without losing much information, and an exponential tail. The deterministic dynamics tells us where the first peak is to be found, but there does not seem to be a simple intuitive argument that yields information about the size of the peak.

In the next subsection we will outline an approximation for the time it takes to go extinct if we condition on survival of the bottleneck, i.e., if the outbreak makes it beyond the first peak in the density of the extinction time and a neighborhood of the endemic steady state is reached. The derivation uses rather sophisticated probabilistic methods. In Section 4.7.5 we will distinguish community sizes depending on whether the time to extinction is very short or extremely long. The results in the beginning of that particular section can be understood with more basic probability theory and readers unfamiliar with probability may wish to go directly to this section.

4.7.2   An approximation for the expected time to extinction

Using some fairly advanced probabilistic methods we obtain an approximation for the (expected) time as follows.[25] First we show that, if the epidemic process is started in quasi-stationarity $Q = \{q_{s,i}\}$ (the distribution conditioned on not having gone extinct during a long period; recall the 'baby example' presented at the end of Section 4.6), the time to extinction T_Q is exponentially distributed with mean $(\alpha + \mu)q_{.,1} := (\alpha + \mu)\sum_s q_{s,1}$.

The job remaining is to find an approximation for $q_{.,1}$, which we deal with in two additional steps, both relying on N being large. The first of these steps shows that the (two-dimensional) epidemic process, properly scaled, converges to a Gaussian process as N tends to infinity (we say that the epidemic process *converges weakly* to a Gaussian process). This limiting process has an explicit two-dimensional normal distribution as its stationary distribution (i.e., when $t \to \infty$). This means that, conditioned on non-extinction, $(S(t), I(t))$, properly scaled, should be approximately distributed according to this normal distribution if N and t are large. We hence use this distribution as an approximation for the quasi-stationary distribution. The second additional step in the approximation is that we use the normal approximation far from the central parts of the quasi-stationary distribution. The normal approximation typically works quite well in the central parts, where numbers are sufficiently large, but not as well in the tails.

We start by showing that T_Q is exponentially distributed. This is quite simple, as it only uses that the exponential distribution is defined by its lack of memory. Take $t > 0$ and $u > 0$ and recall that the quasi-stationary distribution is defined by: $q_{s,i} := \lim_{t \to \infty} P((S(t), I(t)) = (s, i)|I(t) > 0)$. Then, starting in the quasi-stationary distribution, i.e., $(S(0), I(0))$ is distributed as $Q = \{q_{s,i}\}$,

$$P(T_Q > t + u|T_Q > t, (S(0), I(0)) \sim Q)$$
$$= P(T_Q > t + u|T_Q > t, (S(t), I(t)) \sim Q)$$
$$= P(T_Q > u|(S(0), I(0)) \sim Q),$$

where the first equality is true because we are still in the quasi-stationary distribution if we have not gone extinct by t, and the second, since then we can restart the clock at time t. This implies that T_Q, the extinction time when starting in quasi-stationarity, is exponentially distributed. The parameter is simply the rate at which the process experiences extinction from the quasi-stationary distribution. This rate is the sum of the (quasi-stationary) probabilities $q_{s,i}$ multiplied by the jump rate from (s, i) into the absorbing states. The extinction states $(\cdot, 0)$ can only be reached from the states with $i = 1$, and from any such state, the jump into the absorbing class happens if the remaining infective recovers or dies, hence with rate $\alpha + \mu$. The rate parameter of the exponential distribution is therefore $(\alpha + \mu)\sum_s q_{s,1} =: (\alpha + \mu)q_{.,1}$. We hence have the following result:

$$T_Q \sim Exp((\alpha + \mu)q_{.,1}) \quad \text{so} \quad E(T_Q) = \frac{1}{(\alpha + \mu)q_{.,1}} = \frac{1}{\mu}\frac{\varepsilon}{q_{.,1}} \qquad (4.56)$$

---

[25]cf. I. Nåsell: On the time to extinction in recurrent epidemics. *J. Roy. Statist. Soc. B*, **61** (1999), 309–330.

with ε defined by (4.55). The reason for writing the rightmost expression is that the first factor is related to the unit of time. The expected time to extinction is hence the average life-length multiplied by $\varepsilon/q_{.,1}$, and the latter ratio does not depend on the choice of time unit.

We now move to the remaining part, i.e., to obtain an approximation of the quasi-stationary distribution $\{q_{s,i}\}$. For this we need more advanced probability theory. Here we state the results without much explanation and refer to Nåsell and references therein for details. We simplify by assuming that we start in the endemic state: $(\bar{S}(0), \bar{I}(0)) = (\hat{s}, \hat{i})$, which implies that the corresponding deterministic system (4.52, 4.53) will have $(s(t), i(t)) = (\hat{s}, \hat{i})$ for all t. As mentioned in Section 4.7.1, the scaled process $(\bar{S}(t), \bar{I}(t))$ will be close to the solution of the deterministic equations (4.52) and (4.53) when N is large. This can be shown rigorously by investigating the limit $N \to \infty$. It is also possible to show that the fluctuations around the deterministic functions $(s(t), i(t))$ will, after proper scaling and in the limit $N \to \infty$, be a Gaussian process. More precisely, using theory for Markov population processes (Ethier and Kurtz 1986) it can be shown that the process $\{(\tilde{S}_N(t), \tilde{I}_N(t)), \ t \geq 0\}$, defined by $(\tilde{S}_N(t), \tilde{I}_N(t)) = (\sqrt{N}(\bar{S}(t) - \hat{s}), \sqrt{N}(\bar{I}(t) - \hat{i}))$, converges to a certain Gaussian process $\{(\tilde{S}(t), \tilde{I}(t)), \ t \geq 0\}$ as $N \to \infty$. Further, using theory for Gaussian processes (e.g., Karatzas and Shreve 1991), this two-dimensional Gaussian process will as $t \to \infty$, converge to a two-dimensional normal distribution having mean vector $(0, 0)$ and variance-covariance matrix

$$V = \frac{1}{R_0^2} \begin{pmatrix} \frac{1}{\varepsilon} + R_0 & -R_0 \\ -R_0 & R_0 - 1 + \varepsilon R_0^2 \end{pmatrix}.$$

If N is large and the epidemic has evolved long enough without having gone extinct, the combination of these two results suggests that $(S(t), I(t))$ is approximately normally distributed with mean vector $(N\hat{s}, N\hat{i})$ and variance-covariance matrix NV. In particular, $I(t)$ is approximately normally distributed with mean μ_I and standard deviation σ_I given by

$$\mu_I = N\varepsilon \left(1 - \frac{1}{R_0}\right), \tag{4.57}$$

$$\sigma_I = \sqrt{N} \frac{\sqrt{R_0 - 1 + \varepsilon R_0^2}}{R_0}. \tag{4.58}$$

As mentioned before, in the second step of our approximation we use this approximate distribution away from the mean $N\hat{i}$. More precisely, we use this normal distribution to approximate $P(I(t) = 1)$. If we use this distribution to approximate the quasi-stationary distribution we should also condition on not having gone extinct. This suggests the following approximation of $q_{.,1} = \lim_{t\to\infty} P(I(t) = 1 | I(t) > 0)$ for large N:

$$q_{.,1} \approx \frac{\frac{1}{\sigma_I} \varphi\left(\frac{1 - \mu_I}{\sigma_I}\right)}{\Phi\left(\frac{\mu_I - 0.5}{\sigma_I}\right)} \approx \frac{\frac{1}{\sigma_I} \varphi\left(\frac{\mu_I}{\sigma_I}\right)}{\Phi\left(\frac{\mu_I}{\sigma_I}\right)} \tag{4.59}$$

where $\varphi(x) = (2\pi)^{-1/2} e^{-x^2/2}$ is the density function of the standardized normal distribution, and $\Phi(x) = \int_{-\infty}^{x} \varphi(y) dy$ is the corresponding distribution function. The '1' appearing in the argument of the density derives from our focus on $I(t) = 1$. The denominator takes care of the conditioning, using a continuity correction as we are approximating a discrete distribution by a continuous distribution:

$P(I(t) > 0) = P(I(t) > 0.5) \approx 1 - \Phi((0.5 - \mu_I)/\sigma_I) = \Phi((\mu_I - 0.5)/\sigma_I)$. The final approximation relies on $\mu_I \gg 1$.

Using the expressions for μ_I and σ_I we get

$$\frac{\mu_I}{\sigma_I} = \frac{\sqrt{N}\varepsilon(R_0 - 1)}{\sqrt{(R_0 - 1 + \varepsilon R_0)}} \approx \sqrt{N\varepsilon^2(R_0 - 1)},$$

where the last approximation assumes that $\varepsilon R_0^2 \ll R_0 - 1$ (in the numerical illustration in Exercise 4.33 we had $\varepsilon = 1/3750$ and $R_0 = 10$ implying that $\varepsilon R_0^2 = 0.027$ and $R_0 - 1 = 9$, which hence satisfies the assumption). Inserting this expression into (4.56) we obtain an approximation for $E(T_Q)$:

$$E(T_Q) = \frac{1}{\mu} \frac{\varepsilon}{q_{\cdot,1}} \approx \frac{1}{\mu} \frac{\varepsilon \Phi\left(\frac{\mu_I - 0.5}{\sigma_I}\right)}{\frac{1}{\sigma_I}\varphi\left(\frac{1-\mu_I}{\sigma_I}\right)}$$

$$\approx \frac{1}{\mu} \frac{\varepsilon \Phi\left(\sqrt{N\varepsilon^2(R_0 - 1)}\right)}{\frac{R_0}{\sqrt{N(R_0-1)}} \frac{1}{\sqrt{2\pi}} e^{-N\varepsilon^2(R_0-1)/2}} \qquad (4.60)$$

$$\approx \frac{1}{\mu} \sqrt{2\pi} \frac{\sqrt{N\varepsilon^2(R_0 - 1)}}{R_0} e^{N\varepsilon^2(R_0-1)/2} \Phi\left(\sqrt{N\varepsilon^2(R_0 - 1)}\right) \qquad (4.61)$$

$$\sim e^{N\varepsilon^2(R_0-1)/2}.$$

The last line does not correspond to an approximation, but specifies the leading term in the expression as N grows. Quantitatively we expect to get the best approximation by using as few approximation steps as possible, that is, when using the expression on the first line, inserting (4.57) and (4.58) for μ_I and σ_I, respectively. The reason for making further approximations is to increase our understanding of how $E(T_Q)$ depends on the model parameters.

From the approximation (4.61) we see that the time to extinction is increasing with the population size N and also with the length of the infectious period ε. Note that we have parametrized such that we can change ε without affecting R_0; indeed doubling ε means that we consider an infection having an infectious period that is twice as long, while the infectivity per unit of time is halved. From (4.61) it is also seen that $E(T_Q)$ is constant if N and ε are varied such that $N\varepsilon^2$ is kept constant. For example, for two diseases A and B having the same R_0, the expected time to extinction is the same in a community of size N for disease A having infectious period ε (measured as part of the life time) as in another community of size $9N$ for disease B having only a third of that infectious period $(\varepsilon/3)$. Another conclusion that can be drawn by performing a sensitivity analysis of the expressions we derived, is that ε (the length of the infectious period) has the greatest impact on $E(T_Q)$, followed by the community size N.

The effect of R_0 on our approximation (4.61) for $E(T_Q)$ is not as clear because of the factor R_0 in the denominator of (4.61). In fact the expression is not monotone in R_0 for all parameter values, but in some regions of the parameter space $E(T_Q)$ increases with R_0, although this effect is less strong than the effect of increasing ε and N.

Exercise 4.34 Plot $E(T_Q)$ as a function of N using the approximation (4.61), with $R_0 = 10$, $\mu = 1/75$ (corresponding to a life-expectancy of 75 years with years as the unit of time) and $\varepsilon = 1/3750$ (corresponding to an infectious period of one week). Do the same thing, but with $\varepsilon = 1/(2 \cdot 3750)$ (an infectious period half as long) and with $\varepsilon = 2/3750$ (double length of the infectious period). Do the same thing but with $R_0 = 5$ and $R_0 = 20$ (with $\varepsilon = 1/3750$).

4.7.3 The time to extinction when vaccination is introduced

Consider still the stochastic epidemic model with demography, but now assume that a fraction v of all new-born individuals is vaccinated, and assume that the vaccine gives complete and life-long immunity. The only effect this will have on the model is that the birth rate (of susceptibles, which is our only interest!) is changed to $\mu N(1-v)$, because a fraction $1-v$ remains unvaccinated. The reproduction number in a completely susceptible population R_0 is then reduced to $R_v = R_0(1-v)$. Other than that, the model is unchanged.

From this we can conclude the following. If $R_v \leq 1$, which is equivalent to $v \geq 1 - 1/R_0$, the infectious agent cannot become endemic, and if it was when vaccination started, the infectious agent will go extinct since even the deterministic model predicts extinction. The limit $v_c = 1 - 1/R_0$ is hence called the *critical vaccination coverage*, above which herd immunity is obtained. Note that v_c is the same as for the epidemic model for a closed population, treated in Exercise 1.37 and Section 3.5.4.

If $R_v > 1$ the infectious agent may remain endemic, and the formula for the expected time to extinction is the same as without vaccination, only replacing R_0 and N by $R_v = R_0(1-v)$ and $N' = N(1-v)$. We hence get the following expression for $E(T_Q)$:

$$E(T_Q) \approx \frac{1}{\mu} \sqrt{2\pi} \frac{\sqrt{N(1-v)\varepsilon^2 (R_0(1-v)-1)}}{R_0(1-v)} e^{N(1-v)\varepsilon^2 (R_0(1-v)-1)/2} \Phi(x) \quad (4.62)$$

with $x := \sqrt{N(1-v)\varepsilon^2 (R_0(1-v)-1)}$. From this we see that vaccination has two effects: the susceptible community is reduced in size, but also the *effective reproduction number* is now $R_0(1-v)$ as compared with R_0 when no vaccination was in place. This implies that more is gained than only protecting the vaccinated. Of course, when $v \geq v_c$ the infectious agent will go extinct, but even when $v < v_c$ the probability of extinction in a fixed time interval is increased.

Exercise 4.35 Plot $E(T_Q)$ as a function of v using the approximation (4.62), with $R_0 = 10$ and $\varepsilon = 1/3750$, and for $N = 100,000$, $N = 1,000,000$ and $N = 10,000,000$. As before, set $1/\mu = 75$ so that the time unit is a year.

4.7.4 The critical community size

It was seen in Sections 4.7.2 and 4.7.3 that the time to extinction was increasing in the community size. This fact has been observed also in real life. For example, prior to the introduction of vaccination, measles was endemic in the United Kingdom and other countries with large populations, whereas it always went extinct after each outbreak in smaller countries like Iceland. In the context of our model this should be reflected by the fact that $E(T_Q)$ is (very) large for countries like the United Kingdom and (very) small for countries like Iceland.

Exercise 4.36 Compute $E(T_Q)$ for measles in England and Iceland, respectively, using the approximation (4.61). Choose for example $1/\mu = 75$ (although life-expectancy was probably a bit shorter in the 1950s), $\varepsilon = 1/3750$ (one week infectious period) and $R_0 = 15$ (a common estimate of R_0 for measles). Assume that $N = 50{,}000{,}000$ for England and $N = 150{,}000$ for Iceland.

Exercise 4.37 Repeat the previous exercise, but now use the 'more exact' approximation in (4.60).

Also within countries the same type of phenomenon has been observed: in big cities measles have been endemic, but not in smaller towns. This motivates the introduction (and definition) of the *critical community size* N_c separating an endemic situation from a situation where the disease will go extinct. Of course, there will be no clear-cut threshold value, but the idea is that if a community is much smaller than N_c (which will depend on the disease under consideration and community factors!) the disease will not persist, but if the community is much larger than N_c the disease will stay endemic for a long time, relative to host life expectancy. We will 'define' the critical community size in three ways: the first (denoted by $N_c^{(1)}$) being very simple, the second, $N_c^{(2)}$, based on slightly more refined arguments, and the third, $N_c^{(3)}$, deriving from the results of Section 4.7.2 concerning the expected time to extinction $E(T_Q)$.

Before embarking on the derivation of plausible formulas, yielding actual numerical values for the critical community size upon specification of parameter values, we present a robust qualitative argument that, in our opinion, captures the key point. Recall (4.55)

$$\varepsilon := \mu/(\mu + \alpha) \approx \mu/\alpha$$

and that ε is the ratio between the average length of the infectious period and the life expectancy of the host. For fixed N and very small ε, the supply of new susceptibles by births will be too slow, and the infective agent will be doomed to go extinct after an outbreak triggered by introduction. If, however, we fix ε at some positive value and consider very large N then, even though the agent will still go extinct eventually, just as any bounded stochastic population process, it may very well persist for an astronomically long period.

Hence, the limits $\varepsilon \to 0$ and $N \to \infty$ are not interchangeable. One can then look for a 'phase transition,' i.e., a way to let *simultaneously* $\varepsilon \to 0$ and $N \to \infty$ so that the expected time till extinction stays bounded, as well as bounded away from zero. A brief look at the derivation of (4.61) should suffice to conclude that the relevant condition is that $N\varepsilon^2$ stays bounded and bounded away from zero. We conclude that the critical community size is of the order ε^{-2} for $\varepsilon \to 0$.[26]

We now embark on the 'derivation' of our three 'operational definitions' of critical community size. We start with the simplest approach that ignores the results of Sections 4.7.2 and 4.7.3. In Section 4.7.1 (and in fact already in Section 4.2) the number of infectives at the endemic level was seen to equal $\mu_I = N\varepsilon(1-1/R_0)$. The fluctuations around this level will be of the order $\sqrt{N}$. This heuristic conclusion is based on the fact that in nearly all population models the standard deviation is of order $\sqrt{N}$, recall (4.57) and (4.58). If the mean is of the same order as the standard deviation, extinction should be common, whereas if the mean is of higher

---

[26]Alternatively one can consider ε as fixed and introduce $R_0 - 1$ as a small parameter. This is closer to the point of view advocated by I. Nåsell.

order than the standard deviation, it should be rare and in particular it should take a very long time before an excursion from the endemic level to the infection-free state occurs. This motivates the following 'definition': $N_c^{(1)}$ is the solution to the equation

$$k = \frac{\mu_I}{\sqrt{N}} = \sqrt{N}\varepsilon(1 - 1/R_0),$$

where k is some arbitrarily chosen constant. Here we consider R_0 to be independent of N, as in Section 4.7.1. For now let us simply choose the constant $k = 1$. This implies that the critical community size $N_c^{(1)}$ equals

$$N_c^{(1)} = \frac{1}{\varepsilon^2 \left(1 - \frac{1}{R_0}\right)^2}. \tag{4.63}$$

This expression tells us that the critical community size is inversely proportional to the square of the average infectious period, as suggested above. Further, as long as R_0 is not close to 1, $N_c^{(1)}$ is hardly affected by R_0: two infectious agents with the same infectious period, one having $R_0 = 10$ and the other having $R_0 = 20$, have nearly the same $N_c^{(1)}$ (they differ by approximately 5%).

If we include vaccination by assuming, as before, that a fraction v of all newborns are vaccinated with a perfect vaccine, N changes to $N(1-v)$ and R_0 to $R_0(1-v)$. As a direct consequence, the critical community size becomes

$$N_c^{(1)} = \frac{1}{(1-v)\varepsilon^2 \left(1 - \frac{1}{R_0(1-v)}\right)^2}. \tag{4.64}$$

Hence, increasing v leads to an increase in $N_c^{(1)}$. For a fixed community (i.e., fixed N) the implication is that the infectious agent will not persist if the vaccination coverage v is large enough.

We now refine the arguments above by taking into account the results from Sections 4.7.2 and 4.7.3. The simplest refinement is to insert our expression for σ_I, given by (4.58), when looking at the ratio of the mean μ_I, given by (4.57), and the standard deviation. Also, since we are considering excursions and want excursions leading to extinction to be rather unlikely we can try to fine-tune the choice of the constant k. Motivated by the fact that only approximately 0.26% of the mass of the Gaussian distribution is located at a distance from the mean that exceeds three standard deviations, we now choose to characterize the critical community size by the equation $\mu_I/\sigma_I = 3$. In Section 4.7.2 it was shown that $\mu_I/\sigma_I \approx \sqrt{N\varepsilon^2(R_0 - 1)}$, so, also considering vaccination as before, this leads to the formula

$$N_c^{(2)} = \frac{9}{(1-v)\varepsilon^2 (R_0(1-v) - 1)} \tag{4.65}$$

or after rewriting, to

$$N_c^{(2)} = \frac{9}{R_0(1-v) - 1} \left(1 - \frac{1}{R_0(1-v)}\right)^2 N_c^{(1)}.$$

The dependence on ε is hence as before, but R_0 has greater influence on the critical community size when using $N_c^{(2)}$: it is almost inversely proportional to R_0.

Our third 'definition' of N_c is more elaborate. Perhaps the most natural definition of the critical community size is to say that the probability for the infection to persist over a given time horizon t exceeds a given value p whenever $N \geq N_c$. This gives two arbitrary parameters. Here we choose the values $t = 1/\mu$ and $p = 1/2$, so the probability that the infection persists for a period longer than the life expectancy of the host should exceed 0.5 for the community size to be considered as being above criticality.

So we now define the critical community size by the equation $P(T_Q > 1/\mu) = 1/2$. We know from Section 4.7.2 that T_Q is exponentially distributed with parameter $\mu q_{.,1}/\varepsilon$. This implies that $P(T_Q > x) = e^{-(\mu q_{.,1}/\varepsilon)x}$. The defining equation hence becomes $e^{-q_{.,1}/\varepsilon} = 1/2$. Taking logarithms and inserting the approximation (4.59) we find that the critical community size is the solution to

$$\frac{\varepsilon \Phi\left(\frac{\mu_I - 0.5}{\sigma_I}\right)}{\frac{1}{\sigma_I}\varphi\left(\frac{1-\mu_I}{\sigma_I}\right)} = \frac{1}{\ln 2}.$$

In order to gain insight and to simplify the expression somewhat, we use the additional approximations leading to (4.61), and we also consider vaccination as before (as in (4.62)), and define the third critical community size proxy $N_c^{(3)}$ as the solution to

$$\sqrt{2\pi}\frac{\sqrt{N(1-v)\varepsilon^2(R_0(1-v)-1)}}{R_0(1-v)}e^{N(1-v)\varepsilon^2(R_0(1-v)-1)/2}\Phi(x) = \frac{1}{\ln 2} \qquad (4.66)$$

with $x := \sqrt{N(1-v)\varepsilon^2(R_0(1-v)-1)}$. For given parameter values the critical community size $N_c^{(3)}$ is easily determined by solving (4.66) numerically, but there is no explicit expression for $N_c^{(3)}$. Note that, because T_Q is exponentially distributed, the requirement $P(T_Q > 1/\mu) > 1/2$ is equivalent to the requirement $E(T_Q) > 1/(\mu \ln 2)$, and that one can, accordingly, also derive (4.66) by using the approximation of $E(T_Q)$ given in (4.62). The qualitative conclusions are as before: $N_c^{(3)}$ is inversely proportional to the square of the average length of the infectious period, and $N_c^{(3)}$ decreases with R_0 and increases with the fraction vaccinated v.

We have now derived three different expressions for the critical community size N_c. All have their merits, but if we were to recommend the use of one of them in a specific situation, i.e., for a given set of parameter values, it might be $N_c^{(3)}$. The reason for choosing $N_c^{(3)}$ is its simple interpretation through its defining property: if $N \ll N_c^{(3)}$ the infectious agent should go extinct fairly soon, whereas it will persist for quite a long period if $N \gg N_c^{(3)}$). Here 'soon' and 'long' are relative to the life-length of the host.

Exercise 4.38 Compute the critical community size for measles in the absence of vaccination (taking say, $R_0 = 15$ and $\varepsilon = 1/3750$ without vaccination) using all three characterizations $N_c^{(1)}$, $N_c^{(2)}$ and $N_c^{(3)}$.

Exercise 4.39 Do the same thing as in the previous exercise, but now assuming a 90% vaccination coverage (note that with this choice the situation is still above threshold, i.e., $R_v > 1$).

4.7.5   Beyond homogeneous mixing: sub-communities

Our approximating formulas for the expected time to extinction and the critical community size, derived in Sections 4.7.2 and 4.7.4, rely on the assumption that the community consists of identical individuals that mix homogeneously, meaning that there are no social structures in the community and every pair of individuals is equally likely to have a close contact. This is of course a simplification of the real world. In reality individuals are different both in terms of susceptibility and infectivity as well as in their social activity, all of which are important for the spread of infection. In addition, there are social structures in the community. For example, individuals contact each other at higher rate within a household, and larger communities often consist of smaller sub-communities like villages. It is then interesting to see how such deviations from the simplified description affect the time to extinction and the critical community size. In what follows we address the question: How is the expected time to extinction affected when the community consists of sub-communities? We will only give a brief treatment and refer to Lindholm and Britton (2007),[27] and references therein, for a more thorough analysis.

We now describe a model where individuals have a higher chance of infecting individuals belonging to the same sub-community. The model is closely related to the stochastic epidemic model with demography defined in Section 4.7.1. Suppose the community consists of k sub-communities, and assume, for simplicity, that all have equal size n. The total population size is hence $N = kn$. Furthermore, suppose as in the homogeneous model that there is a constant birth rate μN, and that the birth rate is the same, μn, in all sub-communities; individuals have an exponentially distributed life span with mean $1/\mu$; an individual who gets infected immediately becomes infectious and remains so for an exponentially distributed period with mean $1/\alpha$; during the infectious period the individual has 'close' contacts at rate γ. The difference with the homogeneous model is that now a fraction p of all contacts are exclusively with individuals of the same sub-community and the remaining fraction $(1 - p)$ are with individuals chosen at random from the entire community. In other words, for each contact taking place, the person being contacted is chosen uniformly within the same sub-community with probability p, and chosen uniformly from the community at large with probability $1 - p$.

The reproduction number is independent of p and equals $R_0 = \gamma/(\alpha + \mu)$ as before. This is true because the rate γ of contacting other individuals does not depend on p; p only affects *who* will be contacted. Furthermore, the endemic steady state does not depend on p, and in this steady state the fractions susceptible, infectious and removed are the same in all sub-communities.

Clearly for $p = 1$ the community actually consists of k isolated sub-communities. When $p = 0$, the community mixes homogeneously as a whole and there are no sub-communities (from the point of view of infection spread). In what follows we will study the time to extinction for these two extremes. To find approximations for

[27]M. Lindholm and T. Britton: Endemic persistence or disease extinction: the effect of separation into sub-communities. *Theor. Pop. Biol.*, **72** (2007), 253–264; see also M.J. Keeling and B.T. Grenfell: Disease extinction and community size: modeling the persistence of measles. *Science*, **275** (1997), 65–67; M.J. Keeling and B.T. Grenfell: Understanding the persistence of measles: reconciling theory, simulation and observation. *Proc. R. Soc. B*, **354** (2002), 335–343; M. Jesse, P. Ezanno, S.A. Davis and J.A.P. Heesterbeek: A fully coupled, mechanistic model for infectious disease dynamics in a metapopulation: movement and epidemic duration. *J. Theor. Biol.*, **254** (2008), 331–338.

the time to extinction for intermediate values $0 < p < 1$ is more difficult and we refer again to Lindholm and Britton (2007) in footnote 27, and references therein.

Let $\tau_N(0) = E(T_Q)$ for the homogeneously mixing community, so an approximation for $\tau_N(0)$ is given by (4.61). Further, let $\tau_N(p)$ denote the expected time to extinction, starting in the quasi-stationary distribution, for our new model with sub-communities, where we explicitly show the dependence on both the total population size $N = kn$ and the 'mixing parameter' p, but hide the dependence on the other parameters, including k (assumed to be strictly larger than 1). Note that we make the (debatable) assumption that when $p = 1$ we start from the situation that in each of the k isolated sub-communities the quasi-stationary distribution pertains. From the observation above we immediately have

$$\tau_N(0) = \tau_N \approx \frac{1}{\mu}\sqrt{2\pi}\frac{\sqrt{kn\varepsilon^2(R_0-1)}}{R_0}e^{kn\varepsilon^2(R_0-1)/2}\Phi\left(\sqrt{kn\varepsilon^2(R_0-1)}\right)$$

where we have replaced N by kn in the right-hand side of (4.61). This is true because when $p = 0$ we have a homogeneously mixing community of size N. On the other hand, when $p = 1$ we have k isolated sub-communities that hence go extinct independently, all having the same distribution for the time to extinction. This distribution is also the one considered in Section 4.7.2, but with N replaced by n. The times to extinction of the k sub-communities are independent and exponentially distributed, each having mean τ_n that can be approximated using (4.61). The time until the infection goes extinct for the first time in any of the sub-communities is then exponentially distributed with mean τ_n/k, since there are k 'competing' sub-communities. When this happens, the remaining communities are still in quasi-stationarity, implying that the additional time until the infection goes extinct in another sub-community is exponential with mean $\tau_n/(k-1)$. This argument can be continued and as a consequence we have

$$\tau_N(1) = \tau_n\sum_{i=1}^{k}\frac{1}{i} \approx \frac{\sum_{i=1}^{k}\frac{1}{i}}{\mu}\sqrt{2\pi}\frac{\sqrt{n\varepsilon^2(R_0-1)}}{R_0}e^{n\varepsilon^2(R_0-1)/2}\Phi\left(\sqrt{n\varepsilon^2(R_0-1)}\right).$$

We now compare the expected time to extinction for the two extremes $p = 1$ (corresponding to k isolated sub-communities) and $p = 0$ (corresponding to a homogeneously mixing large community). We do this by looking at $\tau_N(1)/\tau_N(0)$ for which we find, using the approximations above,

$$\frac{\tau_N(1)}{\tau_N(0)} \approx \frac{\sum_{i=1}^{k}\frac{1}{i}}{\sqrt{k}}e^{-(k-1)n\varepsilon^2(R_0-1)/2} \qquad (4.67)$$

where we have removed the factor $\Phi\left(\sqrt{kn\varepsilon^2(R_0-1)}\right)/\Phi\left(\sqrt{n\varepsilon^2(R_0-1)}\right)$ because it is close to 1. In (4.67) we see that the ratio is smaller than 1 when n is large.

As a consequence, the time to extinction is longer if the community mixes homogeneously ($p = 0$) as opposed to the case when the community consists of isolated sub-communities ($p = 1$), a result that agrees with intuition. All existing approximations for intermediate values of p indicate that $\tau_N(p)$ decreases monotonically with p, the 'degree of separation.'

Intuition is varied and it is formed by training. So far we took as our starting point that in all sub-communities the situation is described by the quasi-stationary distribution. What if, instead, we introduce the agent in a virgin community ? It may very well be that extinction after the first huge outbreak in the well-mixed community is more likely than extinction after a series of somewhat less correlated first outbreaks in the various sub-communities.

In general, fluctuations in relatively isolated communities will be out of phase. After extinction occurs in one, re-introduction from another may occur. Thus the lack of synchrony may buffer against large fluctuations at the total population scale. In the ecological literature, such considerations pop up when the question is whether a country should have many small national parks or one big one instead.

These final comments have no other aim than to trigger our readers to be critical and curious, rather than to take what we write as the final word.

4.7.6   Waiting forever: on the time to extinction of a reactivating virus

In Exercise 2.5 we considered a herpes virus, of which a characteristic feature is that it may become 'dormant' after the initial infectious period, in order to resurge at a later time (for example in times of stress for the host) to initiate a second infectious period, etc. As a consequence, the virus is not yet extinct if there are no more infectious individuals, but only if there are, in addition, no more latently infected individuals.

In the case of bovine herpes virus in cattle, the aim of control measures is to eradicate the virus. Vaccination is relatively easily applied on farms, and also the effects of vaccination can in principle be assessed without much difficulty. Feral cattle herds, however, living in natural areas, are a different matter.

In the following exposition[28] we introduce many debatable simplifying assumptions, the 'reward' being that we get quite far analytically, leaving only some integrals that need to be evaluated numerically in order to arrive at numbers. In a subsequent step one might test the robustness of the conclusions by performing simulation studies for more complex situations.

We choose to denote the sub-population of infected individuals with a dormant infection by R, normally indicating 'removed' or 'recovered' individuals, i.e., individuals that are no longer infected. We ignore true recovery from infection and note that the R-class now consists of infected, but not infectious, individuals. One could also denote such 'dormant' individuals as 'latently' infected, but the biological mechanisms underlying the period preceding shedding of the agent after the initial infection are different from the mechanisms determining dormancy and resurgence. It is hence better to avoid thinking in terms of 'latent' individuals (the word 're-covered' does apply, in the sense of 'clinical recovery' and immunity to reinfection, from a source outside the host itself). If we incorporate the possibility to return from the R-class to the I-class, the analog of (4.3) is

$$
\begin{aligned}
\frac{dS}{dt} &= B - \beta SI - \mu S, \\
\frac{dI}{dt} &= \beta SI - \mu I - \alpha I + \nu R, \\
\frac{dR}{dt} &= -\mu R + \alpha I - \nu R.
\end{aligned}
\tag{4.68}
$$

---

[28]This section is based on A.A. de Koeijer, O. Diekmann and M.C.M. de Jong: Calculating the time to extinction of a reactivating virus, in particular bovine herpes virus. *Math. Biosci.*, **212** (2008), 111–131.

The total population size $N = S + I + R$ obeys $dN/dt = B - \mu N$ and hence N converges to the steady state value B/μ. To simplify the presentation, we assume that right from the start

$$S + I + R = N = \frac{B}{\mu}. \tag{4.69}$$

We consider resurgence of infectiousness to be indistinguishable from 'first-time' infectiousness, both in infectivity and duration (but note that in fact these resurgent episodes may well be shorter and result in reduced infectiousness.) We are going to disentangle system (4.68) into three separate parts on the basis of time-scale separation. The idea is that when βN and α are very large relative to B, μ and ν, we can pretend that an outbreak is instantaneous at the longer time scale of demographic turnover. So we take out the part

$$\frac{dS}{dt} = -\beta SI, \tag{4.70}$$
$$\frac{dI}{dt} = \beta SI - \alpha I$$

(which we encountered before as (1.12)) and use it only to compute the size of a major outbreak. At the demographic time scale, a major outbreak manifests itself as an instantaneous jump in which S is decreased and R is increased (by the same amount), such that (4.69), in the form

$$S + R = N \tag{4.71}$$

remains valid. The dynamics in between jumps is generated by

$$\frac{dS}{dt} = B - \mu S. \tag{4.72}$$

But how is a jump generated? It is here that a stochastic element enters our otherwise deterministic description. We interpret, as in (4.68), the quantity νR as the probability per unit of time that, by a return from the latency class, an infectious individual appears on the scene. But this is an *event* and we do not, as in (4.68), assume that because of large numbers such events happen all the time. (We neglect for this discussion the biologically possible scenario that resurgence of the infection within an individual is induced by stress and that it may well be that the same type of stress, for example starvation, disturbance, fear, may act on all individuals simultaneously, preventing the events from being independent.) If such an event happens, there are still three possibilities:

- The size of the sub-population of susceptibles is below criticality and, to a very good approximation, nothing happens.

- The size of the sub-population of susceptibles is above criticality, but the outbreak is, by chance, a minor one; in this case something happens, but not much; to keep calculations simple we assume that whatever happens is negligible.

- The size of the sub-population of susceptibles is above criticality and a major outbreak takes place: a jump occurs.

As long as no major outbreak occurs, the R-class of 'dormant' infected individuals is gradually depleted through death. So, given that no major outbreak occurs, the virus goes extinct. To determine how long this time to extinction is likely to be, we have to determine how many major outbreaks occur and how much time elapses between consecutive outbreaks. (Note that if such inter-outbreak times are independent, we can simply 'add' what we compute for a single inter-outbreak interval; note also that when outbreaks are stress-induced, then inter-outbreak times could be regular when stress is a regular occurrence, e.g., starvation in winter.)

Now let us turn to the technical aspects of specifying and analyzing this hybrid dynamical system. By 'hybrid' we mean that continuous deterministic dynamics are combined with instantaneous events generated by a stochastic process. We can fruitfully use some results that we derived earlier.

First recall from Exercise 1.22-i that the basic reproduction number for (4.70) — which we shall call R_1 here to emphasize that it only counts secondary cases during a *single* infectious period of an infected individual — is given by

$$R_1 = \frac{\beta N}{\alpha} \tag{4.73}$$

(you may rightfully argue that, in view of (4.68), the denominator should actually be $\alpha + \mu$, but remember that we assume that $\mu \ll \alpha$). Also recall from Exercise 2.5 that

$$R_0 = \frac{\nu + \mu}{\mu(\alpha + \nu + \mu)} \beta N. \tag{4.74}$$

It may of course happen that $R_1 < 1$ but $R_0 > 1$. However, we are only interested in the situation where $R_1 > 1$ such that a major outbreak can develop even if no individual shows resurgence to infectivity.

Next recall Exercise 1.22-ii and -iv and convince yourself that, in the current situation — in which some non-negative fraction of the population may be in the R-class at the moment that an infective individual appears on the scene — the analog of (1.13) is

$$\ln s_\infty - R_1 s_\infty = \ln s_0 - R_1 s_0 \tag{4.75}$$

where s_0 and s_∞ are fractions (of the total population N), respectively at the start of the major outbreak and at the end (so we require that $R_1 s_0 > 1$, and will find that $R_1 s_\infty < 1$). In the elaboration of Exercise 1.20 it was suggested to rewrite (4.75) in the form

$$R_1 s_\infty e^{-R_1 s_\infty} = R_1 s_0 e^{-R_1 s_0}, \tag{4.76}$$

to draw the graph of $\xi \longmapsto \xi e^{-\xi}$, and to follow a graphical procedure to determine $R_1 s_\infty$ for given $R_1 s_0$ (and hence s_∞ for given s_0 and R_1). Given $R_1 > 1$ we shall write

$$s_\infty = \phi(s_0) \tag{4.77}$$

to denote the root $s_\infty < 1/R_1$ of (4.76) for some specified $s_0 > 1/R_1$.

Finally, recall formula (1.8) for the probability of a minor outbreak in a supercritical situation, when the infectious period follows an exponential distribution. If we define

$$h(s_0) = \begin{cases} 1 - \frac{1}{s_0 R_1}, & s_0 \geq 1/R_1; \\ 0, & s_0 \leq 1/R_1 \end{cases} \tag{4.78}$$

then h is the probability that an individual that returns from the R-class to the infectious class, triggers a major outbreak if a fraction s_0 of the population is susceptible at the moment of returning.

If we work with fractions and scale time such that $\mu = 1$, then (4.72) transforms into

$$\frac{dx}{dt} = 1 - x \qquad (4.79)$$

where

$$x = \frac{S}{N}. \qquad (4.80)$$

In view of (4.78), the point

$$x = \bar{x} := \frac{1}{R_1} \qquad (4.81)$$

is of central importance. From the discussion related to (4.77) we recall that immediately after an outbreak necessarily $x < \bar{x}$. From (4.78), on the other hand, it follows that immediately before an outbreak necessarily $x > \bar{x}$. So in between outbreaks the variable x has to pass the value $\bar{x}$. We say that $\bar{x}$ is a renewal point (meaning that at the moment of passing $\bar{x}$ we can forget all prior history). We shall give $\bar{x}$ a central role in our bookkeeping.

In between jumps, the dormant infected fraction of the population is precisely $1 - x$. So, given x, the probability per unit of time that some individual regains infectiousness is given by $\nu N(1 - x)$. The probability per unit of time that a major outbreak is initiated is therefore $\nu N(1 - x)h(x)$. Let us reset the time variable t to zero at the moment that x passes the value $\bar{x}$. Let $f(t)$ be the 'survival' probability, by which we mean the probability that no major outbreak occurs before t. Then

$$\frac{df}{dt}(t) = -\nu N(1 - x(t))h(x(t))f(t) \qquad (4.82)$$

where $x(t)$ is the solution of (4.79) with initial condition

$$x(0) = \bar{x}. \qquad (4.83)$$

Because x is a monotone function of time there exists a function $F = F(x)$ such that

$$F(x(t)) = f(t). \qquad (4.84)$$

Using (4.79) we may now rewrite (4.82) in the form

$$\frac{dF}{dx}(x) = -\nu N h(x)F(x) \qquad (4.85)$$

and consequently conclude that

$$F(x) = e^{-\nu N H(x)} \qquad (4.86)$$

with

$$H(x) = \int_{\bar{x}}^{x} h(\xi)d\xi = x - \bar{x} + \frac{1}{R_1}(\ln x - \ln \bar{x}). \qquad (4.87)$$

After all these preparations, we are now ready to draw conclusions.

- The probability that no more outbreak will occur equals

$$F(1) = e^{-\nu N H(1)}. \qquad (4.88)$$

- If we condition on the occurrence of the next outbreak, then the susceptible fraction of the population immediately before the outbreak is a random variable with distribution function

$$\frac{1 - F(x)}{1 - F(1)}. \tag{4.89}$$

So if, for $x > \bar{x}$, $T(x)$ denotes the time it takes to let the susceptible fraction grow from $\bar{x}$ to x and next from $\phi(x)$ to $\bar{x}$ (to complete one cycle), then the expected time between any two consecutive passings of $\bar{x}$ is

$$\bar{T} = \frac{\nu N}{1 - F(1)} \int_{\bar{x}}^{1} T(x)h(x)F(x)dx. \tag{4.90}$$

Exercise 4.40 Show that
$$T(x) = \ln \frac{1 - \phi(x)}{1 - x}.$$

Exercise 4.41 The very first outbreak is triggered from outside the system. Start a clock the first time that x passes $\bar{x}$ and stop that clock the last time that x passes $\bar{x}$. Let us call the time at the moment of stopping the clock: 'the time to extinction.' What is the expected value of this random variable?

4.8 BEYOND A SINGLE OUTBREAK: SUMMARY

This chapter has focused on the influence of the demography on the persistence of an infective agent, and vice versa on the influence of the agent on host population growth and persistence.

When new 'fuel' is provided through the replacement of immune individuals by newborn susceptibles, the infective agent may strike again at a later time or even persist and become endemic. Whether we observe repeated outbreaks or relatively small fluctuations around a steady endemic level depends on the temporal and spatial scale that we, the investigators of the system, choose to monitor and/or to model. And likewise it depends on the degree of isolation of the (sub)population on which we focus (e.g., measles on Iceland showed repeated outbreaks). Data on measles provide a rich source to study the above phenomena. For aspects of data collection and analysis in relation to measles, we refer to Cliff, Haggett and Smallman-Raynor (1993), Keeling (1997) and especially Bryan Grenfell and co-workers, reviewed in Grenfell and Harwood (1997), and the references given there.[29]

In Section 4.7.4, we suggested that two parameters, rather than just one, are needed to distinguish between a single outbreak (which may re-occur much later by re-introduction of the agent from the outside) and the endemic situation. In addition to population size N, we consider the ratio α/μ of the time scale of life expectancy and the duration of the infectious period. When $N \to \infty$ for fixed α/μ, we expect an astronomical time until extinction (of the order e^{+vN} for some positive constant v). When $\alpha/\mu \to \infty$ for fixed N, we expect immediate extinction

---

[29]A.D. Cliff, P. Haggett and M. Smallman-Raynor: *Measles: an Historical Geography of a Major Human Viral Disease from Global Expansion to Local Retreat, 1840-1990.* Blackwell, London, 1993; B.T. Grenfell and J. Harwood: (Meta)population dynamics of infectious diseases. *TREE,* **12** (1997), 395–399; M.J. Keeling: Modelling the persistence of measles. *Trends Microbiol.,* **5** (1997), 513–518.

after an outbreak. The 'phase transition' occurs when both N and α/μ tend to infinity such that $\alpha/\mu = O(\sqrt{N})$ and so, in the area of parameter space that is somewhat imprecisely characterized by this relation, we expect extinction on the demographic time scale $1/\mu$.

In Section 4.2 we took the population birth rate B as a constant and found a stable steady endemic state, characterized by damped fluctuations. As the damping requires many periods, the oscillations should be visible in data. But in fact one can easily imagine that periodic temporal heterogeneity (e.g., seasons and the school system) drives the system to truly periodic motion. This is indeed what Schenzle (1984)[30] and others found (and more: even chaotic fluctuations).

Even when analyzing simplified caricature models, it pays to write expressions for quantities such as the level of prevalence, the time scale at which disturbances decay, the period of oscillations etc., in terms of observable quantities such as the mean age $\bar{a}$ at infection. First of all, this helps to estimate parameters from data. But secondly, it makes the conclusions more robust, that is, usable perhaps as rules of thumb in situations where clearly the caricature is an inappropriate description. We emphasize this as a general principle (without repeating the examples we gave in Section 4.2).

Vaccination is, in essence, a way to reduce the rate at which new 'fuel' is provided to the infectious agent. In the context of simple models one can, without much difficulty, characterize the critical coverage that leads to herd immunity.

On a longer time scale, the spread of an infective agent may have its repercussions on population growth. Indeed, in Section 4.4 we considered a host population that grows exponentially and found that an infective agent may reduce the population growth rate. In fact, it may reduce it to the extent that growth is stopped completely.

Competition between different strains of essentially the same infectious agent is mediated by the immune system of the host. Some researchers focus on a particular agent and on known features of the way in which immunity works. Others try to deduce general insights by analyzing caricatures. The hope is that ultimately the two approaches can be combined. Motivated by this hope we have illustrated the adaptive dynamics approach for analyzing evolutionary change by applying it to a caricatural nested model, while incorporating superinfection.

Hospitals provide a situation of special interest, as numbers of hosts are relatively small. Turnover derives from admission and discharge, rather than from birth and death. Increasingly, antibiotic resistance prospers in hospital settings world-wide. So far we have merely introduced the Markov chain formalism for bookkeeping, but in Chapter 14 we shall use the formalism to quantify the spread of infections in hospitals.

---

[30]D. Schenzle: An age-structured model of pre- and post-vaccination measles transmission. *IMA J. Math. Appl. Med. Biol.*, **1** (1984), 169–191.

Chapter Five

Inference, or how to deduce conclusions from data

5.1 INTRODUCTION

A general scientific aim is to draw conclusions from data. Models may be useful along the way, as a tool. Or it may be that the conclusion concerns the choice of the most appropriate model from a family of candidates that differ, qualitatively or quantitatively or both, in the way a phenomenon is related to underlying mechanisms (so that, in the end, the conclusion amounts to pointing out the mechanism that is likely to underlie the phenomenon that is captured by the data).

Rarely, if ever, do we have absolute confidence in our conclusions. In the spirit of Popper we should continue playing the devil's advocate and scrutinize the evidence with an open mind to alternative explanations, as well as an open eye for complementary data. Formal ways to quantify our confidence are very helpful in this respect.

These remarks touch upon a subject, statistical inference or data analysis, to which an immense literature is devoted (see e.g., Rice 1995 for a gentle introduction and Pawitan 2001 for somewhat more advanced general theory; and e.g., Becker 1989, or Andersson and Britton 2000 specifically oriented towards infectious disease data). Here our aim is to illustrate some of the key notions and techniques by way of examples that display their use and usefulness in the context of the epidemiology of infectious diseases.

5.2 MAXIMUM LIKELIHOOD ESTIMATION INTRODUCED IN THE CONTEXT OF THE ICU EXAMPLE

In Section 4.6 we discussed the spread of antibiotic resistant bacteria in intensive care units (ICUs) and in that context there was a need to quantify the length of stay (LOS) of patients in the ICU. As a rule, a hospital can provide for each patient the date of admission to the ICU and the date of discharge from the ICU. In Chapter 14 we shall describe a rather flexible Markov chain model in which such data can be incorporated directly. In Section 4.6, however, we assumed that LOS has an exponential distribution with mean δ, and only this parameter δ entered into the ODE description (4.46). So we adopted a statistical description of LOS and took for granted that the parameter δ can somehow be estimated from available data. Here we shall focus on this estimation.

Suppose that we have observed the sojourn time in the ICU, i.e., the LOS of n patients. The data then consist of n positive numbers $T_1, \ldots, T_n$ (we consider these as real numbers and ignore that they may actually be integers if the time unit of bookkeeping by the hospital is one day).

It is mathematically convenient to conceive of the LOS T as a stochastic variable following the exponential distribution. The convenience mainly derives from two aspects:

- The lack of memory of the exponential distribution allows us to talk about the probability per unit of time of discharge, and to formulate simple Markov chain models in terms of ODE such as (4.46).

- The distribution is parameter scarce: there is just one parameter to estimate.

We choose to adopt this assumption and the n numbers T_i are then considered as realizations of independently identically distributed random variables with distribution

$$P(T \leq t) = 1 - e^{-\lambda t} \tag{5.1}$$

with a positive parameter λ yet to be specified. This distribution has *density*

$$f(t; \lambda) = \lambda e^{-\lambda t} \tag{5.2}$$

meaning that

$$P(a \leq T \leq b) = \int_a^b f(t; \lambda) dt, \tag{5.3}$$

and mean $E(T) = \int_0^\infty t f(t; \lambda) dt = 1/\lambda (= \delta)$.

So even though the probability that T takes the value t, where t is any positive real number, equals zero (for the very simple reason that there are so very many real numbers), we can say somewhat more informally that the *likelihood* that, given λ, T takes the value t is given by $f(t; \lambda)$, as a kind of abridgment of (5.3). By independence the likelihood that, given the value of λ, we would have obtained the n data points that we have, is then given by

$$\prod_{i=1}^n f(T_i; \lambda) = \lambda^n e^{-\lambda \sum_{i=1}^n T_i}. \tag{5.4}$$

However, because we observed the T_i and do not know the value of λ, we turn the interpretation around and call the quantity (5.4) the likelihood of λ, given the data $T_1, \ldots, T_n$. That is, we consider the quantity (5.4) as a function of λ. A very natural idea is now to use as an *estimate* for the unknown parameter that value of λ for which (5.4) is maximal. Indeed, for that value of λ the data have maximum likelihood, and accordingly one refers to this value for λ as the *maximum likelihood estimate* (also called ML-estimate or MLE) and denotes it by λ_{mle}. (Note that λ_{mle} is a random variable, for the simple reason that the T_i are random variables.)

Maximizing (5.4) is equivalent to maximizing the *log-likelihood*

$$n \log \lambda - \lambda \sum_{i=1}^n T_i. \tag{5.5}$$

The log-likelihood has derivative

$$\frac{n}{\lambda} - \sum_{i=1}^n T_i$$

and so to verify that

$$\lambda_{\text{mle}} = \frac{n}{\sum_{i=1}^n T_i} \tag{5.6}$$

we only have to verify that the log-likelihood has indeed a maximum in this critical point. Because the second derivative $-n/\lambda^2$ is negative, this is indeed the case.

If we rewrite (5.6) in the form

$$\lambda_{\mathrm{mle}}^{-1} = \frac{1}{n} \sum_{i=1}^{n} T_i \tag{5.7}$$

we see that the ML-estimator (MLE) is such that the expected LOS ($= 1/\lambda$) is equated with the observed average LOS, exactly as common sense would suggest.

If the (log)likelihood has a sharp peak at $\lambda = \lambda_{\mathrm{mle}}$ we are inclined to have much confidence in our estimate, whereas a rather flat (log)likelihood brings uncertainty. Intuitively, the second derivative at $\lambda = \lambda_{\mathrm{mle}}$, i.e., the quantity

$$-\frac{1}{n}(\sum_{i=1}^{n} T_i)^2 \tag{5.8}$$

should relate to how reliable the MLE is (how much confidence it deserves).

There exist theoretical results that make this intuitive idea precise. These are based on asymptotics for $n \to \infty$ and so are particularly relevant for large sample sizes and less so for relatively small values of n. Moreover, a key assumption is that we believe the model or, in other words, that the data 'obey' the exponential distribution for some true, albeit unknown, value λ_0. The first result is like the law of large numbers. It ascertains the consistency of the MLE, by which we mean that the estimate converges to the true value λ_0 when sample size n goes to infinity. The second result is more like the central limit theorem. It states that for large n the distribution of λ_{mle} is approximately Gaussian, with mean λ_0 and variance $1/nI(\lambda_0)$, where

$$I(\lambda) := -E\left(\frac{\partial^2}{\partial \lambda^2} \log f(T; \lambda)\right). \tag{5.9}$$

Note that with (5.2) we obtain that $\partial^2/\partial \lambda^2 \log f(T; \lambda) = -1/\lambda^2$ for all T so that the expectation also equals $-1/\lambda^2$ and accordingly the asymptotic variance equals λ_0^2/n (by which we mean that λ_0^2/n is the leading term in an asymptotic expansion of the variance for $n \to \infty$.) We add that the second result is based on smoothness assumptions for f.

A difficulty with the second result is that λ_0, the point at which we evaluate I, is not known. The first result suggests a pragmatic solution to this: replace λ_0 by the known λ_{mle}. In our specific case this leads to $nI(\lambda_{\mathrm{mle}}) = 1/n(\sum_{i=1}^{n} T_i)^2 = n(1/n \sum_{i=1}^{n} T_i)^2 = n\lambda_{\mathrm{mle}}^{-2}$.

Define $z(\alpha/2)$ by the requirement

$$\frac{1}{\sqrt{2\pi}} \int_{-z(\alpha/2)}^{z(\alpha/2)} e^{-\frac{1}{2}\xi^2} d\xi = 1 - \alpha \tag{5.10}$$

or, in words, by the requirement that $100(1 - \alpha)\%$ of the mass of the standard normal distribution is contained in the interval $(-z(\alpha/2), z(\alpha/2))$. To actually approximate/compute $z(\alpha/2)$ one uses existing tables or a computer, e.g., $\alpha = 0.05$ leads to $z(\alpha/2) = z(0.025) \approx 1.96$. We now know that

$$P(-z(\alpha/2) \le \sqrt{nI(\lambda_{\mathrm{mle}})}(\lambda_{\mathrm{mle}} - \lambda_0) \le z(\alpha/2))$$

equals $1 - \alpha$ to a good approximation for large values of n. We therefore call

$$\left(\lambda_{\text{mle}} - \frac{z(\alpha/2)}{\sqrt{nI(\lambda_{\text{mle}})}}, \lambda_{\text{mle}} + \frac{z(\alpha/2)}{\sqrt{nI(\lambda_{\text{mle}})}}\right) = \left(\lambda_{\text{mle}} - \frac{z(\alpha/2)}{\sqrt{n}\lambda_{\text{mle}}^{-1}}, \lambda_{\text{mle}} + \frac{z(\alpha/2)}{\sqrt{n}\lambda_{\text{mle}}^{-1}}\right)$$

(5.11)

the $100(1-\alpha)\%$ *confidence interval* for λ_{mle}. The suggested implication is that the true value λ_0 lies with probability $1 - \alpha$ in this interval (here 'probability' refers to the randomness of the data). We have to be careful though, because we evaluated I in λ_{mle} and we determined λ_{mle} on the basis of some finite sample.

Exercise 5.1 Here we consider a discrete-time variant where we replace the exponential distribution by the geometric distribution. So T is now an integer-valued random variable and

$$P(\text{LOS} = t) = \lambda(1 - \lambda)^{t-1}$$

(5.12)

with λ the (unknown) probability of discharge per day. Derive the maximum likelihood estimator for observed LOS data $T_1, \ldots, T_n$. Next determine an expression for the asymptotic variance in terms of the data.

5.3   AN EXAMPLE OF ESTIMATION: THE ICU MODEL

Critically ill patients in an ICU often suffer from infections.[1] A common finding is that, on average, the patients that become infected have a longer length of stay (LOS) than patients who do not become infected. But how should we interpret such an observed correlation?[2]

A tempting first idea is to consider the infection as the cause of an increase in LOS. However, the risk of getting infected increases with LOS, so it may also be the other way around: a long LOS may have as an effect that one gets infected. In addition, there might be heterogeneity among patients such that some will with high probability both stay long and become infected. As these mechanisms do not exclude each other, all three of them may play a role simultaneously. But in order to decide about the potential effect of prevention measures one would like to disentangle how they contribute to the observed correlation.

Reliable conclusions can be obtained from detailed case-control studies (based on a careful matching of patients on the basis of a multitude of traits). Such studies are labor intensive and therefore rather costly, so they are not performed routinely. Is it possible to deduce a first 'guestimate' of attributable extra LOS from readily available information?

If one is willing to ignore heterogeneity, the answer is 'yes' and the aim of the present section is to explain how. The outcome provides quantitative guidance for the thinking about (the need for, and benefit of) prevention measures. But clinicians should separately wonder whether or not the importance of heterogeneity necessitates more costly statistical investigations.

---

[1]In a one-day prevalence study including 10,038 patients in 1417 ICUs in 17 European countries, 45% of the patients were found to suffer from one or more infections, notably ventilator-associated pneumonia (VAP) and bacteremia; J.L. Vincent et al.: The prevalence of nosocomial infection in intensive care units in Europe, results of the EPIC study. *J. Am. Med. Assoc.*, **274**, (1995) 639–644.

[2]This section is based on unpublished joint work of O. Diekmann, M.C.J. Bootsma, M.J.M. Bonten, C.A.M. Schurink, S. Nijssen and I. Pelupessy.

In this section we assume that newly admitted patients are not colonized or infected (and we shall speak about 'infected' even when perhaps 'colonized' would be the better term, cf. Section 4.6).

There is another simplifying assumption that we are going to make: the force of infection is constant in time (but unknown, so this is one of the parameters that we need to estimate). Recall that the force of infection is the probability per unit of time for a non-infected patient to become infected, so by assuming it to be a constant, albeit of unknown value, we eliminate all dependence between patients and thus greatly simplify the problem. In Chapter 14 we shall introduce more sophisticated models that do properly take the dependence into account.

The event 'discharge' happens to every patient (recall from Section 4.6 that in the case of death we do speak about discharge as well; a more sophisticated analysis would make a distinction between death and real discharge). The event 'infection' may or may not happen, but if it happens, it happens before 'discharge.' We do *not* know when 'infection' exactly took place, but we assume that by the time of 'discharge' it is known whether or not it took place. We label a patient I if infection occurred and U otherwise. So the data consist of a list of patients with the LOS and a label, I or U for each patient.

The Markov model involves three *unknown* parameters λ, μ and σ. Here λ and μ are the probability per unit of time of discharge of, respectively, uninfected and infected patients, and σ is the force of infection. So for $\sigma = 0$ the sojourn time in the ICU is exponentially distributed with mean $1/\lambda$. For $\sigma > 0$ the same is true if we condition on the patient to avoid infection. If, on the other hand, a patient becomes infected, the remaining sojourn time is also exponentially distributed with mean $1/\mu$. Hence one can call $1/\mu - 1/\lambda$ the attributable LOS, i.e., the expected prolongation of LOS due to the infection.

Exercise 5.2 Explain the logic underlying the last sentence in more detail. Next, suppose our estimates will yield a value for μ which *exceeds* the value of λ. How would you interpret such an outcome?

The statistically sound way to estimate λ, μ and σ is by maximizing the likelihood. However, we first present a 'deterministic' alternative in terms of simple explicit formulas.

The data consist of the following information: a list of a total of $M + N$ patients numbered by $i = 1, 2, \ldots, M + N$. The first N of these are labeled 'not-infected' whereas the last M are labeled 'infected.' For each i we know the LOS T_i of the patient numbered i. Our task is to estimate λ, μ and σ on the basis of this information.

Exercise 5.3 Let, for a particular patient, τ denote the time elapsed since this patient entered the ICU. Show that the probability that the patient is still in the ICU *and* is not infected, equals, according to our assumptions,

$$e^{-(\lambda+\sigma)\tau}.$$

Also show that the probability that the patient is discharged with label 'infected' is given by

$$\frac{\sigma}{\lambda + \sigma}.$$

Exercise 5.4 Show that the distribution of LOS, *given* that the patient is discharged with label 'not-infected,' has density

$$(\lambda + \sigma)e^{-(\lambda+\sigma)\tau}$$

and hence mean $1/(\lambda + \sigma)$, which is smaller than $1/\lambda$.

Exercise 5.5 Consider a patient who becomes infected. The length of the period *before* the patient is infected is a random variable, with a distribution that has a density. What is the density?

Exercise 5.6 Show that the distribution of LOS, *given* that the patient is discharged with label 'infected,' has density

$$\frac{\mu(\lambda + \sigma)}{\lambda + \sigma - \mu}\left(e^{-\mu\tau} - e^{-(\lambda+\sigma)\tau}\right)$$

and mean

$$\frac{1}{\lambda + \sigma} + \frac{1}{\mu}.$$

The 'deterministic' approach to parameter estimation (also known as the method of moments) is to require that certain means computed from the data should equal the corresponding means computed from the assumptions, and then to choose the parameters such that these requirements are fulfilled. In particular we require, on the basis of Exercise 5.3, that

$$\frac{\sigma}{\lambda + \sigma} = \frac{M}{M + N}, \tag{5.13}$$

on the basis of Exercise 5.4 that

$$\frac{1}{\lambda + \sigma} = \frac{1}{N}\sum_{j=1}^{N} T_j, \tag{5.14}$$

and on the basis of Exercise 5.6 that

$$\frac{1}{\lambda + \sigma} + \frac{1}{\mu} = \frac{1}{M}\sum_{j=N+1}^{N+M} T_j. \tag{5.15}$$

As it happens, we can solve these three equations explicitly:

$$
\begin{aligned}
\mu &= \left(\frac{1}{M}\sum_{j=N+1}^{N+M} T_j - \frac{1}{N}\sum_{j=1}^{N} T_j\right)^{-1}, \\
\sigma &= \left(\frac{1}{N}\sum_{j=1}^{N} T_j\right)^{-1}\frac{M}{N + M}, \\
\lambda &= \left(\frac{1}{N}\sum_{j=1}^{N} T_j\right)^{-1}\frac{N}{N + M}.
\end{aligned}
\tag{5.16}
$$

The fact that the three equations in (5.16) are explicit expressions for the parameters in terms of the data is an attractive feature of this ad hoc approach. As we will show next, the systematic MLE approach leads to equations for the parameters that have to be solved numerically.

Exercise 5.7 Show that, according to our assumptions, the data have likelihood

$$= \binom{N+M}{M} \left(\frac{\sigma}{\sigma+\lambda}\right)^M \left(\frac{\lambda}{\sigma+\lambda}\right)^N \prod_{j=1}^N (\lambda+\sigma)e^{-(\lambda+\sigma)T_j} \quad (5.17)$$

$$\times \prod_{j=N+1}^{N+M} \frac{\mu(\lambda+\sigma)}{\lambda+\sigma-\mu} \left(e^{-\mu T_j} - e^{-(\lambda+\sigma)T_j}\right).$$

Exercise 5.8 Derive the three equations

$$\frac{N}{\lambda} - \frac{M}{\lambda+\sigma-\mu} - \sum_{j=1}^{N+M} T_j + \sum_{j=1}^{N+M} \frac{T_j}{1-e^{-(\lambda+\sigma-\mu)T_j}} = 0,$$

$$\frac{M}{\sigma} - \frac{M}{\lambda+\sigma-\mu} - \sum_{j=1}^{N+M} T_j + \sum_{j=1}^{N+M} \frac{T_j}{1-e^{-(\lambda+\sigma-\mu)T_j}} = 0, \quad (5.18)$$

$$\frac{M}{\mu} + \frac{M}{\lambda+\sigma-\mu} - \sum_{j=1}^{N+M} \frac{T_j}{1-e^{-(\lambda+\sigma-\mu)T_j}} = 0,$$

by computing the derivatives of $\log L$ with respect to λ, σ and μ and next requiring that these are zero.

By subtracting the first two equations of (5.18) we deduce that

$$\frac{\lambda}{\sigma} = \frac{N}{M}, \quad (5.19)$$

while by adding the last two equations one obtains

$$\frac{1}{\sigma} + \frac{1}{\mu} = \frac{1}{M} \sum_{j=1}^{N+M} T_j. \quad (5.20)$$

Both (5.19) and (5.20) also hold for μ, σ, λ defined by (5.16). These equations allow us to express two of the parameters explicitly in terms of the data and the third parameter. The equation that determines the third parameter can, however, not be solved explicitly and one has to resort to numerical methods (in fact it is not at all straightforward to show analytically that this equation has a unique solution, and so one is tempted to take this for granted).

By calculating second order terms one can determine confidence regions for the MLE and get some idea about the robustness of the deterministic estimates (5.16). But the calculations are rather messy and we omit them.

The formula (5.16) was applied to three data sets obtained at ICUs of the University Medical Center Utrecht in 2000–2002. These concerned:

- Colonization with Gram-negative bacteria resistant to third generation cephalosporins (G);

- Bacteremia (defined as a blood culture growing Gram-negative bacteria, *Staphylococcus aureus* or yeasts) (B);

- Ventilator-associated pneumonia (VAP).

The results were (using one day as the unit of time):[3]

- Dataset G: 58 cases among 209 patients, $\sigma = 0.043, 1/\lambda = 9.83, 1/\mu = 7.74$.

- Dataset B: 52 cases among 490 patients, $\sigma = 0.013, 1/\lambda = 9.55, 1/\mu = 15.99$.

- Dataset VAP: 137 cases among 993 patients, $\sigma = 0.017, 1/\lambda = 9.52, 1/\mu = 15.54$.

A first rough conclusion (based on comparing values of $1/\lambda$ to values of $1/\mu$) is that there is no reason to assume that colonization with resistant Gram-negative bacteria prolongs the LOS, but that both bacteremia and ventilator-associated pneumonia lead to substantially longer LOS. For the Gram-negative bacteria and bacteremia this confirmed what was already known and so these studies mainly served to test the method. For ventilator-associated pneumonia there is no generally accepted view concerning its influence on LOS.

5.4   THE PROTOTYPE STOCHASTIC EPIDEMIC MODEL

The estimation procedures of Sections 5.2 and 5.3 turned out to be quite simple because they were based on observations of *independent* events. When estimating parameters from observations concerning *dependent* events, as is the case with infections, estimation is more complicated as we shall now see.

In Section 3.1 the prototype stochastic epidemic model was defined. In the present section we will make inferences concerning the model parameters, and in particular R_0, based on observations from one epidemic outbreak. First we do this when data are restricted to knowing the final size of the epidemic, and next for the case that the epidemic is observed continuously, i.e., where at each time point we know how many individuals are susceptible, latent, infectious and recovered. The community size is also assumed to be known. The results are based on the community size being large and a major outbreak having occurred, which implicitly assumes $R_0 > 1$ (if only very few were infected the outbreak would most likely have gone unnoticed).

5.4.1   Final-size data

In Section 3.5.3 it was shown that, in case of a major outbreak, the final proportion infected $\bar{Y}_N = Y_N/N$ converges in probability (as $N \to \infty$) to the unique positive solution x^* of the equation

$$1 - e^{-R_0 x} = x, \tag{5.21}$$

where $R_0 = cpE(T_I)$. It was also mentioned that, in case of a major outbreak, $\bar{Y}_N$ is asymptotically normally distributed around this value with variance

$$V(\bar{Y}_N) = \frac{1}{N} \frac{x^*(1 - x^*)}{(1 - (1 - x^*)R_0)^2} \left(1 + r^2(1 - x^*)R_0^2\right), \tag{5.22}$$

where $r^2 = V(T_I)/(E(T_I)^2)$ is the squared coefficient of variation (i.e., the square of the standard deviation divided by the mean). This result can be used to derive

---

[3]A more detailed account of (variants of) the data analysis is presented in an unpublished draft manuscript by M.C.J. Bootsma, M.J.M. Bonten, C.A.M. Schurink, S. Nijssen, I. Pelupessi and O. Diekmann: Nosocomial infection and length of stay in intensive care : A simple observation on cause and effect. The manuscript is available on request from the first author.

an estimator for R_0 based on the observed final size Y_N. Since $\bar{Y}_N$ converges to x^* satisfying (5.21), a natural estimator is to replace x^* by $\bar{Y}_N$ and to solve (5.21) for R_0. The leads to the following estimate:

$$\hat{R}_0 = \frac{-\log(1 - \bar{Y}_N)}{\bar{Y}_N}. \tag{5.23}$$

(In Section 5.4.3 we derive a very similar estimate using different methods.) Since $\bar{Y}_N$ converges in probability to x^* as $N \to \infty$, we see from (5.21) that $\hat{R}_0$ converges to R_0. In statistical jargon this means that the estimator is *consistent*. Furthermore, we also know that $\hat{R}_0$ is approximately normally distributed around the true value R_0. In order to find the asymptotic variance we apply the so-called δ-*method* (see for example, Rice 2006, Chapter 4), which we explain now.

Let X be a random variable with finite mean $E(X) = \mu$ and variance $V(X) = \sigma^2$, and let $f(x)$ be a continuous function. The δ-method gives approximations for the mean and variance of the random variable $g(X)$ by Taylor expansion of $g(x)$ around $x = \mu$, and keeping only the constant and linear terms. So we use $g(X) \approx g(\mu) + (X - \mu)g'(\mu)$ and find

$$E(g(X)) \approx g(\mu),$$
$$V(g(X)) \approx (g'(\mu))^2 \sigma^2.$$

The approximation holds better the smaller the variance $\sigma^2 = V(X)$ is. A common application of the δ-method is when $X = (X_1 + \cdots + X_n)/n$ is a mean value and the variance of the mean is small, for n is large.

When we apply the δ-method to obtain a variance approximation for $\hat{R}_0$ (so using $g(x) = -\log(1 - x)/x$), the approximation is good since the variance gets smaller with larger N. The result of the δ-method is that the estimate $\hat{R}_0$ has approximate variance

$$V(\hat{R}_0) \approx \left(\frac{1}{x^*(1 - x^*)} - \frac{R_0}{x^*} \right)^2 \frac{1}{N} \frac{x^*(1 - x^*)}{(1 - (1 - x^*)R_0)^2} (1 + r^2(1 - x^*)R_0^2)$$
$$= \frac{1}{N} \frac{1 + r^2(1 - x^*)R_0^2}{x^*(1 - x^*)}.$$

The standard deviation is the square root of this expression. In the expression the values for the quantities x^*, r and R_0 are unknown. If we replace these by estimates (i.e., only depending on data) we obtain what is called a *standard error*. In order to do this we estimate the limiting fraction infected x^* by the observed fraction infected $\bar{Y}_N$, and of course we estimate R_0 by $\hat{R}_0$. As for r, the coefficient of variation of the infectious period, the final size carries no information about this quantity. Instead r has to be estimated from some other source of information, or else some conservative upper estimate has to be used. Recall that $r = \sqrt{V(T_I)}/E(T_I)$ is the standard deviation divided by the mean (of the infectious period). It is very rare that the standard deviation of an infectious period exceeds the mean length of the infectious period; as a consequence $r = 1$ is an upper bound for most infections. Once a value for r is obtained, we arrive at the following standard error

$$s.e.(\hat{R}_0) = \frac{1}{\sqrt{N}} \sqrt{\frac{1 + r^2(1 - \bar{Y}_N)\hat{R}_0^2}{\bar{Y}_N(1 - \bar{Y}_N)}}. \tag{5.24}$$

The estimate $\hat{R}_0$ defined in (5.23) is not the exact maximum-likelihood (ML) estimator but the estimator is consistent with the ML-estimator because of the result that $\bar{Y}_N$ converges, for $N \to \infty$, in probability to the positive solution of (5.21), given there was a major outbreak. To obtain the exact ML-estimator is difficult even for small populations, because the expression for the likelihood function quickly becomes cumbersome (cf. Exercises 12.23 and 12.24 in Section 12.5.3 where the 'population' is a household of size 3).

From final-size data we can estimate the aggregated quantity R_0 but, since we have no information about time evolution, we cannot say anything about the distributions of the latency and infectious periods, nor about the contact rate c. Finally, since we don't observe the actual contacts we cannot estimate p, the probability that a contact with a susceptible results in infection. The only additional thing we can estimate from observing the final fraction infected $\bar{Y}_N$ is the critical vaccination coverage $v_c = 1 - 1/R_0$, defined as the fraction to vaccinate in order to prevent a major outbreak (see Section 3.5.4).

Of course, the point estimate of $v_c = 1 - 1/R_0$ is given by

$$\hat{v}_c = 1 - \frac{1}{\hat{R}_0} = 1 - \frac{\bar{Y}_N}{-\log(1 - \bar{Y}_n)}.$$

In order to obtain a standard error we can apply the δ-method once more. Using our expression for $V(\hat{R}_0)$, and choosing $g(\hat{R}_0) = 1 - 1/\hat{R}_0$, which has derivative $g'(R_0) = 1/R_0^2$, implying that $V(\hat{v}_c) \approx V(\hat{R}_0)/R_0^4$. Replacing unknown quantities by their estimates as was done with $\hat{R}_0$, we arrive at the following standard error for $\hat{v}_c$:

$$s.e.(\hat{v}_c) = \frac{1}{\sqrt{N}} \sqrt{\frac{1 + r^2(1 - \bar{Y}_n)\hat{R}_0^2}{\hat{R}_0^4 \bar{Y}_n(1 - \bar{Y}_n)}}. \tag{5.25}$$

Note that this standard error was obtained by using the δ-method twice, first writing $\hat{R}_0$ as a function of $\bar{Y}_N$, and then $\hat{v}_c$ as a function of $\hat{R}_0$. A more direct way would be to write $\hat{v}_c$ as a function of $\bar{Y}_N$ and apply the δ-method only once. We encourage the reader to do this. When N is fairly large, the difference between the results obtained by these two methods is small.

Exercise 5.9 Consider a very small community of 30 individuals to which one newly infected individual is added. Suppose the end result is that 10 of the 30 individuals get infected. Estimate R_0 and v_c and their standard errors (do this for the case that $r = 0$ and the case that $r = 1$). Do the same thing for a community of 300 and assume that 100 out of the 300 get infected (i.e. the same fraction infected but now in a larger community). Finally, estimate v_c with standard error for the case $N = 300$ and $r = 0$.

Exercise 5.10 Suppose that, for a certain infection, initially only a proportion s are susceptible (the rest are immune to the disease) and that s is known, for example from serological screening in a period when there is no outbreak. Assume that an outbreak takes place resulting in a proportion $\bar{Y}_n$ of the whole community getting infected. Note that necessarily $\bar{Y}_n \leq s$ since a fraction $1-s$ was assumed to be immune. The proportion among the initially susceptible is $\bar{Y}_n/s$. i) Derive an estimate of R_0 in this situation. ii) Consider the case $N = 60$ and $s = 0.5$, and assume that 10 individuals get infected. Estimate R_0 and the standard error, assuming first $r = 0$ and next $r = 1$. Contrast the results with those of Exercise 5.9, and explain the difference. Hint: Neglect the initially immune individuals completely by rescaling the parameters such that just the initially susceptible individuals are considered.

Exercise 5.11 In Japan an influenza outbreak occurred among racetrack horses in the year 1971.[4] Among 640 horses a total of 580 were infected during the outbreak. Assume that one of the horses was infected from outside and initiated the outbreak.

i) Estimate R_0 under the assumption that horses mix homogeneously and are equally susceptible to the infection.

ii) Give standard errors for the estimate under two different assumptions: the infectious period is non-random (so the coefficient of variation $r = 0$), and the coefficient of variation $r = 0.5$.

Exercise 5.12 Between August and December 2005 an outbreak of kerato-conjunctivitis occurred in a nursing home in Madrid.[5] At the end of the outbreak 46 out of 151 residents were infected so the fraction infected was 30.5% (in reality some of the infecteds were staff members; the incidence among staff members was lower than among residents — we neglect this here). Assume that one person was externally infected and initiated the outbreak. Estimate R_0 and give a 95% confidence interval for the estimate assuming that the coefficient of variation of the infectious period is $r = 1$ (corresponding to an exponentially distributed infectious period, i.e., corresponding to the 'general' stochastic epidemic model).

5.4.2 Continuous time data

In the previous subsection inference was based on final-size data, i.e., assuming that the final number infected was observed. We now consider another specific situation, where we assume that information is collected continuously during the outbreak (reality is quite often somewhere in between: that some temporal information is available albeit not continuously measured). By this we mean that we observe the exact time points at which each individual is infected, when his/her latency period ends and when he/she recovers and becomes immune. The fact that these data are more detailed than when only observing the final number infected, gives us hope that one can estimate more parameters and with higher precision. This is true. The price we have to pay is that the inference procedures are more involved since we now observe a stochastic process (the epidemic process) rather than only an outcome of a random variable (the final size). For this reason, the remainder of Section 5.4 is more statistically advanced. We start by presenting some quite technical general statistical inference procedures when observing a certain class of stochastic process. More about this theory can for example be found in Andersen et al. (1993). Then we move on to applying the methods on our epidemic process. We will focus on the 'general' stochastic epidemic, in which there is no latency period

---

[4]K. Satou and H. Nishiura: Basic reproduction number for equine-2 influenza virus a (H3N8) epidemic in racehorse facilities in Japan, 1971. *Journal of Equine Veterinary Science*, **26** (2006), 310–316.

[5]M.F. Dominingues-Berjon, P. Hernando-Briongosa, P.J. Miguel-Arroyo, J.E. Echevaria and I. Casas: Adenovirus transmission in a nursing home: Analysis of an epidemic outbreak of kerato-conjunctivitis. *Gerontology*, **53** (2007), 250–254.

and the infectious periods are exponentially distributed, and end with a few words about the prototype epidemic model.

Martingales Recall that $S(t)$, $I(t)$, and $R(t)$, respectively, denote the (random) number of susceptible, infectious and recovered individuals at time t. Because we have a closed community and no latency period we have $S(t) + I(t) + R(t) = N + 1$.

In Section 3.2.2 an equivalent definition of the 'general' stochastic epidemic was given in terms of the two jump rates:

$$\lambda_1(t) = cpI(t-)S(t-)/(N+1), \tag{5.26}$$
$$\lambda_2(t) = \alpha I(t-). \tag{5.27}$$

These are jump rates for the two counting processes $N_1(t) = N + 1 - S(t)$ and $N_2(t) = R(t)$, counting the number of infections and the number of recoveries, respectively. The first jump rate is the rate at which a given infectious individual has a successful contact with a given susceptible individual (cp/N) multiplied by the number of such pairs. The second jump rate (5.27) is the individual recovery rate multiplied by the number of infectious individuals at this time.

The probability that one of the counting processes increases by one (i.e., jumps one step) in a small interval $[t, t+h)$ equals h multiplied by its jump rate plus terms of higher order. Since the probability that one of these counting processes increases by two or more is of the order h^2, and hence very small, this is also the expected value of the counting process increment. The results stated rely on knowing the present value of the variables of the epidemic process. Since our model is Markovian, knowing the present values of the process is equivalent to knowing the whole history of the process. We denote the history of the process up to time t by $\mathcal{H}_t$ (in mathematical terminology this is called a 'filtration'). We hence have that

$$E(\Delta N_1(t)|\mathcal{H}_t) = E(N_1(t+h) - N_1(t)|\mathcal{H}_t) = \lambda_1(t)h + o(h) \tag{5.28}$$
$$E(\Delta N_2(t)|\mathcal{H}_t) = E(N_2(t+h) - N_2(t)|\mathcal{H}_t) = \lambda_2(t)h + o(h) \tag{5.29}$$

To obtain the increments over the whole time interval $[0, t]$ we simply accumulate jumps: $\Delta N_i(0) + \Delta N_i(h) + \Delta N_i(2h) + \cdots + \Delta N_i(t-h) = N_i(t) - N_i(0)$. If we take conditional expectation of each term and compare the sum with the corresponding sum on the right hand side of (5.28) or (5.29) we get an integral as h tends to 0. We now define new stochastic processes as the counting processes with the integrated rates subtracted:

$$M_1(t) := N_1(t) - N_1(0) - \int_0^t \lambda_1(s)ds, \tag{5.30}$$

$$M_2(t) := N_2(t) - N_2(0) - \int_0^t \lambda_2(s)ds. \tag{5.31}$$

These new 0-mean processes are so-called *martingales*.

The property that characterizes a martingale is $E(M_i(t+s)|\mathcal{H}_t) = M_i(t)$, i.e., that given the present value, the expected value at some future point equals the present value (which clearly holds for the present processes: use (5.28) and (5.29) and let h tend to 0). As a direct consequence for all martingales we have that $E(M(t)) = E(M(0))$. Quite often the martingales are 'centered' such that $E(M(0)) = 0$ (or even $M(0) \equiv 0$ if the initial value is deterministic). We then speak of a zero-mean martingale.

It is important to know whether or not a stochastic process is a martingale, because there exist central limit theorems for a wide class of martingales. In particular, when properly normalized, the martingale converges in some limit to a Gaussian process. It hence follows that the martingale at a specific time point is approximately normally distributed, an approximation we will make use of later.

Another property of martingales is that if you integrate a left-continuous (also known as predictable) stochastic process with respect to a martingale you obtain a new martingale; a stochastic process is said to be left-continuous if the process as a function of t is left-continuous with probability 1. So, if $f(t)$ is a left-continuous process and $M(t) = N(t) - N(0) - \int_0^t \lambda(s)ds$ is a martingale, it follows that

$$\tilde{M}(t) = \int_0^t f(s)\,(dN(s) - \lambda(s)ds)$$

is a martingale. Recall that $N(s)$ is a counting process, i.e., a stochastic process that increases by one at random time points. If the time-points in $[0, t]$ at which $N(s)$ makes its jumps are denoted $0 < s_1 < \cdots < s_k < t$, then $dN(s)$ will equal 1 at these time-points and 0 elsewhere. As a consequence, $\int_0^t f(s)dN(s) = \sum_{i=1}^k f(s_i)$, i.e., an integral with respect to a counting process is nothing complicated: it is simply a sum.

The so-called optional variation process of $\tilde{M}$ is defined by

$$\left[\tilde{M}\right]_t = \int_0^t f^2(s)dN(s) = \sum_{i=1}^k f^2(s_i), \qquad (5.32)$$

where the last equality is true if we, as before, let s_i with $0 < s_1 < \cdots < s_k < t$ denote the jump points of the counting process $N(s)$.

The most important use of the variation process is based on the following result (that can be derived in a similar way as above where we showed that M_i is a martingale):

$$Var(\tilde{M}(t)) = E([\tilde{M}]_t). \qquad (5.33)$$

We hence know that these martingales have mean 0 and we have an expression for the variance. It can also be shown that linear combinations of martingales are martingales, and that martingales based on separate counting processes not making simultaneous jumps are uncorrelated.

In the next subsections we will use martingales, and these properties, in order to construct estimators and to derive properties of the estimators. Before this we first look at the likelihood for the case of continuous time data, i.e., when observing the epidemic process continuously.

The likelihood As mentioned previously, we now consider the 'general' stochastic epidemic in which there is no latency period and where the infectious periods are exponentially distributed. To heuristically derive the likelihood for the continuous time data, we discretize time into many small time intervals $[t_i, t_{i+1}) = [ih, (i+1)h)$ of length h (so $t_i = ih$). Look at one such a time interval $[t_i, t_{i+1})$ and condition on the state of the epidemic process at a time just before the start of the interval. What can happen in such a (short) interval? Recall that $N_1(t)$ counts the number of infections up to time t and $N_2(t)$ the number of recoveries. We let $\Delta N_1(t_i) = N_1(t_{i+1}) - N_1(t_i)$ denote the number of infections in the interval $[t_i, t_{i+1})$, and similarly $\Delta N_2(t_i) = N_2(t_{i+1}) - N_2(t_i)$ is the number of recoveries in the same interval. If the interval length h is small (which we assume), most likely there will not be many infections or recoveries occurring in the interval. In fact, as we saw before, the probability that one infection occurs in the interval (i.e., that $\Delta N_1(t_i) = 1$) is $\lambda_1(t_i)h + o(h)$ and the probability that one recovery occurs (i.e., that $\Delta N_2(t_i) = 1$) equals $\lambda_2(t_i)h + o(h)$. By far the most likely event (when h is small!) is that there is no jump, so neither infection nor recovery takes place; this event has probability $1 - (\lambda_1(t_i) + \lambda_2(t_i))h + o(h)$ and when this occurs we have $1 - \Delta N_1(t_i) - \Delta N_2(t_i) = 1$. That there is more than one jump (infection and/or recovery) has probability $o(h)$ because each jump has probability proportional to h when h is small.

The probability to observe a specific realization of the epidemic in $[0, t)$ (with infections and recoveries at specified moments in time) then becomes

$$P(N_1(s), N_2(s); 0 \leq s \leq t) \approx \prod_i (\lambda_1(t_i)h)^{\Delta N_1(t_i)} (\lambda_2(t_i)h)^{\Delta N_2(t_i)}$$

$$\times (1 - (\lambda_1(t_i) + \lambda_2(t_i))h)^{1 - \Delta N_1(t_i) - \Delta N_2(t_i)}.$$
(5.34)

This might look very complicated, but it is in fact only a convenient way of writing a common formula for all possible events: from this reasoning exactly one of the three exponents $\Delta N_1(t_i)$, $\Delta N_2(t_i)$ and $1 - \Delta N_1(t_i) - \Delta N_2(t_i)$ will equal 1, and the other two will equal 0 thus not contributing to the product. For each factor in the product, corresponding to a specific time-interval $[t_i, t_{i+1})$, the formula 'picks' the right event: that there was an infection, a recovery or none of the two.

Recall that when h is small most intervals will have no jumps. In fact, for a given realization, the number of intervals having jumps is fixed (and equal to the number of infections plus the number of recoveries); only the number of intervals with no jumps increases as h gets smaller. This implies that we can, in the limit as h tends to 0, replace $1 - \Delta N_1(t_i) - \Delta N_2(t_i)$ in the last exponent of (5.34) by 1 since this will only add a finite number of factors of order $1 - o(h)$. The contribution from such a factor approximately equals $e^{-(\lambda_1(t_i) + \lambda_2(t_i))h}$. As for the first two factors to the right of the product, they become sums if we take logarithms, and when h becomes small these sums will contribute $\log \lambda_k(t_i)$ whenever process k makes a jump. A different way of writing this is as an integral with respect to the counting processes. This completes the heuristic argument for what can in fact be rigorously shown, i.e., that as $h \to 0$, the logarithm of the probability in (5.34) divided by the normalizing factor $h^{N_1(t) + N_2(t)}$ converges to the log-likelihood defined by $\ell_t = \ell_t(c, p, \alpha)$

$$\ell_t(c, p, \alpha) = \int_0^t \log(\lambda_1(s))dN_1(s) + \int_0^t \log(\lambda_2(s))dN_2(s) - \int_0^t (\lambda_1(s) + \lambda_2(s))ds.$$
(5.35)

By replacing $\lambda_1(t)$ and $\lambda_2(t)$ by their expressions and writing $\bar{S}(t) = S(t)/N$, the dependence on c, p and α becomes clear:

$$\ell_t = \int_0^t \log(cpI(s-)\bar{S}(s-))dN_1(s) + \int_0^t \log(\alpha I(s-))dN_2(s)$$

$$- \int_0^t (cpI(s-)\bar{S}(s-) + \alpha I(s-))ds.$$
(5.36)

The first two integrals in (5.36) are sums, as explained earlier: whenever $N_1(s)$ increases by one we add the value $\log(cpI(s-)\bar{S}(s-))$ just prior to s, and similarly for $N_2(s)$. The third integral is the more usual Lebesgue integral and it makes no difference if we integrate the values at s or just prior to s. For this reason we write s instead of $s-$ in what follows.

Viewed as a function of the parameters c, p and α this is the log-likelihood of the data. We first note that, as long as we don't have information about the actual contacts, it is impossible to disentangle the effect of the contact rate c from that of the probability p of a contact resulting in transmission. From now on we hence only study the product of the two, which we denote by $\gamma = cp$, and interpret as the rate of close enough contacts such that infection occurs if one is infectious and the other is susceptible. Note that this notation is in agreement with Chapters 1 and 2 where we have taken the convention of denoting the transmission rate constant by β if the transmission rate is written as βSI, and by γ if it is written as $\gamma SI/N$. To conclude, the log-likelihood is given by

$$\ell_t(\gamma, \alpha) = \int_0^t \log(\lambda_1(s))dN_1(s) - \int_0^t \lambda_1(s)ds + \int_0^t \log(\lambda_2(s))dN_2(s) - \int_0^t \lambda_2(s)ds,$$
(5.37)

where the rates $\lambda_1(s)$ and $\lambda_2(s)$ were defined in (5.26) and (5.27).

Maximum likelihood estimation We continue to focus on the 'general' stochastic epidemic for its mathematical tractability. At the end of the present section we briefly return to the prototype stochastic epidemic model in which the latency and infectious period have arbitrary distributions. In Section 5.6 we devote some words to model choice.

Suppose that we observe the general stochastic epidemic continuously up to some time t. Most often we consider the case where the epidemic is observed up to the end of the outbreak; we will call the time when the outbreak ends $t = \omega$. We now derive parameter estimates of γ, α (or $1/\alpha$) and $R_0 = \gamma/\alpha$ using the maximum likelihood principle, i.e., the parameter values that maximize the likelihood (or equivalently the log-likelihood), or in other words, that best 'explain' the observed data. To obtain maximum-likelihood estimates we differentiate the log-likelihood and equate the expressions to 0. First we differentiate with respect to $\gamma = cp$:

$$\frac{\partial \ell_t}{\partial \gamma} = \int_0^t \frac{1}{\gamma}dN_1(s) - \int_0^t I(s)\bar{S}(s)ds = \frac{N_1(t) - N_1(0)}{\gamma} - \int_0^t I(s)\bar{S}(s)ds.$$

Note that $\partial \ell_t/\partial \gamma = M_1(t)/\gamma$ using (5.30) and (5.26), from which we conclude that $\partial \ell_t/\partial \gamma$ is a zero-mean martingale. By equating $\partial \ell_t/\partial \gamma$ to zero (i.e., to its mean) we get the ML-estimate:

$$\hat{\gamma}_t = \frac{N_1(t) - N_1(0)}{\int_0^t I(s)\bar{S}(s)ds}.$$

The most common case is that the epidemic is observed to its end $t = \omega$:

$$\hat{\gamma} := \hat{\gamma}_\omega = \frac{Y}{\int_0^\omega I(s)\bar{S}(s)ds},$$
(5.38)

because then $N_1(\omega) - N_1(0) = Y$, the total number infected excluding the initially infected, and $\int_0^\omega I(s)\bar{S}(s)ds = \int_0^\infty I(s)\bar{S}(s)ds$, since $I(s) = 0$ for $s > \omega$. From the observation that $\partial \ell_t/\partial \gamma$ is a zero-mean martingale we conclude that $\hat{\gamma}$ is also a moment estimator (a moment estimator is obtained by equating a random variable to its mean).

It is worth studying $\hat{\gamma}$ to see whether or not it is a sensible estimator. The numerator in $\hat{\gamma}$ is the number of individuals who got infected during the course of the epidemic. The denominator is the accumulated rate at which susceptible and infectious individuals have had contact, except for the rate factor γ. From this we see that the estimator makes the number of infections equal to the accumulated close contact rate, i.e., to the expected number of infections. Hence, it is a sensible estimator.

Similarly, we differentiate $\ell_t(\gamma, \alpha)$ with respect to α to obtain the ML-estimate of α:

$$\partial \ell_t / \partial \alpha = \int_0^t \frac{1}{\alpha} dN_2(s) - \int_0^t I(s)ds = \frac{N_2(t) - N_2(0)}{\alpha} - \int_0^t I(s)ds.$$

The term $N_2(0)$ can be omitted as it equals 0 because we assumed that there were no initially removed individuals. From $\partial \ell_t / \partial \alpha = M_2(t)/\alpha$ we conclude that $\partial \ell_t / \partial \gamma$ is a martingale.

If we equate $\partial \ell_t / \partial \alpha$ to 0 we get,

$$\hat{\alpha}_t = \frac{N_2(t)}{\int_0^t I(s)ds}$$

and

$$\hat{\alpha} := \hat{\alpha}_\omega = \frac{Y+1}{\int_0^\infty I(s)ds}, \tag{5.39}$$

in the case that, as before, the epidemic is observed to its end. Recall that α is the rate to recover from the infection, in the sense of reaching the end of the infectious period (recall that we have assumed that the infectious period is exponentially distributed). This means that $1/\alpha$ is the expected length of the infectious period, and if the epidemic is observed continuously to its end, the ML-estimator is hence given by

$$\frac{1}{\hat{\alpha}} = \frac{\int_0^\infty I(s)ds}{Y+1} = \frac{\sum_{i=0}^Y \Delta T_i}{Y+1}, \tag{5.40}$$

where ΔT_i denotes the length of the infectious period of infected individual i. The ML-estimator of the expected length of the infectious period is hence the mean of the observed infectious periods, a very natural estimator.

The ML-estimator of $R_0 = \gamma/\alpha$ is the product of the separate ML-estimators:

$$\hat{R}_0 = \hat{\gamma} \frac{1}{\hat{\alpha}} = \frac{N_1(t) - 1}{\int_0^t I(s)\bar{S}(s)ds} \frac{\int_0^t I(s)ds}{N_2(t)} = \frac{Y \sum_{i=0}^Y \Delta T_i}{(Y+1) \int_0^\infty I(s)\bar{S}(s)ds}. \tag{5.41}$$

The fact that all estimators are based on martingales can be used to show that they are asymptotically Gaussian as $N \to \infty$ and assuming a major outbreak. More precisely, it can be shown that the estimators are consistent, i.e., that $\hat{\gamma} \to \gamma$, $1/\hat{\alpha} \to 1/\alpha$ and $\hat{R}_0 \to R_0$ in probability, and also that $\sqrt{N}(\hat{\gamma} - \gamma)$, $\sqrt{N}(\hat{\alpha}^{-1} - \alpha^{-1})$ and $\sqrt{N}(\hat{R}_0 - R_0)$ converge to zero-mean Gaussian random variables. Furthermore, it is possible to use the variation process to obtain standard errors, due to the previously stated fact that the mean of the variation process equals the variance of the corresponding martingale. We omit the rather technical details and only state the results:

$$s.e.(\hat{\gamma}) = \frac{\hat{\gamma}}{\sqrt{Y}}, \qquad s.e.(\frac{1}{\hat{\alpha}}) = \frac{1/\hat{\alpha}}{\sqrt{Y+1}}, \qquad s.e.(\hat{R}_0) = \hat{R}_0 \sqrt{\frac{1}{Y} + \frac{1}{Y+1}}. \tag{5.42}$$

The above estimates were derived assuming that the 'general' stochastic epidemic model is applicable. When this is not true, but the prototype epidemic model applies, it can be shown that γ and R_0 are still estimated as above (but with different standard errors). Regarding the parameters of the latency and infectious periods, these are estimated by treating the observed latency periods $T_{1,1}, \ldots, T_{1,Y}$ and infectious periods $\Delta T_0, \ldots, \Delta T_Y$ as i.i.d. sequences.

Exercise 5.13 Use $Var(M_1(\omega)) = E([M_1]_\omega) = \int_0^\omega 1^2 dN_1(t) = E(N_1(\tau) - 1) \approx N_1(\tau) - 1$ together with the structure of $\hat\gamma$ to show that $s.e.(\hat\gamma) = \hat\gamma/\sqrt{Y}$. (The standard error for $1/\hat\alpha$ is obtained similarly, as is the standard error for $\hat{R}_0$, because M_1 and M_2 are independent.) Hint: Rewrite $\hat\gamma$ in terms of the true γ, M_1 and Y.

Exercise 5.14 Consider a community consisting of 30 susceptible individuals and assume one newly infected person enters the community at time $t = 0$. The result is that ultimately 10 (out of 30) additional people get infected. The times of infection and recovery of the initial person and the 10 infecteds are (time units are days): (0,4), (1,4), (2,7), (3,5), (3,8), (4,8), (5,6), (6,10), (7,12), (9,13), (10,12). For example, the first to become infected was infected day 1 and recovered day 4. For simplicity we assume all events occur the same time each day (e.g., noon). Assume that the 'general' stochastic epidemic model applies. Estimate γ, $1/\alpha$ and R_0 and their standard errors.

Exercise 5.15 Table 5.1 contains the data of an outbreak of smallpox in a Nigerian village (see Becker 1989, page 112 for further details). Due to a latency period individuals that are infected on a given day do not become infectious the next day, see, for example, the first two lines in the table, and note that a person is infected on day 0 while there still is only one infectious individual (the index case) on day 1. The rate at which individuals get infected is still $\gamma I(t)S(t)/N$ and $N = 119$. Estimate $\gamma = cp$, α and R_0 and give standard errors for the estimates (assuming that the 'general' stochastic epidemic model applies).

5.4.3 Final-size data once again

We return to the case of final-size data, but we will now use martingales. As mentioned before we can only expect to estimate one parameter, $R_0 = cp/\alpha$, from one data point, i.e., the final size Y. If the epidemic has ceased by time ω, the final number infected Y among those initially susceptible equals $N_1(\omega) - 1 = N_2(\omega) - 1$, expressed in terms of the counting processes. The initial configuration with one infectious and the rest susceptible corresponds to $N_1(0) = 1$ and $N_2(0) = 0$, which is also assumed to be known with certainty. Our aim is therefore to obtain a martingale estimator of R_0 that only depends on $N_1(\omega) = N_2(\omega)$ and $N_1(0)$ (=1) and $N_2(0)$ (=0).

From equations (5.30) and (5.31) we know that $M_i(t) = N_i(t) - N_i(0) - \int_0^t \lambda_i(s)ds$, $i = 1, 2$, where $\lambda_1(t) = cpI(t-)\bar{S}(t-)$ and $\lambda_2(t) = \alpha I(t-)$, are martingales. It was mentioned in Section 5.4.2 that linear combinations of martingales are martingales, and that a left-continuous stochastic process that is integrated with respect to martingale increments results in another martingale. To be specific, if

Table 5.1: The spread of smallpox in a Nigerian village. See text for further details.

Day	Number infectives	Number susceptibles	Number infected	Day	Number infectives	Number susceptibles	Number infected
t	I(t)	S(t)	C(t)	t	I(t)	S(t)	C(t)
0	1	119	1	43	5	100	5
1	1	118	0	44	5	99	1
2	1	118	0	45	5	98	1
3	1	118	0	46	4	97	0
4	1	118	0	47	4	97	2
5	1	118	0	48	3	95	1
6	1	118	0	49	3	94	0
7	1	118	1	50	1	94	0
8	1	117	0	51	2	94	0
9	1	117	1	52	3	94	0
10	1	116	0	53	3	94	2
11	1	116	0	54	3	92	0
12	1	116	3	55	2	92	0
13	1	113	1	56	4	92	0
14	1	112	0	57	5	92	0
15	1	112	0	58	5	92	1
16	1	112	0	59	5	91	0
17	1	112	1	60	5	91	0
18	1	111	0	61	7	91	0
19	1	111	0	62	8	91	0
20	1	111	0	63	6	91	1
21	1	111	0	64	5	90	0
22	1	111	1	65	4	90	0
23	2	110	2	66	3	90	0
24	2	110	2	67	5	90	0
25	2	110	2	68	3	90	0
26	5	109	5	69	2	90	0
27	6	109	6	70	2	90	0
28	5	107	5	71	2	90	0
29	5	107	5	72	3	90	0
30	4	105	4	73	3	90	0
31	5	105	5	74	1	90	0
32	5	105	5	75	1	90	0
33	2	105	2	76	1	90	0
34	1	105	1	77	2	90	0
35	1	104	1	78	2	90	0
36	2	104	2	79	1	90	0
37	2	104	2	80	1	90	0
38	1	103	1	81	1	90	0
39	2	102	2	82	1	90	0
40	2	102	2	83	1	90	0
41	4	102	4	84	0	90	0
42	4	102	4				

$f(t)$ and $g(t)$ are two left-continuous processes, then

$$M(t) = \int_0^t f(s)(dN_1(s) - cpI(s-)\bar{S}(s-)ds) + \int_0^t g(s)(dN_2(s) - \alpha I(s-)ds) \quad (5.43)$$

is a martingale. The counting processes increase by 1 at each jump, so those parts of the integrals can be evaluated if $f(s)$ and $g(s)$ are known at the jump points. The integration with respect to the jump rates requires continuous knowledge of these rates — information we don't have when treating final-size data. However, by choosing $f(s)$ and $g(s)$ in a clever way we can get rid of these parts. More specifically, if we choose $f(s) = \alpha/cp\bar{S}(s-) = N/R_0 S(s-)$ and $g(s) = -1$ we get

$$M(t) = \int_0^t \frac{\alpha N}{cpS(s-)} dN_1(s) - \int_0^t dN_2(s)$$

$$= \frac{N}{R_0}\left(\frac{1}{N} + \frac{1}{N-1} + \cdots + \frac{1}{N - (N_1(t) - N_1(0)) + 1}\right) - (N_2(t) - N_2(0)).$$

The second equality holds because at each jump of $N_1(\cdot)$, $S(\cdot)$ decreases by one; and just before the first jump $S(s-) = N$, just before the next jump $S(s-) = N - 1$, and so forth, and $N_1(\cdot)$ makes $N_1(t) - N_1(0)$ jumps in $(0, t]$.

Recall that $N_1(0) = 1$ and $N_2(0) = 0$, and at the end of the epidemic $N_1(\omega) = N_2(\omega) = Y + 1$ (the initially infected plus those who were infected during the outbreak). As a consequence we get

$$M(\omega) = \frac{N}{R_0}\left(\frac{1}{N} + \frac{1}{N-1} + \cdots + \frac{1}{N - Y + 1}\right) - (Y + 1). \quad (5.44)$$

Since $M(t)$ is a zero-mean martingale viewed as a stochastic process in t, it follows that $M(\omega)$ is approximately normally distributed with mean 0 and variance $E([M]_\omega) \approx [M]_\omega$. We derive an estimate for R_0 by equating $M(t)$ to its mean (=0):

$$\hat{R}_0 = \frac{N}{Y+1}\left(\frac{1}{N} + \frac{1}{N-1} + \cdots + \frac{1}{N - Y + 1}\right). \quad (5.45)$$

If N is large this is approximately equal to

$$\hat{R}_0 \approx -\frac{\log(1 - Y/N)}{Y/N}. \quad (5.46)$$

We now derive an expression for the standard error of $\hat{R}_0$ using similar methods as when having continuous time data, i.e., by using that it is based on a martingale. We first rewrite $\hat{R}_0$ defined by (5.45) as follows

$$\hat{R}_0 = \frac{N}{Y+1}\left(\frac{1}{N} + \cdots + \frac{1}{N - Y + 1}\right)$$

$$= \frac{R_0}{Y+1}\left(\frac{N}{R_0}\left(\frac{1}{N} + \cdots + \frac{1}{N - Y + 1}\right) - (Y + 1) + (Y + 1)\right)$$

$$= \frac{R_0}{Y+1}M(\omega) + R_0.$$

This way of writing $\hat{R}_0$ is not meaningful as an estimator since it contains the unknown quantity R_0 — instead we will use it to obtain an expression for the variance. When N is large and assuming a major outbreak (so that $Y + 1 = Nx^*$ plus deviations of order $\sqrt{N}$) we then have that

$$Var(\hat{R}_0) \approx \left(\frac{R_0}{Nx^*}\right)^2 E([M]_\omega) \approx \left(\frac{R_0}{Nx^*}\right)^2 [M]_\omega.$$

Further,

$$
\begin{aligned}
[M]_\omega &= \int_0^\omega f^2(s)dN_1(s) + \int_0^\omega g^2(s)dN_2(s) \\
&= \frac{1}{R_0^2}\int_0^\omega \frac{1}{\bar{S}^2(s-)}dN_1(s) + \int_0^\omega (-1)^2 dN_2(s) \\
&\approx \frac{1}{R_0^2} N\left(\frac{1}{1 - Y/N} - 1\right) + Y.
\end{aligned}
$$

The approximation in the latter step is valid because

$$
\begin{aligned}
\frac{1}{R_0^2}\int_0^\omega \frac{1}{\bar{S}^2(s-)}dN_1(s) &= \frac{1}{1^2} + \frac{1}{((N-1)/N)^2} + \cdots + \frac{1}{((N-Y+1)/N)^2} \\
&\approx N\int_{1-Y/N}^1 \frac{1}{x^2}dx = N\left(\frac{1}{1 - Y/N} - 1\right).
\end{aligned}
$$

Combining these two results we see that a standard error estimate for $\hat{R}_0$ is given by

$$s.e.(\hat{R}_0) = \frac{1}{\sqrt{N}\bar{Y}}\sqrt{\frac{\bar{Y}}{1 - \bar{Y}} + R_0^2\bar{Y}} = \frac{1}{\sqrt{N}}\sqrt{\frac{1}{\bar{Y}(1 - \bar{Y})} + \frac{R_0^2}{\bar{Y}}}. \qquad (5.47)$$

We end this section by comparing our estimator based on martingales with the one obtained in Section 5.4.1 from the central limit theorem for the final size of the epidemic. There the estimator for R_0 was defined by (5.23), which is not exactly the same as (5.45), but it coincides with its approximation (5.46). Since the estimators are nearly identical for large N, it should not come as a surprise that the standard errors are the same: when the infectious period follows the exponential distribution the coefficient of variation r equals 1, and then the standard error (5.24) equals (5.47).

5.5   ML-ESTIMATION OF α AND β IN THE ICU MODEL

Recall from Section 4.6 the situation of an intensive care unit with two beds that are always occupied, but with 0, 1 or both patients colonized by resistant bacteria. The rate to jump from 0 to 1 colonized individual was 2α (each individual gets colonized from an unknown 'outside' source at rate α), and the rate to jump from 1 to 2 colonized was $\alpha + \beta$ (so the uncolonized individual can get colonized 'spontaneously,' with rate α, or by transmission from its already colonized neighboring patient, with rate β. Colonized patients remain colonized (on the present time scale) but are discharged at rate $1/\delta$ and replaced by a new uncolonized patient.

Suppose now that we have data from observing one such intensive care unit over a time interval, and that tests for colonization are frequent enough to say that we continuously observe which patients are colonized and which are not. Let $X(t)$ denote the number of colonized patients (the 'state') at time t, and let $N_{10}(t)$

and $N_{21}(t)$ count the number of jumps from state 0 to state 1 and from 1 to 2, respectively. Further, let $N_{12}(t)$ and $N_{01}(t)$ count the number of jumps from 2 to 1 and 1 to 0, respectively (i.e., the number of discharges when in state 2 and 1, respectively).

We then have four counting processes $N_{10}(t)$, $N_{21}(t)$, $N_{01}(t)$ and $N_{12}(t)$ with jump rates:

$$
\begin{aligned}
\lambda_{10}(t) &= 2\alpha I_{(X(t-)=0)}, & \lambda_{21}(t) &= (\alpha + \beta)I_{(X(t-)=1)} \\
\lambda_{01}(t) &= I_{(X(t-)=1)}/\delta, & \text{and } \lambda_{12}(t) &= 2I_{(X(t-)=2)}/\delta,
\end{aligned}
$$

where for example $I_{(X(t-)=0)} = 1$ if the process is in state 0 just prior to t and 0 otherwise. The reason to have these indicator functions is simply that a specific jump can only occur if the process is in the 'right' state just prior to its occurrence. Note that these rates are exactly those of Figure 4.3 in Section 4.6.1.

By using exactly the same arguments as used when deriving the likelihood for continuous-time data of the 'general' stochastic epidemic, we find that the log-likelihood for our parameters, given time continuous observation of the ICU during the period $[0, t]$ (which contains both $X(s)$ and $N_{ij}(s)$, $0 \leq s \leq t$) equals:

$$
\begin{aligned}
\ell_t(\alpha, \beta, \delta) &= \int_0^t \left(\log(\lambda_{10}(s))dN_{10}(s) + \log(\lambda_{21}(s))dN_{21}(s) + \log(\lambda_{01}(s))dN_{01}(s) \right) \\
&\quad + \int_0^t \log(\lambda_{12}(s))dN_{12}(s) - \int_0^t \left(\lambda_{10}(s) + \lambda_{21}(s) + \lambda_{01}(s) + \lambda_{12}(s) \right) ds.
\end{aligned}
$$

(Recall that $\int_0^t f(s)dN(s) = \sum_{i=1}^k f(t_i)$ if $0 < t_1 < \cdots < t_k < t$ are the jump-times of the counting process $N(s)$.)

In the integrals with respect to the counting processes, which really are sums, there will never be any jumps when the intensity is 0, so $\log(\lambda_{ij}(t))$ need only be defined when the corresponding counting process makes a jump. For instance, $\log(\lambda_{10}(t))$ equals 2α when there is a jump from 0 to 1, so $\int_0^t \log(\lambda_{10}(s))dN_{10}(s) = 2\alpha N_{10}(t)$, i.e., 2α times the number of $0 \rightarrow 1$ jumps. But, $\int_0^t \lambda_{10}(s)ds = 2\alpha T_0(t)$, where $T_0(t) = \int_0^t I_{(X(s-)=0)}ds$ is the time during the interval $(0, t)$ that $X(t)$ is in state 0 ($T_1(t)$ and $T_2(t)$ are defined similarly).

In this manner we find the expression

$$
\begin{aligned}
\ell_t(\alpha, \beta, \delta) &= \log(2\alpha)N_{10}(t) + \log(\alpha + \beta)N_{21}(t) + \log(1/\delta)N_{01}(t) \\
&\quad + \log(2/\delta)N_{12}(t) - 2\alpha T_0(t) - (\alpha + \beta + \frac{1}{\delta})T_1(t) - \frac{2}{\delta}T_2(t)
\end{aligned}
$$

for the log-likelihood, and by straightforward differentiation we obtain:

$$
\begin{aligned}
\frac{\partial \ell_t(\alpha, \beta, \delta)}{\partial \alpha} &= \frac{N_{10}(t)}{\alpha} + \frac{N_{21}(t)}{\alpha + \beta} - 2T_0(t) - T_1(t) \\
\frac{\partial \ell_t(\alpha, \beta, \delta)}{\partial \beta} &= \frac{N_{21}(t)}{\alpha + \beta} - T_1(t) \\
\frac{\partial \ell_t(\alpha, \beta, \delta)}{\partial \delta} &= -\frac{N_{01}(t) + N_{12}(t)}{\delta} + \frac{T_1(t) + 2T_2(t)}{\delta^2}.
\end{aligned}
$$

If we equate these expressions to 0 we get the ML-estimates:

$$
\hat{\alpha} = \frac{N_{10}(t)}{2T_0(t)}, \qquad \hat{\beta} = \frac{N_{21}(t)}{T_1(t)} - \frac{N_{10}(t)}{2T_0(t)}, \qquad \hat{\delta} = \frac{T_1(t) + 2T_2(t)}{N_{01}(t) + N_{12}(t)}.
$$

If we take a closer look at the ML-estimates they are all very sensible. To understand $\hat{\alpha}$ and $\hat{\beta}$ it is probably simplest to see that they satisfy: $2\hat{\alpha}T_0(t) = N_{10}(t)$ and $(\hat{\alpha} + \hat{\beta})T_1(t) = N_{21}(t)$, which means that the estimates are obtained by setting the observed number of jumps equal to its expected value — an application of the method moments, previously encountered in Section 5.3.[6] Concerning $\hat{\delta}$ the numerator equals the accumulated time each colonized patient has spent in the ICU (when $X(t) = 2$, time is counted twice since then two colonized patients are in the ICU). The denominator is simply the number of discharges of colonized patients during the studied interval. As a consequence, $\hat{\delta}$ is simply the observed average time that colonized patients spend in the ICU. This estimator hence agrees with the estimator (5.7), previously derived for ICU duration when colonization was not considered.

5.6   THE CHALLENGE OF REALITY: SUMMARY

The present chapter described methods for making inference about key epidemiological parameters from available data. In Section 5.2 the powerful statistical method called maximum-likelihood (ML) inference was presented and illustrated in the context of a simple transmission model for intensive care units (ICU). This was further developed in Section 5.3 to derive estimators for the parameter length of stay in an ICU. In both these sections the likelihood, and hence the inference procedure, was quite straightforward in that events concerning different individuals were assumed to be independent.

In Section 5.4 we returned to the prototype stochastic epidemic model of Chapter 3 and derived inference methods for key parameters of this model, both for the situation where the epidemic is observed continuously and the situation where only the final size of the outbreak is observed. These inference procedures were more involved and technical due to the underlying dependencies in infectious disease spread: whether or not an individual gets infected is highly dependent on whether or not 'surrounding' individuals get infected. ML-inference methods were extended to stochastic processes (the continuous epidemic process) using so-called martingale methods. Since the likelihood for final-size data is too complicated, two (asymptotically equivalent) estimates of the basic reproduction number R_0 were presented, one using the law of large numbers and the central limit theorem for the final size (presented in Chapter 3), and another using the characterizing property of martingales.

Finally, in Section 5.5 we returned to the ICU situation, but now considering a model with transmission leading to dependencies. Again, we estimated model parameters by ML-inference with the aid of counting processes.

From this short summary of the chapter it is clear that the focus of the chapter has been on *methodology* for making inference from data *given a model*. By 'methodology' we mean that, except for some of the exercises, this chapter (as well as the whole book) is not data-driven. We do not introduce specific sets of data together with focused scientific questions to start off our treatment, but instead we

---

[6] A simple illustration of the moment method is the following. Suppose the mean of a random variable X equals $E(X) = e^\lambda$, where λ is a parameter we want to estimate, and that we observe an outcome x of this random variable. The moment estimator of λ is then simply obtained by equating the mean to the observed value, i.e., by setting $e^\lambda = x$. The moment estimator hence equals $\hat{\lambda} = \log(x)$.

'assume' several possible situations and describe how one would proceed to draw conclusions about parameters of interest *if* such data were available. Secondly, the main focus has been on making inference 'given a model' and also assuming that the data are generated according to this model. We have of course tried to briefly motivate each model, but this has admittedly often been for an idealized situation. In a typical situation that an epidemiologist who has collected data encounters, there are complications that have not been treated in the present chapter. For example, quite often there are shortcomings such as different sorts of bias, large individual variability in key indicators (such as length of infectious period, incubation period or latency period), mismatch between what can be observed and what is needed for inference (such as observing onset of symptoms rather than onset of infection), missing data and/or other uncertainties.

It is far beyond our scope and intentions to try to cover these important problems in a systematic way. Instead we restrict ourselves to mention various types of problems that can occur, and general methods for getting around these, with the hope that the reader shall learn more in other literature, for example in the textbooks on statistics listed at the end of this book.

As for any mathematical/statistical model, epidemic models are always simplifications of the real world they try to mimic. As a consequence, statistical inference (or any other conclusion using models) need not perfectly apply to the real world situation. In each specific situation one has to critically judge the simplifying assumptions and try to understand which deviation from reality might affect the conclusions drawn. The hope is of course that the model studied contains the most relevant features for the problem to be solved, such that the conclusions are approximately true for the real epidemic.

Another important question to ask when making inference, before estimating model parameters from data, is which model to use. Sometimes there is only one 'obvious' candidate, but in many cases there can be several alternative models. A common procedure for selecting a model is to compare likelihoods of different models and to prefer a model with higher (maximum) likelihood, which hence 'fits' better to the available data. In general, the more parameters a model has, the more flexible it is, which implies that models with many parameters will tend to fit data better. A specific example is where a hierarchy of models is considered, where simpler models are special cases of a larger 'encompassing' model, i.e., so-called nested models. Whether or not to prefer a model having more parameters, compared to a sub-model having fewer, can be determined using a likelihood-ratio test that tests the hypothesis that data agree with the smaller sub-model. This method compares the maximized likelihoods of the two models in a way such that the model with more parameters is penalized for this. Other tests for comparing and selecting models are based on Akaike's information criterion (AIC) and the Bayesian information criterion (BIC).

Even when only one model is considered it is of relevance to see how well it explains the observed data. This can for example be done with the R-square (R^2) statistic, which is a measure of how much of the variation in data is explained by the fitted model.

As described above, the level of model complexity to be preferred will depend on how well the various models considered explain the observed data. Two other aspects that will affect the level of model complexity are: the purpose of the analysis, and the level of detail in data. As an example, in the prototype stochastic epidemic model an infectious individual has contact with others at rate c and the probability of disease transmission if the contacted person is susceptible is p. However, unless data contains information about when contacts actually occur (which is extremely rare) there is no hope to estimate c and p separately: they are *unidentifiable* and only their product $\gamma = cp$ might possibly be estimated, thus implying that we have to reduce the number of parameters by one. On the other hand, if one is mainly interested in estimating the basic reproduction number $R_0 = cpE(T)$, there is no need to estimate the parameters separately even if it were possible. Similarly, if the epidemic process is observed continuously (as described in Section 5.4.2) cp and the mean length of the infectious period $E(T)$ may be estimated separately, whereas if only final-size data are available this is not possible and all one can estimate is their product R_0.

Deterministic models are easier to analyze but stochastic models allow for parameter estimates to be equipped with uncertainty measures (e.g., standard errors). For this reason deterministic models are more suited for large communities where uncertainty is negligible. Stochastic models are clearly preferred when analyzing data from smaller communities, but also when analyzing data from large communities when the focus is on features known to exhibit much randomness, for example the initial phase of an outbreak, or outbreaks within small sub-units in the community like households. One should realize that one 'type' of model is not 'better' than the other type. Both approaches need assumptions on many aspects, the fact that a population being studied is small in size is only one of the dimensions of a modeling problem for which choices have to be made. Assumptions on other aspects may influence the outcome much more than the choice between a stochastic or deterministic approach. It then depends on the question being studied and the aim whether one deems population size to be important enough to determine that a stochastic approach is called for.

When analyzing real data, data quality issues might give rise to additional problems. Quite often 'bits and pieces' of information are missing, for example knowledge about prior immunity, vaccination status and also infection status of certain individuals. There is no general method for solving such problems, but one possibility is to 'impute' likely values for such missing observations. One method of doing this lies in the realm of Markov chain Monte Carlo or MCMC (see Section 15.4) where these missing observations are treated as latent variables, and another is the so-called EM-algorithm. The same methodology can be applied to situations where some hypothetical, more detailed, data would have simplified the statistical analysis. Even though this is different from the previous case, where certain observations really were missing, the same type of methods can be adopted by 'pretending' that the more detailed data is missing.

One aspect that is rarely taken into account in models is that behavior of host individuals may change during and due to the outbreak itself: once people start to get infected, others will start taking precautions, and the more serious the symptoms, the earlier such changing behavior starts. If a model without behavior change is fitted to data from an outbreak in which individuals have started to protect themselves then this can lead to very misleading conclusions.

To conclude, modeling and analyzing epidemics can help in increasing the understanding of epidemic outbreaks. Still, there are always aspects not considered in the model that might make conclusions drawn from models questionable. It is hence important to have a critical mind when analyzing real world epidemics. In Part III of the book we will return to inference methodology for epidemic outbreaks.

Part II

Structured populations

Chapter Six

The concept of state

6.1 i-STATES

The basic idea of dynamic structured population models is to distinguish individuals from one another according to characteristics that determine the birth, death and resource consumption rates — more generally, the interaction with the environment — and to describe the rates with which an individual's characteristics themselves change. Since we are mainly interested in infectious diseases, we limit ourselves to those characteristics that influence the force of infection of a given infectious agent, i.e., those traits that influence the rate with which susceptible individuals become infected (encompassing both infectivity, susceptibility and contact pattern).

The first step in building a structured population model in order to investigate a concrete question is to choose those characteristics that are deemed relevant to the problem of interest. In mathematical jargon, this is called choosing the i-state, where 'i' denotes 'individual.' The i-state of an individual at some point in time t is therefore the set of values for the chosen traits for that individual at time t. For example, if we choose age as the only relevant characteristic then the i-state of an individual at any time t is simply the individual's age at t. If the individual is born at some time t_0 then its age at some time $t > t_0$ will be $t - t_0$. Another example is to take sex, partnership status (i.e., single or with steady partner) and age (in discrete classes) simultaneously as individual characteristics. A possible value of this i-state is (male, single, 34).

The concept of 'state' has a more fundamental content than just any collection of individual characteristics. Informally speaking, the *state* of a system — in our case an individual — is the set of precisely that information about the system relevant to predict the system's future development/behavior. In the context of epidemic models, the future 'behavior' we wish to characterize often encompasses the expected infectious output of an infected individual as a function of its i-state; we wish to determine how infectious this individual is to others depending on its characteristics. The state could change according to, for example, a collection of laws describing the individual's life, interaction with other individuals, the transmission of infection, the time-course of the disease inside the individual (if infected), or changes in a broadly defined environment.

It is clear from the first example that the following can hold for an individual's i-state. Given the state, say $x(t_0)$, at some time t_0, the state at time $t_1 + t_0$ is determined by

$$x(t_1 + t_0) = \mathcal{T}(t_1)x(t_0).$$

Here $\mathcal{T}(t_1)$ denotes an operator that maps states into states and that has the semi-group property

$$\begin{aligned} \mathcal{T}(0) &= I, \\ \mathcal{T}(t_2)\mathcal{T}(t_1) &= \mathcal{T}(t_2 + t_1), \quad t_2, t_1 \geq 0. \end{aligned}$$

In words, states have the following property. If we start at time 0 and want to know the state at some future time $t_2 + t_1$, it does not matter whether we go from 0 to $t_2 + t_1$ immediately, $x(t_2 + t_1) = \mathcal{T}(t_2 + t_1)x(0)$, or whether we take the intermediate step of first going from 0 to t_1, and then taking $x(t_1)$ as our new starting value to go time t_2 further: $x(t_2 + t_1) = \mathcal{T}(t_2)\mathcal{T}(t_1)x(0)$.

So, whatever the dynamics in the time interval $[0, t_1]$, if we want to predict the future after t_1, we need only look at the state of the system at time t_1 and can disregard the precise history of the evolution between 0 and t_1. In a way, the state at t_1 carries with it a 'memory' of what happened to the system before t_1, at least of those aspects that are relevant to predicting the future course of the system.

In the context of epidemic models, choosing the i-state ingredients is a double task since we have to deal with both population dynamics and infection transmission, and we shall accordingly refer to those components of the i-state that describe the development of the infection within an individual as the d-state — where 'd' denotes 'disease.' (It would be more appropriate to use the 'i' of 'infection,' as not every infected individual develops (clinical) disease, but we have already used 'i' to denote 'individual.') More precisely, the d-state is that part of the i-state that describes the difference between infected and susceptible individuals. From the point of view of the infectious agent, the rest of the i-state reflects heterogeneity in the population. Accordingly, we will call this part of the i-state the h-state. We will usually call h-state values *types*.

In the presentation above, we have concentrated on so-called 'autonomous' systems, characterized by the absence of time-dependent input. This allows us to work with time as a relative quantity ('the time elapsed since . . .'). If, for instance, seasonal weather conditions do have an impact on the success of transmission or if contact patterns are time-dependent (think of the school system), we should take absolute time into account. This then is an additional technical complication in the bookkeeping, but it does not in any way influence the state concept. The state should contain all information relevant for predicting the future development/behavior, given the environmental input in the intervening period.

6.1.1   d-states

The d-state determines the infectious output of an infected individual. The course of an individual infection is a stochastic process, reflecting among other things the status of the immune system and its 'battle' with the infectious agent. We would like to avoid modeling this complex process in detail. From a system-theoretic point of view, there are but two kinds of d-state: those where the input is a unique event and those where the input is a repeated process.

- *infection age:* Suppose that after infection the disease develops as an autonomous process within the infected individual, and that super-infections (re-infections of an already-infected individual) therefore play no role. We have in mind that the invading organism reproduces within the host at such a rate that further infections with the same agent can be neglected; examples include measles, influenza, rabies and HIV, and fall under the general heading of *micro*parasites. Usually they are viruses or bacteria, hence the name. Irrespective of the precise biological-chemical-physical interpretation of the d-state we can then describe disease-progress — and with it, morbidity, mortality and infectious output — by an infection-age representation (d-age).

We act as if a clock starts ticking the very moment the individual becomes infected. It is convenient in this respect to refer to the h-state at d-age zero as the *state at infection* (or *state at birth*) of the individual — birth meaning recruitment into the infected population. See Section 7.1 for a more detailed exposition.

We will usually denote infection age by the variable τ. In Section 2.1 we gave examples of how to compute the expected infectivity $A(\tau)$ from sub-models for the dynamics of an underlying d-state that can assume finitely many 'values.' See in particular expression (2.3).

- *Infection degree:* Here infection is not a unique event but rather a repeated process. Examples are schistosomiasis and other diseases caused by helminths and they fall under the general heading of *macro*parasites. As a rule, the parasites within a host in this class can be counted, in contrast to microparasites. Sexual reproduction or cloning of a particular life stage of the parasite within the host is allowed to take place, but the full life cycle of the parasite usually involves one or more stages outside the host (possibly in another host species). Since in these cases the morbidity, mortality and infectious output of an infected host depend on the level of infection, the d-state is represented by the number of parasites a host harbors (infection degree or d-degree).

While these seem to be exclusive categories, there is at least one important class of infectious agents, the protozoan parasites (among these are the causal agents of malaria), that belong to both. Superficially speaking, protozoan infections would belong to the first category because protozoans in relevant stages of the life cycle multiply very rapidly within the host. However, the phenomenon of acquired immunity occurs in many protozoan infections. The more additional infections with the parasite (possibly different strains of the same species) that an individual acquires, the higher its level of immunity will rise (leaving aside intricacies that concern the required length of the time period between successive infections). The immunity usually does not protect against re-infection, but individuals with a high level of immunity do not experience the severe disease symptoms. Immune individuals can still be infectious to others for a number of protozoan infections, but usually this infectivity is much reduced compared to that of non-immunes. This phenomenon of acquired immunity through superinfection places the protozoan infections in the second category.

In most of what follows we restrict ourselves to systems allowing for infection *age* as the state representation. Models incorporating infection *degree* will be treated briefly in Chapter 11.

There is a type of distinction in transmission opportunity for which both of the above descriptions of d-state are inadequate, or, more precisely, incomplete. This concerns sexually transmitted infections, for which one should often explicitly take the formation of long-term monogamous partnerships into account. In that case a susceptible can only become infected if its current partner is infected and an infected partner can cause at most one new infection as long as no new partnership is initiated. As a consequence, all contacts between the partners are 'wasted' from the point of view of the infectious agent, once both are infected. Potentially, the infected individual would be capable of causing more infections with the same infectious output. Because of the contact structure, however, these infections are not realized. This implies that the description with d-age, which would be appropriate in the case of, for example, random contacts, is not a good indicator of transmission ability. What we have to take into account is the survival of the partner of the infected individual. Only when the partner dies, or the partnership is dissolved for other reasons, can the infected individual cause new infections. We cannot describe this situation by simply looking at the d-age of the infected, since that quantity does not describe the status of the partner. In Section 7.8 we will show how our general methodology for calculating R_0 easily extends to cover this kind of situation, involving 'super-individuals,' viz., pairs of individuals; see also 6.1.3.

6.1.2   h-states

We now turn to the possible h-states. Characteristics on which these are based can, for a given individual, be *static* (like sex, genetic composition) or *dynamic* (suffering from another disease, stage of development), and they can take *discrete* values (like sexual orientation, partnership status) or *continuous* values (like spatial position of a plant, or age). In particular cases h-states can be very complicated, and contain simultaneously continuous and discrete, static and dynamic components. We will denote by Ω the state space of all possible values of the chosen set of characteristics (i.e., the h-state space).

If the h-component of the i-state has more than one possible value, and the d-state has a d-age representation, we have to take into account that the expected infectivity function A may depend on both the h-state of the susceptible and the h-state of the infected individual taking part in a contact. The major modeling effort involved in addressing a question pertaining to a specific host/infection system is to make precise how A depends on these states. The dependence of A of a given infected individual on the h-state of susceptibles is primarily through the frequency of contacts (for example, if the h-state denotes sex, we could specify different contact rates for homosexual and heterosexual contacts). In Chapter 7, we give one method to obtain A expressed in the parameters (of sub-models) that govern the changes in individual state, the infection transmission and contact pattern, illustrated by a number of extended examples. As is typical for structured population models, the process of obtaining A involves detailed stochastic modeling of events at the individual level (fortunately, on this level one often has possibilities to experimentally measure or estimate parameters). By assuming that our population consists of many individuals, we can then invoke a law-of-large-numbers argument that allows us to describe the changes at the population level by deterministic equations (recall Section 3.4). The theory of Markov chains is tailor-made to compute A from sub-models. The Markov property states that the conditional probability of a given event only depends on the present state of the system and not on the manner in which the present state was reached. This is exactly the property that characterizes i-states.

6.1.3   Various forms of heterogeneity

In the exposition above, our focus was on traits that characterize individuals, i.e., traits that can be described by labeling the individuals. If N denotes the total number of individuals, then one assumes that for $N \to \infty$ the various types of individuals occur in fixed proportions, and that accordingly the population is described by a frequency distribution over the type space (see the next section). Much of Chapter 7 is devoted to analyzing the consequences for defining and computing R_0.

The nature of heterogeneity is manifold and there are several complications. Indeed, often we observe that the frequency of contacts between two individuals depends heavily on whether or not a certain *relationship* exists between the two (such as being a neighbor, belonging to the same family, going to the same school). When $N \to \infty$ the number of individuals with whom a specific individual has such a relationship may very well stay fixed. In that case therefore, the proportion decreases with increasing N, instead of remaining fixed.

When the relationship is an *equivalence* relationship (meaning that i) $x \sim x$; ii) $x \sim y$ implies $y \sim x$ (symmetry); iii) $x \sim y$ and $y \sim z$ imply $x \sim z$ (transitivity); note that in our context the first is rather irrelevant), we can consider the equivalence class as a *super-individual*, and consider transmission at the 'super' level (we present an example in Section 7.4.2 below). Concrete examples include households and, in the veterinary context, herds (or farms) (see for instance Section 12.5).

A further complication may arise if the relationship is *dynamic*. This is highly relevant for STDs that spread in a population in which the majority of the sexual contacts occur between couples that maintain a stable relationship. In Section 7.8 we shall show how to use pair formation (and dissociation) models as a basis for computing R_0.

The distribution over space creates a relationship between individuals that is symmetric but not transitive (my two next-door neighbors find my door in between theirs). In Chapter 10 we investigate some of the consequences of spatial structure in the context of continuum models (as opposed to lattice models, cf. Section 10.7). In Section 12.5 we regard contact heterogeneity by looking at individuals socially structured into households. In Section 12.6 we consider less regular structures that are described by *networks*. In 12.7 we look at a rudimentary form of association between individuals, somewhere 'in between' mass-action contacts and networks, and briefly discuss so-called 'pair approximation.'

Among the many forms of heterogeneity that we do not discuss is heterogeneity created by the infection process itself. Even when 'identical' individuals, for example farm animals, are exposed to an infection, the 'exposure history' of individuals will differ, with some individuals receiving more often or earlier (or higher doses of) infection than others. When the infectious agent induces immunity only upon repeated exposure, the variation in exposure in space and time will then create heterogeneity in response to infection, including heterogeneity in infectivity.

The upshot of this brief discussion is that it is still a major challenge to incorporate dynamic non-equivalence relationships into epidemic models, and that most (but not all) of the material in the rest of this book is concerned with the heterogeneity of individuals, rather than the heterogeneity of their mutual relationships.

6.2   p-STATES

The population state (p-state) is nothing more than the *distribution* of individuals over the i-state space. Changes on the individual level give rise to changes in the composition of the population. The equation for p-state change is obtained basically by bookkeeping, once the dynamics of i-states have been described. We do not go into this issue here, but refer for general background reading on structured population models to Metz and Diekmann (1986), Tuljapurkar and Caswell (1997), Cushing (1998), Diekmann et al. (1998) and Diekmann et al. (2010).[1]

In view of the next chapter, we now focus on the h-state distribution when explaining some concepts and some notation.

When there are finitely many h-states and finitely many individuals, it is simple what we mean by 'distribution': for each h-state we specify the number of individuals that happen to have that state. Or, alternatively, the fraction of the population with that state (in which case we complement the information by one additional number, the total population size).

When we talk about people aged 67, we mean those people who were born between 67 and 68 years ago. True age, in the sense of 'time elapsed since birth,' is actually a continuous variable that can take uncountably many different values. But we are used to subdividing the age axis in one-year intervals and to count accordingly. Other subdivisions are conceivable (and sometimes useful) and they can be taken as a basis for counting.

The object that assigns to every reasonable subset of the age axis the number/fraction of individuals that have their 'true age' in that subset is called a *measure* (the mathematical jargon for 'reasonable' is 'measurable,' and there is a precise technical definition for it). We shall, as much as we can, denote measures by the letter m.

When the measure m is used to describe a population age distribution and ω is an age interval (or union of such intervals, or just a measurable subset) then $m(\omega)$ is the number/fraction of individuals with age in ω. We also write

$$m(\omega) = \int_\omega m(da)$$

where the notation following the integral symbolizes the (mathematical idealization of the) process of adding/counting all individuals with age in ω.

If, for instance, individuals with age a have contact intensity $c(a)$ then the mean contact intensity in the population is given by

$$\bar{c} := \frac{\int_\Omega c(a)m(da)}{\int_\Omega m(da)},$$

where $\Omega = [0, \infty)$. Whenever $m(\Omega) = \int_\Omega m(da) = 1$, we call m a *probability measure*, and we can omit the denominator. Such is the case when m describes fractions. The terminology refers to the fact that the probability that a randomly chosen individual belongs to some specified subgroup equals the fraction of the total population that that subgroup constitutes. The numerator is called the integral of c with respect to the measure m. The operation as a whole corresponds precisely to giving individuals of age a the weight $c(a)$ and then computing the average.

Hidden in the notation are certain mathematical subtleties related to the fact that $[0, \infty)$ is uncountable. For instance, imagine entering a classroom and try to guess what the probability is that one of the students will have true age 19 years, 3 months, 1 week, 2 days, 5 hours, 10 minutes, 57 seconds, 817 milliseconds, ...

---

[1]O. Diekmann, M. Gyllenberg, J.A.J. Metz & H.R. Thieme: On the formulation and analysis of general structured population models. I: Linear theory. *J. Math. Biol.*, **36** (1998), 349–388; O. Diekmann, M. Gyllenberg, J.A.J. Metz, S. Nakaoka and A.M. de Roos: Daphnia revisited: local stability and bifurcation theory for physiologically structured population models explained by way of an example. *J. Math. Biol.*, **61** (2010), 277–318.

Probably you will quickly conclude that already without the '...' being specified, the probability will be low, and that it will be zero if we could specify 'age' with infinite precision. Thus the theory of measure and integration constitutes a highly non-trivial mathematical challenge. In this book, however, all we need is the small conceptual and notational part sketched here.

Sometimes the integral with respect to a measure boils down to the integral that is familiar from calculus. We say that the measure m has a *density* n when, symbolically, $m(da) = n(a)da$, by which we mean that, for all (measurable) ω,

$$m(\omega) = \int_\omega m(da) = \int_\omega n(a)\,da.$$

For instance, when there is a constant population birth rate b, and of all newborns a fraction $\mathcal{F}(a)$ survives until at least age a, then the above holds with

$$n(a) = b\mathcal{F}(a).$$

The point is that the density gives complete information about the measure and, because it is a function for which one can draw a graph, it is a more familiar object. There are, however, at least two points in favor of measures:

- Not all measures have a density.

- Densities are not directly interpretable in terms of numbers, which makes it more dangerous to employ intuitive arguments based on the interpretation.

Returning from age to general h-state, we add two final remarks. We always assume that Ω is measurable, i.e., that one can define measures on Ω. Certainly this is the case when Ω is a nice subset of $\mathbb{R}^k$ for some $k \geq 1$. In that case, one can also introduce the notion of density on the basis of the standard n-dimensional integral. Just as in the case of age ($k = 1$), one requires that, for all ω,

$$\int_\omega m(d\xi) = \int_\omega n(\xi)\,d\xi.$$

For an introduction to the theory of measures see e.g., Kolmogorov and Fomin (1975) or Rudin (1974).

6.3   RECAPITULATION, PROBLEM FORMULATION AND OUTLOOK

Suppose individuals differ from each other with respect to traits that are relevant for the transmission of an infectious agent. How do we describe the spread of the agent? How do we quantify the infectivity? What happens in the initial phase? Can we characterize the final size?

Examples of the 'traits' we have in mind are age, sex, sexual activity level, sexual disposition and spatial position. So a trait may be *static* or *dynamic*, it may be *discrete* or *continuous*. Often the modeler's subjective and pragmatic striving for manageable problems will suggest to take a trait such as 'sexual disposition' as static and discrete, while one may rightfully wonder whether reality isn't more polymorphic and changeable than that.

Despite such doubts, we shall consider the traits as i-states, where 'i' means 'individual' and where 'state' signifies that the current value together with the environmental input in the intervening period completely determines future behavior (although probably in a stochastic sense).

Thus we classify the heterogeneity of individuals in terms of a component, h-state, of their i-state, while the other component, d-state, summarizes all relevant information about output of infectious material. A population of such individuals is no longer characterized by one number, the population size. We need, in addition, to know the composition of the population, i.e., the distribution with respect to i-state. We are dealing with a *structured* population.

Our approach is *top down*. This means that we start with abstract general principles and then gradually become more concrete by being more specific and quantitative. Some readers might conclude that we stay abstract throughout the book, but that is partly a matter of taste and background. We shall sketch computational schemes, but not perform actual computations.

The next chapter concentrates on the definition and the computation of the basic reproduction number R_0 in the context of structured population models. In Chapter 8 we deal much more briefly with the real-time growth rate r, the final size of an epidemic and the probability of a minor outbreak. In particular, we show how the assumption of *separable mixing* allows one to reduce all computations to the situation where there is only a one-dimensional unknown. In Chapter 9 we elaborate on the special case of age-structure, while in Chapter 10 we concentrate on the *asymptotic speed of propagation* c_0 as an important indicator of the spread at the population level in a spatially structured population. Both age and spatial position are determinants of contact structure. Transmission is superimposed on the dynamic contact network that one may use to represent the host population. In Chapter 12 we try to gather together several of the kaleidoscopic facets of contact structure that are most relevant in the context of infectious diseases. In between, in Chapter 11, we present models where 'structure' derives from parasite load, relevant for parasitic worms for example.

Chapter Seven

The basic reproduction number

7.1 THE DEFINITION OF R_0

The basic reproduction number (or ratio) R_0 is arguably the most important quantity in infectious disease epidemiology. It is among the quantities most urgently estimated for infectious diseases in outbreak situations, and its value provides insight when designing control interventions for established infections. From a theoretical point of view R_0 plays a vital role in the analysis of, and consequent insight from, infectious disease models. There is hardly a paper on dynamic epidemiological models in the literature where R_0 does not play a role. R_0 is defined as the average number of new cases of an infection caused by one typical infected individual, in a population consisting of susceptibles only. The meaning of the word 'typical,' which is there to emphasize the subtlety that the word 'average' needs to be interpreted in the right way, will be explained below. The aim of the present chapter is to formally show how R_0 can be characterized mathematically, and to provide detailed examples of its calculation in terms of parameters of epidemiological models, culminating in a set of algorithms (or 'recipes') for the calculation for compartmental epidemic systems. We postpone to Chapter 13 a review of methods to estimate a numerical value for R_0 from data.

In epidemic models, individuals can typically be in a number of different states, reflecting both differences in traits and differences in infection status or stage. Some of these states are simply labels that specify the various traits of individuals. Of these, some will be changing with time, such as age class, and others will be fixed, such as sex or species. Other states indicate the progress of an infection: for example, an individual can, upon becoming infected, typically first enter a state of latency, then progress to a state of infectiousness, and then lose infected status to progress to a recovered/immune state. We have attempted to bring some structure into thinking about these states in Chapter 6, referring to individual traits as the h-state or type of an individual, and to infection status as d-state. From the states that apply to infected individuals, we now single out those states that individuals can be in *immediately* after they have been infected. We call such a state a *state-at-infection* (and sometimes 'state-at-birth'). They play a special role in the definition and calculation of R_0. We will introduce the mathematical definition of R_0 first, and following that show how to identify the states-at-infection from among the total set of states in a given epidemiological system.

We start by considering the situation in which there are only finitely many states-at-infection $1, 2, \ldots, n$. Characterizing R_0 relates to addressing the 'introduction' issue: given a population in demographic steady state, with no history of a given infection, will the introduction of the infectious agent cause an outbreak? We adopt a deterministic point of view — that is, we only consider expected values — and we linearize, that is, we neglect that the agent itself diminishes the availability of susceptibles. The assumption of a demographic steady state means that the agent experiences the world as constant, i.e., environmental conditions do not vary. We will show that R_0 is mathematically characterized by regarding infection transmission as a 'demographic process'; where producing offspring is not seen as giving birth in the demographic sense, but as causing a new infection through transmission (we will refer to this as an *epidemiological birth*). In a natural way this leads to viewing the infection process in terms of consecutive 'generations of infected individuals,' in complete analogy to demographic generations. Subsequent generations growing in size then reflect a growing population (i.e., an epidemic), and the growth factor per generation indicates the potential for growth. In a natural way this growth factor is then the mathematical characterization of R_0. In the spirit of the demographic analogy we will call a state-at-infection also a *state-at-birth*, where 'birth' is interpreted in the epidemiological sense described above: the individual is 'born' from an epidemiological point of view with that state.

Define k_{ij} to be the expected number of new cases that have state-at-infection i, caused by one individual with state-at-infection j, during its entire period of infectiousness, where all contacts of this individual are with susceptibles. With this definition of the k_{ij}, we obtain n^2 non-negative numbers. The question is how these numbers should be averaged. More precisely, we look for a single summarizing number that has two desired properties: i) the introduction of the infectious agent succeeds if and only if the number is greater than 1, and ii) the number has a biological interpretation similar to that in the homogeneous case.

Example 7.1 (cf. Section 2.2) Consider the matrix

$$K = (k_{ij}) = \begin{pmatrix} 0 & 100 \\ 10 & 0 \end{pmatrix},$$

which could, for example, relate to a heterosexually transmitted infection with female and male as states-at-infection (or to host-vector transmission). An individual with state at infection 1 produces, on average, 10 new cases, all with state-at-infection 2. Each individual with state-at-infection 2 produces, on average, 100 new cases, all with state-at-infection 1. The multiplication factors 10 and 100 alternate because, starting from, for example, a female (host), the infectious agent has to 'pass through' a male (vector) before it can enter a new female (host). In other words, it takes two generations to get back to the same type, and every two generations numbers are multiplied by $10 \times 100 = 1000$. The average per generation multiplication factor is therefore $\sqrt{1000}$. How do we arrive at such a number for general matrices? How does one measure the 'size' of a matrix with non-negative entries?

We regard *generations* of infected individuals, described by vectors (note that now we use 'vector' in the mathematical sense, whereas above we used the same word to denote a biological carrier; the etymology is the same). The jth component of the vector equals, by definition, the number of cases with state-at-infection j, in that particular generation. The vector describing the next generation is obtained from the vector describing the current generation by applying the matrix K to it, as in the example above:

$$\phi_i^{\text{new}} = \sum_{j=1}^{n} k_{ij} \phi_j^{\text{old}} \tag{7.1}$$

being the number of susceptibles with state i that are infected by infectives with state j, summed over all states j. In short,

$$\phi^{\text{new}} = K\phi^{\text{old}}. \tag{7.2}$$

Let us use a superscript to number the generations and write

$$\phi^{m+1} = K\phi^m. \tag{7.3}$$

We call K the *next-generation matrix*, and will frequently abbreviate the name as 'the NGM' in our exposition.

We observe that the generation process is described by iteratively applying K. If we use a superscript for a matrix to denote powers with respect to matrix multiplication, e.g., $K^2 = KK$, we can write

$$\phi^m = K\phi^{m-1} = K^2\phi^{m-2} = \cdots = K^m\phi^0, \tag{7.4}$$

where ϕ^0 is the vector describing the generation that starts the process. We note that K is a *positive* matrix in that all its elements are necessarily non-negative (we also use 'non-negative matrix' at times to denote the same property). We use the shorthand notation $K \geq 0$, to indicate that all elements $k_{ij} \geq 0$, $1 \leq i,j \leq n$. Likewise, we speak of positive vectors when all components are non-negative and write $\phi \geq 0$.

Exercise 7.2 Let K be a 2×2 matrix that has two distinct real eigenvalues λ_1 and λ_2, with $\lambda_1 > 0$ and $\lambda_1 > |\lambda_2|$. Let $\psi^{(1)}$ and $\psi^{(2)}$ be the associated (right) eigenvectors. Then any vector $x \in \mathbb{R}^2$ can be written as a linear combination of these eigenvectors: $x = c_1\psi^{(1)} + c_2\psi^{(2)}$.

i) Express $K^m x$ as a combination of the eigenvectors.

ii) By rewriting the expression for $K^m x$, conclude that, after many generations, the influence of the $\psi^{(2)}$ component on the growth of $K^m x$ will become negligible and that we have asymptotically for the number of generations $m \to \infty$

$$K^m x \sim \lambda_1^m c_1 \psi^{(1)}. \tag{7.5}$$

Where is the influence of generation zero expressed? Conclude that λ_1 has the desired threshold property to determine growth or decline (on a generation basis) of the infective population.

In the present context a natural *norm* for vectors, to measure their 'size,' is

$$\| \phi \| = \sum_{j=1}^n |\phi_j|, \tag{7.6}$$

i.e., $\| \phi \|$ is the total number of cases in the generation that is described by ϕ (note that the absolute value has no impact in our setting of positive vectors). Now that a norm for vectors has been specified, we can characterize the 'size' of matrices by

$$\| K \| = \sup_{\|\phi\|\neq 0} \frac{\| K\phi \|}{\| \phi \|} = \sup_{\|\phi\|=1} \| K\phi \|. \tag{7.7}$$

In the epidemiological setting we can interpret $\| K \|$ as the maximum multiplication number for the total number of cases, when we allow the distribution with respect to state-at-infection to take all possible forms. For the matrix in our Example 7.1 we find $\| K \| = 100$.

The point is now that we can choose ϕ^0 quite arbitrarily, but that subsequently the distribution will be determined by the dynamics itself. In Example 7.1, if $\phi^0 = (0,1)^\top$ (where $\top$ denotes 'transpose') then $\phi^1 = (100,0)^\top$ and indeed we find the number of cases multiplied by 100. In the next generation, however, the multiplication factor is only 10. Consequently, the norm of K is too coarse a measure for generation growth. This motivates us to look at multiplication in m generations, but on a 'per generation' basis. In other words, we look at the power $1/m$ of the growth in m generations:

$$\| K^m \|^{1/m} . \tag{7.8}$$

In general, we have to be patient and consider the limit for $m \to \infty$ of this quantity. (Which exists! This requires proof of course, but is not really difficult.) This limit is, for reasons explained below, called the *spectral radius* $\rho(K)$ of K. As it is by definition the long-term average per generation multiplication number, we arrive at the requested mathematical characterization of R_0:

$$R_0 = \rho(K) := \lim_{m \to \infty} \| K^m \|^{1/m} \tag{7.9}$$

where K is the NGM.

This may be a mathematically natural definition of R_0, but should we also use this as an algorithm to compute it? Or are there other characterizations that lend themselves more readily to computation? The answer to the latter question is yes. We can in addition obtain more detailed information about the (linearized) generation process, in particular about the asymptotic behavior for $m \to \infty$. Here the fact that $K \geq 0$ plays an important part.

Theorem 7.3 *Let $K \geq 0$. Then the spectral radius R_0 is an eigenvalue of K, which we call the* dominant eigenvalue *since $|\lambda| \leq R_0$ for all other eigenvalues λ of K.*

We next assume that R_0 is *strictly* dominant in the sense that $|\lambda| < R_0$ for all other eigenvalues λ of K and that R_0 is an algebraically simple eigenvalue (i.e., there is only one factor $\lambda - R_0$ in the characteristic equation $\det(\lambda I - K) = 0$). The eigenvector ψ^d ('d' denotes 'dominant') corresponding to R_0 can be chosen in such a way that all its components are non-negative and their sum equals one (we then call ψ^d normalized). Then one can prove, along the lines of Exercise 7.2, that for any initial vector ϕ^0 one has

$$K^m \phi^0 = c(\phi^0) R_0^m \psi^d + o(R_0^m) \quad \text{for } m \to \infty. \tag{7.10}$$

Here $o(R_0^m)$ summarizes the transient behavior in a crude form, it signifies that $[K^m \phi^0 - c(\phi^0) R_0^m \psi^d]/R_0^m \to 0$ as $m \to \infty$.

The dominant eigenvector ψ^d (when normalized) describes the *stable distribution*. This terminology expresses two properties: invariance and attraction. We now elucidate both of these. If we take $\phi^0 = \psi^d$, we have exactly

$$K^m \psi^d = R_0^m \psi^d, \tag{7.11}$$

or, in words, the distribution remains unchanged from generation to generation while numbers are multiplied by R_0. For general initial-generation vectors the situation will more and more resemble this special case in the course of the generations.

Only one aspect of the initial situation, the scalar quantity $c(\phi^0)$, remains manifest. There exists a slightly technical procedure to compute $c(\phi^0)$ from a given ϕ^0, and for the interpretation this procedure does not matter much.[1]

So when R_0 is strictly dominant and algebraically simple, the situation is very clear (if ϕ^0 is such that $c(\phi^0) \neq 0$): if we apply the next-generation matrix K repeatedly, the (normalized) distribution converges to the stable distribution ψ^d and the per-generation multiplication number converges to R_0. The (normalized) stable distribution can be interpreted as the probability distribution for state-at-infection.

Can we determine whether or not R_0 is strictly dominant and algebraically simple without computing all eigenvalues? The following collection of definitions and results shows that indeed we can. Together with Theorem 7.3 above, this is collectively called *Perron-Frobenius theory* of positive matrices (see e.g., Minc 1988).

Definition 7.4 *The non-negative matrix K is called* irreducible *if for every index pair i, j there exists an integer $m = m(i, j) > 0$ such that $(K^m)_{ij} > 0$. And K is called* primitive, *or* aperiodic, *if one can choose one m for all i, j, that is, if there exists m such that K^m has all its entries strictly positive.*

One alternative definition of irreducible, and a means to facilitate checking whether it is satisfied for a given matrix, centers around a directed graph representation of the matrix: for an n-dimensional matrix, the graph consists of n nodes, with a directed edge from i to j if the entry $K_{ij} > 0$. Irreducible means that it is possible to go from any node to any other node, possibly in several steps. For primitivity one can also use the graph representation, but there the emphasis is on loops in the graph. For primitivity the greatest common divisor of the loop lengths in the graph should be 1.

Exercise 7.5 i) Start with one case that has state-at-infection j. Let K be irreducible. Show that eventually there will be cases with state-at-infection k, no matter what combination of k and j we consider. What do we mean by 'eventually'? And how could we interpret 'eventually' when K would be primitive?

ii) The property discussed in i) can serve to define 'irreducible' and 'primitive'. Do you agree?

iii) Consider the example

$$K = \begin{pmatrix} 0 & 100 \\ 10 & 0 \end{pmatrix}.$$

Is K irreducible? Is K primitive? Compute $\| K^m \|^{1/m}$ for $m = 1, 2, \ldots, 10$.

iv) Consider a sexually transmitted disease in a population composed of homosexual and heterosexual males and females. Argue that in the absence of bisexuality we have a reducible situation. What does 'reducible' mean?

Theorem 7.6 *Let K be primitive. Then:*

---

[1] For completeness, we describe it. Let ψ^{d*} denote an eigenvector of the transposed matrix $K^\top$ corresponding to the eigenvalue R_0. Normalize it such that $\psi^{d*} \cdot \psi^d = \sum_{j=1}^{n} (\psi^{d*})_j (\psi^d)_j = 1$. Then $c(\phi^0) = \sum_{j=1}^{n} (\psi^{d*})_j \phi_j^0$.

- R_0 *is strictly dominant.*

- ψ^d *and* ψ^{d*} *have strictly positive components (and as a consequence* $c(\phi^0) > 0$ *for any non-trivial positive* ϕ^0*).*

- R_0 *is an algebraically simple eigenvalue.*

- *No other eigenvalue has a positive eigenvector.*

When K is merely irreducible, the last three properties still hold, but on the circle of radius R_0 in the complex plane there are other eigenvalues. These have to be roots of the equation $\lambda^m = R_0^m$ for some m, i.e., they have to be of the form $\lambda = R_0 \exp(i\frac{2\pi l}{m})$, $l = 1, \ldots, m-1$.

Exercise 7.7 Consider the situation described in Section 2.2, but now assume that a fraction ρ_1 of the contacts of 1-individuals is with 1-individuals and a fraction ρ_2 of the contacts of 2-individuals is with 2-individuals. Consistency requires that $(1 - \rho_1)c_1 N_1 = (1 - \rho_2)c_2 N_2$, so for given c_i and N_i either ρ_1 or ρ_2 is a free parameter, but not both. For which combination of ρ_1, ρ_2 is R_0 maximal, respectively minimal?

7.2 NEXT-GENERATION MATRIX FOR COMPARTMENTAL SYSTEMS

In this section we restrict ourselves to the most frequently used class of models in the literature: the compartmental models. We add this section because of the current predominance of such models, but at the same time hope that the book convinces the reader that many issues can be addressed equally well in terms of far more flexible models, at hardly any cost in technical difficulty. Characterizing R_0 is one of these issues.

In the previous section we have presented the definition of R_0, and shown that a generation-based approach naturally leads to a characterization that lifts the desired properties from homogenous to heterogeneous systems. Before moving to greater generality, and after that proceeding with examples and specific situations, we take a step back and approach the problem starting from a real-time description of the evolution of the infected sub-populations. In this section, we explain how the generation description of the evolution of infected individuals is related to the real-time description, and we show that three related types of next-generation matrix exist, one of which is the NGM defined in Section 7.1.[2]

In compartmental models, individuals can be in a finite number of discrete states. With each state one can associate the sub-population of individuals that are in that particular state at the given time (e.g., a female in a latent state of infection). Often the same symbol is used as a label for a state and to denote the corresponding sub-population size, either as a fraction or as a number (e.g., I or Y for individuals in an infectious state). The dynamics are generated by a system of nonlinear ODEs that describes the change with time for all subpopulation sizes.

---

[2]This exposition is based on the paper by O. Diekmann, J.A.P. Heesterbeek and M.G. Roberts: The construction of next-generation matrices for compartmental epidemic systems. *J. R. Soc. Interface*, **7** (2010), 873–885. This paper also gives algorithms for the construction of the three types of next-generation matrix that can be associated with compartmental systems.

We have seen in Section 7.1 that the definition of R_0 as the spectral radius of the NGM K, is only based on a subset of all possible states in the system. It is based on those states that can be classified as a state-at-infection, i.e., those states that an individual can have *immediately* after becoming infected. Typically, there will be (many) more infected states than states-at-infection. As an example consider the standard SEIR model (recall Exercise 2.2). There are two states for infected individuals, the latency state E and the infectious state I. Only the E-state is a state-at-infection, however, because all newly infected individuals start their 'infected life' in state E. One cannot be in the I state immediately after becoming infected, but can only enter state I in the course of the infection. Hence the NGM K is a one-dimensional matrix in this example. One could, however, also envisage a description of the generation process based on both infected states, leading to a two-dimensional matrix. This is a general principle and we will show below how these matrices are related. We will start by considering the larger subset of states consisting of all the infected states.

To calculate R_0 one begins with those equations of the ODE system that describe the production of new infections and changes in state among infected individuals. We will refer to the set of such equations as the *infected (or infection) subsystem*. The first step is to linearize the infected subsystem of nonlinear ODEs about the infection-free steady state that, as a rule, exists. Epidemiologically the linearization reflects that R_0 characterizes the potential for initial spread of an infectious agent when it is introduced into a fully susceptible population, and that we assume that the change in the susceptible population is negligible during the initial spread. This linearized infected subsystem is the starting point of our calculations.

Any linear system of ODEs is described by a matrix, usually called the Jacobi matrix when derived by linearization of the original nonlinear ODE system. We will relate the structure of this matrix to the epidemiological interpretation and show how it leads to various next-generation matrices, depending on the subset of infected states to which we restrict.

We have used a matrix description before in Section 2.1 to describe the changes in infected states as a Markov process. Let x be the vector describing the sizes of the infected sub-populations. We write the linearized infection subsystem in the form

$$\dot{x} = (T + \Sigma)\, x. \tag{7.12}$$

The matrix T corresponds to *transmissions* and the matrix Σ to *transitions*. Remember that we include death and removal in the transition matrix to keep the notation simple. Hence, all epidemiological events that lead to new infections are incorporated in the model via T, and all other events (changes of state) are incorporated via Σ. The elements of the matrix T have the following interpretation: T_{ij} is the rate at which individuals currently in infected state j give 'birth' to individuals in infected state i. We have seen in Exercise 2.6 (and will see again in Section 7.5 below) that the elements of the non-negative matrix $-\Sigma^{-1}$ have a clear interpretation: element $-(\Sigma^{-1})_{ij}$ is the expected time that an individual will spend in state i, given that it currently has state j. We can now define a next-generation matrix

$$K_L = -T\Sigma^{-1},$$

that we call the *NGM with large domain*, whose elements $(K_L)_{ij}$ have the following interpretation: the expected number of new infections starting in state i caused by an infected individual in state j. We emphasize that, in general, K_L is not equal to the NGM K; the dimension of K_L is generally higher than that of K because K_L describes the generations of infecteds stratified by all infected states, whereas we have defined the NGM proper to describe generations stratified by states-at-infection only.

Exercise 7.8 Consider a system with the following states: S susceptible; E_1 latently infected of category 1; E_2 latently infected of category 2; I infectious; and R recovered/removed/immune. As usual, the letters for the states also indicate the size of the sub-population in that state, where 'size' in our case is the number of individuals in that state. The idea behind this system might be that category 1 and 2 represent individuals that, once infected, progress to infectiousness at different rates. For this model we assume that the trait that causes this difference in disease progression does not manifest itself as a difference in susceptibility, so there is only one S state. We assume that there is a fixed ratio of the two categories in the population, $p : 1 - p$, hence susceptibles enter the E_1 and E_2 states in that fixed ratio following exposure to infection. Let β be the transmission rate, μ the per capita birth and death rates, ν_1 and ν_2 the rates of leaving the respective latency states and entering the infectious state, and α the rate of leaving the infectious state.

i) Formulate the full model and verify that the equations for the *linearized* infected sub-system are

$$\dot{E}_1 = p\beta I - (\nu_1 + \mu) E_1, \qquad (7.13)$$
$$\dot{E}_2 = (1 - p) \beta I - (\nu_2 + \mu) E_2, \qquad (7.14)$$
$$\dot{I} = \nu_1 E_1 + \nu_2 E_2 - (\alpha + \mu) I \qquad (7.15)$$

(or the variant with β written as γ/N).

ii) Verify that T and Σ are given by

$$T = \begin{pmatrix} 0 & 0 & p\beta \\ 0 & 0 & (1-p)\beta \\ 0 & 0 & 0 \end{pmatrix} \quad \Sigma = \begin{pmatrix} -(\nu_1 + \mu) & 0 & 0 \\ 0 & -(\nu_2 + \mu) & 0 \\ \nu_1 & \nu_2 & -(\alpha + \mu) \end{pmatrix}.$$

iii) Show that the NGM with large domain, and the NGM are given by

$$K_L = \begin{pmatrix} \frac{p\beta\nu_1}{(\nu_1+\mu)(\alpha+\mu)} & \frac{p\beta\nu_2}{(\nu_2+\mu)(\alpha+\mu)} & \frac{p\beta}{\alpha+\mu} \\ \frac{(1-p)\beta\nu_1}{(\nu_1+\mu)(\alpha+\mu)} & \frac{(1-p)\beta\nu_2}{(\nu_2+\mu)(\alpha+\mu)} & \frac{(1-p)\beta}{\alpha+\mu} \\ 0 & 0 & 0 \end{pmatrix}$$

and

$$K = \begin{pmatrix} \frac{p\beta\nu_1}{(\nu_1+\mu)(\alpha+\mu)} & \frac{p\beta\nu_2}{(\nu_2+\mu)(\alpha+\mu)} \\ \frac{(1-p)\beta\nu_1}{(\nu_1+\mu)(\alpha+\mu)} & \frac{(1-p)\beta\nu_2}{(\nu_2+\mu)(\alpha+\mu)} \end{pmatrix}.$$

iv) Show that the dominant eigenvalue R_0 of K is equal to the dominant eigenvalue of K_L and given by

$$R_0 = \left(\frac{p\nu_1}{\nu_1 + \mu} + \frac{(1-p)\nu_2}{\nu_2 + \mu} \right) \frac{\beta}{\alpha + \mu}. \qquad (7.16)$$

The exercise above illustrates the general principle. From the matrix description of the linearized infection subsystem, one can compute the NGM with large domain K_L, and this matrix determines the evolution of generations of individuals in all infected states. The NGM K is based on a subset of these states and basically corresponds to K_L, restricted to this subset. From the epidemiological interpretation and the set up of the model, it is possible to distinguish those infected states that are also states-at-infection from those that are not. One can, however, also see this formally by observing that T has a special structure. In the system given in Exercise 7.8 we see that the third row of T consists of zeros only. Individuals can therefore not be in the third state (in this case state I) *immediately* after infection. Hence the system has only two states-at-infection: all individuals start their infected life in either E_1 or E_2. The NGM in the exercise is therefore a two-dimensional matrix, whereas the NGM with large domain is three-dimensional.

The formal approach to obtaining K from K_L is then as follows. We pre- and post-multiply K_L by an auxiliary matrix $\mathcal{E}$ that singles out the rows and columns relevant for the reduced set of states. Specify $\mathcal{E}$ as consisting of unit column vectors e_i, for all i such that the ith row of T is not identically zero (in other words: the columns of $\mathcal{E}$ span the range of T). In short, create a matrix $\mathcal{E}$ whose columns consist of unit vectors relating to non-zero rows of T only. In the above case this leads to

$$\mathcal{E} = \begin{pmatrix} 1 & 0 \\ 0 & 1 \\ 0 & 0 \end{pmatrix}.$$

To find the NGM we then perform the matrix multiplication

$$K = \mathcal{E}^{\top} K_L \mathcal{E} = -\mathcal{E}^{\top} T \Sigma^{-1} \mathcal{E} \tag{7.17}$$

where $\mathcal{E}^{\top}$ indicates, as before, the transpose of $\mathcal{E}$.

For a 2×2 matrix the dominant eigenvalue can be obtained from the trace $tr(K)$ and the determinant of the matrix as

$$R_0 = \rho(K) = \frac{1}{2}\left(tr(K) + \sqrt{tr(K)^2 - 4\det(K)} \right). \tag{7.18}$$

Note that, in Exercise 7.8, $\det(K) = 0$, i.e., K is a singular matrix. Because K is a 2×2 matrix we can conclude right away that $R_0 = tr(K)$. The resulting expression is as in (7.16).

Apart from resulting in a simplified expression for R_0 in the two-dimensional case, an NGM with the property that $\det(K) = 0$ has the added feature that we can achieve further reduction in dimension of the matrix. Typically this situation arises when the incidences corresponding to two or more different states-at-infection occur in a fixed (i.e., time independent) ratio. One way of describing this property is by saying that there is then only one state-at-infection in a stochastic sense, even though formally there are still two states-at-infection. By 'stochastic sense' we mean that the probability distribution of state-at-infection is fixed, i.e., does not depend on the infectious individual responsible for the transmission. We call the lower-dimensional matrix the *NGM with small domain*, and denote it by K_S. To determine K_S from K, in cases where a reduction is possible, we again examine the transmission matrix T, but instead of only examining the rows we now also examine the columns. For the example we see that T has two columns containing only zeros, and only one column that is a non-zero vector. All three columns are therefore multiples of the same vector $C := (p, 1-p, 0)^\top$, the first two columns being zero times this vector, the third column being β times this vector. Similarly, the rows of T are all multiples of one row vector $R := (0, 0, \beta)$, the first row is p times this vector, the second row is $(1 - p)$ times this vector, and the third row is zero times this vector. Actually, R and C constitute a (multiplicative) decomposition of the transmission matrix T, in the sense that $T = CR$, i.e., $T_{ij} = C_i R_j$. We define the NGM with small domain by

$$K_S = -R\Sigma^{-1}C. \tag{7.19}$$

Exercise 7.9 For the system introduced in Exercise 7.8, derive the NGM with small domain and show that its dominant eigenvalue is equal to the dominant eigenvalue R_0, given by (7.16).

A less general way of looking at the NGM with small domain is to define another subset of infected states: the *states-of-infectiousness*, defined as those states in which the infected individual is able to reproduce (in the epidemiological sense, i.e., to cause new cases). In principle, one could do the bookkeeping, in the generation sense, based on this later phase in 'epidemiological life.' One can compare this to the situation in demography where one could count newborns that arise from a newly-born individual throughout its life, or alternatively count the number of individuals reaching adulthood that arise from one individual that has just reached adulthood. In the example from Exercise 7.8, there are two states-at-infection, but only one state-of-infectiousness (I). In this case we can reduce the dimension of the generation matrix to the number of states-of-infectiousness, and describe the generation process and calculate R_0, by using K_S. More generally, such states are called 'renewal points' in a life cycle: they are states that any individual who will ever reproduce will necessarily visit. Bookkeeping can be based on any such point, but for epidemiological application we can only imagine basing this on the moment of becoming infected (state-at-infection), or the moment of starting transmission (state-of-infectiousness).

Even if the number of states-at-infection and states-of-infectiousness is equal, special circumstances or assumptions can cause a reduction in dimension from the NGM to the NGM with small domain. The most common of these assumptions is separable mixing, to be introduced in various forms in Section 7.4.

In summary, we see that we can associate three related next-generation matrices with a compartmental system, the NGM K, the NGM with large domain K_L, and the NGM with small domain K_S. By definition, the basic reproduction number is the largest eigenvalue of the NGM, $R_0 = \rho(K)$. In Exercise 7.10 below we show, in general, that the largest eigenvalues of K, K_L and K_S are the same. In most situations only K and K_L will be relevant, and the computation of R_0 can be based on either description. However, because generally there are fewer states-at-infection than infected states, the dimension of K is often lower than that of K_L. Therefore, using K helps to arrive at an explicit expression for R_0.

Exercise 7.10 Show that the largest eigenvalues of K, K_L, and K_S are the same.

Although the route to R_0 via the formal linear algebra approach given in this section is appealing, we want to emphasize that the epidemiological interpretation of the elements of the NGM allow for a more direct approach that is biologically sound and leads to the same result. By looking at the specification of the compartmental system, possibly guided by a flow diagram, one can write down expressions for all k_{ij} in the following way:

- Determine the states-at-infection.

- For each state-at-infection j imagine one individual who has just started its infected life in state j, follow this individual through its subsequent life and determine how many new cases it is expected to cause for each state-at-infection i.

In the final series of exercises of this section we provide examples to allow the reader to practice the direct approach, as well as the linear algebra approach.

Exercise 7.11 Derive the NGM K for system (7.13–15) in Exercise 7.8 directly from the biological interpretation of the elements of K and the ingredients of the system.

Exercise 7.12 Imagine that plants are grown in a field (flowers for example) where cuttings are taken at some per capita rate γ. The cuttings are planted in a nursery, where they will mature. Plants in the nursery are replanted in the field with rate ζ. The grower chooses these rates in such a way as to maintain a fixed population size of plants in the field (N_1) and plants in the nursery (N_2), taking into account natural mortality. Suppose that a fungus spreads in this host population and that this fungus can be directly transmitted between plants in the field (assuming mass action with transmission rate constant β_1), and between plants in the nursery (assuming mass action with transmission rate constant β_2). In addition, there is a probability p that a cutting taken from an infected plant is itself infected. Let μ_1 and μ_2 be the natural ('background') death rates of field and nursery plants, respectively. Finally, let ρ_1 and ρ_2 be the per capita infection-induced death rates of field and nursery plants, respectively.

Give an explicit expression for R_0.

Exercise 7.13 The bovine viral diarrhea virus (BVDV) in cattle has a complicated epidemiology. We base the example on a model by Cherry et al.[3] The system has both horizontal and vertical transmission at different rates β_1 and β_2. Horizontally infected animals can be in an exposed (E), infectious (I) or several immune/recovered states. Animals that have been pregnant for less than 150 days when becoming infected may, following recovery into one particular immune state Z, give birth to an infected calf. These offspring are classified as *persistently infected* (P state): they transmit infection, give birth at a lower rate and die at a higher rate than cattle that were infected by the horizontal route. Let γ be the recovery rate, and let the constant p_1 be the probability that an infected animal enters the immune state Z upon recovery. Let $1/\alpha$ be the average time spent carrying an infected fetus, and let p_2 be the probability that an infected fetus survives to enter the herd. Finally, ν is the rate of leaving the exposed class to become infectious, μ is the 'natural' death rate of cows, and a and b represent the reduction in birth rate and increase in death rate of persistently infected animals, respectively.

---

[3]B.R. Cherry, M.J. Reeves and G. Smith: Evaluation of bovine viral diarrhea virus control using a mathematical model of infection dynamics. *Prev. Vet. Med.*, **33** (1998), 91–108.

With a change in notation from Cherry et al., the model is described by

$$\begin{aligned}
\dot{E} &= (\beta_1 I + \beta_2 P) S - (\nu + \mu) E, \\
\dot{I} &= \nu E - (\gamma + \mu) I, \\
\dot{Z} &= p_1 \gamma I - (\alpha + \mu) Z, \\
\dot{P} &= p_2 \alpha Z + (\mu - a) P - (\mu + b) P,
\end{aligned}$$

where, as before, we restrict ourselves to the (linearized) infection sub-system.

i) Convince yourself that there are indeed four infected states and explain what is different about the recovered/immune state Z compared to recovered states we have encountered so far. Argue that there are just two states-at-infection E and P.

ii) Show, by working directly from the interpretation, that the next-generation matrix K is given by

$$K = \begin{pmatrix} \dfrac{\nu}{\mu+\nu}\dfrac{\beta_1}{\mu+\gamma} & \dfrac{\beta_2}{\mu+b} \\ \dfrac{\nu}{\mu+\nu}\dfrac{\gamma p_1}{\mu+\gamma}\dfrac{\alpha p_2}{\mu+\alpha} & \dfrac{\mu-a}{\mu+b} \end{pmatrix},$$

and compute R_0 explicitly by using equation (7.18).

iii) Derive K by the linear algebra route outlined in this section (i.e., give T and Σ, compute K_L and then compute K).

Exercise 7.14 Consider the within-host model described in Section 4.5.2. Assume that the susceptible target cells are in a steady state, and denote the steady state level and the infected target-cell population level by $\hat{X}$ and X^*, respectively (where we change notation from the original model to avoid confusion with our transmission matrix T). So, consider the linear system

$$\begin{aligned}
\frac{dX^*}{dt} &= k\hat{X}V - (\mu + d)X^*, \\
\frac{dV}{dt} &= pX^* - (k\hat{X} + c)V.
\end{aligned}$$

Decide what you would like to call reproduction/transmission, and transition (including death). There is more than one option, so this exercise also serves to make you aware of a certain inherent ambiguity. Next specify the matrices T and Σ, and derive an expression for R_0.

We end with a cautionary remark. We have gone into some detail to explain the approach to R_0 when starting from a compartmental system of ODEs because many researchers are used to writing down systems of ordinary differential equations as a starting point for deriving an expression for R_0. A popular approach is then to derive a threshold condition from stability analysis of the infection-free steady state. We will see in Section 8.2, where we study the relation between R_0 and the real-time growth rate r of the infected sub-population, that indeed R_0 directly relates to the (asymptotic) stability of the infection-free steady state. The opposite of this statement is, however, not true in the sense that not every condition that describes the stability of the infection-free steady state has a direct relation to R_0. Generally speaking, the condition derived from stability analysis shares only the threshold property with R_0, but not the biological interpretation. This is, for example, often the case when the so-called Routh-Hurwitz criteria (see e.g., Edelstein-Keshet 1988) are used to investigate stability. This is illustrated in the next exercise.

Exercise 7.15 Consider the same system as in Exercise 7.13, but assume $\beta_1 = 0$. Convince yourself that R_0 is given by

$$R_0 = \frac{1}{2}B + \frac{1}{2}\sqrt{B^2 + 4A},$$

with

$$B = \frac{\mu - a}{\mu + b},$$

$$A = \frac{\nu}{\mu + \nu}\frac{\gamma p_1}{\mu + \gamma}\frac{\alpha p_2}{\mu + \alpha}\frac{\beta_2}{\mu + b}.$$

Cherry et al. (1998), see footnote 3, derive, starting from their full compartmental system of eight nonlinear coupled ODEs, that $A + B$ relative to the threshold 1 describes the stability of the infection-free steady state. Note that $A + B$ does not equal R_0, but show that

$$R_0 > 1 \Leftrightarrow A + B > 1.$$

7.3   GENERAL h-STATE

When the h-state is not purely discrete-valued but is continuous, or has, in case it consists of several traits, continuous elements, all we have to do is to replace summation by integration in the considerations in Section 7.1. Let Ω denote the h-state space. One can think of some suitably nice subset of $\mathbb{R}^n$ for some $n \geq 1$.

In a susceptible population, let

$k(\xi, \eta) = $ expected number (per unit of h-state space) of new cases
with state-at-infection ξ,
caused by one individual with state-at-infection η,
during its entire period of infectiousness.

Again, as with k_{ij} in Section 7.1, one can have an identity $k(\xi, \eta) = \int_0^\infty A(\tau, \xi, \eta)\, d\tau$ in mind, see Section 7.5, but this is not essential.

The generations are now described by functions ϕ on Ω such that

$$\int_\omega \phi(\eta)\, d\eta \tag{7.20}$$

equals the expected number of cases with state-at-infection belonging to the subset $\omega \subset \Omega$. The *next-generation operator*, again denoted by K, tells us how the situation changes from one generation to the next:

$$(K\phi)(\xi) = \int_\Omega k(\xi, \eta)\phi(\eta)\, d\eta. \tag{7.21}$$

The function $k(\xi, \eta)$ is called the *kernel* of K; kernels are our main modeling ingredient. The norm:

$$\| \phi \| = \int_\Omega | \phi(\eta) |\, d\eta \tag{7.22}$$

is called the L_1-norm of ϕ, and the norm of K is, exactly as before, defined by

$$\| K \| = \sup_{\|\phi\| \neq 0} \frac{\| K\phi \|}{\| \phi \|} = \sup_{\|\phi\|=1} \| K\phi \| . \tag{7.23}$$

By definition, R_0 is the spectral radius of K:

$$R_0 = \lim_{\ell \to \infty} \| K^\ell \|^{1/\ell} . \tag{7.24}$$

(As in Section 7.2, equation(7.8), it can be proved that this limit exists. One should interpret the product of two operators K and M with kernels k and m as an operator with kernel $\int_\Omega k(\xi, \zeta) \, m(\zeta, \eta) \, d\zeta$, as can be seen from the identity

$$((KM)\phi)(\xi) = \int_\Omega \left(\int_\Omega k(\xi, \eta) \, m(\eta, \zeta) \, d\eta \right) \phi(\zeta) \, d\zeta, \tag{7.25}$$

which follows directly from the definitions and an interchange of the order of the two integrations.)

Often (but not always) R_0 is an eigenvalue. The corresponding (normalized) eigenvector ψ^d is called the *stable distribution* and, provided that R_0 is strictly dominant and algebraically simple, the asymptotic behavior for $\ell \to \infty$ is completely analogous to that in the finite dimensional setting. Technical conditions for this to be the case often involve the notion of compactness (of both Ω and K).

We now present a third variant of the basic ideas that unifies the preceding two. Let ω denote a subset of Ω (we have a measurable subset in mind, but our presentation will emphasize the interpretation and intuition while being deliberately sloppy about technical aspects). Then $\int_\omega k(\xi, \eta) \, d\xi$ is the expected number of cases with state-at-infection in ω caused by one individual with state-at-infection η, during the entire period of infectiousness. (Note that k itself is not a number, but that we do indeed obtain a number after integration with respect to ξ.) The third variant takes these quantities, with arbitrary η and ω, as its starting point (and forgets about k). We denote the quantities by $\Lambda(\eta)(\omega)$. So, rather than using k, we now take as the main modeling ingredient

$\Lambda(\eta)(\omega) \quad := \quad$ the expected number of cases with state-at-infection in ω

caused by one individual with state-at-infection η,

during its entire period of infectiousness.

Likewise, we describe the generations by measures m, which assign to every $\omega \subset \Omega$ the number of cases with state-at-infection in ω. (If one can alternatively describe the generations by functions ϕ, the relation between ϕ and m is given by

$$m(\omega) = \int_\omega \phi(\eta) \, d\eta. \tag{7.26}$$

Since measures cannot always be related to functions, this third formulation is more general.) The next-generation operator is now given by

$$(Km)(\omega) = \int_\Omega \Lambda(\eta)(\omega) \, m(d\eta). \tag{7.27}$$

Concerning norms, we confine ourselves to the remark that for non-negative measures $\| m \| = m(\Omega)$. As before we define R_0 to be the spectral radius of K.

The handiness of formulations in terms of measures will be demonstrated by (classes of) examples in Sections 7.4.2 and 7.4.3.

We end our presentation of the definition of R_0 with a cautionary remark. On the one hand, we linearize (i.e., we neglect the diminishing of susceptibles), which we justify by saying that we consider the initial phase only. On the other hand, we define R_0 by looking after many generations, arguing that the details of how precisely the agent is introduced affect only the transient behavior in a relatively short period. The upshot of these somewhat contradictory arguments is that one is not sure that indeed data, when organized on a generation basis, should exhibit multiplication by R_0 in the initial phase. Since looking at data in a generation perspective is somewhat artificial anyhow, we emphasize that the remark carries over to the real-time growth rate r. Whether or not we can observe r in data will depend on such factors as the difference between r and the next eigenvalue (ordered according to real part), and the quantitative aspects of irreducibility (measured, for instance, by the ratio of the maximum and the minimum component of the dominant eigenvector). This admonition detracts nothing from the key property of R_0:

$$\{\text{The infective agent will be able to grow}\} \;\Leftrightarrow\; R_0 > 1.$$

We will return to this property in Section 8.2, where we present an elementary proof for the finite-dimensional case.

Exercise 7.16 The aim of this exercise is to illuminate the remark about 'quantitative aspects of irreducibility.' Consider the reducible matrix

$$\begin{pmatrix} 10^3 & 0 \\ 0 & 1 \end{pmatrix}$$

with dominant eigenvalue 10^3 and corresponding eigenvector $\begin{pmatrix} 1 \\ 0 \end{pmatrix}$. If we make the anti-diagonal terms slightly positive, the matrix becomes irreducible (even primitive). Use perturbation analysis to show that the second component of the normalized eigenvector is of the same order of magnitude as the coupling coefficients in the matrix. Now explain this remark (formulate your conclusions in terms of the interpretation rather than in mathematical terms).

7.4   CONDITIONS THAT SIMPLIFY THE COMPUTATION OF R_0

There are efficient numerical methods for the computation of the dominant eigenvalue of a non-negative matrix, but that is not what will concern us here. Instead we try to establish analytical procedures. Quite naturally these require extra conditions. In this section we discuss three of such conditions (and their interpretation) that facilitate the computation of R_0.

7.4.1   One-dimensional range

When the range of an operator is one-dimensional, no matter what operator you consider, there is only one relevant candidate eigenvector and a formula for the one and only non-zero eigenvalue follows at once. Following the three variants of the formulation of the next-generation operator presented in the previous sections, the assumptions leading to a one-dimensional range and the resulting formulas for R_0 take the following form:

$$k_{ij} = a_i b_j \quad \Rightarrow \quad R_0 = \sum_{j=1}^{n} b_j a_j, \tag{7.28}$$

$$k(\xi, \eta) = a(\xi) b(\eta) \quad \Rightarrow \quad R_0 = \int_{\Omega} b(\eta) a(\eta) \, d\eta, \tag{7.29}$$

$$\Lambda(\eta)(\omega) = \alpha(\omega) b(\eta) \quad \Rightarrow \quad R_0 = \int_{\Omega} b(\eta) \alpha(d\eta). \tag{7.30}$$

Exercise 7.17 Verify at least one of the expressions for R_0. Hint: Write down the respective eigenvalue problems for the special kernels, remark that there is only one relevant candidate eigenvector and find the only non-zero eigenvalue.

Exercise 7.18 This exercise extends and continues the example introduced in Section 2.2 and it gives an introduction to 'sexual activity'-based models as presented at the end of the present section. Assume that $k_{ij} = a_i b_j$ with $b_j = c_j$ and $a_i = c_i N_i / (\sum_l c_l N_l)$ (which guarantees that $k_{ij} N_j = k_{ji} N_i$, which is a consistency condition required by a particular context as explained in Section 2.2). Show that

$$R_0 = \text{mean} + \frac{\text{variance}}{\text{mean}}$$

where the right-hand side refers to the activity c.

But what is the interpretation of the one-dimensional range assumptions given above? Can we understand when and why they are reasonable?

The assumptions express that the probability distribution for the state-at-infection of a newly infected individual is *independent* of the state-at-infection of the individual that is responsible for the infection. So in a stochastic sense, there is just one possible state-at-infection. Stated in yet another manner: the states-at-infection of the two individuals involved influence transmission independently. We argued before (Section 2.2) that straightforward averaging works unless there is correlation between infectivity and susceptibility and that even if such correlation exists within one individual, we can still average (in the right way as expressed by an explicit formula for R_0) provided there is no correlation between the infectivity of the 'donor' and the susceptibility of the 'receiver.'

The condition is called *separable mixing*. When a is proportional to b (or α to the integral of b) we speak of *proportionate mixing*, which is in fact a weighted form of random mixing. Proportionate mixing was introduced in Barbour (1978).[4]

Let us discuss an example. Let ξ be some indicator of sexual activity and consider an STD (sexually transmitted disease). It is reasonable to assume that both susceptibility and infectivity are proportional to ξ, because we have in mind that the number of contacts per unit of time is proportional to ξ. We consider a homosexual population (but see Section 7.4.3 for the slightly more complicated heterosexual case). We take $\Omega = [0, \infty)$ and introduce a measure m on Ω to describe the population composition. In other words, for $\omega \subset \Omega$ the probability that a randomly chosen individual has $\xi \in \omega$ is $m(\omega)$. Let $E(\xi) = \int_{\Omega} x m(dx)$ denote the

---

[4]A.D. Barbour: Macdonald's model and the transmission of bilharzia. *Trans. Roy. Soc. Trop. Med. Hyg.*, **72** (1978), 6–15.

mean level of sexual activity and $V(\xi) = \int_\Omega x^2 m(dx) - E(\xi)^2$ the variance, where we write x for realizations of the stochastic variable ξ. One could think that R_0 is simply proportional to $E(\xi)$, but this is the wrong way of taking averages. As we shall show, the formula above leads to the conclusion that R_0 is proportional to

$$E(\xi) + \frac{V(\xi)}{E(\xi)}.$$

When the variance is large (which it is whenever there are some very active and some very inactive individuals) the second term contributes a lot, and the wrong way of taking averages leads to a much too optimistic (i.e., too low) estimate of R_0.

If we sample individuals by choosing a *contact* at random, we will not observe the distribution m but the distribution obtained from m by applying a weight ξ. We denote this distribution by α. So,

$$\alpha(\omega) = \left(\int_\omega x\, m(dx) \right) / E(\xi)$$

(the factor $1/E(\xi)$ serves to turn α into a probability distribution, i.e., to ensure that $\alpha(\Omega) = 1$). We *assume* that the partners of any individual follow the distribution α, irrespective of the sexual activity level of the individual itself. Furthermore, we take $b(x) = \pi x$, where the constant π involves the probability of transmission, given a contact between a susceptible and an infective. The formula then yields

$$R_0 = \frac{\pi}{E(\xi)} \int_\Omega x^2\, m(dx) = \pi \left(E(\xi) + \frac{V(\xi)}{E(\xi)} \right), \qquad (7.31)$$

which is exactly what we already discussed above.

7.4.2   Additional within-group contacts

Suppose contacts basically follow the pattern just described, but that by an additional mechanism there are extra contacts with individuals having the same h-state (it may be helpful, for the time being, to think of the h-state as being static). Then we can no longer derive an explicit expression for R_0, but we can deduce a rather simple nonlinear equation that R_0 has to satisfy. From this equation we can derive a quantity, Q_0 say, with the property that $R_0 > 1$ if and only if $Q_0 > 1$. (Here we assume that we have a subcritical situation when we restrict contacts to individuals having exactly the same h-state, since otherwise we will always have $R_0 > 1$.) Again we list three variants of the assumption and the expression for the threshold quantity Q_0 that these result in:

$$k_{ij} = a_i b_j + c_j \delta_{ij} \qquad \Rightarrow Q_0 = \sum_{j=1}^{n} \frac{b_j a_j}{1 - c_j}, \qquad (7.32)$$

$$k(\xi, \eta) = a(\xi)b(\eta) + c(\eta)\delta(\xi - \eta) \Rightarrow Q_0 = \int_\Omega \frac{b(\eta)a(\eta)}{1 - c(\eta)}\, d\eta, \qquad (7.33)$$

$$\Lambda(\eta)(\omega) = \alpha(\omega)b(\eta) + c(\eta)\delta_\eta(\omega) \quad \Rightarrow Q_0 = \int_\Omega \frac{b(\eta)}{1 - c(\eta)}\alpha(d\eta). \qquad (7.34)$$

Here δ_{ij} is Kronecker's delta, i.e., $\delta_{ij} = 1$ if $i = j$ and zero otherwise. Furthermore, $\delta(\cdot - \eta)$ and δ_η both represent Dirac's delta 'function,' i.e., the unit point measure concentrated in η. In fact the formulation of $k(\xi, \eta)$ has debatable mathematical underpinning, and the more precise formulation is exactly the one in terms of $\Lambda(\eta)(\omega)$.

Exercise 7.19 Verify at least one of the expressions for Q_0. Hint: For the special kernels write down the eigenvalue problems.

One can understand Q_0 in terms of 'super-individuals,' which here correspond to subcritical epidemics within sub-populations consisting of individuals with one particular h-state. Indeed, if one individual is expected to infect c individuals, these collectively are expected to infect $c \cdot c = c^2$ individuals, etc. So the size of the 'clan' of one case is $1 + c + c^2 + c^3 + \cdots = (1 - c)^{-1}$ (see Exercise 1.41). At this clan level we are back to separable mixing and we can use the formula of the preceding subsection, which gives us the corresponding expression for Q_0. Note that we did implicitly assume that $\max c < 1$.

7.4.3   Finite-dimensional range

When the dimension of the range of the next-generation operator K is finite, but exceeds one, we may still reduce the determination of R_0 to the determination of the dominant eigenvalue of a matrix. This is analogous to the reduction from K to K_S in Section 7.2.

Exercise 7.20 Assume that $\Lambda(\eta)(\omega) = \sum_{i=1}^{n} \alpha_i(\omega) b_i(\eta)$. Show that R_0 is the dominant eigenvalue of the matrix L with entries

$$l_{ij} = \int_\Omega b_i(\eta) \alpha_j(d\eta).$$

Hint: Show first that the range of K is spanned by the α_i. Next compute how the coefficients with respect to this basis transform under K.

In Section 7.5 we shall consider an example having this form. But here we wish to understand the meaning of the assumption. So, what we are after is a version of a finite-dimensional range condition that allows an interpretation. Therefore assume that ξ has two components: a discrete one (which we indicate by i or j but also by m or f, for male and female respectively) and a continuous one (which we indicate by ζ and for which we have sexual activity as the motivating concrete example). Because ζ is a stochastic variable, we will denote specific realizations by z here as in the previous subsection. We denote by $\widetilde{\Omega}$ the set over which the continuous variable ranges and let $\widetilde{\omega} \subset \widetilde{\Omega}$. For $\omega = (i, \widetilde{\omega})$ and $\eta = (j, z)$ we use the notation

$$\Lambda(\eta)(\omega) = \Lambda_i(j, z)(\widetilde{\omega}).$$

So $\Lambda_i(j, z)(\widetilde{\omega})$ is the expected number of new cases with discrete component i and continuous component in $\widetilde{\omega}$, caused by one individual with h-state (j, z). We assume that

$$\Lambda_i(j, z)(\widetilde{\omega}) = \alpha_i(\widetilde{\omega}) b_{ij}(z),$$

which means that, conditional on the discrete component being i, the probability distribution of the continuous component is α_i, independently of the h-state of the individual that causes the infection. But dependence in the discrete component is still allowed, as described by b_{ij}. We will also refer to the finite-dimensional range condition as multi-group separable mixing.

Exercise 7.21 Deduce that, under this assumption, R_0 is the dominant eigenvalue of the matrix L with entries

$$l_{ij} = \int_{\widetilde{\Omega}} b_{ij}(z)\alpha_j(dz).$$

As a concrete example, consider a population structured according to sex and sexual activity level. The population composition is described by the two probability distributions μ_f and μ_m. As before, we denote the mean by E and the variance by V. Moreover, we introduce weighted distributions

$$\alpha_i(\widetilde{\omega}) = \frac{1}{E(\zeta)_i}\int_{\widetilde{\omega}} z\mu_i(dz)\,,\quad i = f, m.$$

Assuming strict heterosexuality, we are led to define

$$b_{ij}(z) = c_{ij}z,\quad \text{with } c_{ij} = 0 \text{ if } i = j$$

(note that it is not necessarily the case that $c_{fm} = c_{mf}$, since the probability of transmission may be asymmetric).

The matrix L is then given by

$$\begin{pmatrix} 0 & \frac{c_{fm}}{E(\zeta)_m}\int_0^\infty z^2\mu_m(dz) \\ \frac{c_{mf}}{E(\zeta)_f}\int_0^\infty z^2\mu_f(dz) & 0 \end{pmatrix},$$

and we conclude that

$$R_0 = \sqrt{c_{fm}c_{mf}\left(E(\zeta)_m + \frac{V(\zeta)_m}{E(\zeta)_m}\right)\left(E(\zeta)_f + \frac{V(\zeta)_f}{E(\zeta)_f}\right)}.$$

We will study two examples of multi-group separable mixing in Section 7.7. In Section 12.6.4 we will study an STD model for heterosexual transmission in a network.

7.5   SUB-MODELS FOR THE KERNEL

In this section, we address the dynamics of h-state change, the contact structure and the probability of transmission. We will then put the pieces together to give a more specific form of the kernel k.

A fruitful and often applicable way of modeling change in an individual's h-state is to regard h-state change as a Markov process. For this we introduce

$$
\begin{aligned}
P(\tau, \omega, \eta) \;=\; & \text{probability that an individual originally} \\
& \text{with h-state } \eta \text{ has an h-state with} \\
& \text{value in } \omega \subset \Omega,\ \tau \text{ units of time later.}
\end{aligned}
$$

(Note that the possibility $P(\tau, \Omega, \eta) < 1$ accounts for death.)

Examples:

- *Static* h-*state*: $P(\tau, \omega, \eta) = \delta_\eta(\omega)$ (if death can occur one has to multiply with the survival probability as a function of τ and η).

- *Age:* Let

$$P(\tau, \omega, \eta) = \frac{\mathcal{F}(\eta + \tau)}{\mathcal{F}(\eta)} \delta_{\eta + \tau}(\omega),$$

  where $\mathcal{F}$ is the survival probability as a function of age. (Note that $\mathcal{F}(\eta + \tau)/\mathcal{F}(\eta)$ is the conditional probability to survive until at least age $\eta + \tau$, given that the individual was alive at age η.)

- *Finite Markov chains*: Let v_η denote the unit vector whose ηth component equals 1 while all other components are zero. Then the probability vector $P = P(\tau, \eta)$ changes according to

$$\frac{dP}{d\tau} = \Sigma P,$$

  where Σ describes 'jumps' between states, including jumps to 'states' not explicitly accounted for in our bookkeeping, such 'no longer infectious' or 'dead.' We have encountered Σ already in Section 2.1 and 7.2 (as the matrix describing the transitions between states, as distinct from state changes caused by transmission). To the equation for change we have to add the initial condition that $P = v_\eta$ at $\tau = 0$. So formally,

$$P(\tau, \eta) = e^{\Sigma \tau} v_\eta \,,$$

  but of course the actual computation of $\exp[\Sigma \tau]$ from the 'data' Σ is a laborious task. With forethought, we note already now that

$$\int_0^\infty e^{\Sigma \tau} d\tau = -(\Sigma)^{-1}.$$

  The point is that the left-hand side makes clear that the ijth component of this matrix equals the expected time that an individual will spend in state i in the rest of its life after we have observed the individual in state j. The right-hand side tells us how to compute the matrix from the data Σ. The matrix Σ will be called the Markov transition matrix (although actually it describes both transitions and death).

Concerning contacts, there is very little that can be said in any generality. So let us just observe that we need to specify

$$
\begin{aligned}
c(\xi, \zeta) \quad = \quad & \text{probability per unit of time that an individual} \\
& \text{with current h-state } \zeta \text{ has contact with an} \\
& \text{individual with current h-state } \xi.
\end{aligned}
$$

With regard to the probability of transmission, one has to keep in mind that whenever one is decomposing a quantity into multiplicative factors there is a certain degree of arbitrariness. If the probability of transmission is allowed to depend on the h-states of the individuals involved in a contact, it is logical to include that dependence in the ingredients we now discuss. However, since we have already included a general function of two variables in the preceding paragraph, this gives a redundancy, which we prefer to avoid. Therefore we choose to describe the probability of transmission as a function h of the time τ elapsed since infection took place.

We now put the various pieces together. Being more specific entails a proliferation of modeling ingredients. Indeed, we have arrived at the identity

$$k(\xi, \eta) = \int_0^\infty h(\tau) \int_\Omega c(\xi, \zeta) P(\tau, d\zeta, \eta) \, d\tau,$$

which expresses k in terms of more easily mechanistically interpretable quantities at the expense of having to deal with three functions instead of one. In a bottom up approach one starts with a multitude of ingredients and aspires to combine the relevant information in a few computable numbers, such as R_0. Here we started from R_0, found that we need k to determine it, and have now found that we need to specify h, c and P to determine k and could go on to anatomize these. We will, however, do so only in more specific contexts and not in general. One example is treated in detail in Section 7.7.

Exercise 7.22 Assume the h-state space is a finite set of n states. Let the h-state dynamics be described by the Markov transition matrix Σ. Let $h(\tau) = h$ be constant. Show that

$$k_{ij} = -h \sum_{l=1}^n c_{il}(\Sigma)_{lj}^{-1} \, .$$

Hint: Recall the remarks for a finite Markov chain.

What remains to be done is to specify c. We do this in terms of the steady demographic state of susceptibles that the infective agent invades. This steady state can, for example, be obtained from a system of equations describing the dynamics of the susceptibles, distributed according to h-state, in the absence of infection. In applications, one could have the possibility to specify the demographic steady state from data (think of the age pyramid for example) or estimates (or even experimental design for certain problems related to animal or plant infections).

7.6   SENSITIVITY ANALYSIS OF R_0

In the ecological literature, matrix models are often used to describe the growth of populations of a species in the life cycle of which one can distinguish stages or classes (see Caswell 2001). The dominant eigenvalue of the transition and reproduction matrix describing the life cycle is interpreted as the growth rate of the population in an environment that does not change. Methods that are able to quantify the relative contributions to this growth rate of the various life-history ingredients (including in particular the transition rates between stages/classes in the life-cycle, as well as the production of offspring), have produced substantial ecological insight for the species and populations studied. Most notably, sensitivity and elasticity analysis play an important role.

In epidemiology we have a similar set-up, but sensitivity and elasticity analysis has (so far) been used only occasionally. The next-generation matrix describes the transmissions (of an infectious agent) between the states-at-infection and the dominant eigenvalue R_0 describes the growth (in terms of generations) of the population (of infecteds). The right eigenvector corresponding to R_0 has a similar interpretation as the stable distribution in ecology, but now concerns the distribution of newly infected individuals in type space. The left eigenvector is called the reproductive value, and has an interpretation in epidemiology similar to that in ecology (i.e., contribution to future 'reproduction'). Elasticity analysis can highlight which entries of the matrix K, or even which components in the entries, have the largest relative influence on the value of R_0. This can highlight and quantify the relative importance of different transmission routes (e.g., vertical transmission, co-feeding in tick-borne infections, or different vectors transmitting the same agent), different infected types (e.g., identifying key species in a wildlife system) or different ingredients (e.g., identifying both parameters that have little and parameters that have a lot of influence, indicating where efforts to secure better estimates could best be directed). One could also compare the influence of a specific parameter for two agents transmitted by the same insect vector (e.g., the influence of the vector's life history parameters could be different for two different viruses transmitted by that vector). Finally, the analysis can be used to predict the effects of postulated future changes in ingredients, for example due to changes in the environment or climate.

Regard the situation of a system with n types. Denote by v and w the left and right eigenvectors of K corresponding to the dominant eigenvalue R_0 (recall that the left eigenvector of a matrix is equal to the right eigenvector of the transpose of that matrix). The sensitivity of entry R_0 for a unit change in entry k_{ij} of K is given by Caswell (2001)

$$\frac{\partial R_0}{\partial k_{ij}} = \frac{v_i w_j}{v \cdot w}$$

where $v \cdot w = \sum v_i w_i$ is the inner product of vector v and w.

Often, it is better to look for relative instead of absolute change, especially when there are orders of magnitude differences between the entries. A related reason is that the components of the entries may have a physical dimension, and so the choice of (physical) units matters (e.g., contacts per day versus contacts per year). The more important reason is that a unit change in a small element can be expected to have a much larger impact than a unit change in a very large element. In ecology, the difference in magnitude of the entries is frequently large because in life history descriptions there are ingredients for which the value is naturally restricted between 0 and 1 (e.g., probability of evolving to the next stage in the life cycle), whereas others are comparatively unrestricted (e.g., number of offspring produced). The elasticity with respect to entry k_{ij} is defined by

$$\frac{k_{ij}}{R_0} \frac{\partial R_0}{\partial k_{ij}}.$$

These elasticities of the matrix elements sum to 1. This has led to the interpretation that the elasticity gives the relative contribution of a matrix entry,[5] or of a group of matrix entries (for example all entries related to a certain route of transmission), to the (generation) growth factor R_0. The useful fact that the elasticities sum to 1 can be seen most easily by looking at homogeneous functions. A function $f = f(x_1, \ldots, x_n)$ is homogeneous of degree s if $f(cx_1, \ldots, cx_n) = c^s f(x_1, \ldots, x_n)$ for all $c > 0$. Now regard $R_0 = R_0(k_{11}, k_{12}, \ldots, k_{nn})$ as a function of the n^2 entries of K. If ψ_d is the (right) eigenvector of K corresponding to R_0, we have that ψ_d is also an eigenvector of $K_c := (ck_{ij})_{i,j}$ with eigenvalue cR_0 because $K_c \psi_d = cK\psi_d = cR_0\psi_d$. We conclude that R_0 is a homogeneous function of degree 1, in terms of the entries of K. The fact that the elasticities sum to 1 then follows from

---

[5]See for an example: A. Matser et al.: Elasticity analysis in epidemiology: an application to tick-borne infections. *Ecol. Letters*, **12** (2009), 1298–1305.

the fact that a homogeneous function f of degree s is characterized by the relation $\sum_{i=1}^{n} x_i \partial f / \partial x_i(x_1, \ldots, x_n) = sf(x_1, \ldots, x_n)$ (this is known as Euler's Theorem and can be seen immediately by taking the derivative to c on both sides of the property defining a homogeneous function, and choosing $c = 1$). Using this relation for R_0 (with $s = 1$), we find the desired result.

For completeness we also give the sensitivity to lower-level ingredients of the entries k_{ij}. Let θ be a parameter involved in a mechanistic description of the transmission process underlying K. Then $k_{ij} = k_{ij}(\theta)$ is a function of θ. Then, the change in R_0 resulting from an infinitesimal change in θ is given by

$$\sum_{i,j=1}^{n} \frac{\partial R_0}{\partial k_{ij}} \frac{\partial k_{ij}}{\partial \theta} \quad \text{and} \quad \frac{\theta}{R_0} \frac{\partial R_0}{\partial \theta} = \sum_{i.j=1}^{n} \frac{k_{ij}}{R_0} \frac{\partial R_0}{\partial k_{ij}} \frac{\theta}{k_{ij}} \frac{\partial k_{ij}}{\partial \theta}.$$

7.7   EXTENDED EXAMPLE: TWO DISEASES

It has been observed in, for example, Africa that certain genital ulcer diseases, particularly chancroid and syphilis, can increase the risk of HIV infection. The damage that these ulcerative sexually transmitted diseases (we write USTDs for short) cause to the genital skin and membranes may facilitate both transmission and acquisition. AIDS has clearly been able to establish itself in the heterosexual population in Africa, as opposed to the situation in Europe and North America. There transmission is highest in other sub-populations. Because of the higher prevalence of other STDs in Africa, one can pose a few obvious questions. To what degree can USTDs that are endemic in a population facilitate the spread of HIV into that population (for example the heterosexual population in Europe or in the United States)? What is the efficacy of control measures aimed at these USTDs in halting the spread of HIV in that population? At which aspects of the USTD should control measures then be aimed to be most effective? We will not attempt to answer these questions; we will, however, illustrate the building of sub-models for the kernel — involving a major and a (relatively) minor disease — with which these questions could be studied. We do so in the form of a series of exercises. We give a few additional exercises where one can study the relative efficiency of various control measures aimed at the minor disease, using the results derived. We concentrate on STDs for our exposition, but other diseases, such as malaria and tuberculosis, also have been indicated as influencing the spread of HIV in Africa.

We denote the major (in the sense of being incurable) disease by 'D' and the minor disease by 'd.' We ask how the R_0 for invasion of D into a population where d is endemic depends on the parameters that govern the spread of d and D. Let us first look at (7.35) as a very simple model for D. For a change, we choose absolute numbers of individuals instead of densities as our variables. We consider

$$\frac{dI}{dt} = \frac{\beta SI}{N} - (\mu + \sigma)I, \tag{7.35}$$

where S and I are population sizes of D-susceptibles and D-infecteds respectively, and σ describes the increased mortality rate due to disease D. For our purpose it is convenient to split β into two factors, $\beta = pc$, where c is the number of new contacts per individual per unit of time, assumed to be common to d and D transmission, and where p is the probability that D-transmission is successful upon contact.

Exercise 7.23 Let R_0^D be the basic reproduction number for the invasion of D into a homogeneous population. Give an expression for R_0^D.

Assume that disease d is in an endemic steady state. We next want to calculate R_0 for the disease D in a heterogeneous population, assuming that the susceptibility to D is, for individuals having d, v times as large as for individuals without d. What we have in mind is that meetings between individuals are totally random, but that the success ratio for infection transmission, given that contact takes place, is enlarged by a factor v. We denote by $w > 1$ the factor by which the success ratio is enlarged when a D-infectious individual is also suffering from d, and let p be the success ratio when both individuals involved in the contact are free from d. Consistency demands that $pvw \leq 1$. For our h-state space we take $\Omega = \{0, +\}$, where '0' means free of d, and '+' means having d. So from the point of view of infection D there are two infected states, which are both also states-at-infection. In this example, the NGM with large domain and the NGM are equal. We assume that meetings occur independently of the h-state of the individuals involved and that the contact rate (i.e., the number of contacts per individual per unit of time) is given by c. Since we are interested in calculating R_0 for D-invasion, we start — as usual — with a population consisting of D-susceptibles only, and can therefore write $S \approx N$.

In Section 7.5 we noted the arbitrariness in assigning multiplicative factors to either infectivity or contacts. Write $T_{il} = hc_{il}$ for the combined factors describing the (constant) infectivity and contact pattern in the kernel as given in Section 7.5. Describe by N_0 and N_+ the steady (with respect to d) state population sizes of '0' and '+' individuals, in the absence of D.

Exercise 7.24 Check that the transmission matrix $T = (T_{il})_{1 \leq i, l \leq 2}$ is given by

$$T = \frac{pc}{N} \begin{pmatrix} N_0 & N_0 w \\ N_+ v & N_+ v w \end{pmatrix}.$$

Before proceeding, we derive expressions for N_0 and N_+ in terms of N and parameters describing the spread of d. Let ζ denote the (constant) force of d-infection in the steady state and let γ be the probability per unit of time that d is cured by treatment (whereupon susceptibility to d returns). We do not concern ourselves with the question of how these parameters arise, we assume that they completely describe the endemic dynamics of d phenomenologically. Furthermore, let μ be the natural death rate.

Exercise 7.25 Give a system of two ODEs for N_0 and N_+, based on the assumptions above (also recall Section 4.2), and check that the steady state of this system is characterized by

$$N_0 = \frac{\gamma + \mu}{\gamma + \mu + \zeta} N, \quad N_+ = \frac{\zeta}{\gamma + \mu + \zeta} N,$$

where $N = N_0 + N_+$.

Exercise 7.26 Give the Markov transition matrix Σ, as introduced in Sections 7.2 and 7.5. Check that $-\Sigma^{-1}$ is given by

$$\frac{1}{(\mu + \sigma)(\mu + \sigma + \gamma + \zeta)} \begin{pmatrix} \mu + \sigma + \gamma & \gamma \\ \zeta & \mu + \sigma + \zeta \end{pmatrix}.$$

Argue that the next-generation matrix K is equal to the NGM with large domain and is therefore given by

$$K = -T \, \Sigma^{-1}.$$

Exercise 7.27 Show that $\det K = 0$ and that K has a one-dimensional range. How is the answer connected to a specific assumption we have made for the entries of the matrix T?

Exercise 7.28 Give the matrices R and C, as defined in Section 7.2, that span the rows and columns of T, respectively. Show that the NMG with small domain K_S is a one-dimensional 'matrix,'

$$K_S = -\frac{pc}{N}\begin{pmatrix}1\\w\end{pmatrix} \cdot \Sigma^{-1}\begin{pmatrix}N_0\\vN_+\end{pmatrix},$$

where '·' denotes the inner product.

Exercise 7.29 Check that R_0 is given by

$$R_0 = pc\frac{(\gamma+\mu)(\mu+\sigma+\gamma+\zeta w) + \zeta vw(\frac{\gamma}{w}+\mu+\sigma+\zeta)}{(\gamma+\mu+\zeta)(\mu+\sigma)(\mu+\sigma+\gamma+\zeta)}. \qquad (7.36)$$

Before doing the algebra, what do you expect the special case $w = v = 1$ to give?

In the following exercises we take a brief look at one way to extract information from the complex function of the parameters in (7.36). For this we note that R_0^D is a factor in the right-hand side and that we can therefore rewrite (7.36) as $R_0 = R_0^D F$, with

$$F := \frac{(\gamma+\mu)(\mu+\sigma+\gamma+\zeta w) + \zeta vw(\frac{\gamma}{w}+\mu+\sigma+\zeta)}{(\gamma+\mu+\zeta)(\mu+\sigma+\gamma+\zeta)}.$$

We now study only the multiplication factor F. The way in which F depends on its ingredients can give us a first idea of the relative influence that the various parameters determining d have on the ability of D to invade.

Exercise 7.30 i) First show that, since all parameters are positive, $F \geq 1$.

ii) Show that for large γ or small ζ, or for v and w close to one, $F \approx 1$. Is this obvious from the biological interpretation?

iii) Show that F is a strictly decreasing function of γ and a strictly increasing function of ζ.

iv) Show that for ζ relatively large (i.e., high infective pressure for disease d), we have $F \approx vw$.

One can draw graphs of F as a function of γ for various fixed values of ζ, or as a function of ζ for fixed γ (keeping the other parameters constant). One can then study questions such as: in attempts to lower F (and thereby to decrease R_0 for D), is it more efficient to aim control measures at increasing γ (improving medical care) or at decreasing ζ (e.g., by campaigning against unprotected sex)? Under what conditions is one preferable to the other? We will not go into that much detail.

The previous exercise hints that notably the product vw can be an important determinant of the size of F. In the next exercise we look at the special case $v = w$ in more detail.

Exercise 7.31 i) Show that, for $v = w$, we can write

$$F = \left(1 - \frac{N_+}{N}\right) \frac{\mu + \sigma + \gamma + \zeta v}{\mu + \sigma + \gamma + \zeta} + \frac{N_+}{N} v^2 \frac{\frac{\gamma}{v} + \mu + \sigma + \zeta}{\mu + \sigma + \gamma + \zeta}.$$

ii) Make, in addition, the following assumptions:

$$\frac{\gamma}{\gamma + \mu} \approx 1, \qquad \frac{\sigma}{\gamma + \mu} \approx 0.$$

How could these be motivated biologically?

iii) Show that, with these assumptions, we have

$$F \approx 1 + \frac{N_+}{N}(v - 1)\left(\frac{N_+}{N}(v - 1) + 2\right).$$

We conclude that, under the given assumptions, F approximately increases quadratically with both $v - 1$ and the prevalence N_+/N of d in the population.

In order to get even more familiar with the techniques from this chapter, we provide a further series of exercises for the d/D-setting, with some added complications.

Exercise 7.32 Let us incorporate into our model the influence of disease D on the cure rate of disease d by introducing a factor z ($0 \le z \le 1$) that describes to what extent the probability per unit of time to recover from d is decreased by the presence of D. Retaining all assumptions in the previous calculations, how do the steady-state population sizes N_0 and N_+ change; how do T and the transition rate matrix Σ change? How does R_0 change? In the case that $v = w = 1$, will z matter?

In the remaining exercises we regard two extensions of these derivations to illustrate the idea of multi-group separable mixing (see Section 7.4.3). The first allows the individuals to be identified by sex, in addition to the marker 0 or +. Let the index f denote females and the index m males. Letting some parameters depend on sex, allows us to incorporate, for example, a higher probability of D-transmission from males to females than vice versa, or a higher contact rate for females when concentrating on a sub-population of prostitutes.

There are four D-infected states, which we indicate by $(f, 0)$, $(f, +)$, $(m, 0)$ and $(m, +)$, and which we shall always consider in that order. Because individuals can start infected life in each of these states and are infectious in each of these states, the sets of states-at-(D)-infection and states-of-(D)-infectiousness coincide with the set of D-infected states. The NGM with large domain and the NGM are therefore identical and have dimension four. We will see, however, that the separable mixing assumption leads to an NGM with small domain that has dimension two. The four D-infected states naturally separate into two pairs $(f, 0)$, $(f, +)$ and $(m, 0)$, $(m, +)$ since the (f, m) distinction is static, while the $(0, +)$ distinction is dynamic. The dynamics of state transitions not involving transmission are then described by

$$\Sigma = \begin{pmatrix} \Sigma_f & 0 \\ 0 & \Sigma_m \end{pmatrix},$$

with both Σ_f and Σ_m as the matrix in Exercise 7.26, but with the parameters γ and ζ different for f and m, and with '0' denoting a 2×2-matrix containing only zeros. Consequently, we have

$$-\Sigma^{-1} = \begin{pmatrix} -\Sigma_f^{-1} & 0 \\ 0 & -\Sigma_m^{-1} \end{pmatrix}.$$

The contact rates are c_f and c_m respectively, with consistency requiring that $c_f N_f = c_m N_m$ if all contacts are heterosexual, where N_f and N_m are the population sizes of females and males respectively. In self-explanatory notation, we have that $N_f = N_{f,0} + N_{f,+}$ and $N_m = N_{m,0} + N_{m,+}$ where $N_{f,0}$ and $N_{f,+}$ can be related to N_f and the parameters γ, μ and ζ with index f exactly as in Exercise 7.25, and similarly $N_{m,0}$ and $N_{m,+}$ can be related to N_m and the parameters γ, μ and ζ with index m.

Let p_{fm} be the D-transmission probability during a contact of a D-infectious male with a D-susceptible female, both of which are free from d. If the D-infectious individual is female while the D-susceptible individual is a male, the corresponding quantity is denoted by p_{mf}. The enhancement factors are v_f and v_m for d-infected female/male D-susceptibles and w_f and w_m for d-infected female/male D-infectious individuals.

Exercise 7.33 Convince yourself that the analog of the matrix T from Exercise 7.24 is now given by

$$T = \begin{pmatrix} 0 & T_{fm} \\ T_{mf} & 0 \end{pmatrix},$$

where 0 is a 2×2 matrix containing only zeros, and where

$$T_{fm} = \begin{pmatrix} c_m \frac{N_{f,0}}{N_f} p_{fm} & c_m \frac{N_{f,0}}{N_f} p_{fm} w_m \\ c_m \frac{N_{f,+}}{N_f} p_{fm} v_f & c_m \frac{N_{f,+}}{N_f} p_{fm} v_f w_m \end{pmatrix}$$

and T_{mf} is similar with f and m interchanged throughout. Convince yourself that the columns of T belong to the subspace spanned by the two vectors

$$\frac{c_f}{N_m} p_{mf} \begin{pmatrix} 0 \\ 0 \\ N_{m,0} \\ N_{m,+} v_m \end{pmatrix}, \qquad \frac{c_m}{N_f} p_{fm} \begin{pmatrix} N_{f,0} \\ N_{f,+} v_f \\ 0 \\ 0 \end{pmatrix}$$

and that the rows of T belong to the subspace spanned by the two vectors $(0, 0, 1, w_m)^\top$ and $(1, w_f, 0, 0)^\top$. Conclude that the separable mixing assumption leads to a two-dimensional NGM with small domain, and that we can therefore derive an explicit expression for R_0.

Exercise 7.34 Convince yourself that the NGM with small domain is given by

$$K_S = -R\Sigma^{-1}C = \begin{pmatrix} 0 & k_{fm} \\ k_{mf} & 0 \end{pmatrix},$$

with $k_{fm} = -\binom{1}{w_m} \cdot \Sigma_m^{-1} \psi_2$ and $k_{mf} = -\binom{1}{w_f} \cdot \Sigma_f^{-1} \psi_1$, where

$$\psi_1 = \frac{c_m}{N_f} p_{fm} \begin{pmatrix} N_{f,0} \\ N_{f,+} v_f \end{pmatrix}$$

and

$$\psi_2 = \frac{c_f}{N_m} p_{mf} \begin{pmatrix} N_{m,0} \\ N_{m,+} v_m \end{pmatrix}.$$

Derive an explicit expression for R_0 in terms of $\Sigma_f^{-1}, \Sigma_m^{-1}, c_m, c_f, p_{fm}, p_{mf},$ $v_f, v_m, w_f, w_m, N_{f,0}/N_f, N_{f,+}/N_f, N_{m,0}/N_m$ and $N_{m,+}/N_m$.

As a second extension, we regard (fixed) sexual activity level as the first component of the h-state. For exposition purposes, we let this trait take discrete values $i \in \{1, 2, \ldots\}$. Assume

$$\frac{dN(i, +)}{dt} = i\zeta N(i, 0) - \gamma N(i, +) - \mu N(i, +)$$

and write $\phi = (\phi_0, \phi_1, \ldots)^\top$, with

$$\phi_i := \begin{pmatrix} \phi(i, 0) \\ \phi(i, +) \end{pmatrix}.$$

The next-generation matrix is now an infinite matrix acting on ϕ, where ϕ is an infinite array of two-vectors. Because the set of states-at-infection is again equal to the set of infected states, the NGM is equal to the NGM with large domain. Let the Markovian h-state dynamics for the ith two-vector be described by the matrix Σ_i.

Exercise 7.35 Write $N_i = N(i, 0) + N(i, +)$. Give $N(i, 0)$ and $N(i, +)$. Give Σ_i operating on the ith two-vector ϕ_i.

Exercise 7.36 Give a proportionate mixing expression for the meeting rate between $(i, \cdot)$-individuals and $(j, \cdot)$-individuals.

Exercise 7.37 Convince yourself that the ijth element of K is given by

$$K_{ij} = \frac{-pcij}{\sum_k kN_k} \begin{pmatrix} N(i, 0) & wN(i, 0) \\ vN(i, +) & wvN(i, +) \end{pmatrix} \Sigma_j^{-1}.$$

Calculate $(K\phi)_i$. Show that the range of K is spanned by the infinite array of the two-vectors

$$i \begin{pmatrix} N(i, 0) \\ vN(i, +) \end{pmatrix} \propto i \begin{pmatrix} \gamma + \mu \\ iv\zeta \end{pmatrix}.$$

Exercise 7.38 Show that

$$R_0 = \frac{-pc}{\sum_k kN_k} \begin{pmatrix} 1 \\ w \end{pmatrix} \cdot \sum_{j=0}^{\infty} j^2 \Sigma_j^{-1} \begin{pmatrix} N(j, 0) \\ vN(j, +) \end{pmatrix}.$$

We leave it to those readers who are not yet exhausted to combine both extensions, that is, discuss heterosexual transmission taking sexual activity levels into account.

7.8   PAIR FORMATION MODELS

In deterministic models in which, by assumption, all sub-population sizes are infinite, two individuals never have contact twice.

Exercise 7.39 Digest this statement. Do you agree? Reflect on the following arguments: i) the expected number of contacts of some given individual in a given finite time interval is finite; ii) there are infinitely many candidates for a contact and they are 'chosen' with equal probability; iii) ergo, for any particular *couple* of individuals the probability of having contact in that time interval is zero.

Such an 'idealization' seems questionable when, for instance, STDs are considered, but also when one thinks of families, schools, offices, ... As an alternative, we may conceive of a population as a network, with connections that form and break but exist for some extended period of time. There we think of transmission as being restricted to individuals that are connected. When every individual can be connected to at most one other individual, the resulting 'network' is simply a collection of singles and pairs, which makes the bookkeeping simple. It is this class of pair formation models that we consider in the present section.

What kind of difference do we expect? The key point is that, from the point of view of the infective agent, contacts between two infected individuals are wasted. Let us try to determine R_0 and look for such an effect.

As a warm-up, we consider the pair formation and dissociation process by itself, in the absence of infection transmission. We use a Markov process description as summarized in Figure 7.1.

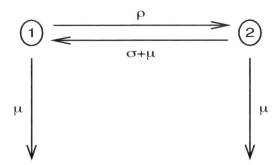

Figure 7.1: A setup with two states and the transitions allowed between them. State 1 signifies that an individual is single, while in state 2 the individual is paired to one other individual.

We assume an exponential lifetime distribution with parameter μ, a probability per unit of time ρ of acquiring a partner and a probability per unit of time σ that a partnership ends in a divorce. Since an individual can also return to the single state due to the death of the partner, the rate from 2 to 1 equals $\sigma + \mu$. The death of the individual that we consider takes it to a hypothetical state outside the system or, in other words, our probability distributions will be defective.

Exercise 7.40 i) Show that the transition matrix

$$\Sigma = \begin{pmatrix} -\rho - \mu & \sigma + \mu \\ \rho & -\sigma - 2\mu \end{pmatrix}$$

has determinant $D = \mu(\rho + \sigma + 2\mu)$ and that

$$-\Sigma^{-1} = \frac{1}{D} \begin{pmatrix} \sigma + 2\mu & \sigma + \mu \\ \rho & \rho + \mu \end{pmatrix}.$$

ii) Use these results to check that any living individual has an expected future lifespan $1/\mu$ and that, if it is single, it will spend a fraction $(\sigma + 2\mu)/(\rho + \sigma + 2\mu)$ of this time in that state, while this fraction equals $(\sigma + \mu)/(\rho + \sigma + 2\mu)$ if it is in a pair.

iii) Show in two different ways that the expected number of partners of a single individual equals $\rho(\sigma + 2\mu)/D$: a) divide the expected time spent in a pair by the expected duration of a partnership; b) denote the quantity we wish to determine by Q and use first (and second) step analysis and the Markov property to derive the equation

$$Q = \frac{\rho}{\rho + \mu}\left(1 + \frac{\sigma + \mu}{\sigma + 2\mu}Q\right).$$

Next, let us distinguish between susceptibles and infectives. To make life simple at first, assume that a newly infected individual is infectious right away and that it has constant infectivity for the rest of its life. We also assume that the disease does not give rise to an extra death rate. We want to focus on a newly infected individual and to compute the expected number of new cases that it will make, i.e., R_0. Necessarily, a newly infected individual will be in a pair with another infective. Before it can infect anybody, it has to pass through the single state. After it has acquired a new partner, the infected individual that we consider may or may not infect the new partner. Note that we assume that the new partner is a susceptible. We are only interested in the initial phase of the process when the infectives still form a negligible fraction of the population. We conclude that the states of the Markov chain need some modification; in particular the previous state 2 has to be split into two states, one for each possible state of the individual's partner. More precisely, we now consider Figure 7.2, where h is the probability per unit of time that the agent is transmitted from an infective to its susceptible partner.

We claim that

$$R_0 = \frac{\rho h(\sigma + \mu)}{\mu(\rho + \sigma + 2\mu)(h + \sigma + 2\mu)}.$$

Exercise 7.41 i) Show that the probability that a susceptible partner is infected equals $h/(h + \sigma + 2\mu)$.

ii) A newly infected individual has state 3. Show that the probability that it will enter state 1 is given by $(\sigma + \mu)/(\sigma + 2\mu)$.

iii) Argue that the expected number of partners of an infected single individual is, as calculated in Exercise 7.40, equal to $\rho(\sigma + 2\mu)/D$. Hint: Does it matter that we consider an infective? Does it matter, as far as we restrict our attention to partnership status, whether the individual is in state 2 or in 3?

iv) Derive the expression for R_0.

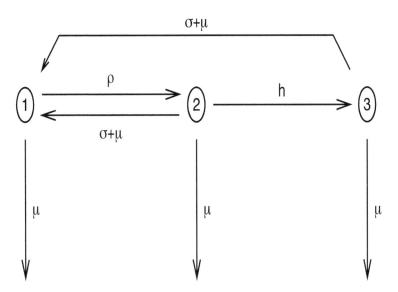

Figure 7.2: A setup with three states and the transitions allowed between them. State 1 signifies an infected individual that is single, in state 2 the infected individual is paired to a susceptible individual, while in state 3 the infected individual is still paired, but has infected the partner.

Exercise 7.42 Here we give an alternative derivation of the expression for R_0.

i) Let P_{13} be the probability to ever arrive in 3 starting from 1. Use the Markov property to deduce that

$$P_{13} = \frac{\rho}{\rho + \mu} \left(\frac{h}{h + \sigma + 2\mu} + \frac{\sigma + \mu}{h + \sigma + 2\mu} P_{13} \right),$$

and solve for P_{13}.

ii) Let P_{33} be the probability to ever return to 3 when starting from 3. Show that

$$P_{33} = \frac{\sigma + \mu}{\sigma + 2\mu} P_{13}.$$

iii) Show that $R_0 = P_{33}(1 + R_0)$, and determine R_0.

iv) Observe that $R_0 > 1 \Leftrightarrow P_{33} > 1/2$ and try to interpret the latter condition directly as a threshold condition.

Motivated by the preceding exercise, we make the following useful observation. A newly infected individual is in state 3 and can itself infect some other individual only when in state 2. From state 3 to state 2, it has to pass through state 1. So we can also base our bookkeeping on passage through 1. Of course, an infected individual in state 1 may re-enter state 1 after having infected someone else, i.e., from state 3. When we define

$$\tilde{R}_0 = \text{expected number of individuals entering state 1 from state 3}$$
$$\text{due to one individual that just entered state 1 from state 3,}$$

then we also count such a re-entering individual. Note that clearly $\widetilde{R}_0$ has threshold value 1 but that it is, in general, different from R_0.

Exercise 7.43 Show that

$$\widetilde{R}_0 = 2\frac{\sigma+\mu}{\sigma+2\mu}P_{13} = 2P_{33}.$$

This different way of doing the bookkeeping is of much help when dealing with different types of partnerships, as in Exercise 7.47. Before turning to further complications, however, let us look at simplifications and, more particularly, at the relation with R_0 for models that disregard pair formation.

Exercise 7.44 i) In the limit $\sigma, h \to \infty$ with $h/\sigma \to p$ we arrive at a model with contact rate ρ and success ratio $q = p/(p+1)$ for which $R_0 = \rho q/\mu$. Verify this statement.

ii) Take $h = p\sigma$ for fixed p. Show that R_0 is an increasing function of σ.

iii) Contemplate the difficulty of gauging epidemic models when comparing them.

Exercise 7.45 i) Suppose the disease causes an additive death rate κ. Modify the model and compute R_0.

ii) Formulate a model in which an individual enters a state of mourning after the death of its partner, from which it may jump with constant probability per unit of time to the single state.

Exercise 7.46 Suppose an infected individual has an exponentially distributed infectious period with parameter α, during which it has constant infectivity. Modify the original model and compute R_0.

Exercise 7.47 Suppose there are two types of pairs: long-term ones (with divorce rate σ_1 and transmission probability/time h_1) and casual ones (with divorce rate σ_2 and transmission probability/time h_2). Assume that a single still acquires a partner with rate ρ but that it will then form a long-term pair with probability f and a casual pair with probability $1 - f$. Elaborate the model formulation and compute R_0. Hint: Use the 'decoupling' argument about passage through the single state. This makes the problem one-dimensional, despite the fact that there are now two different states for newly infected individuals.

Exercise 7.48 So far we have not taken the sex of an individual into account. If the sex ratio is 1/2, does it matter? Can you think of situations in which it does matter?

7.9 INVASION UNDER PERIODIC ENVIRONMENTAL CONDITIONS

In most of our exposition we assume that environmental conditions are constant over time. A fluctuating environment can, however, influence the dynamics. One can think of seasonality in contact opportunities (e.g., caused by the school system, in connection to childhood infections), in host presence or abundance (e.g., in vector-borne infections, where the emergence of new vector individuals is influenced by environmental variables such as temperature and humidity), or in the value of important epidemiological determinants such as the length of the latency period or infectious period (these are, for vectors in vector-borne infections, strongly influenced by temperature). Also, environmental stages in the life cycles of macroparasites entail that temporal variation in the environment influences, for example, developmental rates and death rates, and thus can have a marked influence on the transmission dynamics and hence on the ability to invade a naive population.

In general, few methods are available to analyze the effect of temporal variation. If the variation is not periodic, success of invasion could be studied by calculating dominant Lyapunov exponents. These exponents characterize the rate at which initially close trajectories of the dynamical system separate in phase space. If variation is periodic (e.g., seasonal), we can make use of the Floquet theory for stability of the trivial steady state of systems of ordinary differential equations with periodically varying rate constants. We only regard the periodic case here. We provide some details about this theory and refer to books on ODEs, e.g., Hale (1969), for an in-depth treatment.

The basic idea of Floquet theory is simple. Consider a linear system of ordinary differential equations with periodic coefficients. Let p denote the period. Choose an initial time t_0 and prescribe an initial condition x_0 at this time. The time-p-map assigns to x_0 the value of the solution (with initial condition x_0 at t_0) at time $t_0 + p$. This map is represented by a matrix (called the monodromy matrix). The eigenvalues of this matrix are called the Floquet multipliers. It turns out that these eigenvalues: i) are independent of the choice of t_0, and ii) govern the stability of the zero (in our case: infection-free) steady state. The latter property means that all solutions of the linear ODE system converge exponentially to zero if all Floquet multipliers are, in absolute value, strictly smaller than one; if at least one multiplier has absolute value greater than one, there exists an exponentially growing solution.

So, from a theoretical point of view, the situation is as clear cut as in the case of constant coefficients. Yet from a practical point of view major difficulties arise: how to compute the Floquet multipliers? Before illustrating the difficulties by way of concrete examples, we mention one more positive conclusion: the principle of linearized stability holds for periodic solutions. More precisely, to test whether or not a periodic solution of a nonlinear periodic ODE system is stable, one linearizes and determines the Floquet multipliers; these govern the local stability in essentially the sense described above, but with the subtlety that necessarily there is a Floquet multiplier equal to one, reflecting the invariance with respect to translation along the periodic orbit.

Exercise 7.49 Regard the following simple SIR model

$$\frac{dS}{dt}(t) = -\gamma(t)\frac{S(t)I(t)}{N}, \tag{7.37}$$

$$\frac{dI}{dt}(t) = \gamma(t)\frac{S(t)I(t)}{N} - \alpha I(t), \tag{7.38}$$

where

$$\gamma(t) = \gamma_0(1 + \gamma_1 \cos(2\pi t)) \tag{7.39}$$

with γ_0 and γ_1 positive constants. So the contact rate 'parameter' is periodic, with period normalized to 1 (keeping in the back of our mind that the time unit may be chosen as one year). Show that in this particular case the stability of the infection-free steady state $(S, I) = (N, 0)$ is determined by γ_0/α, which is equal to the time average of $\gamma(t)/\alpha$. Hint: Use the result that the solution of a first-order ODE with period 1 can be written as $x(t) = \Phi(t)x(0)$ where $\Phi(t+1) = \Phi(t)E$, with $E := \Phi(1)$, and that, in the one-dimensional case, the stability of the trivial steady state is determined by the absolute value of E.

It is tempting to conjecture that one can more generally decide about the stability of the infection-free steady state on this basis of considering averages. We shall now show, by way of an example, that this idea is flawed. In Exercise 2.2 we considered the SEIR-system, characterized by an exponentially-distributed latency period. After linearization, the relevant variables are the number of exposed individuals E, and the number of infectious individuals I. The changes in the course of time of these variables are generated by the differential equations

$$\frac{dE}{dt} = \beta N I - \theta E, \tag{7.40}$$

$$\frac{dI}{dt} = \theta E - \alpha I. \tag{7.41}$$

When β, θ and α are constant, the basic reproduction number is given by $R_0 = \beta N/\alpha$. What changes if these coefficients are periodic? Consider the following special case, admittedly rather contrived for exposition purposes. We let α be constant, but let the values for β and θ vary in a special way: periods where $\beta = \beta_0, \theta = 0$ (a 'β-period') alternate with periods where $\beta = 0, \theta = \theta_0$ (a 'θ-period'). To proceed we observe that the solution operator for a θ-period is given by

$$\exp[t \begin{pmatrix} -\theta_0 & 0 \\ \theta_0 & -\alpha \end{pmatrix}] = \begin{pmatrix} e^{-\theta_0 t} & 0 \\ \frac{\theta_0}{\alpha-\theta_0}(e^{-\theta_0 t} - e^{-\alpha t}) & e^{-\alpha t} \end{pmatrix}$$

and the solution operator for a β-period by

$$\exp[t \begin{pmatrix} 0 & \beta N \\ 0 & -\alpha \end{pmatrix}] = \begin{pmatrix} 0 & \frac{\beta N}{\alpha}(1 - e^{-\alpha t}) \\ 0 & e^{-\alpha t} \end{pmatrix}.$$

The plan is now to compare the following two situations:

- A β-period of length $1/2$ is followed by a θ-period of length $1/2$, and this interval of length 1 is periodically repeated;

- A β-period of length $1/4$ is followed by a θ-period of length $1/4$, and this interval of length $1/2$ is periodically repeated.

Note that also in the second situation we have that in an interval of length 1, the total length of the β- and θ-periods equals $1/2$. In particular, the average values of β and θ are the same in the two situations. Yet, as we will demonstrate, the stability criteria for the two situations are not the same.

From these auxiliary results we deduce that the product of the solution operators for the θ-period and the β-period is given by

$$\begin{pmatrix} 0 & \frac{\beta N}{\alpha}e^{-\theta_0 s}(1 - e^{-\alpha s}) \\ 0 & \frac{\beta N}{\alpha}\frac{\theta_0}{\alpha-\theta_0}(e^{-\theta_0 s} - e^{-\alpha s})(1 - e^{-\alpha s}) + e^{-2\alpha s} \end{pmatrix}.$$

In the first situation, the time-1-map is given by this expression evaluated at $s = 1/2$. In the second situation, the time-$1/2$-map is given by the same expression evaluated at $s = 1/4$. Next note that the eigenvalues of the matrix above are its diagonal elements. So, according to Floquet theory, the condition for the stability of the infection-free steady state is

$$\frac{\beta N}{\alpha} \frac{\theta_0}{\alpha - \theta_0}(e^{-\frac{1}{2}\theta_0} - e^{-\frac{1}{2}\alpha})(1 - e^{-\frac{1}{2}\alpha}) + e^{-\alpha} < 1$$

in the first situation, while in the second situation the condition reads

$$\frac{\beta N}{\alpha} \frac{\theta_0}{\alpha - \theta_0}(e^{-\frac{1}{4}\theta_0} - e^{-\frac{1}{4}\alpha})(1 - e^{-\frac{1}{4}\alpha}) + e^{-\frac{1}{2}\alpha} < 1.$$

(If you prefer to also formulate the conditions for the second case in terms of the time-1-map, even though the minimal period equals $1/2$, you should require that the *square* of the expression on the left-hand side of the last inequality is bounded by one, which is clearly equivalent to the condition given above.) For completeness, we observe that it may happen that one condition holds while the second is violated. Indeed, this is easy to see in the limit $\alpha \to \infty, \beta \to \infty$ with β/α^2 constant.

We conclude that it matters how we distribute β-periods of total length $1/2$ over the interval of length 1. The reason is that, in general, matrix products do not commute: their order matters.

There are additional questions that we need to address: is it possible to reformulate the stability condition in such a way that it can be interpreted in a generation perspective? Is it possible to define the basic reproduction number as introduced in this chapter, in a natural generation-based setting, in the case of a periodic system? Initially, one would be inclined to answer the first question with a firm 'no.'

Exercise 7.50 Explain why a quantity derived from stability arguments in conditions of temporal variation is not likely to have the individual-level interpretation of R_0.

Indeed, in the generation way of looking at population growth, we presume that it does not matter *when* offspring is produced. In case of varying environmental conditions, however, it *does* matter.

Fortunately, there is a trick that can save us. Note that the only problem is that it now matters when exactly, during a period, an individual is infected. We can take care of that problem by declaring the phase in the cycle at the moment of becoming infected to be part of the (h-)state-at-infection (i.e., we label infected individuals by the time (modulo the period) at which they became infected). We have to pay a price for this trick. The problem of characterizing R_0 becomes infinite-dimensional, even when it is finite-dimensional in the corresponding case of a constant environment.

To illustrate the key points in the simplest possible setting, we consider the SIR-model of Exercise 7.49, but now also allow α to be a periodic function of time, with period one. Our first aim is to specify the kernel k, as introduced in Section 7.3, using the type of considerations explained in Section 7.5. We want to determine $k = k(t, s)$ such that the next-generation operator K is given by

$$(K\phi)(t) = \int_0^1 k(t, s)\phi(s)ds.$$

Here, ϕ is a periodic function (of period one) describing the density of the distribution with respect to the state-at-infection (i.e., time in the cycle at infection) of a generation, and $K\phi$ corresponds to this density for the next generation. As usual, τ denotes the time elapsed since infection, and our aim is to compute k from $A = A(\tau, t, s)$ by way of

$$k(t,s) = \int_0^\infty A(\tau, t, s)d\tau.$$

Here, A denotes the rate at which an individual, itself infected at time s in the cycle, produces, τ time units after its infection, secondary cases with state-at-infection t. It follows directly from this interpretation that $A(\tau, t, s)$ can only be different from zero if $t = \tau + s \pmod 1$. This means that A cannot be a function in the usual sense, but is a 'generalized function' (or more precisely, a distribution) involving the Dirac delta 'function.' In fact, using the notation of Exercise 7.49, we have

$$A(\tau, t, s) = \gamma(s+\tau)e^{-\int_s^{s+\tau} \alpha(\sigma)d\sigma} \delta(t - [s+\tau])$$

where we write $[s+\tau] := s+\tau \pmod 1$, normalized such that $0 \leq [s+\tau] < 1$, and we also normalize t and s such that they take values in $[0, 1)$. Hence,

$$\int_0^\infty A(\tau, t, s)d\tau = \begin{cases} \gamma(t)\sum_{j=0}^\infty e^{-\int_s^t \alpha(\sigma)d\sigma - j\bar\alpha} & s < t; \\ \gamma(t)\sum_{j=1}^\infty e^{-\int_s^t \alpha(\sigma)d\sigma - j\bar\alpha} & s > t \end{cases}$$

where $\bar\alpha := \int_0^1 \alpha(\sigma)d\sigma$. After evaluating the sums and recalling that the left-hand side is, by definition, $k(t, s)$, we find

$$k(t,s) = \begin{cases} \left(\gamma(t)e^{-\int_s^t \alpha(\sigma)d\sigma}\right)(1 - e^{-\bar\alpha})^{-1} & s < t; \\ \left(e^{-\bar\alpha}\gamma(t)e^{-\int_s^t \alpha(\sigma)d\sigma}\right)(1 - e^{-\bar\alpha})^{-1} & s > t. \end{cases}$$

Hence

$$(K\phi)(t) = \frac{\gamma(t)}{1 - e^{-\bar\alpha}}\left(\int_0^t e^{-\int_s^t \alpha(\sigma)d\sigma}\phi(s)ds + \int_t^1 e^{-\bar\alpha - \int_s^t \alpha(\sigma)d\sigma}\phi(s)ds\right).$$

After having derived this explicit expression for K, we want to check that a spectral analysis of K indeed yields the anticipated conclusion that, in this essentially one-dimensional situation,

$$R_0 = \frac{\bar\gamma}{\bar\alpha}$$

where $\bar\gamma := \int_0^1 \gamma(\sigma)d\sigma$. For this analysis consider $K\phi = \lambda\phi$ and put

$$\phi(t) = \gamma(t)e^{-\int_0^t \alpha(\sigma)d\sigma}\psi(t).$$

Then,

$$\lambda\psi(t) = \frac{1}{1 - e^{-\bar\alpha}}\left(\int_0^t \gamma(s)\psi(s)ds + \int_t^1 e^{-\bar\alpha}\gamma(s)\psi(s)ds\right).$$

Differentiation of both sides leads to the identity

$$\lambda\psi'(t) = \frac{1}{1 - e^{-\bar\alpha}}\left(\gamma(t)\psi(t) - e^{-\bar\alpha}\gamma(t)\psi(t)\right) = \gamma(t)\psi(t)$$

from which we deduce that

$$\psi(t) = \psi(0)e^{\frac{1}{\lambda}\int_0^t \gamma(\sigma)d\sigma}.$$

Therefore

$$\phi(t) = \psi(0)\gamma(t)e^{-\int_0^t \alpha(\sigma)d\sigma + \frac{1}{\lambda}\int_0^t \gamma(\sigma)d\sigma},$$

and in particular

$$\phi(1) = \psi(0)\gamma(1)e^{-\bar{\alpha}+\frac{1}{\lambda}\bar{\gamma}} = \phi(0)e^{-\bar{\alpha}+\frac{1}{\lambda}\bar{\gamma}}.$$

From this we see that periodicity of ϕ is guaranteed if and only if

$$\lambda = \frac{\bar{\gamma}}{\bar{\alpha} + 2\ell\pi i}$$

for some integer ℓ, and we conclude that indeed, as expected, the eigenvalue of K with the largest absolute value is exactly $R_0 = \bar{\gamma}/\bar{\alpha}$.

In the final part of this section we expand this example in the more general context of compartmental epidemic systems, as in Section 7.2, with linearized infection sub-systems of the form

$$\frac{dx}{dt}(t) = (T(t) + \Sigma(t))\, x(t), \tag{7.42}$$

where T and Σ are now periodic matrix-valued functions of t, with period p, describing transmissions and changes in infected states, respectively. In the remaining pages of this section we shall

- Describe how to generalize A, k and K to this far more general setting.

- Deduce from these results a general characterization of R_0 that is amenable to numerical calculation.[6]

Recall (see e.g., Hale 1969) that the *fundamental matrix solution* of

$$\frac{dy}{dt}(t) = \Sigma(t)y(t) \tag{7.43}$$

is the matrix $\Phi(t, s)$ satisfying

$$\begin{cases} \frac{\partial}{\partial t}\Phi(t, s) = \Sigma(t)\Phi(t, s), \\ \Phi(s, s) = I. \end{cases}$$

Note that i) $\Phi(t + p, s) = \Phi(t, s - p)$; and ii) the solution of (7.43) with initial condition $y(s)$ at time s is given by $y(t) = \Phi(t, s)y(s)$.

We next define

$$A(\tau, t, s) = T(s + \tau)\Phi(s + \tau, s)\delta(t - [s + \tau])$$

and

$$k(t, s) = \int_0^\infty A(\tau, t, s)d\tau = \begin{cases} T(t)\sum_{m=0}^\infty \Phi(t + mp, s) & \text{if } t > s; \\ T(t)\sum_{m=1}^\infty \Phi(t + mp, s) & \text{if } t < s. \end{cases}$$

---

[6]This part is inspired by the pioneering paper N. Bacaër and S. Guernaoui: The epidemic threshold of vector-borne diseases with seasonality. *J. Math. Biol.*, **53** (2006), 421–436. The use of $L_1 L_2$ in our exposition is a useful refinement introduced in W. Wang and X-Q Zhao: Threshold dynamics for compartmental epidemic models in periodic environments, *J. Dyn. Diff. Equat.*, **20** (2008), 699–717.

We can then compute that

$$
\begin{aligned}
(K\phi)(t) &= \int_0^p k(t,s)\phi(s)ds \\
&= T(t)\left(\int_0^t \sum_{m=0}^\infty \Phi(t+mp,s)\phi(s)ds + \int_t^p \sum_{m=1}^\infty \Phi(t+mp,s)\phi(s)ds\right) \\
&= T(t)\left(\int_0^t \Phi(t,s)\phi(s)ds + \int_0^p \Phi(t+p,s)\phi(s)ds \right.\\
&\qquad \left. + \int_0^p \Phi(t+2p,s)\phi(s)ds + \cdots\right) \\
&= T(t)\left(\int_0^t \Phi(t,s)\phi(s)ds + \int_{-p}^0 \Phi(t,\sigma)\phi(\sigma)d\sigma \right.\\
&\qquad \left. + \int_{-2p}^{-p} \Phi(t,\sigma)\phi(\sigma)d\sigma + \cdots\right) \\
&= T(t)\int_{-\infty}^t \Phi(t,\sigma)\phi(\sigma)d\sigma.
\end{aligned}
$$

We conclude: R_0 is the spectral radius of the operator K defined by

$$
(K\phi)(t) = T(t)\int_{-\infty}^t \Phi(t,\sigma)\phi(\sigma)d\sigma
$$

acting on the space of locally integrable p-periodic vector-valued functions ϕ. A proof that indeed the sign of $R_0 - 1$, with this definition of R_0, tells us whether or not the infection-free steady state of (7.42) is stable, is provided by Wang and Zhao (2008, see footnote 6 in this section). The reasoning is similar to that of the proof of the corresponding result for the constant coefficient case, which we give in Section 8.2.

But, one may well ask, how now to compute R_0? For this, first note that $K = L_2 L_1$ where

$$
(L_1\phi)(t) := \int_{-\infty}^t \Phi(t,\sigma)\phi(\sigma)d\sigma
$$

and

$$
(L_2\phi)(t) := T(t)\phi(t).
$$

It is a general result that the spectrum of $L_2 L_1$ equals the spectrum of $L_1 L_2$ if we disregard $\lambda = 0$ (in other symbols: $\sigma(L_2 L_1) \setminus \{0\} = \sigma(L_1 L_2) \setminus \{0\}$). In particular, $L_2 L_1$ and $L_1 L_2$ have the same spectral radius.

Now suppose that $L_1 L_2 \phi = \lambda\phi$, then, by differentiation, it follows that

$$
\lambda\phi'(t) = T(t)\phi(t) + \Sigma(t)(L_1 L_2\phi)(t) = T(t)\phi(t) + \lambda\Sigma(t)\phi(t)
$$

which, for $\lambda \neq 0$ is equivalent to

$$
\phi'(t) = \frac{T(t)}{\lambda}\phi(t) + \Sigma(t)\phi(t). \tag{7.44}
$$

Conversely, if ϕ is p-periodic and satisfies this differential equation then, by using the variation-of-constants formula and the fact that (under appropriate conditions on $\Sigma(t)$, cf. Section 8.2) $\Phi(t,s) \to 0$ exponentially for $t - s \to \infty$,

$$\phi(t) = \int_{-\infty}^{t} \Phi(t,\sigma) \frac{T(\sigma)}{\lambda} \phi(\sigma) d\sigma,$$

i.e.,

$$\lambda\phi = L_1 L_2 \phi.$$

We conclude that non-zero eigenvalues of K are in one-to-one correspondence with $\lambda \in \mathbb{C}$ for which (7.44) has a non-trivial p-periodic solution. By invoking positivity arguments we can, in fact, say a bit more.

Let $\Psi(t,s,\lambda)$ denote the fundamental matrix solution of (7.44). For positive λ, the matrix $\Psi(p,0,\lambda)$ is positive. The spectral radius $\rho(\Psi(p,0,\lambda))$ is a monotonically decreasing function of λ. If $R_0 > 0$, then R_0 is the unique root of the equation $\rho(\Psi(p,0,\lambda)) = 1$. In principle it may happen that $\rho(\Psi(p,0,\lambda)) < 1$ for all $\lambda > 0$, and in that case necessarily $R_0 = 0$; we refer once again to Wang and Zhao (2008), see footnote 6, for a detailed proof. The upshot of it is that a numerical procedure for solving $\rho(\Psi(p,0,\lambda)) = 1$ yields an operational characterization of R_0.

As we discussed at length in Section 7.2, the range of $T(t)$ is often spanned by fewer than n t-independent vectors, where n denotes the number of infected states (i.e., the dimension of system (7.42)). As a consequence, the operator K has a reduced range too. Yet we did not exploit that fact. The reader may wonder why. The reason is that, after reduction to the range, the problem is still infinite-dimensional. So the reduction does not help much, neither from a theoretical point of view, nor from a practical, computational, point of view. Therefore, we emphasized instead the characterization of R_0 as a root of $\rho(\Psi(p,0,\lambda)) = 1$, as this *does* actually help to compute R_0 for concrete settings.

7.10   TARGETED CONTROL

We assume throughout this section that $R_0 > 1$. We consider the situation that a control method is implemented in a heterogeneous population, but that the effort is targeted at a subset of the different types of individuals, rather than at all individuals indiscriminately. In Section 1.4, Exercise 1.37, we saw that in a homogeneous population, vaccination against a given agent with a perfect vaccine administered to a fraction v of the population, will prevent outbreaks of that agent in the vaccinated population if and only if $R_v < 1$, i.e., if $v > 1 - 1/R_0$. One can say that the *critical control effort*, or critical vaccination effort in this case, v_c, defined here as the minimum effort required to prevent outbreaks of the agent in that population, is $1 - 1/R_0$. This result was first derived by C.E.G. Smith in 1964.[7] In Section 4.3, Exercise 4.22 we then looked at the changes in critical control effort when the vaccine is not perfect, but still for the case of a homogeneous population.

In a heterogeneous population, the same relation between R_0 and the critical control effort holds when the control method is imposed on the individuals regardless of their type. Consider the case of a finite number of types and a next-generation matrix K, with spectral radius R_0. If we vaccinate a fraction v of the individuals of each type in the population, we effectively regard a new next-generation matrix K_v with elements $(1-v)k_{ij}$. The spectral radius R_v of K_v is given by $R_v = (1-v)R_0$.

---

[7]C.E.G. Smith: Factors in the transmission of virus infections from animal to man. *Scientific Basis of Medicine Annual review* (1964), 125–150; see also J.A.P. Heesterbeek: A brief history of R_0. *Acta Biotheoretica*, **50** (2002), 189–204.

Exercise 7.51 Show that for any positive matrices K and $K_v = (1-v)K$ we have that $R_v = (1-v)R_0$.

When control is targeted at certain types, however, the situation changes, and the simple relation between critical control effort and R_0 breaks down. In this section we discuss a threshold quantity T that replaces R_0 in case of non-uniform vaccination of a heterogeneous population.[8] Please note with care that the meaning of the symbol T is completely different in the current section, as compared to its meaning in Sections 7.2 and 8.2, where it represented the transmission matrix as part of the NGM.

As an example, think of a vector-transmitted infectious agent. There are at least two types of infected individuals in such a system: the host species and the vector species (note that this is a largely anthropocentric distinction as in fact both species are a 'host' of the pathogen). In control, one could wish to contrast control effort of vaccinating hosts with control effort of chemically attacking the vector population. First consider the special case of pure vector transmission.

Exercise 7.52 Let type 1 denote a host and type 2 a vector and consider a purely vector-transmitted agent, i.e., the entries k_{11} and k_{22} of the next-generation matrix K are zero. Suppose we have a perfect vaccine to protect hosts. Show that the critical control effort for hosts is given by $v_c = 1 - 1/R_0^2$, i.e., that vaccination of a fraction $v > 1 - 1/R_0^2$ is required to prevent outbreaks in the system as a whole. Note that this result derives of course from our definition of 'generation,' and hence of R_0; in particular recall our discussion of Example 7.1.

Exercise 7.53 Consider the case of Exercise 7.52 where, additionally, there is vertical transmission in the vector species, i.e., $k_{22} > 0$. Again vaccinate hosts and show that critical control effort is given by

$$v_c = 1 - \frac{1 - k_{22}}{k_{12}k_{21}} = 1 - \frac{1}{R_0}\left(\frac{1 - k_{22}}{R_0 - k_{22}}\right) = 1 - \frac{1}{R_0^2}\left(\frac{1 - k_{22}}{1 - \frac{k_{22}}{R_0}}\right).$$

From this expression we deduce two things. First of all, but this can already be seen from the original expression, the calculation — as well as the attempt to prevent invasion by controlling only type 1 — does not make sense when $k_{22} \geq 1$. Second, given the condition $k_{22} < 1$, the term in square brackets is smaller than 1, meaning that, as to be expected, more control effort is needed in the case of the additional transmission route in the vector population.

Exercise 7.54 Explain why the calculation only makes sense from a biological point of view if $k_{22} < 1$.

Readers wanting yet another example can consider the case where also the host can directly cause cases in hosts, i.e., also $k_{11} > 0$. In exactly the same manner as

---

[8]The quantity T was introduced in M.G. Roberts and J.A.P. Heesterbeek: A new method to estimate the effort required to control an infectious disease. *Proc R. Soc. B*, **270** (2003), 1359–1364; and J.A.P. Heesterbeek and M.G. Roberts: The type-reproduction number T in the control of infectious diseases. *Math. Biosci.*, **206** (2007), 3–11. We base our exposition largely on these papers.

in Exercise 7.53 one then shows that for

$$K_v = \begin{pmatrix} (1-v)k_{11} & (1-v)k_{12} \\ k_{21} & k_{22} \end{pmatrix}$$

the critical control effort is

$$v_c = 1 - \frac{1-k_{22}}{k_{11}+k_{12}k_{21}-k_{11}k_{22}} = 1 - \frac{1-k_{22}}{k_{11}+R_0(R_0-k_{11}-k_{22})}.$$

What we see from these simple examples is that we retain a similar relation between critical control effort needed and R_0, but that critical control effort can no longer be expressed in terms of R_0 alone, as in the case of a uniform application of the control method to all types. In addition, these relations break down for systems where the number of types exceeds two. Except for very special cases there is then no longer an explicit expression for the dominant eigenvalue of the next-generation matrix, and therefore an explicit relation for control effort can no longer be derived in this way.

By concentrating on the secondary cases caused by individuals of the type at which control is targeted, one can define a new threshold quantity T, similar in spirit to R_0, that does not have these problems. We can express the critical control effort as an explicit expression in terms of T alone for any model involving a discrete number of types.

If control is targeted at just one type, out of n types, we can without loss of generality assume that this is type 1 (we will treat the case where control is targeted at a larger subset of the n types later). To define T, start with one individual of type 1 that has just become infected. We trace infection chains initiated by this individual, cycling through all future generations and all other types, and we count the number of secondary cases of type 1 that arise along the way. Note that we do not continue a chain after it has produced a new type-1 infection. T is therefore defined as the number of secondary cases of type 1 arising in any future generation, without allowing these secondary cases to reproduce (in epidemiological sense). Note the difference between this and the number of secondary cases produced by one type-1 individual directly. In the latter case these secondary cases are all in the next generation from the generation to which the infector belongs. In the case of T, the secondary cases are in principle in all future generations. Because it counts the future reproduction capability of a specific type, we call T the *type-reproduction number*.

How can we constructively define this quantity mathematically? Iteration of the NGM K takes into account all future infection chains arising from a given generation. First note that the discontinuation of future infection chains in the population whenever a type 1 case is produced can be taken into account by projecting the matrix K onto the subset consisting of the other types 2, ... , n after each generation, before continuing the remaining chains of infection (i.e. all chains that have so far not produced a new type 1 individual). We can cumulatively calculate the number of type 1 infections arising along the way. For example, we start with one individual of type 1. Let $e = (1,0,\ldots,0)^\top$ be the first unit vector in $\mathbb{R}^n$ (where as usual $\top$ denotes 'transpose'). Convince yourself that the expected number of type 1 cases in the next generation, if we start from the generation described by e, is given by $e^\top K e$. Let I be the identity matrix in $\mathbb{R}^n$, and let P be the projection matrix with entries $p_{11} = 1$ and $p_{ij} = 0$ for all $(i,j) \neq (1,1)$.

Exercise 7.55 Show that the total number of cases of type 1 defined in the way
described above in all future generations is given by

$$e^\top K \sum_{j=0}^{\infty} \left((I - P)K)^j \right) e.$$

Let a matrix M be defined by

$$M := PK \sum_{j=0}^{\infty} \left((I - P)K)^j \right)$$

whenever the infinite sum converges and define a quantity T by

$$T := e^\top M e.$$

The infinite sum converges when the spectral radius ρ of the matrix $(I - P)K$ is
less than 1. For later use we remark that, by the structure of the matrix M, one
can also interpret T as the spectral radius of the matrix M:

$$T = \rho(M).$$

This is because M has only non-zero entries in the first row. This means that
$n-1$ of the n eigenvalues of M are zero, and the dominant eigenvalue, and spectral
radius, is equal to the element m_{11}, which is also the definition for T given above.

Exercise 7.56 What is the biological interpretation of the situation where the
spectral radius $\rho((I - P)K) > 1$?

Given the existence of T we obtain the following explicit expression (by noting
that for a matrix X with $\rho(X) < 1$ one has that $\sum_{j=0}^{\infty} X^j = (I - X)^{-1}$ as in the
scalar case):

$$T = e^\top PK(I - (I - P)K)^{-1}e = e^\top K(I - (I - P)K)^{-1}e$$

where we used $e^\top P = e^\top$ in the last step. The reason we give this version is that
later we will study the case where control effort is aimed at larger subsets of types,
rather than just a single type. The type-reproduction number T has the following
properties:

- Given that $\rho((I - P)K) < 1$, a value of $T < 1$ precludes outbreaks in the
population as a whole:
$$T < 1 \Longleftrightarrow R_0 < 1.$$

- If we vaccinate (with a perfect vaccine) exclusively type-1 individuals, then
the minimum control effort v_c (to prevent outbreaks in the population as a
whole) is
$$v_c = 1 - \frac{1}{T}.$$

Exercise 7.57 Show that the type-reproduction number T indeed has these two
properties.

We conclude that we obtain a similar expression for critical control effort as in the case where the same control effort is imposed on all types of infected individuals, but with R_0 replaced by T: the type-reproduction number T takes over the role of R_0 in situations where we have more than one type and control effort is targeted at one type only.

Exercise 7.58 Give the expression for T for the host-vector systems of Exercises 7.52 and 7.53, as well as for the case where $k_{22} > 0$.

It is clear that one would also like to have the possibility of calculating critical control effort when targeting more than just a single type. The entire method outlined here carries over to the case where control is targeted at the first l types, among the n types. In the generalized definition T then is the spectral radius

$$T = \rho(M_l)$$

of the $l \times l$-matrix M_l

$$M_l = E_l^\top P_l K (I - (I - P_l)K)^{-1} E_l$$

where P_l is the $n \times n$-projection matrix on the subspace spanned by the first l unit vectors in $\mathbb{R}^n$, and E_l is the $n \times l$-matrix obtained by aligning the first l unit vectors in $\mathbb{R}^n$. For $l = 1$ we recover the situation discussed earlier. The two basic properties of T are proven in the same way as in the one-dimensional case. The matrix again represents a sum of infinitely many terms. This converges, and T is finite, whenever $\rho((I - P_l)K) < 1$, i.e., when the remaining $n - l$ types alone cannot sustain the infectious agent.

Apart from estimating control effort for a given type or set of types, one could use T to compare or rank the effort needed when targeting different types alone or different sets of types. For ranking one would need an additional criterion such as quantification of costs involved in the different options.

7.11 SUMMARY

In this section we summarize the definition of R_0 for host populations consisting of various types of individuals. A key aspect of the characterization is that we regard *generations* of only the infected individuals as they are distributed (at the moment of becoming infected) over all possible individual types (collected in the h-state space Ω). We then construct a *linear positive* operator that provides the next generation of infecteds (both the size and its distribution over Ω) when applied to the present generation. In the case that there are only finitely many types, the operator is (represented by) a matrix. Iteration of the next-generation operator describes the initial progression of infection within the heterogeneous population. Hence the value of the dominant eigenvalue/spectral radius of the operator relative to the threshold 1 determines whether generations of infecteds increase in size or decrease to zero. Therefore, R_0 is defined to be the dominant eigenvalue of the linear next-generation operator. The interpretation is the expected number of new cases produced by a *typical* infected individual during its entire infectious period, in a population consisting of susceptibles only. By 'typical' we mean an individual whose state is distributed over Ω according to the eigenvector corresponding to R_0. Once we suitably normalize this eigenvector, the vector can be interpreted as the probability distribution for the state at the moment of becoming infected (state-at-birth).

In general, the dominant eigenvalue (spectral radius) of a positive operator is not easy to compute. We have seen, however, that under additional assumptions (but for arbitrary Ω), R_0 can be expressed as a formula, rather than as the solution of an equation. The theory of linear positive operators, moreover, presents various approaches for the approximation of the spectral radius (see e.g., the book by Krasnosel'skij et al. 1989).

The basic steps to compute R_0 from first principles are:

- *Identification*: Specify the relevant h-states and transmission routes and single out the states-at-infection from among them; we denote the set of states-at-infection by Ω_b.

- *Modeling*: Construct the elements $k(\xi, \eta)$ of the kernel, defined as the expected number[9] of new cases of state-at-infection ξ that are produced by one infected with state-at-infection η.

- *Calculation*: Compute the dominant eigenvalue of the next-generation operator K defined by

$$(K\phi)(\xi) = \int_{\Omega_b} k(\xi, \eta)\phi(\eta)\, d\eta.$$

Connected to the identification step is that one chooses the h-states in such a way that all relevant information can be incorporated. For example, adding transmission routes can have as a consequence that one has to consider more h-states, in particular as states-at-infection. In general we can distinguish indirect transmission (via a vector, an intermediate host, contaminated food or environment), direct horizontal transmission between individuals, vertical transmission (i.e., from mother to unborn offspring) and *diagonal* transmission. We propose the latter term for situations where horizontal transmission is also in a sense vertical since it is strictly related to birth (e.g., contact of individuals with infected afterbirth or aborted fetuses as in respectively scrapie and brucellosis). It may happen (see Exercise 7.13) that an individual born by vertical transmission ('born' in a double sense) behaves differently from a horizontally infected individual.

With regard to modeling the kernel k, one could differentiate between h-states that are characterized by one number and h-states that are higher-dimensional. Furthermore, one can differentiate between h-states that are static (for a given individual) and those that are dynamic. We have presented an approach for dynamic h-states in Section 7.5. As part of the construction of the kernel, one also has to specify the demographic steady state of susceptibles in the absence of infection.

With regard to the calculation of the dominant eigenvalue of K, we have seen various simplifying assumptions in Section 7.4 that lead to explicit expressions or to explicit criteria. We have also seen that one assumption (multi-group separable mixing) can be relevant for higher dimensional h-state space and that this assumption can lead to a reduction of the dimension of the space on which the operator acts.

---

[9]Using the word 'number' here is not precise enough. The point is that the dimensions should match. The more precise formulation is that $k(\xi, \eta)d\xi$ is the number in the volume element $[\xi, \xi + d\xi)$. One should really speak of a 'number per unit of h-state-at-birth space.'

Chapter Eight

Other indicators of severity

This chapter is devoted to the initial real-time growth rate r, the probability of a major outbreak, the final size, and the endemic level, in structured populations, with special attention for computational simplifications in the case of separable mixing.

In the previous chapter we studied the basic reproduction number R_0 for epidemic models in populations manifesting various forms of heterogeneity. According to the definition, R_0 is the per-generation growth factor of the number infected during the early stages of an outbreak in a previously unaffected community. It was illustrated that R_0 depends on the transmission parameters, contact rates, the infectious period and on the community structure. The importance of R_0 lies in the fact that an epidemic can, and will in the deterministic setting, take off only if $R_0 > 1$, a characteristic referred to as supercritical. In a community having births or immigration of susceptibles, this also means that the disease can become endemic. If the parameters and community are such that $R_0 < 1$ (or $R_0 = 1$), we are in the subcritical (critical) regime and an epidemic outbreak cannot occur.

In the present chapter we study important supplementary characteristic features and see how they depend on the different parameters of the model.

8.1 THE PROBABILITY OF A MAJOR OUTBREAK

In this section we focus on stochastic models in finite populations and see how heterogeneities affect the *probability* of a major outbreak. As before, we consider the case when the population size N is large — otherwise a distinction between 'minor outbreak' and 'major outbreak' cannot be made. In Chapter 3 it was shown that in the prototype stochastic epidemic model, the final proportion infected in a large community could be either close to 0 or, if $R_0 > 1$, close to some positive deterministic value x^* depending only on R_0. This type of result, denoted 'threshold limit theorem' and saying that the final proportion infected converges to a two-point distribution, is valid for a very large class of stochastic epidemic models.

In Chapter 3 it was also shown that, for the prototype stochastic epidemic model in a homogeneous community, the probability of a major outbreak depends on the model parameters (in particular on the number of initially infectious and the randomness of the infectious period). We will now show, by means of a simple example, that when the community is heterogeneous, the probability of a major outbreak not only depends on how many are initially infectious, but also on characteristics these individuals have.

To this end we introduce a simplistic model for the spread of a sexually transmitted infection (STD) in a heterogeneous community. In this model it is assumed that individuals have spontaneous heterosexual contacts over time and that, each time, the partner is chosen independently at random among the individuals of opposite sex (in Section 12.6.4 we study a more realistic model based on networks, allowing for steady partnerships).

8.1.1   A simplistic model for an STD in a heterosexual community

Suppose the community of N individuals consists of N_f females and N_m males and denote the corresponding proportions by $\pi_f = N_f/N$ and $\pi_m = N_m/N$, respectively. Individuals that get infected, females as well as males, have infectious periods ΔT that are independently distributed with the same distribution (having mean $E(\Delta T) = 1$ without loss of generality). We ignore a latency period. Whether infectious or not, a male individual is assumed to engage in contact with any given female independently at rate c/N. If the male is infectious and if a contacted female is susceptible she becomes infectious with probability p_{fm}. Similarly, an infectious female has contact with males independently, each at the rate c/N, and any such contact results in infection with probability p_{mf} if the male is susceptible (note that contacts are symmetric but the transmission probabilities need not be, and usually are not for STDs). Contacts with non-susceptibles have no effect. Given the initial number of infected males m_m and females m_f, the epidemic goes on until there are no infectious individuals anymore — then the epidemic stops. The numbers of males Y_m and females Y_f ultimately infected specify the final size of the epidemic.

The exact distribution of (Y_m, Y_f) is complicated — it can be derived using recursive formulas in a similar (but more involved) way as for the homogeneous community (Section 12.5.3). We now assume that the community size N is large and study the initial stage of an outbreak. Just as for the prototype stochastic epidemic model for a homogeneous community, during the initial stages of an outbreak the process corresponding to this model behaves like a branching process, in this case a two-type branching process. This is the case because, during the early stages of the epidemic in a large community, nearly all individuals will still be susceptible, so the probability to have a contact with someone already infected is negligible. As a consequence, infected individuals infect new individuals more or less independently, and this independence is the crucial assumption underlying branching processes. It remains to determine how many new individuals an infected typically infects, and how these new infections are distributed over the two types of individuals.

An infected male has infectious contacts with a given female at rate cp_{fm}/N and, since there are $N_f = \pi_f N$ females, an infected male infects females at rate $cp_{fm}\pi_f$. Given the length of the infectious period $\Delta T = t$ it follows that the male infects a random number of females having Poisson distribution $Po(cp_{fm}\pi_f t)$. If we remove the conditioning on ΔT, the number infected, X_{fm}, follows a mixed Poisson distribution. Similarly, conditioning on the infectious period $\Delta T = t$, an infected female will infect a $Po(cp_{mf}\pi_m t)$ number of males during the early stages of the epidemic. Removing the conditioning, the number of males that a female will infect during the early stages of an outbreak, X_{mf}, follows a mixed Poisson distribution too. In general, for a two-type branching process, each type of individual can give birth (infect) to both other types of individual, but for our model this is not the case since we are assuming a heterosexual community.

Exercise 8.1 What is the offspring distribution X_{mf} if

    i) $\Delta T \equiv 1$,

    ii) $\Delta T \sim Exp(1)$.

Our approximating branching process is hence a two-type branching process where each type only produces offspring of the other type, and the offspring distributions are mixed Poisson: $X_{fm} \sim Po(cp_{fm}\pi_f\Delta T)$ and $X_{mf} \sim Po(cp_{mf}\pi_m\Delta T)$, where ΔT is random. The reproduction number R_0 for this model depends only on the expectations of the offspring distributions. Since we have assumed that $E(\Delta T) = 1$ the reader can check that it follows that $\mu_{fm} = E(X_{fm}) = cp_{fm}\pi_f$ and $\mu_{mf} = E(X_{mf}) = cp_{mf}\pi_m$. The basic reproduction number for a multi-type branching process equals the largest eigenvalue to the matrix of mean offspring distribution (recall Section 7.1). In our case the mean offspring distribution equals

$$\begin{pmatrix} 0 & \mu_{fm} \\ \mu_{mf} & 0 \end{pmatrix}$$

and consequently,

$$R_0 = \sqrt{\mu_{mf}\mu_{fm}} = c\sqrt{p_{mf}p_{fm}\pi_f\pi_m}. \tag{8.1}$$

How about the probability of a major outbreak? For this the actual offspring distribution matters, and not only its mean, but also who are initially infected. We now determine q_m (and q_f), the probability of having no major outbreak starting with one initially infected male (one initially infected female). Similar to the case of a homogeneous community we derive equations for q_m and q_f by conditioning on the number of offspring in the first generation. If the first infected male infects k females, each such female must avoid causing a major outbreak, and each such female avoids causing a major outbreak independently with probability q_f. Similar reasoning applies when starting with one infected female. It follows that q_m and q_f must satisfy

$$q_m = \sum_{k=0}^{\infty} q_f^k P(X_{fm} = k), \tag{8.2}$$

$$q_f = \sum_{k=0}^{\infty} q_m^k P(X_{mf} = k). \tag{8.3}$$

Given the model parameters (in particular the distribution of the infectious period ΔT), the offspring distribution consists of two fully defined mixed Poisson distributions, so equations (8.2) and (8.3) actually characterize q_m and q_f. It is clear that $q_m = q_f = 1$ is always a solution, but if $R_0 > 1$ there is one more solution in $(0, 1] \times (0, 1]$ and this is the solution defining q_m and q_f. In this particular case we can write $q_m = f_1(q_f)$ and $q_f = f_2(q_m)$, with f_1 and f_2 defined by the right-hand sides of (8.2) and (8.3), respectively. So then $q_m = g(q_m)$ with $g = f_1 \circ f_2$. Because $f_i' > 0$ and $f_i'' > 0$, for $i = 1, 2$, it follows that $g' > 0$ and $g'' > 0$, and we are in the situation considered in Section 1.2.2. Having derived q_m and q_f we can use these to express outbreak probabilities also for other initial configurations. The probability to avoid a major outbreak starting with m_m males and m_f females simply equals

$$P(\text{no major outbreak}|m_m, m_f) = q_m^{m_m} q_f^{m_f}. \tag{8.4}$$

We now determine q_m and q_f for a specific example. Suppose that $\Delta T \equiv 1$, $\pi_m = \pi_f = 0.5$, $c = 10$, $p_{fm} = 0.4$, $p_{mf} = 0.2$. In words this means that all infectious periods are equally long, there are equally many males and females, individuals on average have $c\pi_m = c\pi_f = 5$ sex contacts (with different partners) over a period corresponding to the length of the infectious period, and the transmission probability from male to female is 0.4 and 0.2 for transmission in the opposite direction. It then follows that $X_{fm} \sim Po(2)$ and $X_{mf} \sim Po(1)$. Equations (8.2) and (8.3) then become

$$q_m = \sum_{k=0}^{\infty} q_f^k \frac{e^{-2} 2^k}{k!} = e^{-2(1-q_f)},$$

$$q_f = \sum_{k=0}^{\infty} q_m^k \frac{e^{-1} 1^k}{k!} = e^{-(1-q_m)}.$$

Solving these equations numerically leads to $q_m \approx 0.410$ and $q_f \approx 0.554$. This means that, in this example, the probability of a major outbreak equals $1 - q_m \approx 0.590$ or $1 - q_f \approx 0.446$ depending on who was the initially infected. Since the male has a higher chance of transmitting the disease onwards it is in fact quite intuitive that the epidemic has higher probability of taking off if the initial infective is male. If there is more than one initially infected individual, say $m_m = 1$ males and $m_f = 2$ females, the probability of a major outbreak increases to $1 - q_m^1 q_f^2 \approx 0.874$. This hence illustrates that the *probability* of a major/minor outbreak is rather different depending on who and how many are initially infectious. However, *if* the epidemic takes off, it can be shown that the final number infected will be more or less the same irrespective of who are initially infected, as long as the *fraction* of initially infected remains negligible — these individuals can hence trigger the outbreak, but once the outbreak takes off, it continues independently from the index cases. See, for instance, Exercise 8.37 and the discussion in Section 12.4.2.

With this simple model we have illustrated that who and how many are initially infectious greatly affects the probability of a major outbreak, but not the final size in case there is a major outbreak. The same type of result holds for many epidemic models for heterogeneous communities.

Exercise 8.2 Consider the same numerical example as above with the exception that $\Delta T \sim Exp(1)$. Elaborate equations (8.2) and (8.3) for this infectious period distribution. Compute q_m and q_m from the equations.

8.1.2 Partially vaccinated populations (a second example)

To further illustrate various aspects of definitions and calculations for heterogeneous populations, we shall treat in some detail a relatively simple example that is of some interest by itself, viz., the case of a population in which a fraction v is vaccinated. We assume that the description of Section 2.1 applies to the population before vaccination. To avoid the complications associated with formula (2.6), we assume that infectivity is deterministic (so notably the infectious period is the same for all individuals). In particular we have that

$$R_0 = c \int_0^{\infty} A(\tau)\, d\tau,$$

which we assume to be bigger than one, that r is the unique positive root of

$$1 = c \int_0^{\infty} e^{-r\tau} A(\tau)\, d\tau,$$

that the probability z_∞ of a minor outbreak is, in the situation we consider, the unique root in $(0,1)$ of the equation

$$z = e^{R_0(z-1)},$$

and that the fraction $s(\infty)$ that escapes from a major outbreak is likewise the unique root in $(0,1)$ of the equation

$$s(\infty) = e^{R_0(s(\infty)-1)},$$

so that actually $z_\infty = s(\infty)$. We have in mind that this description concerns a local population in a metapopulation (i.e., a population consisting of spatially separated local populations that are connected by migration of individuals; see Hanski and Gilpin 1997) and we are also interested in transmission at the metapopulation level. A relevant quantity in this respect is

$$(1 - z_\infty)(1 - s(\infty)),$$

the probability of a major outbreak times the size of a major outbreak. The idea is that R_0 at the metapopulation level is proportional to the expected outbreak size and that $(1 - z_\infty)(1 - s(\infty))$ is a convenient proxy for this expected size.

Now assume that, as a result of a vaccination campaign, in every local population a fraction v is vaccinated. We shall assume that vaccination reduces the susceptibility to a fraction f (of the original susceptibility), and correspondingly the infectivity to a fraction ϕ, while leaving the time course of infectivity unaltered. By this we mean that a contact between an infected individual with infection age τ and a susceptible leads to transmission with probability

$fA(\tau)$ if the susceptible is vaccinated while the infective is not;

$\phi A(\tau)$ if the susceptible is not vaccinated but the infective is;

$f\phi A(\tau)$ if both are vaccinated,

where $0 \le f, \phi \le 1$. By allowing $f\phi$ to be positive, we acknowledge that the vaccine may not give full protection.

The question we now ask is: how are the analogs of R_0, r, z_∞ and $s(\infty)$ characterized for this new situation and what replaces $(1 - z_\infty)(1 - s(\infty))$? In particular, we want to know how the answer depends on v, f and ϕ, the three parameters characterizing the vaccination coverage and the vaccine efficacy.

We now deal with a structured population, with two types of individuals: unvaccinated ones, which we denote by the index 1, and vaccinated ones, to be denoted by the index 2. Transmission by infecteds of infection age τ is then described by the matrix

$$\begin{pmatrix} c(1-v)A(\tau) & c(1-v)\phi A(\tau) \\ cvfA(\tau) & cvf\phi A(\tau) \end{pmatrix}$$

during the early rise of an epidemic. In particular R_v, the reproduction number for the vaccinated population, is the dominant eigenvalue of the next-generation matrix

$$R_0 \begin{pmatrix} 1-v & (1-v)\phi \\ vf & vf\phi \end{pmatrix}.$$

Note that this matrix has a one-dimensional range spanned by the vector $(1-v, vf)^\top$ that gives the proportions in which unvaccinated and vaccinated individuals will occur in early generations of the infected sub-population.

Exercise 8.3 Why did we say 'early' generations?

Exercise 8.4 Show that the non-zero eigenvalue is

$$R_v = (1 - v + vf\phi)R_0.$$

Hint: For every matrix the trace is the sum of the eigenvalues, or, if this hint does not mean anything to you, use the fact that you know the eigenvector.

In the following we assume that $R_v > 1$, i.e., large outbreaks are still possible despite the vaccinations.

Exercise 8.5 Convince yourself that r_v, the real-time growth rate for the population with vaccination, is characterized by the condition that the matrix

$$\begin{pmatrix} c(1-v)\int_0^\infty e^{-r_v\tau}A(\tau)\,d\tau & c(1-v)\phi\int_0^\infty e^{-r_v\tau}A(\tau)\,d\tau \\ cvf\int_0^\infty e^{-r_v\tau}A(\tau)\,d\tau & cvf\phi\int_0^\infty e^{-r_v\tau}A(\tau)\,d\tau \end{pmatrix}$$

has dominant eigenvalue one. Next show that this amounts to the condition that

$$1 = R_v\frac{\int_0^\infty e^{-r_v\tau}A(\tau)\,d\tau}{\int_0^\infty A(\tau)\,d\tau}$$

by exploiting once more the one-dimensional range property.

Next we want to derive equations that characterize (determine) extinction probabilities. As we explained in the preceding subsection, the plural is justified since we have to deal with the probability π_1 of extinction when starting with one un-vaccinated infected individual, and the probability π_2 of extinction when starting with one vaccinated infected individual. An infected individual is making contacts according to a Poisson process with intensity c, and with probability $1 - v$ the contacted individual will be unvaccinated, while with probability v it will be vacci-nated. If the original infectious individual is itself unvaccinated, the transmission probabilities for these two cases are $A(\tau)$ and $\phi A(\tau)$ respectively, while if the orig-inal individual is actually vaccinated, these are $fA(\tau)$ and $f\phi A(\tau)$. With the same reasoning as in Sections 1.2.2 and 8.1.1, we arrive at the system of equations

$$\begin{aligned} \pi_1 &= e^{R_0(1-v)(\pi_1-1)}e^{fR_0v(\pi_2-1)}, \\ \pi_2 &= e^{\phi R_0(1-v)(\pi_1-1)}e^{\phi fR_0v(\pi_2-1)}. \end{aligned} \tag{8.5}$$

Exercise 8.6 Give the derivation in some detail.

Exercise 8.7 Do you expect any special structure in this system? Do you see any special structure? Hint: The infectivity of the two types differs by a factor ϕ.

Exercise 8.8 Define $\xi := (1 - v)(1 - \pi_1) + fv(1 - \pi_2)$. Show that ξ is determined by the equation

$$\xi = (1-v)(1 - e^{-R_0\xi}) + fv(1 - e^{-\phi R_0\xi}) \tag{8.6}$$

and that, once ξ is known, we have explicitly

$$\pi_1 = e^{-R_0\xi}, \quad \pi_2 = e^{-\phi R_0\xi}.$$

Thus we have reduced the system to a scalar problem. How would you inter-pret ξ? Hint: Victims are made in the proportions $1 - v : fv$.

It remains to characterize the final sizes. Let σ_1 denote the fraction of the unvaccinated individuals that escapes a major outbreak and let σ_2 be the same quantity for the vaccinated individuals. Then consistency requires that

$$
\begin{aligned}
\sigma_1 &= e^{-R_0\{(1-v)(1-\sigma_1)+\phi v(1-\sigma_2)\}}, \\
\sigma_2 &= e^{-fR_0\{(1-v)(1-\sigma_1)+\phi v(1-\sigma_2)\}}.
\end{aligned}
\tag{8.7}
$$

Exercise 8.9 Provide the arguments underlying these equations. Hint: Reread the derivation in Section 1.3.1 of the final-size equation in terms of a probabilistic consistency consideration.

Exercise 8.10 Define $\theta := (1-v)(1-\sigma_1) + \phi v(1-\sigma_2)$. Show that θ is determined by the equation

$$
\theta = (1-v)(1-e^{-R_0\theta}) + \phi v(1-e^{-fR_0\theta})
\tag{8.8}
$$

and that, once θ is known, we have explicitly

$$
\sigma_1 = e^{-R_0\theta}, \quad \sigma_2 = e^{-fR_0\theta}.
$$

Note the analogy with equation (8.6) for ξ and, in particular, the role of reversal of ϕ and f in these equations. How would you interpret θ?

We now claim that the relevant quantity to consider when thinking about transmission at the metapopulation level is $\xi\theta$. Indeed, if a local population is challenged from outside with a standard 'amount' of infectious material, the probability that an unvaccinated or a vaccinated individual is infected is proportional to respectively $1-v$ and fv, with the same constant of proportionality. So, the probability that a major outbreak occurs is proportional to ξ. If a major outbreak occurs, it will affect a fraction $1-\sigma_1$ of the fraction $1-v$ of unvaccinated individuals and a fraction $1-\sigma_2$ of the fraction v of vaccinated individuals. Since we are interested in the output of infectious material towards other local populations that may result from this outbreak, we give the vaccinated individuals a weighting factor ϕ to account for the reduction in infectivity. So, θ is indeed a measure for the output of infectious material that results from a major outbreak.

Thus we have completed the task we set out to perform. As a brief excursion into applied issues, we address the question whether it is more efficient to reduce ϕ or to reduce f, when there is a choice (admittedly this will not happen very often; yet it may be that several vaccines exist and that one has to make a comparison). More precisely, we ask which combination of f and ϕ, satisfying the constraint $f\phi = k$, a constant less than one, minimizes the product $\xi\theta$. We shall put $\phi = k/f$ and use f as a free parameter in the interval $[k, 1]$.

Exercise 8.11 Analyze how the product $\xi\theta$ depends on f when $\phi f = k$. Interpret the results. Hint: Exploit the symmetry related to interchanging f and ϕ. Moreover, it is helpful to concentrate on the extreme situations $R_0 \to \infty$ and $R_v = (1-v+vk)R_0$ (see Exercise 8.4) only slightly bigger than one. *Warning*: This is a rather technical exercise.

We conclude that the extreme cases $f = k$ and $f = 1$ are equally good (or bad?), but that they show a trade-off between the likeliness of a major outbreak and the size of a major outbreak. Moreover, we see that what is considered optimal depends on quantitative aspects, no generally valid criterion emerges. Note that we have neglected several complicating issues that could influence what is 'optimal.' For example, differences in f and θ may be linked to differences in the clinical manifestation of disease in individuals, with possible individual and public health consequences.

At the end of Section 8.3 we present a few additional exercises concerning the probability of a major outbreak in a more general setting, using notation to be introduced in that section.

8.2   THE INTRINSIC GROWTH RATE

In Section 1.2.3, our starting point was (a special case of) the equation

$$i(t) = c \int_0^\infty A(\tau)i(t-\tau) \, d\tau \tag{8.9}$$

for the incidence i in the initial phase of an epidemic. A natural question is therefore:

> What is the analog of this equation when we account for heterogeneity in the host population?

The details of the answer depend on the choices we make (in particular concerning the factoring of the expected number of transmissions, like the product $cA(\tau)$ above; recall Section 7.5), but the overall structure is determined by the bookkeeping, that we hope by now is becoming familiar. Our choice is to write

$$i(t,\xi) = N(\xi) \int_0^\infty \int_\Omega A(\tau,\xi,\eta)i(t-\tau,\eta)d\eta d\tau \tag{8.10}$$

where $N(\xi)A(\tau,\xi,\eta)$ is the rate at which an individual that was itself infected while having h-state η is expected to generate new cases with h-state ξ at time τ after its own infection, and where $i(t,\cdot)$ is the h-state specific incidence at time t. The h-state specific population density $N(\xi)$ describes the virgin steady state. The word 'choice' in the above refers to the fact that we have written N as a factor, even though we confine ourselves to stating the interpretation of the product $N(\xi)A(\tau,\xi,\eta)$. The point is that the precise interpretation of $A(\tau,\xi,\eta)$ depends on features of the contact process, as considered in Sections 1.3.2 and 1.3.3. These features, in fact, only matter if we compare different host populations or consider what happens after the initial phase, when susceptible individuals are depleted. We will refer to (8.10) as the equation for the (linearized) real-time evolution (or real-time growth).

In Section 1.2.3, we proceeded by substituting the ansatz $i(t) = ke^{\lambda t}$ into (8.9), which led to the characteristic equation

$$1 = c \int_0^\infty A(\tau)e^{-\lambda \tau} \, d\tau. \tag{8.11}$$

The real solution for λ, if it exists and is unique, is called the Malthusian parameter, or intrinsic or exponential growth rate, and will be denoted by the symbol r. If the host population is heterogeneous, the corresponding ansatz is

$$i(t,\xi) = e^{\lambda t}\Psi(\xi), \tag{8.12}$$

or, in words, that i factorizes into the product of a function of t (which then actually has to be an exponential since its translates should be multiples of the function itself) and a function of h-state ξ. We find that λ and Ψ should be such that

$$K_\lambda \Psi = \Psi \tag{8.13}$$

has a positive solution Ψ, where

$$(K_\lambda \phi)(\xi) = N(\xi) \int_\Omega \int_0^\infty A(\tau, \xi, \eta) e^{-\lambda \tau} \, d\tau \, \phi(\eta) \, d\eta. \tag{8.14}$$

In other words, K_λ should have eigenvalue 1. Since the kernel is positive when λ is real and because Ψ should be positive, we can say slightly more if we assume irreducibility: K_r should have dominant eigenvalue one.

Exercise 8.12 Verify that R_0 is the dominant eigenvalue/spectral radius of K_0.
   Hint: Convince yourself that $k(\xi, \eta) = N(\xi) \int_0^\infty A(\tau, \xi, \eta) \, d\tau$ (for notation see Section 7.3).

We expect, of course, that[1]

$$\mathrm{sign}(R_0 - 1) = \mathrm{sign}\, r. \tag{8.15}$$

The key element for a proof is a lemma that states that the spectral radius of K_λ is a continuous and strictly decreasing function of λ when λ is real. A precise formulation requires precise assumptions on Ω, A and N. We refrain from a general elaboration here and refer to the literature,[2] and to the rest of this section for a special case.

The next question then is:

Does $r < 0$ imply that all solutions of (8.10) decay to zero exponentially?

In fact, we have not even addressed this question for the homogeneous case of Section 1.2.3. There are two aspects that need attention:

- For what $\lambda \in \mathbb{C}$ does (8.10) have a non-trivial solution of the form $i(t, \xi) = e^{\lambda t} \Psi(\xi)$?

- When all such λ have negative real part, can we conclude that all solutions decay to zero exponentially?

For the second aspect we refer to the literature.[3] The key observation related to the first aspect is that $|e^{\lambda t}| \le e^{\mathrm{Re}\lambda t}$, which has as a corollary that the spectral

---

[1]The function sign is defined in the usual way: $\mathrm{sign}(y) = y/|y|$ if $y \ne 0$, and $\mathrm{sign}(0) = 0$.

[2]See e.g., H. Inaba: Threshold and stability results for an age-structured epidemic model. *J. Math. Biol.*, **28** (1990), 411–434; H.J.A.M. Heijmans: The dynamical behaviour of the age-size distribution of a cell population. In: J.A.J. Metz and O. Diekmann (1986), pp. 185–202. H.R. Thieme: Spectral bound and reproduction number for infinite-dimensional population structure and time heterogeneity. *SIAM J. Appl. Math.*, **70** (2009), 188–211.

[3]O. Diekmann, S.A. van Gils, S.M. Verduyn Lunel and H.-O. Walther: *Delay Equations: Functional-, Complex-, and Nonlinear Analysis*. Springer-Verlag, Berlin, 1995; G. Gripenberg, S-O. Londen and O. Staffans: *Volterra Integral and Functional Equations*. Cambridge University Press, Cambridge, 1990; O. Diekmann, P. Getto and M. Gyllenberg: Stability and bifurcation analysis of Volterra functional equations in the light of suns and stars. *SIAM J. Math. Anal.*, **39** (2007), 1023–1069.

radius of K_λ is less than or equal to the spectral radius of $K_{\mathrm{Re}\lambda}$. So all such λ lie to the left of r in the complex plane. Or, in other words, the sign of r determines whether or not there are such λ in the right-half plane.

The observation that r is the rightmost of all such λ is also the key to the conclusion that, when $r > 0$, every positive solution of (8.10) exhibits exponential growth with exponent r. In summary, the intrinsic growth rate r is the unique real root of the 'equation'

$$\text{spectral radius of } K_\lambda = 1. \tag{8.16}$$

Just as in the case of R_0, one can derive computational simplifications of this rather implicit characterization, by adopting assumptions that give K_λ a special structure. We do not elaborate on this theme, but simply refer back to Section 7.4 for inspiration.

Exercise 8.13 Assume that $A(\tau, \xi, \eta) = a(\xi)b(\tau, \eta)$. Derive an equation for r.

Why does R_0 have such a predominant position in epidemic theory, when one could just as well characterize success of invasion in terms of the sign of r, where, furthermore, r is a quantity that relates more clearly to something that can be observed? One reason is that the characterization of R_0 is *explicit*, which often helps to derive *explicit* formulas, or approximations, and to compute numerical values.

Many modelers are used to writing down systems of ordinary differential equations and to deriving a threshold condition by determining the stability of the infection-free (or virgin) steady state (e.g., by verifying the Routh-Hurwitz criteria; see the book by Edelstein-Keshet 1988 for these criteria). These modelers are not used to introducing infectivity kernels A and calculating R_0 from those. Some might even distrust whether indeed the threshold value one of R_0, as defined in the preceding chapter, corresponds exactly to their stability condition. Often the appearance of the two conditions is so different that the correspondence cannot be decided by just staring at them (see Exercise 7.15 for an example of correspondence). Even more often, the Routh-Hurwitz criteria remain implicit and do not yield an explicit formula straight away. The remainder of this section is intended for modelers who recognize themselves in the above description.

Earlier (in particular in Sections 2.1, 7.2 and 7.4) we have considered the situation in which infected individuals could be in a finite (say n) number of different states, with state transitions (including death) occurring according to a rate matrix Σ, and transmission (i.e., reproduction in an epidemiological sense), described by a matrix T, so the ijth element of T is the rate at which an infected individual with current h-state j produces secondary cases with state-at-infection i. We shall show that the zero steady state of the linear system

$$\frac{dx}{dt} = (T + \Sigma)x \tag{8.17}$$

is asymptotically stable if and only if $R_0 < 1$. We have seen in Section 7.2 that R_0 is the dominant eigenvalue of the NGM and equal to the dominant eigenvalue of the NGM with large domain

$$K_L = -T\Sigma^{-1}. \tag{8.18}$$

As already explained in Section 7.5, the fact that K_L is a next-generation matrix follows once we realize that the matrix exponential

$$e^{\tau\Sigma}$$

applied to the ith unit vector e_i yields the probability distribution over the various h-states at time τ after being infected in state i, and that consequently

$$K_L = \int_0^\infty Te^{\tau \Sigma} \, d\tau = -T\Sigma^{-1}.$$

Here we assume that Σ is such that the integral converges, which is equivalent to $-\Sigma^{-1}$ being positive (see Lemma 8.14 below).

Before formulating the key hypotheses concerning T and Σ we introduce some notation. For a square matrix H we denote by $s(H)$ the *spectral bound* and by $\rho(H)$ the *spectral radius*:

$$s(H) := \sup \left\{ \mathrm{Re}\,(\lambda) : \lambda \in \sigma(H) \right\},$$
$$\rho(H) := \sup \left\{ |\lambda| : \lambda \in \sigma(H) \right\}$$

where $\sigma(H)$ denotes the *spectrum* of H, that is the set of eigenvalues. All matrices that we consider have real entries. As customary, we call a non-zero matrix H *positive*, and write $H \geq 0$, if all entries are non-negative; and *positive-off-diagonal* if all entries are non-negative except possibly those on the diagonal. We start by showing that for a positive-off-diagonal matrix H we have that: $s(H) < 0$ if and only if H is invertible and $-H^{-1}$ is a positive matrix. In the following proofs we will use the notation $\mathbf{I}$ for the identity matrix (to distinguish from the population size of the infectious sub-population I); $\mathbf{0}$ for the vector consisting of zeros; and $\mathbf{1}$ for the matrix with all entries equal to 1.

Lemma 8.14 *Let H be a real matrix with non-negative off-diagonal elements (i.e., $h_{ij} \geq 0$ for $i \neq j$). Then $e^{\tau H}$ is a positive matrix. Moreover, for the spectral bound $s(H)$ we have the equivalence*

$$s(H) < 0 \Leftrightarrow \left(\det H \neq 0 \text{ and } -H^{-1} \geq 0 \right). \tag{8.19}$$

Proof: For suitably large $\theta > 0$ we have that $H + \theta\mathbf{I} \geq 0$ (where $\mathbf{I}$ is the identity matrix) and hence it follows from the Taylor series definition of the matrix exponential (see e.g., Hirsch and Smale 1974) that

$$e^{\tau(H+\theta\mathbf{I})} \geq 0.$$

Hence

$$e^{\tau H} = e^{-\tau\theta} e^{\tau(H+\theta\mathbf{I})} \geq 0.$$

Now assume that $s(H) < 0$. Then $\det H \neq 0$, since $\det H = 0$ would imply that $\lambda = 0$ is an eigenvalue of H and hence that $s(H) \geq 0$, a contradiction. It follows from $s(H) < 0$ that the integral

$$\int_0^\infty e^{\tau H} \, d\tau$$

converges. Clearly, $\int_0^\infty e^{\tau H} \, d\tau \geq 0$ since $e^{\tau H} \geq 0$. Moreover, using the Taylor series once again, we see that

$$H \int_0^t e^{\tau H} \, d\tau = \int_0^t e^{\tau H} \, d\tau \, H = e^{tH} - \mathbf{I},$$

which, by taking the limit $t \to \infty$, implies that

$$\int_0^\infty e^{\tau H} \, d\tau = -H^{-1}.$$

We conclude that $-H^{-1} \geq 0$.

To prove the converse, assume that H^{-1} exists and that $-H^{-1}$ is positive. Let ϕ be such that $H\phi = r\phi$ with $\phi \geq 0$ and r real. Applying H^{-1} to this identity, we see that necessarily r must be negative. The fact that such a combination of ϕ and r must exist and that $r = s(H)$ follows from Theorem 7.3 applied to $H + \theta \mathbf{I}$. This concludes the proof.   □

In the following we assume that T is a positive matrix, and that Σ is a positive off-diagonal matrix with $s(\Sigma) < 0$, hence $-\Sigma^{-1}$ is a positive matrix. These assumptions reflect the biological meaning of both matrices; the condition $s(\Sigma) < 0$ reflects that one cannot remain (potentially) infectious for ever.

In the remainder of this section we will substantiate the following result.

Theorem 8.15 *Let T be a positive matrix and let Σ be a positive off-diagonal matrix with $s(\Sigma) < 0$. Let $R_0 = \rho(-T\Sigma^{-1})$ and $r = s(T + \Sigma)$. Then the following equivalence holds:*

$$\operatorname{sign}(r) = \operatorname{sign}(R_0 - 1).$$

We recall that the basic reproduction number R_0 is defined by $R_0 = \rho(K)$, where K is the NGM, and that we have shown in Section 7.2 that $\rho(K) = \rho(K_L) = \rho\left(-T\Sigma^{-1}\right)$. This observation justifies that we chose $R_0 = \rho(-T\Sigma^{-1})$ in the formulation of Theorem 8.15.

The stability of the zero steady state of the linear system (8.17) is determined by the sign of the Malthusian parameter r, which is defined as

$$r = s(T + \Sigma).$$

This criterion extends to the nonlinear system by the Principle of Linearized Stability if, in addition, the demographic dynamics make the infection-free steady state stable in the invariant subspace corresponding to the absence of the infectious agent.

We first prove the result under the extra assumptions that $T + \Sigma$ is irreducible and $R_0 > 0$, and then employ an *approximation and continuity* argument to establish the result in general.[4]

Lemma 8.16 *If $R_0 > 0$ then $s\left(R_0^{-1}T + \Sigma\right) = 0$.*

Proof: First assume that $T + \Sigma$ is irreducible. Let v be the non-negative left eigenvector of $K_L = -T\Sigma^{-1}$ corresponding to the eigenvalue R_0. Hence $vK_L = R_0 v$, which can be rearranged to obtain

$$v\left(R_0^{-1}T + \Sigma\right) = 0 \tag{8.20}$$

---

[4]The proof is based on ideas in C.K. Li and H. Schneider: Applications of Perron-Frobenius theory to population dynamics. *J. Math. Biol.*, **44** (2002), 450–462, who addressed a similar problem in population dynamics in a discrete-time setting — building, in turn, on ideas in J.M. Cushing and Zhou Yicang: The net reproductive value and stability in matrix population models. *Nat. Res. Mod.*, **8** (1994), 297–333. In P. Van den Driessche and J. Watmough: Reproduction numbers and sub-threshold endemic equilibria for compartmental models of disease transmission. *Math. Biosci.*, **180** (2002), 29–48, a proof is presented in terms of M-matrices, and we refer to H.R. Thieme: Spectral bound and reproduction number for infinite-dimensional population structure and time heterogeneity. *SIAM J. Appl. Math.*, **70** (2009), 188–211, for the analogous result for the infinite dimensional case.

The irreducibility of $T + \Sigma$ implies that $R_0^{-1}T + \Sigma$ is irreducible. By adding a large positive multiple of the identity, $k\mathbf{I}$, to $R_0^{-1}T + \Sigma$, we obtain a positive irreducible matrix, and since v is non-negative it must be the eigenvector corresponding to the spectral radius of that matrix. It follows that all the other eigenvalues have smaller real parts. By subtracting $k\mathbf{I}$ again all eigenvalues shift to the left in the complex plane, but the order relation between their real parts remains intact. Hence we conclude from equation (8.20) that $s\left(R_0^{-1}T + \Sigma\right) = 0$.

Next consider the case that $T + \Sigma$ is reducible. Regard the irreducible matrix $T + \epsilon\mathbf{1} + \Sigma$, where $\mathbf{1}$ is the matrix with all entries equal to one. Denote the spectral radius of the matrix $-(T + \epsilon\mathbf{1})\Sigma^{-1}$ by ρ_ϵ. For $\epsilon \downarrow 0$, we have that $\rho_\epsilon \to R_0$ and hence $\rho_\epsilon > 0$ for ϵ small. So, by the above proof for the irreducible case, $s((T + \epsilon\mathbf{1})/\rho_\epsilon + \Sigma) = 0$. Finally, for $\epsilon \downarrow 0$ we have, as noted above, that $\rho_\epsilon \to R_0$, and hence $s(T/R_0 + \Sigma) = \lim_{\epsilon \to 0} s((T + \epsilon\mathbf{1})/\rho_\epsilon + \Sigma) = 0$. $\qquad\square$

Lemma 8.17 *If $T + \Sigma$ is irreducible then*

$$y \mapsto s\left(y^{-1}T + \Sigma\right)$$

is strictly monotone decreasing.

Proof: We first add $k\mathbf{I}$ to $T + \Sigma$ for some k large enough to obtain a positive matrix. The spectral radius of an irreducible positive matrix strictly decreases (increases) if any entry of that matrix decreases (increases), see Theorem 2.1 in Li and Schneider (2002) and the references given there. Hence the spectral radius of $y^{-1}T + \Sigma + k\mathbf{I}$ is a monotone function of y. For a positive matrix the spectral radius is equal to the spectral bound, and the corresponding point on the real axis remains equal to the spectral bound as the spectrum shifts to the left when we subtract $k\mathbf{I}$. $\qquad\square$

Lemma 8.18 *If $T + \Sigma$ is irreducible and $R_0 > 0$ then $\mathrm{sign}\,(r) = \mathrm{sign}\,(R_0 - 1)$.*

Proof: If $R_0 > 1$ then (by Lemma 8.17) $s\,(T + \Sigma) > s\left(R_0^{-1}T + \Sigma\right)$, but (by Lemma 8.16) $s\left(R_0^{-1}T + \Sigma\right) = 0$, hence $r = s\,(T + \Sigma) > 0$. If $R_0 = 1$ then (by Lemma 8.16) $r = s\,(T + \Sigma) = 0$. If $R_0 < 1$ then (by Lemma 8.17) $s\,(T + \Sigma) < s\left(R_0^{-1}T + \Sigma\right)$, by Lemma 8.16 $s\left(R_0^{-1}T + \Sigma\right) = 0$, hence $r = s\,(T + \Sigma) < 0$. $\qquad\square$

Lemma 8.19 *If $s\,(T + \Sigma) = 0$ then $R_0 \geqslant 1$.*

Proof: By the shifting argument used above, it follows that $s\,(T + \Sigma)$ is an eigenvalue of $T + \Sigma$. Let $u \neq \mathbf{0}$ be a vector such that $(T + \Sigma)\,u = \mathbf{0}$, and define $v = \Sigma u$. As Σ is invertible, $v \neq \mathbf{0}$. Moreover, $\left(T\Sigma^{-1} + \mathbf{I}\right)v = (T + \Sigma)\,u = \mathbf{0}$, hence $K_L = -T\Sigma^{-1}$ has a unit eigenvalue and $\rho(K_L) \geq 1$. $\qquad\square$

Lemma 8.20 *If $s\,(T + \Sigma) = 0$ then $R_0 = 1$.*

Proof: Below we approximate $K_L = -T\Sigma^{-1}$ with a continuous family of matrices, parameterized by ϵ, that have spectral radius less than or equal to one for $\epsilon > 0$, and which converge to K_L as $\epsilon \downarrow 0$. It follows that $R_0 \leqslant 1$. Because from Lemma 8.19 it follows that $R_0 \geqslant 1$, we conclude that $R_0 = 1$.

Define $H(\epsilon) = T + \Sigma + \epsilon\mathbf{1}$. From similar arguments as those used in the proof of Lemma 8.17, it follows that the function $\epsilon \mapsto s\,(H(\epsilon))$ is monotone increasing. So, if we define $\widetilde{H}(\epsilon) = H(\epsilon) - s\,(H(2\epsilon))\,\mathbf{I}$, then $s\left(\widetilde{H}(\epsilon)\right) = s\,(H(\epsilon)) - s\,(H(2\epsilon)) \leqslant 0$. The decomposition $\widetilde{H}(\epsilon) = (T + \epsilon\mathbf{1}) + (\Sigma - s(H(2\epsilon))\mathbf{I})$ motivates us to introduce the matrix

$$M(\epsilon) = -\left(T + \epsilon\mathbf{1}\right)\left(\Sigma - s\,(H(2\epsilon))\,\mathbf{I}\right)^{-1}.$$

Clearly, $M(\epsilon)$ converges to K_L as $\epsilon \downarrow 0$, and as the spectral radius of K_L exceeds one (by Lemma 8.19), the spectral radius of $M(\epsilon)$ must be positive for small positive ϵ. Because $\widetilde{H}(\epsilon)$ is irreducible, we use Lemma 8.18 and deduce $\rho(M(\epsilon)) \leqslant 1$.   □

We are now ready to prove Theorem 8.15.

Proof of Theorem 8.15: By combining Lemma 8.20 and Lemma 8.16 (with $R_0 = 1$) we conclude that

$$s(T + \Sigma) = 0 \Leftrightarrow R_0 = 1.$$

By Lemma 8.17 we have that, at least for small $\epsilon > 0$,

$$s(T + \epsilon \mathbf{1} + \Sigma) < 0 \Leftrightarrow \rho(-(T + \epsilon \mathbf{1}) \Sigma^{-1}) < 1$$

and so, by considering the limit $\epsilon \downarrow 0$, we have that $s(T + \Sigma) < 0 \Rightarrow R_0 \leqslant 1$ and $R_0 < 1 \Rightarrow s(T + \Sigma) \leqslant 0$. Because, as already noted, $s(T + \Sigma) = 0 \Leftrightarrow R_0 = 1$, we conclude that $s(T + \Sigma) < 0 \Leftrightarrow R_0 < 1$. It follows that $s(T + \Sigma) > 0 \Leftrightarrow R_0 > 1$, and the proof is complete.   □

An early version of this theorem is presented in a paper by Nold (1980).[5] The assumption on $s(\Sigma)$ is certainly satisfied when $\Sigma = \widetilde{\Sigma} - D$, where D is a diagonal matrix with strictly positive diagonal elements, and where $\widetilde{\Sigma}$ is a positive off-diagonal matrix for which the elements of each column add to zero. The idea here is that the matrix D represents death/removal/recovery, while $\widetilde{\Sigma}$ represents transitions between states associated with current or potential future infectivity.

The special situation of just one possible state-at-infection corresponds to

$$Tx = (h \cdot x) b,$$

with b a given positive vector (preferably normalized such that its elements sum to one). In this situation we obtain

$$R_0 - h \cdot (-(\Sigma)^{-1} b)$$

as the explicit formula.

We need to point out that linearization near the infection-free steady state of a nonlinear ODE model leads to a system of higher dimension than (8.17), since the (various kinds of) susceptibles and removed individuals also occur in the bookkeeping. However, the invariance of the infection-free situation guarantees that this system decouples. And provided one has stability within the invariant infection-free subspace, the stability is completely governed by (8.17). We therefore emphasize the crucial <u>We conclude</u> that within the context of ODE models, a demographically stable infection-free steady state is asymptotically stable if and only if $R_0 < 1$.

As indicated above this conclusion extends to infinite-dimensional models, but the precise formulation of the assumptions becomes a lot more subtle and cumbersome in that case.

Finally, we mention that also in more general h-state spaces and in the case of age structure, one can find a relation between the formulation in terms of an infectivity function A and a formulation in terms of a system of (ordinary or partial) differential equations. The function A, however, needs to satisfy a restrictive condition, as we see from the next two exercises.

---

[5] A. Nold: Heterogeneity in disease-transmission modeling. *Math. Biosci.*, **52** (1980), 227–240.

Exercise 8.21 Assume
$$A(\tau, \eta, \xi) = \beta(\xi, \eta)e^{-\alpha(\eta)\tau}.$$

Let $I(t, \xi)$ be the total size of the infective sub-population with h-state ξ at time t. Give an expression for I and show by differentiation and manipulation that (8.10), in the nonlinear version with $N(\xi)$ replaced by $S(t, \xi)$, simplifies to
$$\frac{dI}{dt}(t, \xi) = S(t, \xi) \int_{\Omega} \beta(\xi, \eta)I(t, \eta) \, d\eta - \alpha(\xi)I(t, \xi).$$

Exercise 8.22 Let $\Omega = \mathbb{R}_{\geq 0}$, and let the h-state be 'age.' Again regard equation (8.10) for the incidence i in the nonlinear version, but use a as the variable, instead of ξ. Assume
$$A(\tau, a, \eta) = \beta(a, \eta + \tau)e^{-\int_{\eta}^{\eta+\tau} \alpha(\sigma) \, d\sigma}$$

(note that an individual that was infected with h-state η necessarily has h-state $\eta + \tau$ at infection-age τ, given that it is still alive). Define again the total size of the infective sub-population of age a at time t,
$$I(t, a) = \int_0^a i(t - \tau, a - \tau)e^{-\int_{a-\tau}^{a} \alpha(\sigma) \, d\sigma} \, d\tau,$$

and carry out the same manipulation as in the previous exercise to show that I satisfies
$$\frac{\partial I}{\partial t} + \frac{\partial I}{\partial a} = S(t, a) \int_0^{\infty} \beta(a, a')I(t, a') \, da' - \alpha(a)I(t, a),$$

with $a' = \eta + \tau$.

8.3 A BRIEF LOOK AT FINAL SIZE AND ENDEMIC LEVEL

In this section we review aspects of the time course, final size, endemic level and probability of extinction that are relevant for the formulation of general models for the spread of infectious agents in heterogeneous host populations. We do so in the form of exercises. Ideally, these should be elaborated in detail. We hope that the hurried reader — who has no time for that many exercises — will still get some information and gain some understanding by just reading them and checking the main ideas. In contrast to the choice in Section 8.1.2, we absorb the contact intensity c into the function A, to simplify the notation.

Exercise 8.23 Spell out the assumptions that underlie the equation
$$\frac{\partial S}{\partial t}(t, x) = S(t, x) \int_{\Omega} \int_0^{\infty} A(\tau, x, \eta)\frac{\partial S}{\partial t}(t - \tau, \eta) \, d\tau \, d\eta. \tag{8.21}$$

Do they apply to plants in a spatial domain Ω? Do they allow for dynamic h-states? Is inflow of new susceptibles or loss of immunity incorporated?

Exercise 8.24 Derive from (8.21) by integration the final-size equation
$$\ln \frac{S(\infty, x)}{S(-\infty, x)} = \int_{\Omega} \left(\int_0^{\infty} A(\tau, x, \eta) \, d\tau \right) \{S(\infty, \eta) - S(-\infty, \eta)\} \, d\eta. \tag{8.22}$$

For later use, it is convenient to derive, as a first step, the equation

$$\ln \frac{S(t,x)}{S(-\infty,x)} = \int_\Omega \left(\int_0^\infty A(\tau,x,\eta)\, d\tau \right) \{S(t-\tau,\eta) - S(-\infty,\eta)\}\, d\eta. \quad (8.23)$$

Note that $S(-\infty, x) = N(x)$, i.e., the population composition before the infectious agent entered the population.

Exercise 8.25 Derive the alternative form of (8.22)

$$\frac{S(\infty,x)}{S(-\infty,x)} = e^{-\int_\Omega (\int_0^\infty A(\tau,x,\eta)\, d\tau) S(-\infty,\eta)\{1 - \frac{S(\infty,\eta)}{S(-\infty,\eta)}\}\, d\eta} \quad (8.24)$$

by a probabilistic consistency consideration as in Sections 1.3.1 and 8.1.

Exercise 8.26 How would you define R_0? Do you agree that (8.22) should have a nontrivial solution for $R_0 > 1$? (the trivial solution being $S(\infty,\eta) = S(-\infty,\eta)$). How would you prove that this is indeed the case? Hint: Use monotone iteration.

Exercise 8.27 When the aim is to characterize the endemic level in a population with demographic turnover, we should replace (8.10) by

$$\frac{\partial S}{\partial t}(t,x) = B(x) - \mu(x)S(t,x) - i(t,x),$$

$$i(t,x) = S(t,x) \int_\Omega \int_0^\infty A(\tau,x,\eta)i(t-\tau,\eta)\, d\tau\, d\eta$$

where the h-state x is assumed to be static.

i) Check that you know what B, μ and i are supposed to describe.

ii) Calculate explicitly what the infection-free steady state is.

iii) Do you recall how R_0 is defined in this setting?

iv) Derive equations for $\overline{S}$ and $\overline{i}$ characterizing the endemic steady state.

v) For the mathematically inclined: Do you see any relation between $R_0 > 1$ and the existence of a non-trivial steady state? Hint: Do not care too much about technical details at this point. The problem is actually not at all easy and involves the notion of ejective fixed points for nonlinear maps defined on cones.[6]

We turn to the probability that only a minor outbreak occurs when nevertheless $R_0 > 1$. We start by giving a description in terms of

$p_k(x) :=$ the probability that an individual of type x begets k offspring

and

$m(x, \omega) :=$ the probability that the state-at-infection of a child of an
individual of type x belongs to the subset ω of Ω,

---

[6]See e.g., R.D. Nussbaum: *The Fixed Point Index and Some Applications.* Séminaire de Mathématiques Supérieures. les Presses de l'Université de Montreal, Canada, 1985.

while postponing a discussion of how $p_k(x)$ and $m(x,\omega)$ are related to $S(-\infty,x)$ and $A(\tau,x,\eta)$. The basic assumption is that offspring are produced independently, i.e., given that k offspring are produced, the state-at-infection of each of these is drawn from the probability distribution $m(x,\cdot)$.

Let

$\pi(x) :=$ probability of extinction when starting with one individual of type x.

Exercise 8.28 Convince yourself that consistency requires that

$$\pi(x) = \sum_{k=0}^{\infty} p_k(x) \left(\int_\Omega \pi(\eta) m(x, d\eta) \right)^k .\qquad(8.25)$$

We previously encountered several (very) special cases of (8.25) as (3.4), (8.2) and (8.3), and (8.5).

Exercise 8.29 Let $f(y,x)$ be defined as

$$f(y,x) := S(-\infty,y) \left(\int_0^\infty A(\tau,y,x)\, d\tau \right).$$

Check that the specifications

$$p_k(x) = \frac{1}{k!} \left(\int_\Omega f(y,x)\, dy \right)^k e^{-\int_\Omega f(y,x)\, dy}\qquad(8.26)$$

and

$$m(x,\omega) = \frac{\int_\omega f(y,x)\, dy}{\int_\Omega f(y,x)\, dy}\qquad(8.27)$$

are consistent with transmission being superimposed on a Poisson contact process. Use these specifications to rewrite (8.25) as

$$\pi(x) = \sum_{k=0}^{\infty} \frac{1}{k!} \left(\int_\Omega \pi(\eta) f(\eta,x)\, d\eta \right)^k e^{-\int_\Omega f(y,x)\, dy}.\qquad(8.28)$$

8.4 SIMPLIFICATIONS UNDER SEPARABLE MIXING

The general theme that emerges from the preceding section is that, on a formal level, the addition of structure complicates the bookkeeping, but that it does not fundamentally alter the kind of relations that exist between various quantities. When it comes to doing calculations, however, the difference between the structured and the unstructured cases is enormous. In Section 7.4.1 we showed how the assumption of separable mixing facilitates the computation of R_0 (see also Exercise 8.13). In mathematical terms, the point is that whenever operators have a one-dimensional range we can work with scalar quantities. In terms of the interpretation, we can say that whenever the h-state at the moment of becoming infected is following an a priori given distribution (in particular independently of the h-state of the infecting individual), all individuals are identical in a stochastic sense and therefore we know how to take averages. The aim of this section is to demonstrate that this principle is not restricted to R_0, but extends to other aspects of the spread of infectious agents. The general conclusion is that the assumption of separable mixing allows

us to work with scalar quantities. Note that Section 8.1.2 has already offered a
simple and relatively concrete example.

Throughout this section we assume that

$$A(\tau, x, \eta) = a(x)b(\tau, \eta). \tag{8.29}$$

Exercise 8.30 Formulate in words what (8.29) means. What will be the distribution of the state-at-infection of newly infected individuals?

Exercise 8.31 Using (8.29) and the ansatz

$$\frac{\partial S}{\partial t}(t, x) = -\frac{dw}{dt}(t)a(x)S(t, x), \tag{8.30}$$

for some function w to be determined, rewrite equation (8.21) and integrate
from $-\infty$ to t to obtain

$$w(t) = \int_0^\infty \int_\Omega b(\tau, \eta)S(-\infty, \eta)\left(1 - e^{-a(\eta)w(t-\tau)}\right) d\eta \, d\tau. \tag{8.31}$$

Hint: Note that (8.30) implies that $S(t, x) = S(-\infty, x)e^{-a(x)w(t)}$ (provided
we require $w(-\infty) = 0$).

So once we determine w from the scalar integral equation (8.31), we find S by
substitution:

$$S(t, x) = S(-\infty, x)e^{-a(x)w(t)}. \tag{8.32}$$

To obtain the final-size equation, we simply take the limit $t \to \infty$ in (8.31) to
deduce

$$w(\infty) = \int_\Omega \int_0^\infty b(\tau, \eta) \, d\tau \, S(-\infty, \eta)\left(1 - e^{-a(\eta)w(\infty)}\right) d\eta, \tag{8.33}$$

which is a nonlinear scalar equation for the unknown $w(\infty)$.

Exercise 8.32 Compute R_0 and show that for $R_0 > 1$, (8.33) has a non-trivial
solution.

Exercise 8.33 Derive (8.33) from (8.23) using (8.29) and the appropriate ansatz.

Exercise 8.34 Introduce the assumption (8.29) in the setting of Exercise 8.27,
while making the ansatz

$$i(t, x) = S(t, x)a(x)v(t). \tag{8.34}$$

You should obtain the system of equations

$$\frac{\partial S}{\partial t}(t, x) = B(x) - \mu(x)S(t, x) - S(t, x)a(x)v(t), \tag{8.35}$$

$$v(t) = \int_0^\infty \int_\Omega b(\tau, \eta)S(t - \tau, \eta)a(\eta) \, d\eta \, v(t - \tau) \, d\tau.$$

From these derive the scalar equation

$$1 = \int_0^\infty \int_\Omega b(\tau, \eta)d\tau \frac{a(\eta)B(\eta)}{\mu(\eta) + \bar{v}a(\eta)} d\eta \tag{8.36}$$

for the steady state value of $\bar{v}$. Check that a unique positive solution exists
provided that $R_0 > 1$.

Exercise 8.35 In the setting of Exercise 8.28, assume that

$$m(x, \omega) = v(\omega) \tag{8.37}$$

i.e., assume independence of the distribution of the h-state at birth from the h-state of the mother. Define

$$z = \int_\Omega \pi(\eta) v(d\eta),$$

and derive from (8.25) the scalar equation

$$z = \sum_{k=0}^{\infty} z^k \int_\Omega p_k(\eta) v(d\eta). \tag{8.38}$$

Finally, check that once z is known we can recover π via the formula

$$\pi(x) = \sum_{k=0}^{\infty} z^k p_k(x). \tag{8.39}$$

Exercise 8.36 Exploiting the mild degree of arbitrariness in the multiplicative decomposition (8.29), normalize a such that

$$\int_\Omega a(\eta) S(-\infty, \eta) \, d\eta = 1. \tag{8.40}$$

Show that the formulas of the preceding exercise apply with

$$v(\omega) \;=\; \int_\omega a(\eta) S(-\infty, \eta) \, d\eta, \tag{8.41}$$

$$p_k(x) \;=\; \frac{1}{k!} \left(\int_0^\infty b(\tau, x) \, d\tau \right)^k e^{-\int_0^\infty b(\tau, x) \, d\tau}. \tag{8.42}$$

The purpose of the final exercise below is to illustrate that, for fixed R_0, a more contact-heterogeneous community has a higher *probability* to suffer from a major outbreak, but that the *size* of that outbreak will be smaller, compared to a more contact-homogeneous community. We emphasize that this conclusion derives from the assumption that the index case belongs to the core group (see Section 12.4.2) of very 'contact-active' individuals, which seems a reasonable assumption for STDs. It is this feature that explains the disparity with the conclusion of Exercise 2.11 and Section 3.3.2 concerning the influence of heterogeneity in infectious periods on the probability of a major outbreak.

Exercise 8.37 This exercise aims at comparing a homogeneous community with a community that has a small subgroup with higher transmission than the rest of the community. One such example could be a group of promiscuous individuals when considering a sexually transmitted disease in a homosexual community. Admittedly our model is too simple to be realistic for STDs but still serves our purpose here. Later in the book we will discuss the notion of a 'core group' in more detail (Section 12.4.2) and consider a slightly more realistic model for STDs (Section 12.6.4).

Consider a large community of N individuals, N_s of which are 'standard' individuals and N_c that belong to the core group ($N_s + N_c = N$). Let $\pi_c = 1 - \pi_s = N_c/N$ denote the (typically small) fraction of individuals in the core group. Suppose for simplicity that all infectious periods are non-random (and equal to 1 without loss of generality) for both types of individual. During the infectious period an individual of type j has infectious contacts with a given individual of type i randomly in time at rate $\lambda_{ij}/N = \alpha_i \alpha_j/N$, i, $j = s$ or c, where we assume proportionate mixing. The initial stages of this epidemic can be approximated by a branching process as described in Section 8.1.1 for the simplistic STD in a heterosexual community without core group. Now the two types of individual are 'standard' and 'core' instead of male and female.

i) Give the next-generation matrix and an expression for R_0.

Now fix $R_0 = 1.5$ and also $\pi_c = 0.1$ (so $\pi_s = 0.9$). Note that this is a normalization that we *choose*, that this will have impact on the outcome and that we therefore have to be aware of the choice we made when interpreting the results. We want to compare the probability of a major outbreak, and the size of such an outbreak in case it occurs, for different values of α_s and α_c. Since $R_0 = 1.5$ and $\pi_c = 0.1$ is fixed, this boils down to varying one of the two parameters, say α_c. We look at the two extreme cases and an intermediate case. The first is where there actually is no core group: $\alpha_c = \alpha_s = \alpha = \sqrt{1.5} \approx 1.22$ (since there is then only one type that infects others at rate α^2 during a period of mean 1, so $R_0 = \alpha^2$). The other extreme situation is where core individuals are responsible for all infections (and are the only ones getting infected!). This means that $\alpha_s = 0$, and so everything takes place within the core group implying that the basic reproduction number equals $\alpha_c^2 \cdot \pi_c$ so $\alpha_c = \sqrt{15} \approx 3.87$. Finally, we also look at the intermediate case $\alpha_c = 2.5$ (implying that $\alpha_s = 0.986$, by solving $1.5 = R_0 = 0.9\alpha_s^2 + 0.1\alpha_c^2$).

ii) Compute numerically the probability of a major outbreak for the three situations $(\alpha_s, \alpha_c) = (\sqrt{1.5}, \sqrt{1.5})$ and $(\alpha_s, \alpha_c) = (0.986, 2.5)$ and $(\alpha_s, \alpha_c) = (0, \sqrt{15})$ (all having the same $R_0 = 1.5$). Do this for the case that the outbreak is initiated by one infected individual from the core group.

We end this exercise by also computing the final size in case a major outbreak takes place. We have a two-type epidemic with parameters previously defined. For a single type epidemic the final size of an outbreak in case there is one, starting with initially few infectives, is the unique positive solution x^* to the equation

$$1 - x = e^{-R_0 x},$$

(cf. Sections 1.3 and 3.5.3). When there is more than one type we get a system of equations and a vector of solutions x_i^* giving the fraction infected of the different types. In our case this system of equations is given by a 'discrete' variant of (8.24):

$$1 - x_s = e^{-\alpha_s^2 \pi_s x_s} e^{-\alpha_s \alpha_c \pi_c x_c},$$
$$1 - x_c = e^{-\alpha_c \alpha_s \pi_s x_s} e^{-\alpha_c^2 \pi_c x_c}.$$

These two equations can be simplified to one equation for one unknown by defining a new variable $y = \alpha_s \pi_s x_s + \alpha_c \pi_c x_c$.

iii) Compute x_s^*, x_c^* and the overall fraction infected $x^* = \pi_s x_s^* + \pi_c x_c^*$ for the three cases treated above.

Chapter Nine

Age structure

Especially in the context of infectious diseases among humans, 'age' is often used to characterize individuals. Partly this reflects our system of public health administration (and, perhaps, our preoccupation with age). Indeed, we can exploit that data on the distribution of the random variable 'age at (first) infection' contain information about the prevailing force of infection in an endemic situation.

There is, however, also a more 'mechanistic' reason to incorporate age structure: patterns of human social behavior and sexual activity correlate with age. In addition, the effect that the infective agent has on the host sometimes depends heavily on the age of the host (e.g., in polio) or it may depend on another aspect of the host, such as pregnancy, which correlates with age (e.g., in rubella).

Age is a dynamic variable, but the dynamics are very simple: $da/dt = 1$ by definition. In this short chapter we elaborate some of the material of Chapters 7 and 8 for this special case. We shall deal with endemic steady states, but we postpone a discussion of the related inverse problem of estimating R_0 from data about the average age at infection, seropositivity as a function of age, etc, to Chapter 13. We do discuss, in the final section, vaccination strategies as one of the major applied issues of age-structured epidemic models. To start, we give a very brief introduction to the key notion of mathematical demography: the stable age distribution.

9.1 DEMOGRAPHY

The cohort *survival function* $\mathcal{F}_d(a)$ (d for death, a for age) describes the probability that an arbitrary newborn individual will survive at least until age a. The age-specific *force of mortality* $\mu(a)$, i.e., the per capita probability per unit of time of dying, is related to $\mathcal{F}_d(a)$ by

$$\mu(a) = -\frac{\mathcal{F}_d'(a)}{\mathcal{F}_d(a)} = -\frac{d}{da}\ln\mathcal{F}_d(a) \Leftrightarrow \mathcal{F}_d(a) = e^{-\int_0^a \mu(\alpha)\,d\alpha}.$$

We first give, without derivation, some results. In a density-independent situation (in other words, when a *linear* model applies) the total population size will eventually grow exponentially with a certain rate, which is traditionally denoted by r (but note that this now refers to the growth rate of the *host* population and not, as before, to the sub-population of infected hosts). Moreover, the distribution with respect to age will stabilize to a fixed shape, the normalized *stable age distribution* given explicitly in terms of $\mathcal{F}_d(a)$ and r by

$$N(a) = Ce^{-ra}\mathcal{F}_d(a), \tag{9.1}$$

with $C = (\int_0^\infty e^{-ra}\mathcal{F}_d(a)\,da)^{-1}$. The factor e^{-ra} reflects that the relative contribution of an individual has to be discounted as the individual ages, since meanwhile the total population is changing (cf. the explanation of $R_{0,\text{relative}}$ in Section 4.4, following Exercise 4.23).

Exercise 9.1 i) Show that the mean age at death of a cohort (i.e., a group of individuals born at more or less the same time, say within a certain year) is given by

$$\int_0^\infty a\mu(a)\mathcal{F}_d(a)\,da.$$

ii) Show that the mean age of those dying at more or less the same time, say within a certain year, is given by

$$\frac{\int_0^\infty a\mu(a)e^{-ra}\mathcal{F}_d(a)\,da}{\int_0^\infty \mu(a)e^{-ra}\mathcal{F}_d(a)\,da}.$$

iii) Reflect upon the difference.

iv) Elaborate for $\mu(a) \equiv \mu$. Note that the quantity calculated in i) is often called the *life expectancy* of a newborn individual.

Exercise 9.2 Analyze the meaning of the following statement: 'Fast-growing populations have a steep age pyramid.'

One frequently sees a formulation of age-dependent population growth in terms of a partial differential equation (possibly first derived by McKendrick[1]) to describe aging and dying,

$$\frac{\partial n}{\partial t} + \frac{\partial n}{\partial a} = -\mu n,$$

and a boundary condition $n(t,0) = \int_0^\infty \beta(\alpha)n(t,\alpha)\,d\alpha$ to describe the 'inflow' of newborns. Here $\beta(a)$ is the age-specific fecundity, i.e., the probability per unit of time of giving birth. Let $B(t) := n(t,0)$ be the total birth rate at time t. The equation above can be reformulated as a so-called renewal equation

$$B(t) = \int_0^\infty \beta(a)\mathcal{F}_d(a)B(t-a)\,da.$$

The population growth rate r is then found from the ingredients β and μ as the unique real root of the so-called Euler-Lotka equation

$$1 = \int_0^\infty e^{-\lambda a}\beta(a)\mathcal{F}_d(a)\,da$$

by substituting the ansatz $B(t) = ce^{\lambda t}$ in the renewal equation.

Exercise 9.3 How would you define in words the basic reproduction number R_0 in this demographic context? And can you give a formula for it?

9.2 CONTACTS

How does the expected number of contacts per unit of time that an individual of age a has with individuals of age α depend on the population size and composition? To make such a question meaningful, we have to be more specific about the precise interpretation of contact. For example, do we refer to sexual contacts or

---

[1]A.G. McKendrick: Applications of mathematics to medical problems. *Proc. Edin. Math. Soc.*, **44** (1926), 98–130.

to the inhaling of aerosols just exhaled by someone else? But even after a further specification, the question is a very difficult one![2] Therefore most modelers adopt a very pragmatic approach by simply assuming something that keeps the equations relatively tractable and does not require the estimation of very many parameters. Our later assumptions will be very much in that spirit. At this point we try to keep the model relatively general and flexible by introducing a *contact coefficient* $c(a, \alpha)$ having, by definition, the meaning that an individual of age α has per unit of time $c(a, \alpha)N(a)$ contacts with individuals of age a. An individual of age α therefore has $\int_0^\infty c(a, \alpha)N(a)\, da$ contacts in total. All the difficulties mentioned in Chapter 1 concerning the dependence of c on population size are intensified here, now that we also have to think about dependence on population composition. Since we have nothing to add to what has already been said in Chapter 1, we shall think here of a mild disease in one particular population and not worry about scaling of c.

An obvious question presents itself: are contacts necessarily symmetric or, more precisely, should c and N satisfy the relation

$$c(a, \alpha)N(a)N(\alpha) = c(\alpha, a)N(\alpha)N(a),$$

which amounts to $c(a, \alpha) = c(\alpha, a)$?

Exercise 9.4 i) Contemplate the kind of contacts for which the answer is 'yes' and those for which it is 'no.' Hint: Also think of parents caring for children.

ii) Contemplate the indeterminacy in decomposing 'probability of transmission' and 'probability of transmission, given a contact,' and how this bears upon the symmetry of c. In other words, reflect upon the inherent vagueness of 'contacts.'

9.3 THE NEXT-GENERATION OPERATOR

The dynamics of the h-state are very simple in the case of age: an individual of age a will τ units of time later be of age $a + \tau$ (given that it survives that long). The probability that it has not died in this time interval of length τ equals $\mathcal{F}_d(a + \tau)/\mathcal{F}_d(a)$ (here we assume that the death rate is the same for infected and uninfected individuals; so the formulas below have to be adapted when infection entails a serious risk of death for the host). We need one more ingredient: the probability of transmission given a contact. In principle this could depend on the ages of both individuals involved. Due to Exercise 9.4-ii, we are aware of an element of freedom in our description. Since c is already a general function of two variables, it would be redundant to introduce yet another such function. Let our last ingredient be

$h(\tau, \alpha)$:= probability of transmission of the infective agent,

given a contact between a susceptible (of arbitrary age)

and an individual that was itself infected while having age α.

You may wonder why we allowed for this dependence on α. What we have in mind is that — as a rule and up to a certain point — older individuals are bigger

[2]J. Mossong et al.: Social contacts and mixing patterns relevant to the spread of infectious diseases. *PLoS Med.*, **5** (2008), e74.

and so may distribute (much) larger quantities of the infective agent around them. Or, working in the other direction, the immune system of very young individuals may be less effective in dealing with the agent.

The next-generation operator is given by

$$(K\phi)(a) = \int_0^\infty k(a, \alpha)\phi(\alpha)\, d\alpha.$$

Exercise 9.5 i) Repeat in words the rationale underlying the following formula for the kernel k of the next-generation operator (cf. Section 7.5)

$$k(a, \alpha) = \int_0^\infty h(\tau, \alpha)c(a, \alpha + \tau)N(a)\frac{\mathcal{F}_d(\alpha + \tau)}{\mathcal{F}_d(\alpha)}\, d\tau.$$

ii) Show that under the assumption that, for certain functions f and g,

$$c(a, \alpha) = f(a)g(\alpha),$$

R_0 is given by

$$R_0 = \int_0^\infty \psi(\alpha)f(\alpha)N(\alpha)\, d\alpha, \tag{9.2}$$

with

$$\psi(\alpha) = \int_0^\infty h(\tau, \alpha)g(\alpha + \tau)\frac{\mathcal{F}_d(\alpha + \tau)}{\mathcal{F}_d(\alpha)}\, d\tau. \tag{9.3}$$

Reformulate the assumption on c in words, perhaps starting with the special case $f = g$.

iii) Assume more generally that there are, for some n, certain functions f_k and g_k, with $k = 1, \ldots, n$, so that we can write

$$c(a, \alpha) = \sum_{k=1}^n f_k(a)g_k(\alpha).$$

Specify an $n \times n$ matrix M such that R_0 is the dominant eigenvalue of M. This prepares for Section 9.4.

Many childhood diseases have an infectious period that is three orders of magnitude shorter than the average human lifespan. In that case we might approximate $\mathcal{F}_d(\alpha + \tau)/\mathcal{F}_d(\alpha)$ by 1 and $c(a, \alpha + \tau)$ by $c(a, \alpha)$ and next put $H(\alpha) = \int_0^\infty h(\tau, \alpha)\, d\tau$. We shall call this the 'short-disease approximation.'

Exercise 9.6 Simplify the expressions obtained in Exercise 9.5-ii and 9.5-iii on the basis of the short-disease approximation.

Exercise 9.7 Suppose animals enter a farm at a constant rate v and stay there for a fixed period M during which they 'mix' uniformly with the other animals at the farm. Define age = 0 at the moment of entering the farm.

i) Assume that death does not occur on the farm, but interpret removal from the farm as a type of death at age M (which indeed it most likely is). Specify $\mathcal{F}_d(a)$.

ii) What can you say about $c(a, \alpha)$ in this situation?

iii) Assume that infected individuals become immune at an age-independent rate, while having constant (in particular age-independent) infectivity during the infectious period. Derive an expression for R_0.

Exercise 9.8 Assume that, as in Section 7.4.2, $c(a, \alpha)$ can be split into a part that has separable mixing and a part giving additional contacts with individuals of the same age. Mathematically this would pose no problems. Is it a meaningful assumption from a biological point of view?

In the following three exercises we regard the calculation of R_0 and r for bovine spongiform encephalitis (BSE) in an age-structured cattle population.

Exercise 9.9 We focus on the supposed two ways of transmission for BSE: through consumption of contaminated food and through vertical transmission from mother to fetus. In the time period in which the epidemic started, it was common practice to bring deceased dairy cows to a rendering factory to be turned into meat and bone meal that was subsequently fed back to cattle. We assume an infected cow distributes its infectivity in this way randomly over all N cattle, and infectious 'contacts' therefore scale with $1/N$. We wish to express R_0 in terms of the (mostly age-dependent) functions and parameters that could describe the transmission cycle.

Let $b(a)$ be the per capita birthrate, $\mu(a)$ the 'natural' per capita death rate (by culling), and $\nu(\tau)$ the infection-age-dependent culling rate for symptomatic animals. Regard $\mathcal{F}_d(a)$ and $\mathcal{F}_c(\tau) = \exp(-\int_0^\tau \nu(\sigma)\, d\sigma)$ as the age-dependent demographic survival function and the infection survival function, respectively. We assume, for convenience only, that all infected animals are rendered (completely) and that the rendering process does not influence the infectiousness of the 'agent.' Let $\beta(a)$ be the susceptibility as a function of age. Finally, let $\gamma(\tau)$ describe the infectiousness of an infected individual, irrespective of the route by which it became infected, and let $m = m(\gamma)$ be the probability of maternal transmission for a cow with infectivity function γ. The demographic steady state $\overline{S}(a)N = \mathcal{F}_d(a)N / \int_0^\infty \mathcal{F}_d(\alpha)d\alpha$ represents the stable age distribution of the cattle population (held artificially constant by the farmers). Assume that the infected cattle feed is distributed randomly over the cattle herd.

i) Argue that the age of animals infected by feed will be distributed according to the density function

$$\frac{\beta(a)\mathcal{F}_d(a)}{\int_0^\infty \beta(\alpha)\mathcal{F}_d(\alpha)\, d\alpha}. \tag{9.4}$$

ii) What are the states-at-infection in this system?

iii) Then R_0 is the dominant eigenvalue of

$$K = \left(\begin{array}{cc} k_{11} & k_{12} \\ k_{21} & k_{22} \end{array} \right),$$

with k_{11} the expected number of new feed-infected individuals from a feed-infected individual and with similar interpretations for the other three elements. So

$$R_0 = \frac{1}{2}(k_{11} + k_{22}) + \frac{1}{2}\sqrt{(k_{11} + k_{22})^2 - 4(k_{11}k_{22} - k_{12}k_{21})}.$$

Express the elements of K in terms of the ingredients of the model.

iv) Although direct horizontal transmission between animals is unlikely, one could investigate the influence on R_0 of infection by other than maternal or feed sources. How does the above calculation change if we include a parameter w to summarize direct mass-action transmission?

Exercise 9.10 The real-time evolution of the BSE infection in an age-structured cattle population is described by

$$i(t, a) = \overline{S}(a) N \beta(a) \int_0^\infty \int_0^\infty A(\alpha, \tau) i(t - \tau, \alpha) \, d\tau \, d\alpha$$

if we restrict ourselves to the dominant mode of transmission via feed and neglect all other routes.

i) Give an expression for the kernel $A(\alpha, \tau)$ using the ingredients in Exercise 9.9.

ii) Show that

$$i(t, a) = f(a) e^{\lambda t}$$

is a solution of this integral equation if f is an eigenfunction of an operator M corresponding to eigenvalue 1, where M is defined as

$$(Mf)(a) = \overline{S}(a) N \beta(a) \int_0^\infty \int_0^\infty A(\alpha, \tau) f(\alpha) e^{-\lambda \tau} \, d\tau \, d\alpha.$$

Show that f is indeed the function we are looking for if and only if λ satisfies

$$\int_0^\infty \int_0^\infty A(\alpha, \tau) \overline{S}(\alpha) N \beta(\alpha) e^{-\lambda \tau} \, d\tau \, d\alpha = 1. \tag{9.5}$$

iii) Argue that a unique value $r > 0$ exists if $R_0 > 1$ and that r can be found numerically by computing the unique zero of a function $g(\lambda) - 1$, where g is defined by the left-hand side of (9.5), and hence is monotonically decreasing.

Exercise 9.11 Regard the same problem as in Exercise 9.10, but now considering both feed and vertical transmission. Derive an equation from which one can compute the exponential growth rate r at the start of an outbreak. This construction serves as an example of how to find an equation for r in multi-type systems, which then needs to be solved numerically.

Hint: As in the case of a single state-at-infection in Exercise 9.11, write down the linearized real-time evolution of infecteds and look for an exponential solution.

9.4   INTERVAL DECOMPOSITION

We now turn to *the* example of a finite-dimensional range situation (see Section 7.4.3). The idea is to discretize age by forming age intervals, which together cover precisely all feasible ages, and to discretize c accordingly. This means that a function of two continuous variables is replaced by a matrix. Mathematically we formulate this by introducing intervals I_i, $i = 1, 2, \ldots, n$, which are non-overlapping (i.e., $I_i \cap I_j = \emptyset$ for $i \neq j$) and together cover the positive axis (i.e., $\mathbb{R}_+ = \cup_{i=1}^n I_i$), and by requiring that

$$c(a, \alpha) = c_{ij} \text{ for } a \in I_i \text{ and } \alpha \in I_j$$

and certain given numbers c_{ij}, $1 \le i, j \le n$. The intervals can conveniently be adapted to the school system, the public health administrative system, etc. That the numbers c_{ij} are 'given' is a euphemism, and we shall return to this issue below.

With these conventions c is of the form $\sum_{k=1}^{n} f_k(a) g_k(\alpha)$, introduced in Exercise 9.5-iii. To demonstrate this, it is helpful to introduce the characteristic function χ_I of a set I defined by

$$\chi_I(a) = \left\{ \begin{array}{ll} 1 & \text{if } a \in I, \\ 0 & \text{otherwise.} \end{array} \right.$$

Indeed, with this piece of notation at hand, we can write

$$\begin{aligned} c(a, \alpha) &= \sum_{k,l=1}^{n} c_{kl} \chi_{I_k}(a) \chi_{I_l}(\alpha) \\ &= \sum_{k=1}^{n} \chi_{I_k}(a) \sum_{l=1}^{n} c_{kl} \chi_{I_l}(\alpha), \end{aligned}$$

which is of the indicated form since we can take $f_k = \chi_{I_k}$ and $g_k = \sum_{l=1}^{n} c_{kl} \chi_{I_l}$.

Exercise 9.12 Elaborate for this special case the expression of Exercise 9.5-iii in the short-disease approximation of Exercise 9.6.

9.5 THE ENDEMIC STEADY STATE

Let us now return to the general setting and assume that a steady (= stationary = time-independent) endemic situation exists. By analogy with the situation described in Section 9.1, we can introduce the cohort 'remain uninfected' function $\mathcal{F}_i(a)$ (i for infection, a for age) describing the probability that an arbitrary individual will remain uninfected (in the sense of 'never before being infected') at age a, given that it survives until at least that age. Accordingly, we can introduce the age-specific *force of infection* $\Lambda(a)$, i.e., the per susceptible probability per unit of time of being infected, by

$$\Lambda(a) = -\frac{d}{da} \ln \mathcal{F}_i(a) \Leftrightarrow \mathcal{F}_i(a) = e^{-\int_0^a \Lambda(\alpha)\, d\alpha}. \tag{9.6}$$

In the demographic context, the survival function is a purely descriptive statistical object. In the context of infectious diseases, a postulated mechanism is responsible for the observed 'survival' function $\mathcal{F}_i(a)$, and hence there should be a consistency condition that serves as an equation from which, at least in principle, $\mathcal{F}_i$ (or, equivalently, Λ) can be determined. Indeed, the interpretation of all the ingredients involved requires that

$$\Lambda(a) = \int_0^\infty \int_0^\infty h(\tau, \alpha) c(a, \alpha + \tau) \Lambda(\alpha) S(\alpha) \frac{\mathcal{F}_d(\alpha + \tau)}{\mathcal{F}_d(\alpha)}\, d\tau\, d\alpha, \tag{9.7}$$

where

$$S(a) = N(a) \mathcal{F}_i(a), \tag{9.8}$$

i.e., S is the relative density of susceptibles ('relative' because N is normalized to have integral 1, recall formula (9.1)). This certainly requires an explanation.

The basic idea is that we are dealing with an innocent disease that does not affect population numbers. Moreover, we adhere to the interpretation of c as discussed in Section 9.2 (this is a subtle issue deserving careful attention, in particular when comparing host populations of different sizes).

Let P denote the total population size. Then the age-specific incidence is given by $P\Lambda(a)S(a)$, and, according to the assumptions preceding Exercise 9.5, the number of new cases per unit of time resulting from this steady incidence equals

$$S(a)\int_0^\infty\int_0^\infty h(\tau,\alpha)c(a,\alpha+\tau)P\Lambda(\alpha)S(\alpha)\frac{\mathcal{F}_d(\alpha+\tau)}{\mathcal{F}_d(\alpha)}\,d\tau\,d\alpha$$

where the first factor is due to the fact that a contact with an individual of age a is with probability $S(a)$ with a susceptible (read it as $N(a)\mathcal{F}_i(a)$ or $N(a)\frac{S(a)}{N(a)}$, whichever you prefer). Consistency then requires that this quantity is equal to $P\Lambda(a)S(a)$. If we factor out $PS(a)$, we arrive at (9.7).

Because $\mathcal{F}_i(a)$ depends on Λ, the consistency condition is a nonlinear integral equation for Λ. A rigorous mathematical analysis of this equation[3] leads in particular to the conclusion that for $R_0 > 1$ there is indeed a non-negative solution.

Exercise 9.13 Show that linearization at $\Lambda \equiv 0$ leads us back from the right-hand side of (9.7) to the next-generation operator. Hint: Interpret ΛN as ϕ.

9.6   VACCINATION

As an applied problem that motivates the use of age-structured models, we regard vaccination as a control strategy. We do so in two different situations: i) we investigate under what conditions a given agent can successfully invade a partially vaccinated population (or, more precisely, how large a fraction of the population do we have to vaccinate at what age in order to prevent the agent from establishing); ii) the more realistic situation that a vaccination campaign is started when the agent has become endemic, and we ask the question whether the vaccination effort will be sufficient to eliminate the agent. We consider here, for convenience, perfect vaccines only.

In the first situation the idea is to calculate a reproduction ratio R_v for invasion of a population of susceptibles in a demographic steady state where a vaccination strategy v is in operation. Strategy v prevents the agent from establishing if $R_v < 1$. Let

$$\mathcal{F}_v(a)$$

denote a vaccination 'survival function,' i.e., the conditional probability that an individual of age a that is alive is still susceptible (and has not been made immune by vaccination). To arrive at explicit and manageable criteria, we adopt the one-dimensional range condition of Exercise 9.5-ii.

Exercise 9.14 Argue that R_v is given by

$$R_v=\int_0^\infty\int_0^\infty h(\tau,\alpha)g(\alpha+\tau)\frac{\mathcal{F}_d(\alpha+\tau)}{\mathcal{F}_d(\alpha)}d\tau f(\alpha)N(\alpha)\mathcal{F}_v(\alpha)\,d\alpha.$$

[3]H. Inaba: Threshold and stability results for an age-structured epidemic model. *J. Math. Biol.*, **28** (1990), 411–434; D. Greenhalgh: Threshold and stability results for an epidemic model with an age-structured meeting rate. *IMA J. Math. Appl. Med. Biol.*, **5** (1988), 81–100.

Exercise 9.15 In the easiest case, let a fraction v of the population be vaccinated at birth. Show that in order to prevent establishment we must have

$$v > 1 - \frac{1}{R_0}$$

where R_0 is given by (9.2).

Exercise 9.16 Now suppose we vaccinate a fraction v of the individuals of age a_v. Derive the appropriate inequality for the minimal proportion to be vaccinated in order to prevent establishment.

In the second situation, when the agent is endemic in the population, developing a procedure based on R_0 no longer makes sense. We do still assume that a steady state has arisen. We have to discount the available susceptibles not only by the probability to be unvaccinated, but also by the probability to have escaped infection so far. Therefore, instead of (9.8), we obtain for the steady-state density of susceptibles in the endemic situation

$$S(a) = N(a)\mathcal{F}_v(a)\mathcal{F}_i(a),$$

and furthermore the force of infection Λ still satisfies the nonlinear integral equation (9.7), but now with S defined by this modified expression.

Exercise 9.17 Assume separable mixing $c(a, \alpha) = f(a)g(\alpha)$.

i) Show that necessarily $\Lambda(a) = Qf(a)$ for a constant Q that has to satisfy

$$1 = \int_0^\infty \psi(\alpha)N(\alpha)\mathcal{F}_v(\alpha)e^{-Q\int_0^\alpha f(a)\,da}f(\alpha)\,d\tau\,d\alpha, \qquad (9.9)$$

where $\psi(\alpha)$ is given by (9.3). The same expression, apart from notation, was derived by Dietz and Schenzle (1985).[4]

ii) In the short-disease approximation (Section 9.3), show that (9.9) can be approximated by

$$1 = \int_0^\infty f(\theta)g(\theta)H(\theta)N(\theta)\mathcal{F}_v(\theta)e^{-Q\int_0^\theta f(a)\,da}\,d\theta.$$

If one further assumes the same age dependence in activity level ($f = g$), one can use data about the endemic state to estimate f, Q and the demographic ingredients and subsequently calculate whether or not a given vaccination schedule suffices to eliminate the agent, i.e., causes $R_v < 1$. We refer once more to Dietz and Schenzle (1985) for more information.

Finally, we briefly address how the mean age at first infection relates to control measures, such as vaccination, that influence the force of infection in the population.

In Exercise 4.6 (equation (4.10)), we already encountered a simple formula that expressed the mean age at first infection $\bar{a}$ as a function of a constant force of infection Λ and a constant per capita death rate μ: $\bar{a} = 1/(\mu + \Lambda)$. It follows that $\bar{a}$ will increase when vaccination is implemented in a population.

---

[4]K. Dietz and D. Schenzle: Proportionate mixing models for age-dependent infection transmission. *J. Math. Biol.*, **22** (1985), 117–120.

In Exercise 13.6 we show that, in the age-structured case, the mean age at first infection is given by

$$\bar{a} = \frac{\int_0^\infty a\Lambda(a)\mathcal{F}_i(a)\mathcal{F}_d(a)\,da}{\int_0^\infty \Lambda(a)\mathcal{F}_i(a)\mathcal{F}_d(a)\,da}. \tag{9.10}$$

When a fraction of the population is vaccinated, the force of infection will decrease. Heuristically, we expect that, as a result, the mean age at infection $\bar{a}$ will increase, since with a lowered infection pressure in the population, it would take longer for a given susceptible to meet an infectious individual. The formula (4.10) shows that this relation indeed holds when both the force of infection and the force of mortality are constant. Naively, the authors set out to prove a general result, starting from (9.10). But soon they discovered that the normalization factor (the denominator) has subtle effects and that, in fact, one can construct a counterexample in which the mean age at infection decreases, despite the fact that the force of infection decreases. The key feature is that the force of infection at high ages decreases much more strongly than the force of infection at low ages. We still conjecture that $\bar{a}$ should increase when $\Lambda(a)$ is multiplied by some factor less than one.

The special case of no mortality (i.e., $\mathcal{F}_d(a) \equiv 1$) is much simpler, since then the denominator equals 1. By partial integration, the numerator can be rewritten as

$$\int_0^\infty e^{-\int_0^a \Lambda(\alpha)\,d\alpha}\,da$$

(provided $\exp(-\int_0^a \Lambda(\alpha)\,d\alpha) \to 0$ for $a \to \infty$, i.e., provided no individual escapes for ever from being infected), which clearly shows a monotone decreasing dependence on Λ.

We want to show that a decrease in the force of infection leads to an increase of the risk to become infected at a relatively advanced age. Let $a_2 > a_1 > 0$. The risk of becoming infected at an age between a_1 and a_2, given that one does not die before a_2, is given by

$$\mathcal{F}_i(a_1) - \mathcal{F}_i(a_2) = \left(1 - \frac{\mathcal{F}_i(a_2)}{\mathcal{F}_i(a_1)}\right)\mathcal{F}_i(a_1)$$
$$= \left(1 - e^{-\int_{a_1}^{a_2}\Lambda(\alpha)\,d\alpha}\right)e^{-\int_0^{a_1}\Lambda(\alpha)\,d\alpha}.$$

Clearly, the first factor decreases when Λ decreases, but the second factor *increases*, and that may very well be the dominant effect. Thus vaccination may, for those individuals that were not vaccinated (or for whom vaccination failed), lead to an increase of the risk of becoming infected at an older age than in a comparable unvaccinated population.

This 'side effect' of vaccination can lead to dangerous situations because seriousness of complications arising from childhood infections is often positively correlated with age. An example is vaccination against rubella. The well-known problem of congenital rubella syndrome (CRS) in babies born from females who become infected with rubella during pregnancy is a main reason to vaccinate against rubella. The vaccination strategy and coverage needs to be such that the number of females of fertile age who have not been vaccinated, nor have experienced a natural infection in early childhood, is small. A rise in $\bar{a}$ induced by a sub-optimal vaccination strategy, or by a sub-optimal vaccination coverage, could in theory lift $\bar{a}$ from a young age into the fertile age group. In principle, this could then lead to a larger number of CRS cases in the population, compared to the situation where no vaccination is implemented. An illustration of this has been found in Finland (see Anderson and May 1991 and the references given there), and has been clearly demonstrated in Greece.[5] The data from Greece clearly show a steady rise in the percentage of women of childbearing age susceptible to rubella from 10% at the start of vaccination, to 36% in 1990. Age-structured data of rubella patients from 1986 show a peak in the age group of 5–9-year old children, whereas by 1993 this peak has shifted to the individuals in age group 15–19. In 1993, Greece experienced a severe 'outbreak' of CRS as a result of this effect.

---

[5]See T. Panagiotopoulos, I. Antoniadou and E. Valassi-Adam: Increase in congenital rubella occurrence after immunisation in Greece: retrospective survey and systematic review. *British Medical Journal*, **319** (1999),1462–1466.

Chapter Ten

Spatial spread

10.1 POSING THE PROBLEM

As an example of a situation where spatially structured models are relevant, think of a fungal pathogen affecting an agricultural crop. A farmer having ascertained that his field is affected wants to know: How fast is the infection spreading? What fraction of the yield do I stand to lose if I do not spray with fungicides? The trade-off here could be that spraying is expensive and bad for the environment. Suppose that harvest is three months away. Do I take the loss of plants or do I invest in fungicide and accept the concomitant pollution?

So, the key question is: How fast is the infection spreading?

A student of the preceding chapters might be inclined to answer the farmer by first drawing Figure 10.1 and next, in an attempt to be pragmatic rather than scrupulous, saying that the relevant part of the curve is to a good approximation described by Ce^{rt}, with C determined from the current situation and r from (one hopes) known data about spore production and dispersal. The aim of this chapter is to provide the student with better ingredients for an answer. In particular, we will explain that it is much more likely that the fraction of the crop affected will grow as a quadratic function of time.

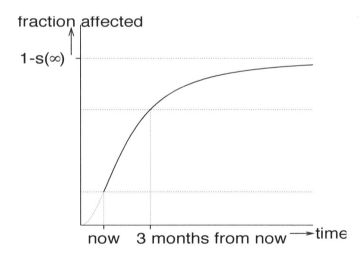

Figure 10.1: Pragmatic sketch of increase in fraction of a plant crop affected by a fungal pathogen as a function of time. See Section 10.1 for context.

The main point is that the infestation is localized in patches, called foci, which expand more or less radially. The population growth parameter r, however, describes population growth that is uniform (in space), as can be concluded from the eigenfunction, which is constant (in fact also when we consider R_0). This will be elaborated further.

Thus we arrive at the following set of questions: Do models predict radial expansion of epidemic fronts? If so, how can we determine the *speed* of the front from the model ingredients? What are the conditions that promote wave-like expansion?

10.2 WARMING UP: THE LINEAR DIFFUSION EQUATION

Let u denote the density of a pest species. Let this species inhabit a very large domain without structure (no hedges, roads, canals, rivers, mountains, ...). We take the plane $\mathbb{R}^2$ as an idealized representation of this domain and let $x \in \mathbb{R}^2$ denote a spatial position. Note that $\mathbb{R}^2$ is homogeneous (i.e., translation-invariant: points are interchangeable) and isotropic (the same structural properties in every direction).

We assume that the species grows at a net per capita rate κ. In addition, it disperses. If movement is completely random, we may use the diffusion equation to describe dispersal (the underlying assumption being that the flux is proportional to the gradient). Hence we postulate that

$$u_t = D\triangle u + \kappa u, \tag{10.1}$$

where $u_t = \partial u/\partial t = \partial u(t,x)/\partial t$ is the partial derivative of u with respect to time, where D is a species-dependent constant (called the diffusion constant) representing the mean square distance covered per unit of time, and where

$$\triangle u = \frac{\partial^2 u}{\partial x_1^2} + \frac{\partial^2 u}{\partial x_2^2}$$

is shorthand for the sum of the second partial derivatives of u with respect to the two coordinate directions.

The chief advantage of (10.1), and the reason to consider it here, is that we can solve this equation explicitly. The explicit expression allows us to pinpoint certain quantitative and qualitative properties, in particular the phenomenon of an *asymptotic speed of propagation* c_0, which equals the *minimal speed* for which *plane traveling-wave solutions* exist. Armed with this knowledge, we then study, in subsequent sections, the robustness of the conclusions: Do the qualitative phenomena extend to the nonlinear realm? And to models involving kernels $A(\tau, x, \xi)$?

The solution of (10.1),

$$u(t,x) = \frac{1}{4\pi D t} e^{-\frac{|x|^2}{4Dt} + \kappa t}, \tag{10.2}$$

describes the effect of a localized (at $x = 0$) disturbance (at $t = 0$) of the unstable steady state $u \equiv 0$ (read as: u is identically zero).

Exercise 10.1 Verify that u given in (10.2) satisfies the equation (10.1).

Exercise 10.2 Verify that $\int_{\mathbb{R}^2} u(t,x)\, dx = e^{\kappa t}$ for $t > 0$.

Exercise 10.3 Show that uniformly for $|x| \geq \varepsilon$ we have

$$\lim_{t \downarrow 0} u(t,x) = 0$$

and conclude that u is indeed concentrated in $x = 0$ at $t = 0$. (In the language of distributions and measures, we have $\lim_{t \downarrow 0} u(t, x) = \delta(x)$, with δ the Dirac 'function' (measure/distribution) concentrated in $x = 0$.)

What can be said about the behavior of $u(t, x)$ for large t? Of course the factor $1/t$ goes to zero, but $e^{\kappa t}$ goes to infinity much faster. So, if, on the one hand, we fix x and let t tend to infinity, we find that u grows exponentially with rate κ. If, on the other hand, we fix t and let $|x| \to \infty$ (that is, we observe far ahead in space) we find that u is negligibly small. Thus it appears that the limits $t \to \infty$ and $|x| \to \infty$ cannot be interchanged: the order matters. In such situations of non-uniform convergence, one expects to see a *transition layer* (in which transition from one extreme, zero, to the other extreme, infinity, is made) once we approach infinity in (t, x) space in a tailor-made fashion. We can immediately infer from the explicit expression what this 'tailor-made fashion' is: we have to avoid the exponent in (10.2) going to either $+\infty$ or $-\infty$. (The factor $1/t$ necessitates that we be a little more precise about the first of these possibilities, viz., we have to avoid approaching $+\infty$ too quickly; it is indeed this factor $1/t$ that makes the precise characterization of the transition layer rather subtle.) We refrain from a more precise study, and restrict ourselves to the observation that, for any $\varepsilon > 0$, for $t \to \infty$

$$u(t, x) \to \begin{cases} 0 & \text{if } |x|^2 > (4D\kappa + \varepsilon)t^2; \\ \infty & \text{if } |x|^2 < (4D\kappa - \varepsilon)t^2. \end{cases}$$

In suggestive words, we could say that we distinguish the 'not yet' region, being the exterior of a disc, the radius of which grows like $2t\sqrt{D\kappa + \varepsilon/4}$, and the 'already over' region, being the interior of a disc, the radius of which grows like $2t\sqrt{D\kappa - \varepsilon/4}$. (Here ε is a positive number that can be taken arbitrarily small. It relieves us from going into the details of the subtle limiting behavior that occurs when x grows as $2t\sqrt{D\kappa} + O(\ln t)$.) Yet another way of expressing this result is to state that

$$c_0 := 2\sqrt{D\kappa}$$

is the *asymptotic speed of propagation* of the disturbance. We conclude that the solution of the linear diffusion equation (10.1) displays radial expansion of a disturbance with a well-defined speed that can be easily computed from the parameters.

Our next aim is to characterize c_0 in a completely different way, viz., as the minimal speed of plane traveling waves. The point is that this characterization carries over much more easily to other situations, in which explicit calculations are often impossible.

A plane wave traveling in the direction specified by a given unit vector ν is described by a solution of the form

$$u(t, x) = w(x \cdot \nu - ct), \tag{10.3}$$

(here $x \cdot \nu$ is the inner product of the vectors x and ν, i.e., $x \cdot \nu = x_1\nu_1 + x_2\nu_2$). We call w the *profile* of the wave, ν the *direction* of the wave, and c its *speed*.

Exercise 10.4 Show that, in order for (10.3) to define a solution of (10.1), the profile w should satisfy

$$Dw'' + cw' + \kappa w = 0. \tag{10.4}$$

Next argue that this requires w to be of the form $w(\xi) = \exp(\lambda\xi)$, with λ satisfying the characteristic equation

$$D\lambda^2 + c\lambda + \kappa = 0. \tag{10.5}$$

Conclude that

$$\lambda = \lambda_\pm = \frac{-c \pm \sqrt{c^2 - 4D\kappa}}{2D}. \tag{10.6}$$

We want *positive* solutions, because of the interpretation. Oscillating solutions are characterized by complex λ. So we should require $\lambda_\pm$ to be real; that is, we should have $c^2 - 4D\kappa \geq 0$, i.e., $c \geq 2\sqrt{D\kappa} = c_0$. We conclude that plane traveling wave solutions exist for all speeds c that exceed a threshold c_0 and that the minimal plane-wave speed c_0 coincides with the asymptotic speed of propagation.

Exercise 10.5 For $c = c_0$ we have $\lambda_\pm = -\frac{c_0}{2D}$ and

$$w(\xi) = e^{\lambda_\pm (x \cdot \nu - c_0 t)} \sim e^{-\lambda_\pm c_0 t} = e^{\frac{c_0^2}{2D} t} = e^{2\kappa t}$$

for large t. What do you conclude from this?

10.3   VERBAL REFLECTIONS SUGGESTING ROBUSTNESS

Consider a steady state (zero/infection-free) that is unstable and such that any physically feasible (in particular *positive*) perturbation triggers a transition towards another steady state ('infinity' in the case of the linear diffusion equation, i.e., 'after the epidemic'). To this local dynamics, add a spatial component and coupling, which means that perturbations at some point generate perturbations at nearby points. How fast do perturbations spread?

Imagine space to be the same in every point in all directions (i.e., homogeneous and isotropic). Traveling plane waves are uniform in all directions but one. So they manifest how disturbances travel in one direction (although this direction is arbitrary because of the isotropy). And the speed tells us how fast the spread will be.

Planc waves do not come with a unique speed, but rather with a continuum of possible speeds, bounded only at one side. Why should the minimal speed be the truly relevant one?

To fix a unique solution, the partial differential equation (PDE) (10.1) has to be supplemented with an initial condition. At least as a thought experiment, we can therefore manipulate the solution.[1] Imagine a series of fireworks placed in a row, with fuses of varying lengths. By lighting the fuses, one can create a 'traveling wave' of explosions. By choosing the lengths of the fuses appropriately, one can achieve any speed one wants. If, however, the fireworks also have the tendency to kindle their nearest neighbors, this process of self-kindling will dominate as soon as one tries to achieve, by manipulation of the fuses, a speed that is too low. Therefore, the minimal plane-wave speed corresponds to the inherent speed of the self-infection mechanism!

Traveling plane-wave solutions are examples of similarity solutions, i.e., solutions depending only on a certain combination of the independent variables. They

---

[1]The thought experiment is from J.A.J. Metz.

show up whenever the dynamics are equivariant under a group of transformations (in this case translations and rotations). They are often the quintessence of intermediate asymptotic behavior, when the transients reflecting the initial conditions have died out, but where the final state has not yet been achieved everywhere and boundary conditions (every real domain is finite) do not yet impinge upon the natural dynamics.[2]

We conclude that whenever

- local dynamics consist of a transition from an unstable steady state to a stable one;

- perturbations spread, i.e., there is some form of coupling of local dynamics;

- space is homogeneous and isotropic;

we are bound to find that:

- Traveling plane-wave solutions exist for all speeds c exceeding a minimal speed c_0;

- The minimal wave speed c_0 is the asymptotic speed of propagation associated with the self-triggering mechanism.

Many of these conclusions remain valid if there is homogeneity but no isotropy.[3] Of course, the minimal plane-wave speed will then depend on the direction ν. When defining what asymptotic propagation means, one should then blow up not a disc but rather another (convex) set defined on the basis of the function $c_0(\nu)$.

Exercise 10.6 Suppose a species spreads by wind-borne propagules. Assume there is a prevailing wind direction. Let the unit vector σ point to this wind direction and let θ be the wind velocity. Then (10.1) should be replaced by

$$u_t = D\triangle u - \theta\sigma \cdot \nabla u + \kappa u,$$

where ∇u is the gradient (i.e., $\nabla u = (\partial u/\partial x_1, \partial u/\partial x_2)^\top$, the vector of partial derivatives), and so $\sigma \cdot \nabla u = \sigma_1 \partial u/\partial x_1 + \sigma_2 \partial u/\partial x_2$ is the directional derivative in the direction σ. In case you wonder about the minus sign, put $D = 0$ and $\kappa = 0$ and check that $u(t, x) = \phi(\sigma \cdot x - \theta t)$ satisfies the equation for any function ϕ, and that such solutions correspond to plane waves traveling in the direction σ. Look at Figure 10.2, turn it into an animation of a traveling wave and verify that the direction of propagation is to the right.

Now let us look for wave solutions

$$u(t, x) = w(\sigma \cdot x - ct)$$

traveling with speed c in the direction σ. Show that such solutions exist for all $c \geq c_0 + \theta$ with $c_0 = 2\sqrt{D\kappa}$ as before. Next look for traveling waves in the opposite direction. What do you conclude?

---

[2]G.I. Barenblatt, *Similarity, Self-similarity and Intermediate Asymptotics*. Plenum, New York, 1979.

[3]H.F. Weinberger: Long-time behavior of a class of biological models. *SIAM J. Math. Anal.*, **13** (1982), 353–396; F. van den Bosch, O. Diekmann and J.A.J. Metz: The velocity of spatial population expansion. *J. Math. Biol.*, **28** (1990), 529–565.

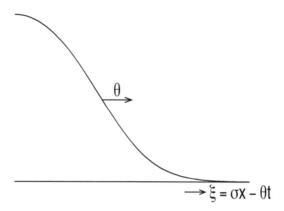

$$\rightarrow \xi = \sigma x - \theta t$$

Figure 10.2: Schematic rendering of a traveling wave solution of a diffusion system, see Exercise 10.6.

This seems an appropriate place for the following remark. Within the general framework, it is possible that a population grows while continuously moving in space to ever more removed regions (i.e., to 'infinity'). The spectral radius criterion would say $R_0 > 1$, so the population grows. A local observer, however, would only notice a very temporary growth, followed by local extinction. Within a more measure-theory-oriented approach, growth is defined in terms of a local indicator, the *Perron root*. In the situation just described, it follows that the Perron root is less than one and this is interpreted as 'the population does not grow.' We refer to Jagers (1995) for a precise definition of the Perron root and note that Shurenkov (1992) has shown that the spectral radius and the Perron root coincide whenever the state space is compact.[4]

10.4   LINEAR STRUCTURED POPULATION MODELS

Let $B(\tau, x, \xi)$ be the expected rate at which an individual born at position ξ in space produces offspring (per unit of space) at position x at age τ. Then straightforward bookkeeping considerations suggest that (10.1) be replaced by the time-translation invariant integral equation

$$u(t,x) = \int_0^\infty \int_\Omega B(\tau, x, \xi) u(t - \tau, \xi) \, d\xi \, d\tau, \qquad (10.7)$$

where Ω represents the area in which the species lives (readers are invited to read the interpretation of the right-hand side of (10.7) aloud to themselves, to check their understanding of it). When $\Omega = \mathbb{R}^2$ and we assume homogeneity, B should be a function of the *relative position* $x - \xi$ rather than of x and ξ separately. When, in addition, we assume isotropy, we can consider B as a function of the *distance*

---

[4]P. Jagers: The deterministic evolution of general branching populations. In: O. Arino, D. Axelrod and M. Kimmel (eds.), *Mathematical Population Dynamics*. Wuerz, Winnipeg, 1995; V.M. Shurenkov: On the relationship between spectral radii and Perron roots. Preprint 1992-17, Department of Mathematics, Chalmers University Göteborg, 1992.

$|x - \xi|$ only. Here we restrict our attention to that situation, that is, we consider (with slight notational abuse of the symbol B)

$$u(t,x) = \int_0^\infty \int_{\mathbb{R}^2} B(\tau, |x - \xi|) u(t - \tau, \xi) \, d\xi \, d\tau. \tag{10.8}$$

Tempted by our analysis of the linear diffusion equation and the robustness considerations of the preceding section, we look for traveling plane-wave solutions, i.e., we set

$$u(t,x) = w(x \cdot \nu - ct) \tag{10.9}$$

and deduce that w should satisfy

$$w(\theta) = \int_{-\infty}^\infty V_c(\zeta) w(\theta - \zeta) \, d\zeta, \tag{10.10}$$

where, by definition,

$$V_c(\zeta) = \int_0^\infty \int_{-\infty}^\infty B(\tau, \sqrt{(\zeta + c\tau)^2 + \sigma^2}) \, d\sigma \, d\tau, \tag{10.11}$$

which we note does not depend on ν.

Exercise 10.7 Derive (10.10) in detail.

Exercise 10.8 Derive the characteristic equation

$$1 = \int_{-\infty}^\infty e^{-\lambda\zeta} V_c(\zeta) \, d\zeta \tag{10.12}$$

by inserting the trial solution $w(\theta) = e^{\lambda\theta}$ into (10.10).

Exercise 10.9 We give the right-hand side of (10.12) a name, i.e., we write $1 = L_c(\lambda)$, where

$$L_c(\lambda) := \int_{-\infty}^\infty e^{-\lambda\zeta} V_c(\zeta) \, d\zeta. \tag{10.13}$$

Show that

$$L_c(\lambda) = \int_{-\infty}^\infty e^{-\lambda\alpha} \int_0^\infty \int_{-\infty}^\infty e^{\lambda c\tau} B(\tau, \sqrt{\alpha^2 + \sigma^2}) \, d\sigma \, d\tau \, d\alpha, \tag{10.14}$$

and show from this that

$$L_c(0) = \int_0^\infty \int_{\mathbb{R}^2} B(\tau, |\eta|) \, d\eta \, d\tau = R_0, \tag{10.15}$$

$$\frac{dL_c}{d\lambda}(0) = c \int_0^\infty \int_{\mathbb{R}^2} \tau B(\tau, |\eta|) \, d\eta \, d\tau > 0 \tag{10.16}$$

for $c > 0$, and that

$$\frac{d^2 L_c}{d\lambda^2}(\lambda) > 0 \tag{10.17}$$

for all λ and all $c \geq 0$. Finally show that for every $\lambda < 0$, $L_c(\lambda)$ is a monotonically decreasing function of c with limit zero for $c \to \infty$.

Conclude from all this that, whenever $R_0 > 1$, the set $\{c : \text{there exists } \lambda < 0 \text{ such that } L_c(\lambda) < 1\}$ consists of a half-line (c_0, ∞).

Exercise 10.10 Establish that c_0 can be characterized, together with the corresponding value of λ, say λ_0, as the solution of the pair of equations (10.12) and

$$\frac{dL_c}{d\lambda}(\lambda) = 0. \tag{10.18}$$

We conclude that, starting from the modeling ingredient $B(\tau, |\eta|)$, one can constructively define a minimal plane-wave speed c_0 by the pair of equations (10.12) and (10.18). It remains to ascertain that c_0 thus defined is also the asymptotic speed of propagation of disturbances. We postpone remarks on this issue to the next section, where we deal with the nonlinear problem.

Exercise 10.11 When the species considered is actually an infectious agent exploiting a host population, and if we assume mass-action contacts, we have

$$B(\tau, x, \xi) = S_0(x)A(\tau, x, \xi), \tag{10.19}$$

where A is our familiar epidemic model ingredient and $S_0(x)$ is the host density as a function of position. Do you agree? If so, check that traveling front solutions require a uniform host density S_0 (as may be an appropriate assumption within fields of agricultural crops, or, if we think of fields as host individuals, for fields within a region).

We refer to Thieme (1979)[5] for estimates of the speed of propagation using only lower bounds for S_0 and to the book by Shigesada and Kawasaki (1997; see footnote 7 of this chapter) for numerical studies of the speed of propagation when high- and low-density host population patches alternate.

10.5   THE NONLINEAR SITUATION

Nonlinearity leads to boundedness, but under suitable assumptions, nothing much changes otherwise.

Models in population genetics, combustion and population dynamics lead to nonlinear diffusion equations

$$u_t = D\triangle u + f(u) \tag{10.20}$$

with a nonlinear function f having properties as displayed graphically in Figure 10.3.

So, forgetting for a moment about space, zero is an unstable steady state and any positive perturbation ultimately leads to some stable steady state $\bar{u}$, say. Hence the considerations of Section 10.3 apply. But how should one compute c_0?

If we linearize (10.20) at $u \equiv 0$, we obtain equation (10.1) with $\kappa = f'(0)$, to which we can associate a speed c_0. Is this the right one?

To show that (10.20) has traveling plane-wave solutions for every $c \geq c_0$ is a matter of phase-plane analysis, for which we refer to the literature.[6] To show that

---

[5]H.R. Thieme: Density-dependent regulation of spatially distributed populations and their asymptotic speed of spread. *J. Math. Biol.*, **8** (1979), 173–187.

[6]K.P. Hadeler and F. Rothe: Travelling fronts in nonlinear diffusion equations. *J. Math. Biol.*, **2** (1975), 251 263; A.I. Volpert, V.A. Volpert and V.A. Volpert: *Traveling Wave Solutions of Parabolic Systems*. AMS Translations of Mathematical Monographs, Vol. 140, 1994; D.G. Aronson and H.F. Weinberger: Nonlinear diffusion in population genetics, combustion, and nerve

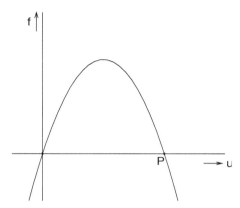

Figure 10.3: Typical shape of a function f in a nonlinear diffusion system in population biology, see equation (10.20).

$u(t,x)$ tends to zero outside a ball that expands with speed larger than c_0, while tending to the stable steady state $\bar{u}$ inside a ball that expands with speed smaller than c_0, is not easy but also not impossible. The intricate proof, for which we refer to Aronson and Weinberger and Diekmann and Temme (see footnote 6), involves comparison arguments based on the maximum principle. Both results involve the condition

$$f(u) \leq f'(0)u, \qquad (10.21)$$

which reflects that the living conditions are optimal at very low densities. When the condition is not satisfied, for example due to an Allee effect (such as when it is more difficult to find suitable mates at low population densities), one may have so-called 'pulled' waves, the speed of which is not determined by the low-density situation. Also for this case many results are known (see the references already quoted).

We conclude that, provided the nonlinearity f satisfies certain interpretable and reasonable conditions, there is a well-defined asymptotic speed of propagation that can be calculated from the appropriate linearization.

In the epidemic context, the starting point is equation (8.23) with $S(-\infty, x) = S_0$, independent of x. Introducing (cf. the elaboration of Exercise 8.26)

$$u(t,x) = -\ln \frac{S(t,x)}{S_0} \qquad (10.22)$$

we arrive at the nonlinear integral equation

$$u(t,x) = \int_0^\infty \int_{\mathbb{R}^2} S_0 A(\tau, x, \xi) g(u(t-\tau, \xi)) \, d\xi \, d\tau \qquad (10.23)$$

with

$$g(u) = 1 - e^{-u}. \qquad (10.24)$$

pulse propagation. In: *Partial Differential Equations and Related Topics*. J.A. Goldstein (ed.), Lecture Notes in Mathematics, Vol. 446, Springer-Verlag, Berlin, 1975, pp 5–49; O. Diekmann and N.M. Temme (eds.): *Nonlinear Diffusion Problems*. Mathematisch Centrum, Amsterdam, 1976.

We observe two things:

- The linearization at $u \equiv 0$ is of the form (10.7).

- $g(u) \leq g'(0)u$ (the 'virgin' situation is optimal for the infective agent).

Motivated by these observations, we expect that one can prove for u satisfying (10.23) that c_0 is the asymptotic speed of propagation, where:

- c_0 is calculated from S_0 and A (which is, of course, assumed to depend on $|x - \xi|$ only) as in the preceding section.

- u tends to the familiar final size within any ball that expands with speed less than c_0.

That these expectations are warranted was shown in detail by Thieme and Diekmann, using comparison methods in the spirit of Aronson and Weinberger; more complicated situations were dealt with extensively in the book by Radcliffe and Rass (2003). Pioneering work had been done much earlier by Fisher, Skellam, Kolmogorov-Petrovski-Piscounov, Kendall and later Mollison (see the references in Metz and Van den Bosch 1995).[7]

We conclude that for our familiar epidemic model, one can compute the asymptotic speed of propagation from the model ingredients S_0 and A, viz., by solving (10.12) and (10.18).

In our top-down approach we are still far from the bottom. As a next step, one should introduce parametrized families of kernels A on the basis of a mixture of mechanistic and pragmatic considerations. In addition, it is useful to derive approximation formulas for c_0 involving moments (both of time-type and of space-type) of the kernel A. Using experimental data to estimate parameters (preferably in an independent manner) and to verify predictions, one can then assess the theory.[8]

10.6   SUMMARY: THE SPEED OF PROPAGATION

Within the context of idealized models, we have unambiguously defined the (asymptotic) speed c_0 of the spatial propagation of an infection, and we have characterized c_0 in terms of the basic model ingredients in such a way that the computation of c_0 from the ingredients is rather simple. Thus we added one more indicator of the infectiousness of an agent to the list (consisting so far of R_0, r, the probability of a major outbreak, the size of a major outbreak, and the endemic level). For many ecological or agricultural systems, this is actually the most relevant indicator!

---

[7]J.A.J. Metz and F. Van den Bosch: Velocities of epidemic spread. In: Mollison (1995), pp. 150–186.

[8]See, for example, F. Van den Bosch, J.A.J. Metz and O. Diekmann: The velocity of spatial population expansion. *J. Math. Biol.*, **28** (1990), 529–565; M.A. Lewis and S. Pacala: Modeling and analysis of stochastic invasion processes. *J. Math. Biol.*, **41** (2000), 387–429; U. Dieckmann, R. Law and J.A.J. Metz (eds.): *The Geometry of Ecological Interactions*. Cambridge University Press, Cambridge, 2000; N. Shigesada and K. Kawasaki: *Biological Invasions: Theory and Practice*. Oxford University Press, Oxford, 1997; J. McGlade (ed.): *Advanced Ecological Theory: Principles and Applications*. Blackwell Science, Oxford, 1999.

10.7 ADDENDUM ON LOCAL FINITENESS

Spatial position creates heterogeneity and, in particular, the contact 'intensity' of a pair of individuals depends on the spatial position of both of them, often by way of the distance between them. It is this feature of spatial structure that took center stage in the present chapter. We systematically ignored another important feature: even if a population as a whole is very, say infinitely, large, it may very well be locally small, in the sense that any individual comes into contact with only a fixed finite set of neighboring individuals. If that is the case, we can ignore neither demographic stochasticity nor dependence (recall Section 2.3), not even in the initial stages of spatial spread.

As always, it helps to look at caricatures first. Imagine that individuals occupy the positions of an integer lattice on a line

o o o o o o o o o o o o o o o o

for instance, cows lined up in a row in a very long stable. Assume that an individual can only infect its two immediate neighbors and that, if infected itself, it does infect a susceptible neighbor with transmission probability $p, 0 < p < 1$. Assume one cow is miraculously infected from outside. Its expected number of offspring equals $2p$, which exceeds 1 if $p > 0.5$. But from then on in every generation, an infected cow has precisely 1 susceptible neighbor, so $R_0 = p < 1$ and, we expect, only minor outbreaks can occur.

Indeed, because $p < 1$ sooner or later the right moving boundary between infected and susceptible cows comes to a halt because transmission fails. Similarly the left moving boundary comes to a halt, leaving us with a finite connected patch of cows that became infected.

The next step is to look at a regular lattice in a plane, such as the square lattice

o o o o o o o o o
o o o o o o o o o
o o o o o o o o o
o o o o o o o o o

where each individual has $N = 4$ neighbors. If an individual is infected by a neighbor, it has at most $N - 1$ susceptible neighbors. Now, however, it may very well happen that: i) it has even fewer than $N - 1$ susceptible neighbors; and ii) it has to compete with other nearby infected individuals of the same generation to make susceptible neighbors into offspring by infecting them. The key point is that there are multiple pathways in the lattice that connect two given points.

As a result, it makes no sense whatsoever to speak about R_0. There does not exist a 'typical' spatial configuration of already infected and still susceptible individuals, unlike in the one-dimensional lattice with only nearest neighbor transmission, where the order structure gave rise to a well-defined boundary. In the case of spatial structure, $R_0 > 1$ is locally still a necessary condition for initial spread, but it is not sufficient to also guarantee such spread.

Even if R_0 doesn't make sense, the distinction between minor and major outbreaks does. Any finite outbreak is called minor and any infinite outbreak major. If again p denotes the transmission probability, we expect that for small values of p only minor outbreaks occur, but that for p close to 1 major outbreaks are possible and do indeed occur. That this intuition is right is proved in a rich mathematical field[9] called Percolation Theory. Moreover, in that field a key problem is to characterize the critical value p_c such that for $p > p_c$ major outbreaks are possible, while for $p < p_c$ they are not (for the square lattice $p_c = 0.5$). So the idea of a threshold does survive, but we lose the easy way to characterize the threshold in terms of a quantity R_0 that is both well-defined and allows a clear interpretation. This phenomenon has been observed in a natural infection-host system.[10]

In Section 12.6 we shall return to network models and discuss a variant that doesn't take the form of a regular spatial lattice. In Section 12.7 we look at approximations that, in a sense, attempt to capture the 'typical' spatial configuration of already infected and still susceptible individuals (these are usually called 'Pair Approximations').

The aim of the present short section has been twofold: i) to draw attention to the fact that local finiteness enhances demographic stochasticity as well as dependence and that, as a result, the approach based on linearization and the concept of R_0 is doomed to fail for certain classes of models; and ii) to refer to the percolation literature for an alternative approach that restores the idea of a threshold between the regime in which all outbreaks are minor and a regime in which major outbreaks do occur.

---

[9]There is an equally rich literature on the subject. See B. Bollobás and O. Riordan: *Percolation Theory*. Cambridge University Press, 2006, and also the many references given there.

[10]S.A. Davis, J.P. Trapman, H. Leirs, M. Begon and J.A.P. Heesterbeek: The abundance threshold for plague as a critical percolation phenomenon. *Nature*, **454** (2008), 634–637.

Chapter Eleven

Macroparasites

11.1 INTRODUCTION

As we have seen in Chapter 6, the defining mathematical distinction between microparasites and macroparasites is that for macroparasites, as a rule, re-infection through the environment is essential to get an increase in individual infectious load and consequent infectious output. In this chapter, we give a brief introduction to the consequences that this distinction has for formulating epidemic models for macroparasites. For the largest part, we concentrate on the definition and calculation of R_0.

Typically, macroparasites are multicellular organisms (e.g., helminths and other worm-like parasites) where definite stages in a life cycle can be distinguished. Several of these stages live outside living hosts. We will regard mostly two stages, adults living within a host and larvae (hatched from eggs produced by the adults and shed by the hosts) living in the environment of the host, since many features can already be illustrated in this minimal setting. Larvae are then infective to hosts and uptake can be by, for example, ingestion or skin penetration.

Exercise 11.1 Reflect on whether, as in the microparasite case, infection age of an infected individual could be adopted as a basis to model infectious output.

The mathematical distinction between micro- and macroparasites has consequences for the way R_0 is defined. What matters for R_0 is the infectious output of infected individuals and the contact patterns of these individuals with susceptibles. In the microparasite case we can often describe the infection within an individual as an autonomous process, disregarding further influence of the individual's environment. The definition of R_0 for microparasites as the expected number of new host individuals infected per infected host is a direct consequence of these considerations, and it makes good sense to follow individual hosts, possibly of various types. In the macroparasite case, however, the influence of the environment is essential in order to describe the infection pressure acting on hosts. In addition, the obligatory environmental stages bring with them the consequence that direct contacts between hosts do not generally provide a good description of the spread of the infectious output to other hosts. Thinking about R_0 as the number of hosts infected per host does not relate as closely to the actual biological processes as in the case of microparasites. It makes little sense to follow individual hosts, since what happens to the parasite in the environment is also crucial. It makes more sense to follow parasites in the traditional demographic spirit: one chooses a reference point in the life cycle (usually 'being newborn,' but here we prefer 'becoming/being adult'), and calculates for one individual at the reference point, the expected number of offspring reaching the reference point. The life cycle then has as a consequence that R_0 could be described as the expected number of new adult parasites (i.e., stages that end up in the major host) produced per adult parasite.

Exercise 11.2 i) R_0 for microparasites refers to the situation that the susceptible hosts are in a demographic steady state and that each contact an infective makes is with a susceptible. What do you think the corresponding assumption will be in the case of macroparasites?

ii) It was pointed out in 1965 by Macdonald and by Nåsell (1985), in their respective books, that for parasites with obligatory sexual reproduction in the main host, the assumption needed to characterize R_0 presents a conceptual difficulty. Can you describe it? Hint: How many parasites will likely be present in an infected host in the early phase of an epidemic?

iii) We expand on this difficulty in the context of a model. The following system of differential equations has been considered as a model for schistosomiasis:[1]

$$\frac{dI}{dt} = -\delta I + C(m)(N - I),$$
$$\frac{dm}{dt} = -\mu m + A\frac{I}{N}.$$

Here I is the density of infected snails (the intermediate host; for details of the life cycle see e.g., Anderson and May 1991) and m is the mean number of adult worms in infected humans. So μ corresponds to the per capita death rate of worms inside the human host, δ to the per capita death rate of infected snails and N to the total density of snails (assumed to be constant). A is proportional to the rate at which an infected snail produces the free-living stage that can penetrate the human skin, and C is proportional to the rate of production of eggs by the female adult worms (which includes the hatching of eggs to produce the free-living stage that can infect snails in water). The dependence of C on the mean number of *paired* worms per human is here incorporated as a dependence on m. The point is, however, that the fact that it takes two to reproduce should be reflected in properties of C. In so-called hybrid models[2] (i.e., mixing stochastic and deterministic components) one takes for C an increasing function that is zero at zero and, most importantly, has a vanishing derivative at zero. This latter assumption reflects that, for small values of m, parasites will not usually be able to find a mate within the human host.

The aim of this exercise is to show, basically via phase-plane analysis of the system given above, that the number of non-trivial steady states is either zero or two, and that in the latter case one needs to introduce a substantial parasite load to escape from the domain of attraction of the infection-free steady state (and to converge to the stable endemic level).

So, assume that the isocline

$$I = \frac{C(m)N}{\delta + C(m)}$$

[1]W.M. Hirsch, H. Hanish and J.P. Gabriel: Differential equation models of some parasitic infections: methods for the study of asymptotic behaviour. *Comm. Pure Appl. Math.*, **38** (1985), 733–753.

[2]See the book by Nåsell (1985); J.P. Gabriel, H. Hamisch and W.M. Hirsch: Worm's sexuality and special function theory. In: Gabriel, Lefèvre and Picard (1990), pp. 137–144.

has a sigmoid shape. First interpret the steady-state condition graphically. Then address the following questions: How does the number of steady states depend on the parameter A? What is R_0? Is there a critical value for A? If so, characterize it and describe the dynamical behavior for A above the critical value.

In Section 11.3 we show how to calculate R_0 for macroparasites in unstructured populations and populations with finite discrete structure. In Section 11.4 we touch upon a subtle mathematical pathology that is a direct consequence of the fact that macroparasite burden in an individual host can, in principle, increase beyond any bound.

We start, however, by describing a typical way in which spread of macroparasites is modeled, where we draw on work by Kostizin, Anderson and May, Hadeler and Dietz, Kretzschmar and Adler, and others (references are provided where they arise in the exposition).

11.2   COUNTING PARASITE LOAD

The derivation of the bookkeeping equations based on infection-degree is similar to a derivation in Exercise 1.34, and serves as another example of a stochastic epidemic model. Let $p_i(t)$ be the number of hosts carrying i adult parasites at time t, with i a natural number (we emphasize that the p_i are definitely *not* probabilities). The term 'susceptibles' could now be reserved for those hosts carrying no parasites, p_0, but actually the term does not make much sense anymore, since every host is susceptible to additional infection. In fact, we will use the customary term 'naive' to denote host individuals that never have experienced parasites. Note, however, that these individuals are not necessarily the same group as those without parasites. If one takes acquired immunity upon re-infection into account, it might be that individuals who have just been cleared of parasites, and so would enter the category p_0, are different from naives, since the ones that had parasites before can have a different immune status because of this previous presence. We neglect this possibility in our exposition.

Note that $N(t) = \sum_{i=0}^{\infty} p_i(t)$ is the total number of hosts at time t, and so $p_i(t)/N(t)$ is the fraction of hosts at time t who carry i adult parasites. In Exercise 11.8 we interpret this fraction as a probability, thus motivating the use of a generating function. As we shall define the generating function in terms of $\{p_i\}$ rather than its normalization, one should keep in mind that the generating function also incorporates information about the total number of hosts. The difference is only minor if N is constant, but crucial when we incorporate host demography.

Let $L(t)$ denote the density of larvae in the environment at time t. We distinguish between the following events/mechanisms:

- Adult parasites die in the host with per capita probability per unit of time μ.

- Larvae die in the environment with per capita rate ν.

- The host experiences a force of infection β, in the sense that the probability per unit of time that a larva enters is β; ultimately, we have to close the feedback loop and relate β to L, for instance as $\beta = \theta L$, for some constant θ.

- Upon entering a host, a larva transforms instantaneously into an adult.

- An adult parasite produces larvae with probability per unit of time λ.

We assume that parasites act independently from each other (yet there is interaction through the death of the host). We disregard host demography and therefore view transmission dynamics as a pure immigration-death process for the parasites. To derive differential equations for the dynamics of the p_i, we consider the events that can lead to a change over an interval of length Δt in a host carrying i adult parasites at the start.

Exercise 11.3 What do we have to assume about Δt and what consequences does this have for the events themselves and for the corresponding probabilities. Hint: The events described above follow exponential probability distributions.

Exercise 11.4 Describe the two key events within a host and give their probabilities in the interval of length Δt.

Exercise 11.5 Use Exercise 11.4 to show that the differential equation for p_i, $i \geq 1$, is given by

$$\frac{dp_i}{dt}(t) = -(\beta + i\mu)p_i(t) + (i+1)\mu p_{i+1}(t) + \beta p_{i-1}(t) \qquad (11.1)$$

and that the differential equation for p_0 is given by

$$\frac{dp_0}{dt}(t) = -\beta p_0(t) + \mu p_1(t). \qquad (11.2)$$

For the larvae, we need some additional assumptions. We need to specify how larvae come into contact with hosts. Assume that this is governed by mass action, with proportionality constant θ, say. Let $\hat{N}$ be the (constant) number of hosts and let $P(t)$ be the total number of adult parasites at time t in the $\hat{N}$ hosts. In the present formulation both p_i and L are numbers, but all relations continue to hold when both are interpreted as densities, i.e., numbers per unit area. We will encounter situations where demography is ignored and the number of hosts is indeed taken as constant, $\hat{N}$. In situations where the host population size is not constant we will use N or $N(t)$.

Exercise 11.6 Convince yourself that an equation incorporating these assumptions and governing the changes in the larval population is

$$\frac{dL}{dt} = \lambda P - \theta \hat{N} L - \nu L, \qquad (11.3)$$

where

$$P(t) = \sum_{i=1}^{\infty} i p_i(t).$$

Moreover, argue that $\beta = \theta L$.

Exercise 11.7 What if we do take host demography into account? We assume a per capita birth rate b of hosts, a natural host death rate d and an additional mortality rate caused by the parasites. For the latter assume that each parasite the host harbors increases the death rate by κ. All newborns are free from parasites. Show that (11.1) and (11.2) become

$$\frac{dp_i}{dt}(t) = -(\beta + d + i(\mu + \kappa))p_i(t)$$
$$+ (i+1)\mu p_{i+1}(t) + \beta p_{i-1}(t), \quad i > 0, \tag{11.4}$$
$$\frac{dp_0}{dt}(t) = b\sum_{i=0}^{\infty} p_i(t) - (d + \beta)p_0(t) + \mu p_1(t). \tag{11.5}$$

This model was first formulated by Kostizin (1934).[3] In Exercise 11.11 we shall return to demography, but until then we shall ignore it.

For the analysis of the infinite system of ordinary differential equations[4] there are two common routes, both making use of the special structure of the system. The first involves generating functions, which we have encountered already in Sections 1.2.2 and 3.3, reducing the study of the infinite system of ODEs to the study of a first-order partial differential equation, which can be solved by the method of characteristics. The second, more common, route involves a brute force approximation that leads to a simplified model of two ODEs for the average number of parasites per host and the density of larvae. We derive both types of equations below, but refrain from a detailed analysis.

Imagine a single host, randomly chosen at time t from the (large) population. Then $p_i(t)/N(t)$ is the probability that this host has i parasites.

Exercise 11.8 Convince yourself that a law-of-large-numbers argument reconciles this interpretation with the original definition of the $p_i(t)$.

We summarize some aspects of the detailed information on parasite distribution $p_i(t)$ into the variable $P(t)$, the number of adult parasites present in the population at time t, and $N(t)$, the total number of hosts at time t:

$$P(t) := \sum_{i=0}^{\infty} i p_i(t), \quad N(t) := \sum_{i=0}^{\infty} p_i(t).$$

Initially, host demography is simply ignored and therefore $N = \hat{N}$ is constant. Some calculation shows that, starting from (11.1) and (11.2), one obtains the following differential equation for P:

$$\frac{dP}{dt} = -\mu P + \beta \hat{N}. \tag{11.6}$$

Let us first give, based on the interpretation, an expression for R_0 for the system (11.3) and (11.6). Consider one adult. Its expected lifespan is $1/\mu$, and during that time it is expected to produce larvae at a rate λ. Larvae have an average lifespan

[3]V.A. Kostizin: *Symbiose, Parasitisme et Évolution.* Hermann, Paris, 1934.

[4]M. Martcheva, H.R. Thieme and T. Dhirasakdanon: Kolmogorov's differential equations and positive semigroups on first moment sequence spaces. *J. Math. Biol.,* **53** (2006), 642–671; and M. Martcheva and H.R. Thieme: Infinite ODE systems modeling size-structured metapopulations, macroparasitic diseases and prion proliferation. In: P. Magal and S. Ruan (eds.): *Structured Population Models in Biology and Epidemiology.* LNiM 1936. Springer-Verlag, Berlin, 2008, pp. 51–114.

given by $1/(\nu + \theta\hat{N})$, and a larva has probability per unit of time $\theta\hat{N}$ of turning into an adult. Therefore

$$R_0 = \frac{\lambda}{\mu}\frac{\theta\hat{N}}{\nu + \theta\hat{N}} \approx \frac{\lambda}{\mu}.$$

Let $g(t, z)$ be the generating function of the $p_i(t)$. That is,

$$g(t, z) := \sum_{i=0}^{\infty} p_i(t)z^i.$$

Exercise 11.9 Show that from the system (11.1)–(11.2) one can derive the linear first-order partial differential equation

$$\frac{\partial g}{\partial t}(t, z) = \beta(z - 1)g(t, z) - \mu(z - 1)\frac{\partial g}{\partial z}(t, z). \tag{11.7}$$

Exercise 11.10 Check equation (11.6). Hint: Differentiate the PDE (11.7) with respect to z and set z equal to 1.

Exercise 11.11 Now reintroduce demography and regard the situation where the host population size is not constant. As a follow-up to Exercise 11.7, show that the differential equations for P and N are given by

$$\frac{dN}{dt} = (b - d)N - \kappa P, \tag{11.8}$$

$$\frac{dP}{dt} = -(\mu + d)P + \beta N - \kappa\sum_{i=0}^{\infty} i^2 p_i, \tag{11.9}$$

if we take (the particular assumptions on) host demography into account. Hint: Either proceed by direct calculation using the definitions of P and N and equation (11.4) and (11.5), or proceed as in Exercise 11.10, starting from the PDE for the generating function related to (11.4) and (11.5):

$$\frac{\partial g}{\partial t}(t, z) = bN(t) - dg(t, z) + \beta(z - 1)g(t, z) + (\mu - (\kappa + \mu)z)\frac{\partial g}{\partial z}(t, z).$$

Readers wishing even more practice can derive this PDE from the system (11.4)–(11.5).

Exercise 11.12 The PDE (11.7) can be solved by the method of characteristics. Let the initial distribution be described by $g(0, z) = g_0(z)$. Show that the solution of (11.7) is

$$g(t, z) = g_0(e^{-\mu t}(z - 1) + 1)\exp\left(\frac{\beta}{\mu}(z - 1)(1 - e^{-\mu t})\right).$$

Deduce that for $t \to \infty$ the following holds: the number of parasites that a randomly chosen individual harbors is Poisson distributed with parameter β/μ. Does this result surprise you?

To link the adults with the dynamics of the larvae in the environment, we note once again that our mass-action assumption of uptake of larvae by hosts implies that $\beta = \theta L(t)$ in (11.6) and (11.9). In the absence of host demography, one can then summarize the model into the two-dimensional system (11.3) and (11.6).

Exercise 11.13 We explore the system from Exercises 11.7 and 11.11 a little further. We see that when we include host demography three differential equations arise (for P, L and N), but that this system is not closed. Anderson and May (1978)[5] initiated the following approximation to close the three-dimensional system. If we have information (e.g., from field studies) on parasite distributions $p_i(t)$, we can try to use this to postulate a formula for the $p_i(t)$ and use the formula to try and express $\sum i^2 p_i$ in terms of P and N, despite the fact that we know that this is incompatible with the assumptions underlying the bookkeeping in the first place. So this is a mixed strategy, in which we combine empirical information and theoretical considerations that are not (necessarily) consistent. In the Anderson and May approach the idea was that aggregation would have a stabilizing influence on the dynamics of the system. We will come back to this at the end of this section.

Typically, data show parasite distributions to be highly aggregated, i.e., a small fraction of the hosts carries a large fraction of the parasite population. First express the parasite distribution as a fraction by defining $r_i(t) := p_i(t)/N(t)$. Let the generating function of the parasite distribution be $g(t, z) = \sum_{i=0}^{\infty} r_i(t) z^i$. Note that $g'(t, 1)$, defined as $g'(t, 1) = \frac{\partial g}{\partial z}(t, z)|_{z=1}$ is equal to the mean of the distribution.

i) Check that for the parasite distribution we have that the mean parasite load is $g'(t, 1) = P/N$.

ii) Show that $\sum_{i=0}^{\infty} i^2 p_i = N \sum_{i=0}^{\infty} i^2 r_i = N(g''(t, 1) + g'(t, 1))$. Note that the variance of the parasite distribution is given by $g''(t, 1) + g'(t, 1) - (g'(t, 1))^2$ and that we can rewrite

$$\frac{1}{N} \sum_{i=0}^{\infty} i^2 p_i = \text{mean} \left(\frac{\text{variance}}{\text{mean}} + \text{mean} \right).$$

iii) Traditionally,[6] a negative binomial distribution is adopted for the r_i, with parameters $m = P/N$, assuming that the mean parasite load is ≤ 1, and $k > 0$, which is a measure of aggregation (a smaller k means that parasites are more aggregated). The generating function is given by

$$g(t, z) = \left(1 + \frac{m}{k}(1 - z) \right)^{-k}.$$

Use this expression to show that

$$\sum_{i=0}^{\infty} i^2 p_i = N \left(\frac{k+1}{k} \frac{P^2}{N^2} + \frac{P}{N} \right),$$

which enables us to close the system of differential equations for P and L by changing (11.9) into

$$\frac{dP}{dt} = P \left(\frac{\theta N L}{P} - (\mu + d + \kappa) - \kappa \frac{k+1}{k} \frac{P}{N} \right). \tag{11.10}$$

[5]R.M. Anderson and R.M. May: Regulation and stability of host-parasite population interactions. *J. of Animal Ecol.*, **47** (1978), 219–267.

[6]R.M. Anderson and R.M. May: Helminth infections of humans: mathematical models, population dynamics and control. *Adv. in Parasitol.*, **24** (1985), 1–101.

Exercise 11.14 If we could give a mechanistic argument to express the variable L in terms of P and N, we could reduce the system even further to a more tractable two-dimensional system. Suppose that the larvae in the environment have a (much) shorter lifespan than the adult parasites living in the host (i.e., during the average lifetime of an adult, many larvae will have come and gone). This difference in time scale could then be exploited by taking the extreme view that, relative to the time scale of adult dynamics, the larvae are in fact in a (quasi) steady state (cf. Section 12.2). Show that, with this assumption, we obtain

$$\beta N = \frac{\lambda \theta P N}{\nu + \theta N}.$$

To summarize, the closing 'trick' of Exercise 11.13 and the time-scale argument of the previous exercise reduce the originally infinite-dimensional system to

$$\frac{dN}{dt} = (b - d)N - \kappa P, \tag{11.11}$$

$$\frac{dP}{dt} = P\left(\frac{\lambda \theta N}{\nu + \theta N} - (\mu + d + \kappa) - \kappa \frac{k+1}{k}\frac{P}{N}\right). \tag{11.12}$$

Exercise 11.15 i) Use the same heuristic argument as that following equation (11.6) to show that for the system (11.11)–(11.12) we have

$$R_0 = \frac{\lambda}{\mu + d + \kappa}\frac{\theta \hat{N}}{\nu + \theta \hat{N}}.$$

When $\theta \hat{N}$ becomes large relative to ν, we obtain

$$R_0 = \frac{\lambda}{\mu + d + \kappa}.$$

ii) Note that we obtain the same expression if we apply the heuristic argument to the three-dimensional system (11.3), (11.8) and (11.10). The time-scale argument for L therefore has no influence on R_0. Is this to be expected?

Exercise 11.16 To study the potential effect of aggregation on stability (see Exercise 11.13), we do some more analysis on the system (11.11)–(11.12) along the lines of Section 4.4. Four cases of different long-term behavior of the system (11.11)–(11.12) can be identified (Anderson and May 1978, see reference in Exercise 11.13) after the parasite is introduced into a naive host population. Assume $b > d$. When no parasites are present, the host population size grows exponentially with rate $b - d$. Four regions of parameter space can be identified, using the rate of production of larvae λ as a bifurcation parameter.

i) Show that for $0 < \lambda < \lambda_0 := \mu + d + \kappa$ the parasite cannot invade successfully and that the host population will asymptotically (i.e., when $t \to \infty$) grow exponentially.

ii) Show that for $\lambda_0 < \lambda < \lambda_1 := b + \mu + \kappa$ the invasion is successful and both the parasite and host populations will grow. Show that the mean parasite load P/N, however, goes to zero, and that N will asymptotically grow exponentially with the same rate as before. Hint: First derive a differential equation for P/N.

iii) Show that for $\lambda_1 < \lambda < \lambda_2 := \mu + d + \kappa + (b-d)\frac{k+1}{k}$ both populations will asymptotically grow exponentially with the same reduced rate which is smaller than $b - d$, but positive. The mean parasite load P/N approaches a constant for $t \to \infty$.

iv) Show that for $\lambda > \lambda_2$ there exists a steady state $(\overline{N}, \overline{P})$ for the system (11.11)–(11.12). One can show that this steady state is locally asymptotically stable. In this situation the parasite is able to regulate the host population by bringing the exponential growth to a standstill.

Kretzschmar and Adler[7] have studied the dynamics of the full infinite dimensional model with host demography and compared this with the two-dimensional approximation by Anderson and May given above. In the infinite-dimensional model the thresholds λ_0 and λ_1, and the concomitant behavior given in Exercise 11.16, are as in the approximation, but the threshold λ_2 is different. The infinite-dimensional model can be shown[8] to have a fixed distribution of parasites, which is approached asymptotically, and which has a variance-to-mean ratio that is larger than one. In other words, a distribution is approached that is over-dispersed. This would justify the Anderson and May approximation (although there one assumes the same distribution at all times, rather than a changing distribution that only asymptotically reaches a fixed over-dispersed shape). In the approximation, λ_2 depends on the aggregation parameter k of the assumed negative binomial distribution. The problem with this is that k is not a biological parameter — it is at best an unknown function of the biological parameters in the system. As a consequence, it can be difficult to draw detailed biological conclusions. Kretzschmar and Adler (1992) (see footnote 7) show that, rather than by aggregation, the system is stabilized by the ratio of the variance to the mean. The infinite-dimensional model and the two-dimensional approximation discussed behave in much the same way. Finally, we note that there is actually no freedom to choose a distribution, because essentially the distribution is generated by the system of the p_i.

If the two models discussed are used as bases to test the influence of additional mechanisms on stability, we find that differences in the predicted dynamics start to occur between the infinite-dimensional and two-dimensional models. For one such additional mechanism, where the parasites are now able not only to influence the death rate of hosts, but also to reduce the host's fertility, it has been shown that periodic solutions are possible.[9]

Exercise 11.17 Let ζ be the per (parasite) capita reduction of host fertility. Two possibilities to model the influence of all the parasites within the host in reducing the host's fertility are to add all contributions (i.e., to take $i\zeta$ for a

---

[7]F.R. Adler and M. Kretzschmar: Aggregation and stability in parasite-host models. *Parasitology*, **104** (1992), 199–205; M. Kretzschmar and F.R. Adler: Aggregated distributions in models for patchy populations. *Theor. Pop. Biol.*, **43** (1993), 1–30.

[8]M. Kretzschmar: Comparison of an infinite-dimensional model for parasitic diseases with a related 2-dimensional system. *J. Math. Anal. Appl.*, **176** (1993), 235–260.

[9]O. Diekmann and M. Kretzschmar: Patterns in the effects of infectious diseases on population growth. *J. Math. Biol.*, **29** (1991), 539–570. For a practical application of such periodicity-inducing fecundity effects, we refer to P.J. Hudson, A.P. Dobson and D. Newborn. *Science*, **282** (1998), 2256–2258. In that paper the interaction is studied between red grouse and a parasitic nematode, where the nematode influences the fecundity of the grouse. The system is a very nice example of regulation of a host population by an infectious agent; the interaction with nematode explains the strong periodic dynamics of the grouse observed in the north of the United Kingdom.

host carrying i parasites), or to multiply them (i.e., to take ζ^i). What is the preferred choice and why?

11.3   THE CALCULATION OF R_0 FOR LIFE CYCLES

In the previous section we have already derived an expression for R_0 by a heuristic argument for parasites with two stages in the life cycle. This argument can be formalized for parasites with n stages, possibly living in different hosts and environments, and for the case that there is heterogeneity in the parasite, the hosts and the environments. The idea is to consider two rates d_i and m_i in the absence of density-dependent effects. Let the stage-i individuals of the parasite give rise to stage-$(i+1)$ individuals at a rate m_i (notably, maturation, of one larval stage into the next, or reproduction). Without loss of generality, we can assume that stage n that determines inflow into stage 1 to close the cycle, is the reproduction stage leading to the birth of new parasites. Therefore, m_n indicates the rate with which new parasites are created ('transmission'), whereas the other m_i, $i = 1, \ldots, n-1$, indicate the rates of development (stage transition) of existing parasites. Let $d_i > 0$ be the rate of leaving stage i (by death or by transition to stage $i + 1$).

Exercise 11.18 i) Give d_1, d_2, m_1, and m_2 in the case of the system (11.3), (11.8) and (11.10).

ii) We can generalize the heuristic argument by considering that there are n links in the life cycle and that each link is characterized by a multiplication factor for its contribution to the propagation of the parasite species. Note that m_i in a sense gives the reproduction rate of stage-$(i+1)$ individuals and that sojourn time in stage i is exponentially distributed with parameter d_i. Argue that

$$R_0 = \prod_{i=1}^{n} \frac{m_i}{d_i}.$$

Exercise 11.19 We can formalize the above using the ideas introduced in Sections 7.2 and 7.5 to calculate R_0. For this we view the progress through the life cycle as a continuous-time Markov chain on $\{1, \ldots, n\}$. Suppose we start with one newborn parasite in stage 1; how many stage-1 parasites is it expected to produce? Reproduction giving rise to new parasites of stage 1 is described by the vector $(0, \ldots, 0, m_n)^\top$ by assumption. Realize that out of the n 'infected states' (stages for the parasite to be in), there is only one state-at-infection. The NGM with large domain is therefore n-dimensional, but the NGM K and the NGM with small domain, K_S, are in this case one-dimensional. Convince yourself that the transition matrix Σ and the 'transmission' matrix T are given by

$$\Sigma = \begin{pmatrix} -d_1 & 0 & \cdots & 0 & 0 \\ m_1 & -d_2 & \cdots & 0 & 0 \\ 0 & m_2 & \ddots & 0 & 0 \\ \vdots & \ddots & \ddots & \ddots & \vdots \\ 0 & \cdots & 0 & m_{n-1} & -d_{n-1} \end{pmatrix}, T = \begin{pmatrix} 0 & 0 & \cdots & 0 & m_n \\ 0 & 0 & \cdots & 0 & 0 \\ 0 & 0 & \ddots & 0 & 0 \\ \vdots & \ddots & \ddots & \ddots & \vdots \\ 0 & \cdots & 0 & 0 & 0 \end{pmatrix}$$

Give the NGM and the NGM with small domain and realize that

$$R_0 = \rho(K_S) = -(0 \cdots 0 \, m_n)\Sigma^{-1} \begin{pmatrix} 1 \\ 0 \\ \vdots \\ 0 \end{pmatrix} = \prod_{i=1}^{n} \frac{m_i}{d_i}.$$

Because R_0 describes the success of invasion into the parasite-free steady state, we see that we can use the same reasoning as in Section 7.2 to characterize it. We write x_i for the mean parasite load for stage-i parasite individuals. By this we mean the average number of parasites per host, for those stages in the life cycle living in hosts, and the density of parasites, for those stages living in another environment (such as in the water or on a pasture). Then, linearized around the steady state $(N, x_1, \ldots, x_n) = (\overline{N}, 0, \ldots, 0)$, we can describe the dynamics by

$$\frac{dx}{dt}(t) = (T + \Sigma)x(t).$$

The same arguments as in Section 8.2 then show that the spectral radius $R_0 = \rho(K_L) = \rho(K) = \rho(K_S)$ determines the stability of the parasite-free steady state.

11.4 A 'PATHOLOGICAL' MODEL

The theory presented in Section 11.2 was based on a description of the parasite population in terms of mean parasite load. How can one create a theory of R_0 for the setting in terms of $p_i(t)$ as introduced in the beginning of Section 11.2? In this short section we look at an atypical model for the spread of macroparasites that shows that the differences between microparasites and macroparasites have a subtle consequence for R_0 (in addition to the subtlety already mentioned in Exercise 11.2-ii). The consequence is that there can be a difference in threshold between the situation where we characterize invasion success of a parasite by an increase in its numbers and the situation where we, in addition, require that also its abundance increases in the sense that the population of infected hosts has to increase. The model originates in the work of Barbour.[10]

In the models treated in Section 11.2 only a single larva could enter the host at any given time (cf. Exercise 11.2). In parasites where clumps of eggs are shed by hosts into the environment (e.g., a common water supply or a pasture) larvae will typically be picked up by the main hosts in clumps as well. Now suppose we model this in the following way, which is akin to microparasites. We pretend that hosts have a certain rate of contact to other hosts and that the number of parasites entering a host depends on the parasite load of the 'infecting' host and that this number is drawn from an 'infection distribution' with mean η per adult parasite in the 'infecting' host (which takes into account the fact that a large proportion of free-living parasites do not make it back into a new host). This is comparable to hosts infecting a common pool and susceptibles drawing from that pool in the following way: first they choose an infected host at random, and then they draw

---

[10] A.D. Barbour: Threshold phenomena in epidemic theory. In: *Probability, Statistics and Optimization*. F.P. Kelly (ed.), Wiley, New York, 1994, pp 101-116; A.D. Barbour, J.A.P. Heesterbeek and C.J. Luchsinger: Thresholds and initial growth rates in a model of parasitic infection. *Ann. Appl. Prob.*, **6** (1996), 1045–1074.

from the infection distribution a number of parasites that enter simultaneously upon infection (depending on the parasite load of the selected infecting host, but without the infecting host experiencing a drop in its own parasite load). More concretely, let c be the 'contact' rate of hosts and let q_{ij} be the probability that the 'contact' of the target host is with an infected host carrying j parasites and that the target host becomes infected with i parasites. We assume that $\sum_{i \geq 0} q_{ij} = 1$ for each j and $\sum_{k \geq 1} k q_{kj} = j\eta$.

Let $r_i(t)$ be the fraction of hosts carrying i parasites at time t. We neglect host demography (which is a crucial assumption) and take μ to be the death rate of adult parasites in the host, as before. Since we are only interested in R_0, we neglect re-infection and assume that all target hosts are naive. Hosts can become infected with arbitrary numbers of parasites (with certain probabilities). Once an individual is infected, we have only death of parasites to deal with, and passing on of infection to others. The dynamics of the r_i are governed by (cf. Section 11.2)

$$\frac{dr_i}{dt}(t) = -i\mu r_i(t) + (i+1)\mu r_{i+1}(t) + c \sum_{j=1}^{\infty} r_j(t) q_{ij}, \qquad i \geq 1. \qquad (11.13)$$

Exercise 11.20 We can derive (11.13) also from the equation for the incidence, involving the infection-age-dependent kernel A, as introduced for direct host-to-host transmission in Chapter 7. Let $x(i, j, \tau)$ be the probability of having i parasites left, τ time units after being infected with j parasites. Let $z(t, k)$ be the incidence at time t (as a fraction of the population) of infected hosts who are 'born' with k parasites. Convince yourself (cf. Section 7.5) that z satisfies

$$
\begin{aligned}
z(t, k) &= \int_0^\infty \sum_{j \geq 1} \left(c \sum_{i=1}^{j} x(i, j, \tau) q_{ki} \right) z(t - \tau, j) \, d\tau \\
&= \int_0^\infty \sum_{j \geq 1} A(\tau, k, j) z(t - \tau, j) \, d\tau
\end{aligned}
$$

and that the fraction of the population carrying i parasites, $r_i(t)$, can be calculated as

$$r_i(t) = \int_0^\infty \sum_{j \geq i} x(i, j, \tau) z(t - \tau, j) \, d\tau.$$

The next-generation operator is given by

$$(K\phi)(k) = \sum_{j \geq 1} \int_0^\infty A(\tau, k, j) \, d\tau \, \phi(j).$$

Now assume that parasites die independently from each other in the host with per capita rate μ and find the binomial distribution

$$x(i, j, \tau) = \binom{j}{i} e^{-i\mu\tau} (1 - e^{-\mu\tau})^{j-i}.$$

Show that with this choice we obtain (11.13).

Exercise 11.21 i) Let us calculate R_0 for the model described in this section. First give a heuristic argument that shows that

$$R_0 = \frac{c\eta}{\mu}.$$

ii) Let a matrix U be defined by the right-hand side of the system (11.13) acting on the infinite-dimensional vector r with elements r_i. Show that (11.13) can be rewritten as

$$\frac{dr}{dt} = Ur,$$

and that the vector $\phi(j) = j$ is an eigenvector of $U^\top$ with eigenvalue $c\eta - \mu$.

iii) Use the notation introduced in Exercise 11.23, and first show that

$$A_{kj} := \int_0^\infty A(\tau, k, j)\, d\tau = \frac{c}{\mu} \sum_{i=1}^{j} \frac{q_{ki}}{i}$$

and next that

$$(K\phi)(k) = \frac{c}{\mu} \sum_{j \geq 1} \sum_{i=1}^{j} \frac{q_{ki}}{i} \phi(j).$$

Hint: The expected sojourn time in the class of hosts carrying i parasites is $(i\mu)^{-1}$, if parasites die independently from each other. This can also be calculated by working out the expectation of $x(i, j, \tau)$ with respect to τ.

iv) Since K is a strictly positive matrix, there is only one positive eigenvalue with a positive eigenvector, and this eigenvalue is defined to be R_0. Since the spectrum of a matrix is the same as the spectrum of its transpose, we can also, at least formally, look for eigenvalues and eigenvectors of $K^\top$ (or alternatively look for a left-eigenvector of K). Show that the vector $\phi(j) = j$ is an eigenvector of the transposed matrix $K^\top$ corresponding to the eigenvalue $c\eta/\mu$.

Analysis of the system (11.13) shows the following threshold behavior. If a parasite enters a naive population, $R_0 = \frac{c\eta}{\mu}$ characterizes the asymptotic behavior of $\sum_{i \geq 1} i r_i(t)$; i.e., the value relative to the threshold 1 determines whether or not the parasite population will grow or decline. For an epidemic, however, we think more in terms of a rise or decline in the number of infected hosts after invasion, and this turns out to be more delicate. Concretely, one can show (using tools that are beyond this text) that for $\eta \leq e$ (where e is the base of the natural logarithm), both parasite and infected host behavior is governed by R_0. For $\eta > e$, however, the behavior of the hosts is characterized by the quantity $\frac{ce \ln \eta}{\mu} \leq R_0$ relative to the threshold 1. In particular, in the case that $\eta > e$, an epidemic among the hosts only occurs if $\frac{ce \ln \eta}{\mu} > 1$. What apparently can happen is that for $R_0 > 1 > \frac{ce \ln \eta}{\mu}$ the invasion is successful in that parasites increase, but they become aggregated in too few hosts to be able to let the infected hosts increase epidemically.

It is an open problem to see what the effect of introducing host demography and, more specifically, mortality caused by the parasites would be on the above phenomenon. Intuitively one would expect an influence if, for example, hosts carrying more parasites have a higher death rate.

Chapter Twelve

What is contact?

12.1 INTRODUCTION

In this section we reflect on the various aspects of contacts and the contact process. The interpretation of the word contact varies for different infectious agents, and only refers to *happenings* where infection transmission could occur — to resurrect a term invented by Ross (1911), who referred to his work on epidemic phenomena as his 'theory of happenings.'

We first note that a contact can refer to two separate types of happenings, i.e., both the actual event of a transmission opportunity and the pairing of two individuals during which several such opportunities can arise. In general, pairings last longer than transmission events. Depending on the question studied and the time scales involved, a situation may call for separate modeling of the two types of contact.

The two most important aspects of contacts for infection transmission are 1) the number of contacts per unit of time and 2) the number of *different* individuals with whom these contacts occur. Without going into too much detail, we reflect on various possibilities concerning both aspects. Aspect 1 is concerned not only with variation in the number of transmission opportunities during a pairing, but also with the duration of the pairing (Section 12.2). We regard three cases, ordered by increasing time scale of contact duration: instantaneous contacts, pairings that are short-lived compared with the other time scales in the system, and finally pairings that are not short-lived on the infection time scale. Aspect 2 concerns spatial or social networks with variation in the set of potential 'contactees.' This could be the entire population, a dynamic subset of the population or a fixed subset of the population. Moreover, the (sub)sets can contain finitely or (for mathematical reasons) infinitely many individuals. In the latter case one may assume that, when contacts are random, the probability to contact the same individual twice is negligible. We give some of the flavor of these aspects in Section 12.6 on graphs and networks. In Sections 12.3 and 12.4 we briefly address the interesting new aspects that arise in heterogeneous populations where different types of individuals are recognized. We deal with consistency conditions, with gauging the effects of subdividing the population into subsets of 'mixing' individuals, and with core groups. We also, in Section 12.5, pay attention to populations that consist of very many small groups, like a community of households, with intense within-group contact. In the final Section 12.7 we briefly describe the technique of pair approximation.

12.2 CONTACT DURATION

Assume that there are infinitely many potential contactees. We only regard happenings involving two individuals.

Assume contact duration is very short. In other words, regard contacts as effectively instantaneous, while the contact intensity (number of contacts per unit of time) is proportional to the density of the two types of individuals involved. One then speaks of mass-action kinetics. As a metaphor one can take gaseous molecules colliding elastically in a reaction vessel (a context where the term 'law of mass action' originated). Pairings have infinitesimally short duration and all transmission opportunities during a pairing are compressed into this one instant. The contact rate between susceptibles and infectives is proportional to the product SI of the respective densities. If S and I represent *numbers* of individuals, it is proportional to SI/N, where N is the total population size. Although it is an extremely simple description of the contact process, there are situations in which mass action can be a good option, for example if contact duration is very short compared with, say, the average length of the infectious period.

Exercise 12.1 Regard a single infected individual. How does the number of contacts per unit of time for this individual rise under mass action as a function of the population density N? Is this a realistic description of the contact process? Can you give examples of happenings where it is certainly not realistic? What would the dependence of the contact rate on N look like for these examples?

We now study more closely the case where contacts have non-negligible duration. This implies that an individual is — purely due to time limitation — not able to engage in more and more contacts per day, say, even if contact opportunities continue to increase. In the following series of exercises we give a way to model these saturation effects mechanistically. For this, let N be the population density and suppose individuals have some form of pairwise contacts of exponentially distributed duration. We introduce the fraction $C(N)$ of the population that is engaged in a contact at any given time. We have that $NC(N)/2$ is the density of pairs in the population at any given time, because pairs consist of two individuals. Because only a fraction $2SI/N^2$ of the random pairs is between a susceptible and an infective, the density of pairs consisting of an infective and a susceptible, at any given time, is given by

$$NC(N)\frac{S}{N}\frac{I}{N} = C(N)\frac{SI}{N}. \tag{12.1}$$

Exercise 12.2 i) Verify that the alternative (which is to consider the fraction of the susceptible population that is engaged in contact at any given time) leads to the same expression.

ii) Give an argument to show that $C(N)$ can also be interpreted as the fraction of the time that an individual will spend on contacts, given that the population density equals N.

Exercise 12.3 Give some properties that the function $C(\cdot)$ will reasonably have. When does (12.1) revert to mass action? Can you give an example of a situation where $C(\cdot)$ would be approximately constant over its entire range?

There are of course many functional forms that have the properties in Exercise 12.3. Often, models are studied without further specification and by only using these properties to derive results.[1] If one studies the contact process per se, it is interesting to have an explicit expression for $C(N)$. Based on Holling's time budget argument (Holling's disc equation) to derive the functional response (i.e., the number of prey eaten per predator per unit of time, as a function of prey density), Dietz[2] suggested the following form:

$$C(N) = \frac{aN}{1 + bN} \tag{12.2}$$

(in chemostat models this is referred to as the Monod equation and in chemical reaction kinetics as the Michaelis-Menten function). We give both a heuristic and a formal derivation of this relation.

Exercise 12.4 The idea of functional response in predator-prey interaction is that the number of prey caught per predator per unit of time as a function of prey density is limited by the time spent handling and eating prey that has been caught. Holling assumed that the number of prey caught per predator is proportional to prey density and search duration. Let T_h be the expected 'handling time' of a single prey that has been caught, and let T be the total time available to a predator for food gathering and eating. Let N be the prey density and let a be the search efficiency (the product of the area covered per unit of time while searching and the probability that a prey in this area is actually detected). If Z is the number of prey caught by a predator in time T, express the actual search time as a function of T, T_h and Z. Express Z as a function of N, a and the search time and derive the formula for the functional response.

Exercise 12.5 A slightly more formal (but not essentially different) derivation makes the time-scale argument explicit. The idea is to distinguish two types of predator ('searching' with index 0, and 'busy handling prey' with index 1) and to assume that the switches in type are fast compared with changes in the density N of prey and in the density of predators due to births and deaths (note that, among other things, this requires that prey greatly outnumber predators). One then regards one predator on the time scale of type change. If h is the rate of return from the handling of prey (i.e., $1/h = T_h$), and if we assume that predators meet prey according to the law of mass action, we can write

$$\frac{dp_0}{dt} = -aNp_0 + hp_1,$$

where p_i is the probability that the predator is in state i, $i = 0, 1$. The non-zero steady state $\bar{p}_0$ is usually referred to as a quasi- (or pseudo-) steady state. Can you explain why? Express $\bar{p}_0$ as a function of N and derive the expression for the functional response.

Exercise 12.6 Suppose we now want to justify Dietz's formula (and to give an interpretation of its parameters) in the case of contacts between individuals.

---

[1]See e.g., H. Thieme: Epidemic and demographic interaction in the spread of potentially fatal diseases in growing populations. *Math. Biosci.*, **111** (1992), 99–130; J. Zhou and H.W. Hethcote: Population size dependent incidence in models for diseases without immunity. *J. Math. Biol.*, **32** (1994), 809–834.

[2]K. Dietz: Overall population patterns in the transmission cycle of infectious disease agents. In: Anderson and May (1982).

Where do we run into trouble? Hint: Does time limitation matter for the prey in the Holling argument?

Instead of trying to rationalize an ad hoc function such as (12.2), we can try and derive an expression from mechanistic assumptions and then show that it has the desired properties. In the exercises below, we give an argument for the unstructured case, which generalizes to structured populations.[3]

Exercise 12.7 Consider a population consisting of singles, with density $X(t)$ (i.e., those individuals that are at time t not engaged in a pairwise contact), and pairs, with density $P(t)$. Consistency requires that $N(t) = X(t) + 2P(t)$. Give a rational underpinning of the following system:

$$\frac{dX}{dt} = -\rho X^2 + 2\sigma P,$$
$$\frac{dP}{dt} = \frac{1}{2}\rho X^2 - \sigma P.$$

Denote by S_1 the density of susceptible singles, by I_1 the density of infective singles, by S_2 the density of pairs consisting of two susceptibles, by M the density of pairs consisting of a susceptible and an infective (so M means 'mixed'), and finally, by I_2 the density of pairs consisting of two infectives.

Assume (for the time being) that infectives stay infectious for ever and that

$$\frac{dS_1}{dt} = -\rho S_1(S_1 + I_1) + 2\sigma S_2 + \sigma M,$$
$$\frac{dI_1}{dt} = -\rho I_1(S_1 + I_1) + 2\sigma I_2 + \sigma M,$$
$$\frac{dS_2}{dt} = \frac{1}{2}\rho S_1^2 - \sigma S_2, \tag{12.3}$$
$$\frac{dM}{dt} = \rho S_1 I_1 - \sigma M - \beta M,$$
$$\frac{dI_2}{dt} = \frac{1}{2}\rho I_1^2 - \sigma I_2 + \beta M,$$

where ρ is the reaction rate constant for pair formation, σ is the probability per unit of time that a pair dissolves into two singles, and β is the probability per unit of time that transmission occurs in a mixed pair. A key point here is that the pair formation/dissolution process proceeds independently of the susceptible/infective distinction. The processes are illustrated in Figure 12.1 (note the symmetry).

Exercise 12.8 Let $X = S_1 + I_1$ and $P = S_2 + M + I_2$. Show that system (12.3) then leads to

$$\frac{dX}{dt} = -\rho X^2 + 2\sigma P, \tag{12.4}$$
$$\frac{dP}{dt} = \frac{1}{2}\rho X^2 - \sigma P.$$

---

[3]Based on: J.A.P. Heesterbeek and J.A.J. Metz: The saturating contact rate in marriage and epidemic models. *J. Math. Biol.*, **31** (1993), 529–539.

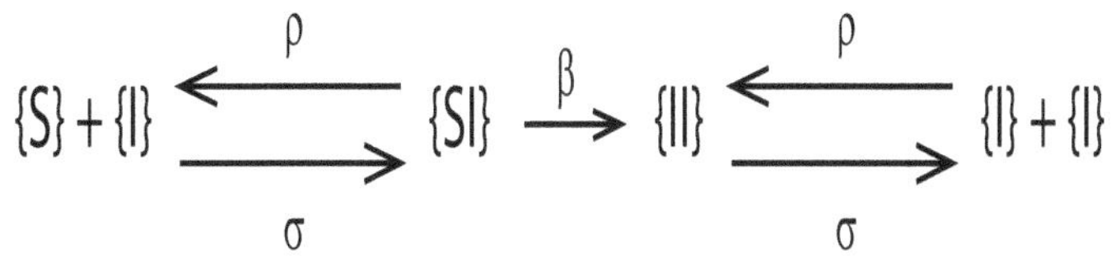

$$\{S\} + \{I\} \quad \begin{array}{c} \rho \\ \longleftarrow \\ \longrightarrow \\ \sigma \end{array} \quad \{SI\} \quad \xrightarrow{\beta} \quad \{II\} \quad \begin{array}{c} \rho \\ \longleftarrow \\ \longrightarrow \\ \sigma \end{array} \quad \{I\} + \{I\}$$

Figure 12.1: Pair formation/dissolution process for susceptible and infected individuals.

Exercise 12.9 Verify that the steady state of the system (12.4) is given by

$$\overline{X} = \frac{\sqrt{1 + 4\nu N} - 1}{2\nu}, \quad \overline{P} = \frac{1 + 2\nu N - \sqrt{1 + 4\nu N}}{4\nu}, \qquad (12.5)$$

where N equals the total population size and where $\nu := \rho/\sigma$. What can you say about the stability and the domain of attraction?

Hint: Use that $X + 2P = N$.

Exercise 12.10 Introduce the density of susceptibles $S = S_1 + 2S_2 + M$ and the density of infectives $I = I_1 + 2I_2 + M$. Show that

$$\frac{dS}{dt} = -\beta M, \qquad (12.6)$$

$$\frac{dI}{dt} = \beta M.$$

Explain how this result could have been obtained directly from the interpretation.

We are now ready to introduce the time-scale argument. The idea is that β is very small relative to both ρ and σ. Consequently, S and I will change slowly relative to the speed at which the pair formation/dissolution process equilibrates. To incorporate and elaborate this technically, we eliminate the βM terms from the systems (12.3) and (12.6), and compute the steady-state values of S_1, I_1, S_2, M and I_2 for the resulting system. These are quasi-steady-state values, since in fact S and I will be slowly changing. To compute how S and I will change on the long time scale, we simply substitute the quasi-steady-state value of M into the system (12.6).

Exercise 12.11 Verify that the quasi-steady state of the system (12.3) is given explicitly by

$$\overline{S}_1 = \overline{X} \frac{S}{N}, \quad \overline{I}_1 = \overline{X} \frac{I}{N}, \quad \overline{S}_2 = \overline{P}\left(\frac{S}{N}\right)^2, \qquad (12.7)$$

$$\overline{M} = 2\overline{P} \frac{S}{N} \frac{I}{N}, \quad \overline{I}_2 = \overline{P}\left(\frac{I}{N}\right)^2,$$

and explain the logic behind these expressions (here $\overline{X}$ and $\overline{P}$ are given by (12.5)).

Exercise 12.12 Write $dS/dt = -\beta M$ in the form $dS/dt = -\beta C(N)SI/N$ and show that

$$C(N) = \frac{1 + 2\nu N - \sqrt{1 + 4\nu N}}{2\nu N} = \frac{2\nu N}{1 + 2\nu N + \sqrt{1 + 4\nu N}}. \qquad (12.8)$$

Show that (12.8) indeed has the desired properties (cf. the elaboration of Exercise 12.3).

Our bookkeeping was relatively simple since we had only two types of individuals, S and I. When we also consider immune individuals and introduce a class R, we get six categories of pairs: {SS}, {SI}, {SR}, {II}, {IR}, {RR}. But we would still have the system (12.4) to describe the dynamics of pairs and singles without paying attention to the S, I or R label that the individuals carry. Instead of the system (12.6), we would have

$$\frac{dS}{dt} = -\beta M, \quad \frac{dI}{dt} = \beta M - \alpha I, \quad \frac{dR}{dt} = \alpha I,$$

where M refers to {SI} pairs. Exactly as in Exercise 12.11, we would obtain the analog of (12.7) and, in particular, the relation

$$\overline{M} = 2\overline{P}\frac{S}{N}\frac{I}{N}.$$

In other words, mathematically hardly anything changes and the resulting system on the long time scale would be

$$\frac{dS}{dt} = -\beta C(N)\frac{SI}{N}, \quad \frac{dI}{dt} = \beta C(N)\frac{SI}{N} - \alpha I, \quad \frac{dR}{dt} = \alpha I, \qquad (12.9)$$

with $C(N)$ given by (12.8) and now $N = S + I + R$.

Likewise, one can incorporate host demography on the long time scale and still work with C as given by (12.8). Our derivation focused on the, from a notation point of view, simplest case, but the result is much more generally valid.

Exercise 12.13 Repeat Exercise 12.9 under the additional assumption that single individuals can revert to a resting state in which they are not active in searching for partners. Let ε_1 and ε_0 be the probabilities per unit of time of entering the resting state and returning to the active state respectively. Pairings dissociate into two active singles (a debatable assumption!). You can check your result by showing that the appropriate limit leads back to (12.5).

Exercise 12.14 With hindsight, we can now also give a Holling-type argument as in Exercise 12.4. This time we take into account that both individuals taking part in a contact are time-limited ('Holling squared,' a procedure and name invented by J.A.J. Metz). We think again of singles and pairs as in the previous exercises and use a mix of notation from 12.9 and 12.4. Let $Y := Z/T$. Give the analogy with the Holling argument that leads to

$$Y = \frac{\rho X}{1 + \nu X},$$

where $\nu = \rho T_h$. In contrast to the prey in the Holling argument, the singles here are time-limited, in the sense that they are not all available for contacts. Convince yourself that

$$X = N\left(\frac{T - ZT_h}{T}\right).$$

Rewrite this in terms of Y. Express the steady-state density $\overline{P}$ of complexes in N, Y and T_h and substitute the expression for Y to find the same expression for $\overline{P}$ as in Exercise 12.9 and therefore the same expression for $C(N)$.

Remark: In this heuristic derivation we have not used the assumption that duration of pairings is exponentially distributed with some parameter σ; we have only used the average duration T_h. This suggests that the assumption of an exponential distribution in the formal derivation — although mathematically convenient — is not necessary.

To conclude, we show that if we do assume instantaneous contacts, the expression (12.8) for $C(N)$ indeed collapses to a constant times N (i.e., to transmission according to mass action), as is to be expected. One has to take care that the right procedure for taking limits is adopted. This exercise makes clear that it is worthwhile to distinguish between pairings and transmission opportunities during pairings. The obvious first thing to do would be to let the average duration of a pairing, $1/\sigma$, tend to zero (i.e., to take the limit $\sigma \to \infty$). Check that if we do this without adapting other quantities then the transmission rate becomes zero. The reason is that if the pairing duration becomes shorter and shorter, there comes a point beyond which the transmission opportunity during a pairing is negligible. The solution is to let the transmission opportunity tend to infinity 'equally fast' as the pairing duration tends to zero.

Exercise 12.15 Argue that the probability of transmission per pairing is given by

$$\frac{\beta}{\beta + \sigma}.$$

Hint: $1 - e^{-\beta t}$ is the probability that transmission occurs within the first t time units of the existence of the pairing; $\sigma e^{-\sigma t}$ is the probability density function for pairing duration.

Exercise 12.16 Show that the limit $\sigma \to \infty, \beta \to \infty$ with β/σ and ρ constant collapses C into a linear function of N.

As a third possibility for variation in contact duration, one has that the time scale of formation and dissociation of pairings is not fast compared with the demographic time scale or the time scale of infection. For this we refer to the literature.[4] For example, with regard to sexually transmitted diseases, it can be relevant to take the formation of longer-lasting (sexual) partnerships into account. We have seen some of the flavor of the resulting pair formation models when we calculated R_0 for examples from this class of models in Section 7.8.

---

[4]K. Dietz and K.P. Hadeler: Epidemiological models for sexually transmitted diseases. *J. Math. Biol.*, **26** (1988), 1–25; For the calculation of R_0 for pair formation models see K. Dietz, O. Diekmann and J.A.P. Heesterbeek: The basic reproduction ratio for sexually transmitted diseases, Part 1: theoretical considerations. *Math. Biosci.*, **107** (1991), 325–339.

12.3   CONSISTENCY CONDITIONS

In earlier chapters, we have repeatedly encountered situations in which the symmetry of contact implied that modeling ingredients should satisfy certain consistency relations. For instance if, according to our model, in a certain period of time 3251 females have sexual contact with a male then the number of sexual contacts of males with females should also equal 3251 in that period. In this subsection we address this consistency issue more systematically.

We consider a population in a demographic steady state. We assume that the h-state is static. This is not essential for the considerations, but it becomes essential if one combines bookkeeping considerations with an infectivity submodel to compute the kernel $A(\tau, x, \xi)$. We distinguish finitely many h-states, which we number $1, \ldots, n$ (this is not at all essential; a continuum of h-states only requires a notational adaptation). We concentrate on *symmetric* contacts (so we exclude 'contacts' such as through blood transfusions, through a common water supply or through infected potato salad at a party). We adopt the counting convention that when two j-individuals have contact, there are *two* contacts (more generally, when a j-individual and a k-individual have contact, we count this as a kj-contact and a jk-contact).

Let

$$\phi_{ij} \quad := \quad \text{total number of contacts per unit time} \qquad (12.10)$$
$$\text{between } j\text{-individuals and } i\text{-individuals.}$$

Then, because of the imposed symmetry, consistency requires that

$$\phi_{ij} = \phi_{ji}. \qquad (12.11)$$

Therefore, out of the n^2 elements of the matrix Φ, only $n(n+1)/2$ can be freely chosen.

In particular when considering sexual contacts, the h-state often refers to a (sexual) activity level. More specifically, assume that j-individuals have, on average, c_j contacts per unit of time. Let N_j denote the number of j-individuals. Then Φ, c and N cannot be completely arbitrary. They should satisfy the relations

$$\sum_{i=1}^{n} \phi_{ij} = c_j N_j, \qquad j = 1, \ldots, n. \qquad (12.12)$$

Therefore, if one considers c and N as 'given,' the number of degrees of freedom in the choice of Φ reduces to $n(n-1)/2$. A particular choice for the elements of Φ is the *proportionate mixing* expression

$$\phi_{ij} = \frac{c_i c_j N_i N_j}{\sum_{k=1}^{n} c_k N_k} \qquad (12.13)$$

(in the terminology of C. Castillo-Chavez and co-workers,[5] this is called a Ross solution). However, other choices satisfying (12.11) and (12.12) are clearly possible. Indeed, the following result derives from work of Castillo-Chavez and various coauthors.

---

[5]C. Castillo-Chavez, J.X. Velasco-Hernandez and S. Fridman: Modeling contact structures in biology. In: *Frontiers in Mathematical Biology*. S.A. Levin (ed.), Springer-Verlag, Berlin, 1994, pp. 454–491.

Theorem 12.17 *Let c and N be given. Let Ψ be a symmetric $n \times n$ matrix. Define*

$$\theta = \left(\sum_{k=1}^{n} c_k N_k \right)^2 - \sum_{i,j=1}^{n} c_i c_j N_i N_j \psi_{ij} \qquad (12.14)$$

and

$$\alpha_j = \sum_{k=1}^{n} c_k N_k - \sum_{i=1}^{n} c_i N_i \psi_{ij}. \qquad (12.15)$$

Then the matrix Φ defined by

$$\phi_{ij} = \frac{c_i c_j N_i N_j}{\sum_{k=1}^{n} c_k N_k} \left(\frac{\alpha_i \alpha_j}{\theta} + \psi_{ij} \right) \qquad (12.16)$$

satisfies relations (12.11) and (12.12). (In order for this to be meaningful in our context, both Ψ and the right-hand sides of (12.14) and of (12.15) should be non-negative. A sufficient condition for this to be the case is $0 \le \psi_{ij} \le 1$ for all i, j.).

Conversely, if Φ is such that (12.11) and (12.12) hold for its elements then

$$\phi_{ij} = \frac{c_i c_j N_i N_j}{\sum_{k=1}^{n} c_k N_k} \psi_{ij}, \qquad (12.17)$$

where the matrix Ψ is symmetric and the elements are such that

$$\sum_{i=1}^{n} c_i N_i \psi_{ij} = \sum_{k=1}^{n} c_k N_k \qquad (12.18)$$

for all j. This shows that a representation of the form (12.16) is possible.

Proof: Clearly, the matrix Φ defined by (12.16) is symmetric. Moreover,

$$\sum_{i=1}^{n} \phi_{ij} = \frac{c_j N_j}{\sum_{k=1}^{n} c_k N_k} \left(\frac{\sum_{i=1}^{n} c_i N_i \alpha_i}{\theta} \alpha_j + \sum_{i=1}^{n} c_i N_i \psi_{ij} \right).$$

By (12.15), the second factor on the right-hand side equals

$$\frac{\sum_{i=1}^{n} c_i N_i \alpha_i}{\theta} \alpha_j + \sum_{k=1}^{n} c_k N_k - \alpha_j.$$

Also by (12.15), we have that

$$\sum_{i=1}^{n} c_i N_i \alpha_i = \left(\sum_{i=1}^{n} c_i N_i \right)^2 - \sum_{i,j=1}^{n} c_i c_j N_i N_j \psi_{ij},$$

which, by (12.14), equals θ. Hence the second factor reduces to $\sum_{i=1}^{n} c_i N_i$ and the product with the first factor to $c_j N_j$, which then yields (12.12).

It is straightforward that θ and α_j are non-negative when $0 \le \psi_{ij} \le 1$.

If, finally, the matrix Φ is given and its elements satisfy (12.11) and (12.12) then a matrix Ψ with elements defined by

$$\psi_{ij} = \frac{\sum_{k=1}^{n} c_k N_k}{c_i c_j N_i N_j} \phi_{ij}$$

is certainly symmetric and, in addition,

$$\sum_{i=1}^{n} c_i N_i \psi_{ij} = \frac{\sum_{k=1}^{n} c_k N_k}{c_j N_j} \sum_{i=1}^{n} \phi_{ij} = \sum_{k=1}^{n} c_k N_k.$$

This ends the proof of Theorem 12.17.   □

On the one hand, Theorem 12.17 provides us with an algorithm to generate, for given c and N, a matrix Φ satisfying the constraints (12.11) and (12.12). It turns out that a symmetric matrix Ψ can be freely chosen, except for sign constraints on the right-hand sides of (12.14) and of (12.15), which themselves depend on c and N. On the other hand, the second part of the theorem tells us that there is no loss of generality in restricting Φ to have elements of the form (12.16).

Exercise 12.18 Let π_{ij} be the probability of transmission, given a contact between a susceptible i-individual and an infectious j-individual. Express the incidence among i-individuals in terms of π_{ij}, ϕ_{ij}, S_i and I_j for $j = 1, \ldots, n$.

The constraint (12.12) derives from the interpretation of c_j as an, on average, *realized* number of contacts per unit of time. If we relax the interpretation and characterize individuals by a *tendency* to make contacts, the rigid constraint disappears. Thus one can use a multi-type version of the time-scale arguments in Section 12.2 to derive ϕ_{ij} from a mass-action sub-model.

Exercise 12.19 Consider a metapopulation of seal colonies. Number the colonies $1, 2, \ldots, n$. Let ρ_{ij} denote the fraction of individuals of colony j that at any particular low tide sunbathe at the haul-out spot of colony i (we assume that such individuals are 'chosen' randomly from among all j-individuals at every low tide). Derive an expression for ϕ_{ij}.

12.4 EFFECTS OF SUBDIVISION

12.4.1 Aggregation

Regard a population where the individuals are divided into n groups $G_1, \ldots, G_n$, of sizes $N_1, \ldots, N_n$ respectively, with different contact patterns. In this subsection we study the effect, on the spread of infection, of decreasing the extent of the heterogeneity in the population. Suppose we combine the n groups into k larger groups $H_1, \ldots, H_k$, with $k < n$. We are interested in the effect that such aggregation has on both R_0 and the final size of an epidemic. For the former we were inspired by results of F.R. Adler;[6] for the latter we refer to work by H. Andersson and T. Britton[7] and to the end of Section 2.2. From Section 2.2, we already know that effects on R_0 and the final size need not be the same since in the heterogeneous case R_0 and the final size are not necessarily related in a monotone way. A small core group with high activity level can contribute substantially to R_0, but may have little impact on the final size.

---

[6]F.R. Adler: The effects of averaging on the basic reproduction ratio. *Math. Biosci.*, **111** (1992), 89–98.

[7]H. Andersson and T. Britton: Heterogeneity in epidemic models and its effect on the spread of infection. *J. Appl. Prob.*, **35** (1998), 651–661.

Let c_{ij} be the per couple contact rate of individuals from group i with individuals from group j. We consider tangible contacts only; we think of, for example, vector-transmitted infections or sexually transmitted infections. Consistency then requires that

$$c_{ij}N_iN_j = c_{ji}N_jN_i$$

(see Section 12.3 where $\phi_{ij} = c_{ij}N_iN_j$), and the contact matrix $C = (c_{ij})_{1 \le i,j \le n}$ should therefore be symmetric.

Now partition the indices $\{1, \ldots, n\}$ into k non-overlapping subsets $s_1, \ldots, s_k$, and define new larger groups and group sizes

$$H_l := \cup_{i \in s_l} G_i, \qquad \widehat{N}_l := \sum_{i \in s_l} N_i.$$

Exercise 12.20 Let mixing between groups be proportional to their size and activity, i.e., assume that $c_{ij} = v_i v_j$. Check that this mixing pattern is preserved between the aggregated groups, i.e., show that $\widehat{c}_{lr} = \widehat{v}_l \widehat{v}_r$, where $\widehat{c}_{lr}$ is the contact rate between group H_l and group H_r. Hint: The total number of contacts between the aggregated groups l and r is $\sum_{i \in s_l} \sum_{j \in s_r} c_{ij}N_iN_j$.

If we include the probability p_{ij} that, upon contact between a susceptible of group i and an infective from group j, transmission is successful, we will as a rule, have $p_{ij} \ne p_{ji}$. Furthermore, let T_j be the average length of the infectious period for infected individuals in group j. In analogy with the homogeneous case, equation (2.2), we can calculate elements m_{ij} of the next-generation matrix, defined as the expected number of new cases in group i caused by an infective from group j, if group i consists of susceptibles only. Instead of a single infection function $A(\tau)$, we then have functions $A_{ij}(\tau)$. We find, if we assume that all variability between infected individuals is in the length of the infectious period, or somewhat more generally if infectivity during the infectious period is independent of the length of the period,

$$m_{ij} = N_i \int_0^\infty A_{ij}(\tau) \, d\tau = N_i c_{ij} p_{ij} T_j.$$

From Chapter 7, we know that R_0 is the dominant eigenvalue of the matrix M.

Now consider the special case that C is symmetric and primitive (i.e., that there is an integer k such that all elements of C^k are strictly positive) and that $p_{ij}T_j = pT$, for all i and j. This effectively means that we take out any influence of variability in infectious output and look at contact structure only. An argument analogous to the one given above gives a matrix $\widehat{M}$ and a dominant eigenvalue $\widehat{R}_0$ for the aggregated groups. We then have the following result by F. Adler (1992) (footnote 6):

$$\widehat{R}_0 \le R_0, \tag{12.19}$$

or, in words, less heterogeneity (more aggregation) leads to a lower value for R_0 if infectivity is the same in all groups and if the contact pattern is symmetric and primitive. Note that in Section 2.2 we illustrated, with a numerical example, this relation for the case of two groups clustered into a single group, in the case that one has proportionate mixing.

Exercise 12.21 i) In Section 2.2 we used c_i to denote the average number of contacts per unit time of a type-i individual. Presently we are working with per-couple contact rates c_{ij}. The aim of this first part of the present exercise is to clarify the relationship between the various contact rates. In particular, show that

$$c_i = \sum_{j=1}^{n} c_{ij} N_j.$$

Moreover, show that a fraction

$$q_i = \frac{c_{ii} N_i}{c_i}$$

of the contacts of a type-i individual is within its own group.

ii) Using the notation from Section 2.2, prove (12.19) for the case of two groups clustered into a single group, in the case of proportionate mixing. Assume, for convenience, that $p_{ij} T_j = 1$ for all i and j.

iii) Consider the case $k = 1$, i.e., ignore heterogeneity completely. Then

$$\widehat{R}_0 = \bar{c} p T$$

where

$$\bar{c} := \frac{\sum_{i=1}^{n} c_i N_i}{\sum_{i=1}^{n} N_i}$$

is the average contact rate, T is the average length of the infectious period, and p is the probability of transmission given a contact between an infectious and a susceptible individual. Now prove relation (12.19).

Hint: Define the re-scaling transformation $S : R^n \to R^n$ by

$$(Sx)_i = \frac{x_i}{\sqrt{N_i}}.$$

Verify that SMS^{-1} is symmetric (the idea underlying the transformation S is that the ℓ_1-norm $\|x\|_{\ell_1} = \sum_{i=1}^{n} |x_i|$ is good for measuring population size, but that the Euclidean norm $\|x\|_{\ell_2} = (\sum |x_i|^2)^{1/2}$ is good for exploiting symmetry). Next recall (or use) that for a symmetric matrix Q the largest eigenvalue is characterized by an extremum property

$$\lambda_d(Q) = \max_{\|w\|_{\ell_2}} (w, Qw)$$

where $(w, v) := \sum_i w_i v_i$ is the Euclidean inner product. Finally, use as an auxiliary tool the vector v with components

$$v_i = \frac{\sqrt{N_i}}{(\sum_{j=1}^{n} N_j)^{1/2}}.$$

iv) Prove relation (12.19) for general k.

Exercise 12.22 For a counterexample in the case that infectivity varies between groups, take the same situation as in Exercise 12.21-ii (i.e., $n = 2$, $k = 1$), and let $p_{ij} T_j = d_j$, where we choose $d_1 = 1$ and $d_2 = 3$ respectively. To make life easier, assume that $N_1 = N_2 = N$ and $c_1 = c_2 = c$ (note that this implies $p_1 = p_2 = p$, equal for both groups). We have to make an assumption about how to define the infectivity for the aggregated group. Here simply take the arithmetic average (so the value is 2). Show that $\widehat{R}_0 > R_0$ for all $p < \frac{1}{2}$.

12.4.2 What is a core group?

From the book *Gonorrhea Transmission Dynamics and Control* by H.W. Hethcote and J.A. Yorke (1984), we quote (p.35): "The population in the groups with high prevalence (with prevalences at least 20%) are lumped together and called the *core*." This is an operational definition (although it presupposes a splitting of the population into subgroups) and the suggestion is that the core causes gonorrhea to remain endemic and, more generally, determines to a large extent the endemic prevalence level. The notion of a core is appealing as well as helpful in organizing our thoughts about the role of heterogeneity in determining overall levels of transmission. Yet, we claim that a good definition is lacking. The present subsection offers an attempt at such a definition. We hope that it serves to stimulate our readers to give the issue some thought and to come up with better ideas.

Consider finitely many h-states numbered $1, 2, \ldots, n$. The next-generation matrix K has as its entries k_{ij}, the expected number of secondary cases of type i from one primary case of type j, when the agent is introduced in a virgin host population. The basic reproduction ratio R_0 is the dominant eigenvalue of the next-generation matrix K (recall Section 7.1). The diagonal elements k_{ii} describe transmission within the sub-population of the same type (the own group). If we were to consider this type in isolation, eliminating all other types, while assuming that this would not increase the 'within-group' contacts, then the corresponding R_0 would be k_{ii}. In general, the contacts outside the own group help to increase R_0 and therefore, as can be proved mathematically as well, $R_0 \geq k_{ii}$ for all i, so $R_0 \geq \max_i k_{ii}$.

It is now tempting to call the sub-population with h-state i a *core group* if k_{ii} and R_0 have the same order of magnitude while k_{jj} is substantially smaller for all $j \neq i$. More generally, we would call the sub-population with h-state $i \in C$, where C is a (small) subset of $\{1, 2, \ldots, n\}$, a core group if for all $i \in C$ the quantity k_{ii} has the same order of magnitude as R_0 while for all $j \notin C$ the quantity k_{jj} is substantially smaller. This definition has (at least) two weak points. The first is that 'order of magnitude' and 'substantially' are subjective and it is a matter of taste how one should compare, for example, 6 with 0.5. The second weak point is that loops at the group level, occurring for example when considering the influence of prostitutes on STD prevalence, are ignored.

The second weakness can be remedied by looking at the diagonal elements of K^2 (and more generally, higher powers of the matrix K). When $n = 2$, we have

$$K^2 = \begin{pmatrix} k_{11}^2 + k_{12}k_{21} & k_{11}k_{12} + k_{12}k_{22} \\ k_{21}k_{11} + k_{22}k_{21} & k_{21}k_{12} + k_{22}^2 \end{pmatrix},$$

and we see the product $k_{12}k_{21}$ appearing on the diagonal. In general (i.e., for all n), the inequality

$$R_0 \geq \sqrt{\sum_{i=1}^{n} k_{ij}k_{ji}}$$

holds for all j. We therefore might use the term *jm-core loop* if $k_{mj}k_{jm}$ is of the same order of magnitude as R_0, while all diagonal elements k_{ii} as well as all products $k_{li}k_{il}$ with $(i, l) \neq (j, m)$ are substantially smaller. Of course we can also consider loops of length three or greater, etc.

12.5   STOCHASTIC FINAL SIZE AND MULTI-LEVEL MIXING

In the epidemic models treated so far there is an underlying assumption that individuals have contact with different individuals at more or less the same rate, i.e., the rates of contact between any two individuals are of the same order. One exception are models for spatial spread where the contact rates decay with the distance between individuals (or for percolation inspired models, where individuals *only* have contact with their 'neighbors,' suitably defined).

In many situations this might be too simplified a description of reality. Quite often one can instead subdivide those contacted into two categories: those that the individual meets often/regularly and those that are more random/occasional, for example fellow bus passengers or fellow shop clients. The prime example of regular contacts are those within a household (one can also think of other community structures playing the role of 'households,' for example day-care centers, class rooms and work places). In the next section we will focus on a household model, but there are other models with similar features. For example, if the N individuals are lined up in a circle and anyone who gets infected has local contacts with its neighbors and global contacts with randomly chosen individuals, this gives an example of so-called small world models.[8] The methods for analyzing household models are quite different from those needed to analyze small world models. The important difference is that the household model has distinct units of completely mixing subunits: the households. This is used in the analysis by treating the households as 'super-individuals' (recall Section 6.1.3). In the great circle model the local mixing is not restricted to distinct units, instead we have *overlapping sub-groups*, which makes the analysis more difficult.

The models we have described above are called *two-level mixing* epidemic models. For finite N it is hard to distinguish between 'two-level' mixing and just mixing at different rates, so the term is based on asymptotic considerations for large N, when the two types of contact occur with rates of different order. The other extreme case compared to household models (which typically have many units of small size) is to have few large units and to assume that the contact rates between individuals depend on the groups to which the two individuals belong. This more resembles the asymptotic situation where the group sizes tend to infinity and the number of groups remains fixed; in that case, the contact rates between any pair of individuals are of the same order $(1/N)$, although different. Then individuals are heterogeneous and a multi-type model (see Chapters 7 and 8) is better suited than a two-level mixing model. For finite N the relevance of talking about two-level mixing models increases with the difference of the individual-to-individual contact rates.

One could of course think of several levels of mixing, multi-level mixing. It is often natural to consider households as the closest unit of mixing but one could also include a second level consisting of day-care centers, schools and workplaces, and on top of that more random type contacts.[9] From now on we restrict ourselves to two levels of mixing: one social structure, typically a household, and then random mixing with anybody in the community. The social structure we are considering is typically a small unit (a household), so an epidemic model for this situation is preferably stochastic. In the next section we present such a model followed by

---

[8]D.J. Watts: *Small Worlds: The Dynamics of Networks Between Order and Randomness.* Princeton University Press, 1999.

[9]See for example, T. Britton, T. Kypraios and P.D. O'Neill: Statistical models for epidemic models with three levels of mixing. *Scand. J. Stat.*, **38** (2011), 578–599.

a heuristic analysis. A more general form of the model as well as rigorous large population asymptotics for the model can be found in Ball, Mollison and Scalia-Tomba (1997).[10]

12.5.1 Modeling transmission within and between households

We now describe a household epidemic model. For this we consider a community of N individuals in which each individual belongs to exactly one household (which of course could be a single individual household). Let n_j denote the number of households of size j, and n the total number of households, so $\sum_j n_j = n$ and $\sum_j j n_j = N$. Further, let $\pi_j = n_j/n$ denote the proportion of households having size i and let $h = \sum_j j \pi_j$ denote the mean household size. In the asymptotic situation we consider a sequence of such epidemics indexed by N, so all variables should be equipped with index N, and the quantities should converge to non-trivial limits at suitable rates — here we skip that level of detail.

Initially there is a specified number of newly infected individuals residing in the households in a specified way (most often it is assumed that there is one initially infected, and that this individual is chosen completely at random among the $N = \sum_j j n_j$ individuals). The remaining individuals are assumed to be susceptible to the disease. Once an individual gets infected he or she immediately becomes infectious (cf. Section 12.5.2) and remains so for a finite random infectious period of duration ΔT. During the infectious period the individual has global contacts according to a Poisson process with rate c_G, each contact being independent and chosen completely at random from the whole community. As a consequence, the global contact rate with a specified individual equals c_G/N. Additional to this, an infectious individual also has local infectious contacts: during the infectious period the individual makes contact independently with each household member at rate c_L. Both types of infectious contacts are such that, if the contacted person is still susceptible, then that person becomes infected with probability p, and with probability $1 - p$ the contact has no effect. Once the infectious period is over, the person stops infecting others and becomes immune and remains so for the remainder of the study period.

We are interested in characterizing the final size of an outbreak. Note that the final size not only counts the total number infected, but also how they are distributed among households. One way to characterize the final size is by using the random vector $\{N_{j,i}; i \leq j\}$, where $N_{j,i}$ denotes the number of households of size j having i infected individuals at the end of the epidemic. This implies that $\sum_{i=0}^{j} N_{j,i} = n_j$. The final (or total) number infected T can be obtained from the final size: $T = \sum_{i,j} i N_{j,i}$.

12.5.2 Comments and extensions of the household epidemic

It is worth pointing out that the distribution of the final size is determined from three different sets of quantities: the community structure $\{n_j\}$, the model parameters: the infectious period ΔT (and its distribution) and c_G, c_L and p, and finally the number and location of the initially infected individuals. The model is over-parameterized when considering the final-size distribution (as opposed to the actual time dynamics) since the final size is not affected by multiplying the infectious periods by any factor and at the same time dividing the contact rates c_G and

---

[10]F. Ball, D. Mollison and G. Scalia-Tomba: Epidemics with two levels of mixing. *Ann. Appl. Prob.*, **7** (1997), 46–89.

c_L by the same factor (and p could be set to 1) — this is simply a change of time unit. For this reason we could set $E(\Delta T) = 1$ without loss of generality. In fact we could generalize the model without affecting the final-size distribution at all. Rather than having a fixed infectious contact rate during the entire infectious period, we can allow this rate to vary, possibly randomly, over time. The generalized model assumes that an infectious individual has global infectious contacts at rate $c_G \iota(\tau)$ and local infectious contacts with each household member at rate $c_L \iota(\tau)$, τ time units after he or she was infected, where $\{\iota(\tau); \tau \geq 0\}$ is a non-negative stochastic process (the 'infectivity process'), being independent and identically distributed between individuals. As a special case we could have a latency period of arbitrary length simply by setting $\iota(\tau) = 0$ during this period.

For this more general model we get exactly the same distribution of the final size as the model defined in the previous section if ΔT (in the present model) has the same distribution as $\int_0^\infty \iota(\tau) d\tau$ (in the extended model).

12.5.3   The distribution of a within-household outbreak

Before determining an approximation for the final-size distribution of the household epidemic we now derive the final-size distribution of an isolated household of size $m + n$, starting with m initially infectious and n susceptible individuals. This distribution is interesting in its own right in that it gives the exact distribution of the final size of the prototype stochastic epidemic model defined in Section 3.1 (with c_G replacing c/N and treating the household as the 'community' with size $m + n$). This section hence also gives a recipe for determining the exact distribution of the prototype stochastic epidemic model in a small community — numerical problems arise when considering larger sizes.

The model we consider in this section is the following. The household initially consists of m infectives and n susceptibles. Any individual who becomes infectious has contacts with each other household member independently at rate c_L during the infectious period having random length ΔT. A contact with a susceptible results in infection with probability p, other contacts have no effect. After the infectious period is over the individual recovers and becomes immune.

In Section 3.4.2 the Sellke construction of the prototype stochastic epidemic model was defined, where each individual was given an exponentially distributed *resistance* Q, and that individual was infected as soon as the cumulative infection pressure $\Lambda_c(t) = c_L p \int_0^t I(s) ds$ (with our new parameters) exceeded Q. The random variable Y, specifying the final number infected among the initially susceptible, was shown (cf. equation (3.17)) to be given by

$$Y = \min\{k \geq 0; \; Q_{(k+1)} > c_L p \sum_{i=-(m-1)}^{k} \Delta T_i\},$$

where $Q_{(1)}, \ldots, Q_{(n)}$ are the order statistics of $Q_1, \ldots, Q_n$, so $Q_{(j)}$ denotes the jth smallest resistance. A heuristic motivation for this equation is that the epidemic stops as soon as the over-all infection pressure caused by those infected does not exceed any further resistance. Let

$$\Lambda_c(\infty) = c_L p \sum_{i=-(m-1)}^{Y} \Delta T_i \qquad (12.20)$$

denote the 'total infection pressure' of the epidemic, sometimes also referred to as the 'total cost' of the epidemic. Then it is possible to show a so-called Wald's identity for the final size (for the prototype stochastic epidemic model) involving the final number of infected (Y) and the total infection pressure ($\Lambda_c(\infty)$). This identity will be the key result for determining the distribution of the final size. Let

$$\phi(\theta) = E(e^{-\theta \Delta T})$$

denote the Laplace transform of the infectious period ΔT. We then have the following theorem:[11]

Theorem 12.23 (Wald's identity) *Consider the prototype stochastic epidemic model using notation from the present section. Then Y, the final number infected among the initially susceptible, and $\Lambda_c(\infty)$, the total infection pressure, satisfy*

$$E\left(\frac{e^{-\theta \Lambda_c(\infty)}}{(\phi(\theta c_L p))^{m+Y}}\right) = 1, \qquad \theta \geq 0. \tag{12.21}$$

Proof: The key to the proof is the Sellke construction, the rest is straightforward. Note that

$$(\phi(\theta c_L p))^{m+n} = E\left[\exp\left(-\theta c_L p \sum_{i=-(m-1)}^{n} \Delta T_i\right)\right]$$

$$= E\left[\exp\left(-\theta\left(\Lambda_c(\infty) + c_L p \sum_{i=Y+1}^{n} \Delta T_i\right)\right)\right]$$

$$= E\left[e^{-\theta \Lambda_c(\infty)}(\phi(\theta c_L p))^{n-Y}\right],$$

where the last identity follows because the $n-Y$ infectious periods $\Delta T_{Y+1}, \ldots, \Delta T_n$, are mutually independent and also independent of $\Lambda_c(\infty)$ which has ϕ as Laplace transform (and which depends on $\Delta T_{-(m-1)}, \ldots, \Delta T_Y$). Dividing by $(\phi(\theta c_L p))^{m+n}$ on both sides completes the proof. $\square$

Wald's identity is the main step for obtaining the final-size distribution of Y. We are now in a position to derive this distribution, which we denote by $P_i^{(m,n)}$, $i = 0, \ldots, n$, where $P_i^{(m,n)} = P(Y^{(m,n)} = i)$ is the probability that i out of the initial n susceptibles become infected starting with m initially infected individuals (the index (m,n) is inserted to show the dependence on m and n).

The second step in deriving the final-size distribution is to consider subsets of individuals in the household and remember that all individuals behave the same, so subsets of the same size are interchangeable. Let $0 \leq i \leq k \leq n$ be given and define the subsets $\mathcal{I} = \{1, \ldots, i\}$, $\mathcal{K} = \{1, \ldots, k\}$ and $\mathcal{N} = \{1, \ldots, n\}$, implying that $\mathcal{I} \subset \mathcal{K} \subset \mathcal{N}$. Define an auxiliary 'set-probability' $P_{\mathcal{I}}^{(m,n)}$ as the probability that the individuals in the set $\mathcal{I}$ get infected and the rest avoid being infected. By interchangeability it follows that $P_i^{(m,n)} = \binom{n}{i} P_{\mathcal{I}}^{(m,n)}$. The event that the epidemic within the household $\mathcal{N}$ results in exactly those in $\mathcal{I}$ getting infected is identical to the event that the epidemic within the sub-group $\mathcal{K}$ results in exactly $\mathcal{I}$ getting

[11]F. Ball: A unified approach to the distribution of total size and total area under the trajectory of infectives in epidemic models. *Adv. Appl. Prob.*, **18** (1986), 289–310.

infected *and* that these i infected together with the initial m infected fail to infect those in $\mathcal{N} \setminus \mathcal{K}$. Further, the $n - k$ individuals in $\mathcal{N} \setminus \mathcal{K}$ escape infection, from those infected, independently of each other (they are assumed to have independent resistances). This hence gives us the following relation between $P_{\mathfrak{J}}^{(m,n)}$ and $P_{\mathfrak{J}}^{(m,k)}$:

$$P_{\mathfrak{J}}^{(m,n)} = P_{\mathfrak{J}}^{(m,k)} E\left(e^{-(n-k)\Lambda_c^{(m,k)}(\infty)}|Y^{(m,k)} = i\right),$$

where we have also equipped $\Lambda_c(\infty)$ with a super-index. This is true because the resistance $Q \sim Exp(1)$, so the probability that an individual avoids infection from the pressure a equals e^{-a}. The conditioning on $Y^{(m,k)} = i$ is needed because the distribution of the total infection pressure depends on how many were infected. In terms of the final-size probabilities we hence have

$$\frac{P_i^{(m,n)}}{\binom{n}{i}} = \frac{P_i^{(m,k)}}{\binom{k}{i}} E\left(e^{-(n-k)\Lambda_c^{(m,k)}(\infty)}|Y^{(m,k)} = i\right). \tag{12.22}$$

Wald's identity, with $\theta = n - k$, for the epidemic in the sub-group gives us

$$E\left(\frac{e^{-(n-k)\Lambda_c^{(m,k)}(\infty)}}{(\phi((n-k)c_{\mathrm{L}}p))^{m+Y^{(m,k)}}}\right) = 1,$$

and if we condition on the final size $Y^{(m,k)}$ we get

$$\sum_{i=0}^{k} \frac{E\left(e^{-(n-k)\Lambda_c^{(m,k)}(\infty)}|Y^{(m,k)} = i\right)}{(\phi((n-k)c_{\mathrm{L}}p))^{m+i}} P_i^{(m,k)} = 1. \tag{12.23}$$

If we combine equations (12.22) and (12.23) we get

$$\sum_{i=0}^{k} \frac{\binom{k}{i} P_i^{(m,n)}}{\binom{n}{i}(\phi((n-k)c_{\mathrm{L}}p))^{m+i}} = 1.$$

If we rearrange this equation by using that $\binom{k}{i}/\binom{n}{i} = \binom{n-i}{k-i}/\binom{n}{k}$ and putting $P_k^{(m,n)}$ on one side, we obtain a formula which determines the final-size probabilities:

$$P_k^{(m,n)} = \binom{n}{k}[\phi((n-k)c_{\mathrm{L}}p)]^{m+k} - \sum_{i=0}^{k-1}\binom{n-i}{k-i}[\phi((n-k)c_{\mathrm{L}}p)]^{k-i} P_i^{(m,n)}, \tag{12.24}$$

for $0 \le k \le n$. Note that this is a recursive formula, so in order to compute $P_k^{(m,n)}$ we must first sequentially compute $P_0^{(m,n)}$ up to $P_{k-1}^{(m,n)}$. As a consequence, the formula is only practically useful for moderately small k. As mentioned in the beginning of this section this distribution is also the distribution of the prototype stochastic epidemic model defined in Chapter 3, if we replace c_{G} above by c/N and treat the household as the 'community' with size $m + n$. This is clear since an outbreak in an isolated household makes the household a community.

Solving equation (12.24) for $k = 0$ gives

$$P_0^{(m,n)} = [\phi(nc_{\mathrm{L}}p)]^m = \left[E\left(e^{-nc_{\mathrm{L}}p\Delta T}\right)\right]^m.$$

The interpretation of this is that each of the n initially susceptible has to escape from the infection force caused by each of the m initially infected, and these infecteds are assumed to act independently. However, the events that initial susceptibles escape the infection pressure of a *given* initially infected are *dependent* as they all depend on the infection shed by that infected individual.

Solving the equation for $k = 1$ then gives us:

$$P_1^{(m,n)} = n[\phi((n-1)c_Lp)]^{m+1} - n\phi((n-1)c_Lp)P_0^{(m,n)}$$
$$= n\phi((n-1)c_Lp)\Big([\phi((n-1)c_Lp)]^m - [\phi(nc_Lp)]^m\Big).$$

For larger k the formula quickly becomes cumbersome. However, with computers this recursive formula can be used to numerically compute the final-size distribution for households of size 10 or more.

Exercise 12.24 Consider the case where the infectious period is constant and equal to 1, i.e., that $\Delta T \equiv 1$. This implies that $\phi(\theta) = e^{-\theta}$. Assume further that $c_Lp = 0.5$ and that the household consists of $m = 1$ initially infectious and $n = 2$ initially susceptible individuals. Derive the final-size distribution, that is, derive $P_0^{(1,2)}$, $P_1^{(1,2)}$ and $P_2^{(1,2)}$.

Exercise 12.25 Consider the same case as in the previous exercise with the exception that the infectious period now is exponentially distributed with mean equal to 1, i.e., that $\Delta T \sim Exp(1)$. This implies that $\phi(\theta) = 1/(\theta+1)$. Derive the final-size distribution, that is, derive $P_0^{(1,2)}$, $P_1^{(1,2)}$ and $P_2^{(1,2)}$.

12.5.4 The final-size distribution of the household epidemic

We now return to the household epidemic model, i.e., to the situation where there are n (assumed to be a large number) households of varying sizes and where there is transmission between households additional to within-household transmission. To rigorously show the exact and asymptotic distribution of the household epidemic is very complicated. Here we give some large community heuristics.

The reproduction number During the early stages of the epidemic most households are unaffected by the outbreak, implying that most global contacts will be with individuals in unaffected households. If we treat the households as 'super-individuals,' this suggests that these super-individuals infect other super-individuals independently, implying that the epidemic among the super-individuals behaves much like a branching process. This reasoning can be made precise in the limit n tending to infinity. We will now derive a reproduction number R_* which determines if a major outbreak is possible or not. In doing this we will make use of the within-household outbreak distribution derived in the previous section.

Consider an individual who has been infected during the early stages of the epidemic. Attribute to him/her all individuals that he/she directly infects globally, but also those infected indirectly in the household outbreaks resulting from these global infections. Do *not* attribute those individuals the person infects within his/her own household. With this new way of attributing infections each infected individual is still 'infected' by exactly one person as long as there are no multiple global contacts to the same household, something that holds with large probability in the initial phase of an outbreak as explained earlier. Define R_* as the *expected* number of individuals an infected individual 'infects' using this new way of attributing infections. Then it is plausible that a major outbreak can occur if and only if $R_* > 1$, since it is only when this average exceeds 1 that the number infected can grow large.

During the early stages the global contacts will (with large probability) be with susceptible individuals residing in different households. As a consequence, an individual who gets infected during the early stages will on average infect $c_{\mathrm{G}} p E(\Delta T)$ individuals through global contacts. Each such contact results in a household outbreak, so it remains to determine the expected value of such an outbreak. The contacted individual who got infected resides in a household of size j with probability $jn_j/N = j\pi_j/h$ (the size-biased household distribution), where $h = \sum_j j\pi_j$ is the mean household size. Given that the contacted person resides in a household of size j, the expected number infected in that household, including the contacted individual, equals μ_j, where μ_j is defined by

$$\mu_j = 1 + \sum_{i=0}^{j-1} i P_i^{(1,j-1)},$$

(using the final-size distribution of the previous section). So μ_j denotes the expected number infected (including the initially infected!) in a household of size j, starting with one initially infected and $j-1$ initially susceptible. The unconditional mean household outbreak size is hence $\mu = \sum_j \mu_j j\pi_j/h$. Each global infection will give the same expected outbreak size, so the reproduction number is hence given by

$$R_* = c_{\mathrm{G}} p E(\Delta T)\mu = c_{\mathrm{G}} p E(\Delta T) \sum_j \mu_j \frac{j\pi_j}{h}. \tag{12.25}$$

It is worth pointing out that, contrary to the case of the prototype stochastic epidemic model without households, the reproduction number now depends on the *distribution* of the infectious period and not only on its expected value $E(\Delta T)$. This is hidden in the notation but follows since the final-size distribution of the within-household outbreak, and its mean, depend on this distribution (see the exercises below for an illustration).

Exercise 12.26 Assume all households are of size 3, that the infectious periods are deterministic and equal to 1, that $c_{\mathrm{L}} p = 0.5$ and that $c_{\mathrm{G}} p = 1$. Determine the reproduction number R_*. (Note that the final-size probabilities are the same as those in Exercise 12.24.)

Exercise 12.27 Assume all households are of size 3, that the infectious periods are exponentially distributed with mean 1, that $c_{\mathrm{L}} p = 0.5$ and that $c_{\mathrm{G}} p = 1$. Determine the reproduction number R_*. (Note that the final-size probabilities are the same as those in Exercise 12.25.)

Exercise 12.28 The situation in Exercise 12.26 and 12.27 are identical except that the distribution of the infectious periods differ even though they have the same average length of 1. Does 'more randomness' in the infectious period seem to increase or decrease R_*?

The final size in case of a major outbreak Assume that the epidemic takes off infecting a positive fraction of the large community of households. Implicitly we are hence assuming that $R_* > 1$ since a major outbreak is only possible in this case. We will now study the distribution of the final size of such a household epidemic. Recall that the final size is specified by the vector $\{N_{j,i}; 0 \leq i \leq j\}$, where $N_{j,i}$ denotes the number of households of size j having i ultimately infected. As before we assume that the number of households n is large. As n tends to infinity it can be shown that, in case of a major outbreak, both $T/N = \sum_{i,j} iN_{j,i}/N$, the final fraction ultimately infected and the proportions $N_{j,i}/n_j$ converge in probability to some fixed constants, u and $p_{j,i}$ say, respectively. Determining the final size hence consists of deriving expressions for these quantities, expressions that will depend on the household epidemic and its parameters.

We now give heuristic arguments to determine the final-size distribution, characterized by $\{p_{j,i}; \ 0 \leq i \leq j\}$. Since $p_{j,i}\pi_j$ is the proportion of all households being of size j and having i infected, and u is the over-all proportion infected, it follows that u must satisfy

$$u = \sum_{i,j} ip_{j,i}\pi_j/h, \tag{12.26}$$

where $h = \sum_j j\pi_j$ is the average household size. The main insight for determining the proportions $\{p_{j,i}; \ 0 \leq i \leq j\}$ lies in noting that individuals of different households are infected more or less independently of each other. Of course, there is a small chance that individuals of two given households have infected each other, but this probability is negligible if the number of households is large. We hence treat the households as being independent. Given that a proportion u of the community are infected, the probability that an individual escapes global infectious contacts equals $e^{-c_{\mathrm{G}}pE(\Delta T)u}$. This follows because the accumulated global infection pressure (acting on a separate individual) caused by all eventually infected equals $c_{\mathrm{G}}p/N$ multiplied by the sum of all infectious periods, and there are Nu such infectious periods, each having mean $E(\Delta T)$. Finally, the probability that a resistance exceeds $c_{\mathrm{G}}pE(\Delta T)u$ (implying that the individual escapes global infection) equals $e^{-c_{\mathrm{G}}pE(\Delta T)u}$.

Individuals are hence globally infected with probability $1 - e^{-c_{\mathrm{G}}pE(\Delta T)u}$, and these events are more or less independent. Another important observation is that, for the final size, it does not matter at what time such global contacts occur, so we can just as well think of them as occurring before the household outbreaks take place. As a consequence, a household of size j will have a binomial number X of globally infected ($X \sim Bin(j, 1 - e^{-c_{\mathrm{G}}pE(\Delta T)u})$). Given this number $X = k$, the outbreak of that household will have a distribution defined by $\{P_r^{(k,j-k)}; 0 \leq r \leq j-k\}$. A household of size j can get i ultimately infected by having any number k between 1 and i globally infected and the household outbreak resulting in another $i - k$ infected. From this it follows that $\{p_{j,i}\}$ must satisfy

$$p_{j,0} = (e^{-c_{\mathrm{G}}pE(\Delta T)u})^j \tag{12.27}$$

$$p_{j,i} = \sum_{k=1}^{i} \binom{j}{k} \left(1 - e^{-c_{\mathrm{G}}pE(\Delta T)u}\right)^k \left(e^{-c_{\mathrm{G}}pE(\Delta T)u}\right)^{j-k} P_{i-k}^{(k,j-k)}, \quad i = 1, \ldots, j. \tag{12.28}$$

Given the model parameters $c_{\mathrm{G}}, c_{\mathrm{L}}, p$ and the distribution of the infectious period ΔT we can derive the within-household outbreak probabilities $\{P_i^{(m,n)}\}$. These quantities together with the community structure of household sizes $\{\pi_j\}$, then determine the large population distribution of the household epidemic, $\{p_{j,i}\}$, and its overall proportion infected u by way of equations (12.26), (12.27) and (12.28).

For finite n the distribution of $N_{j,i}/n_j$ is of course not exactly that of a point mass at $p_{j,i}$. It is random with some complicated distribution around $p_{j,i}$. However, it can be shown using a law-of-large numbers argument that, as n tends to infinity the distribution gets more and more concentrated around $p_{j,i}$. In fact, a central limit theorem has also been shown stating that the distribution tends to a normal distribution with mean $p_{j,i}$ and standard deviation proportional to $1/\sqrt{n}$. This is complicated to show and beyond our scope.

Exercise 12.29 Assume all households are of size 3, that the infectious periods are deterministic and equal to 1, that $c_{\mathrm{L}}p = 0.5$ and that $c_{\mathrm{G}}p = 1$ (as in Exercise 12.27). Determine the final proportions $\{p_{3,i}; i = 0, \ldots, 3\}$ that $\{N_{3,i}/n\}$ will (approximately) equal in case of a large outbreak.

Exercise 12.30 Assume that all households are of size 3, that the infectious periods are exponentially distributed with mean 1, that $c_{\mathrm{L}}p = 0.5$ and that $c_{\mathrm{G}}p = 1$ (as in Exercise 12.27). Determine the final the proportions $\{p_{3,i}; i = 0, \ldots, 3\}$ that $\{N_{3,i}/n\}$ will (approximately) equal in case of a large outbreak.

12.6   NETWORK MODELS (AN IDIOSYNCRATIC VIEW)

12.6.1   Reflections: what do we want and why?

We recall an observation made at the beginning of Section 7.8: in deterministic models in which, by assumption, all sub-population sizes are infinite, two individuals never have contact twice. A quite different type of model conceives a population as a *network*, with connections (called *edges*) between individuals (called *nodes* or *vertices*) that are either fixed or that can be formed and broken in the course of time. Each connection symbolizes a relationship between individuals that involves repeated contacts, and therefore transmission of an infective agent proceeds along connections. In pure network models, transmission is restricted to connections, but of course one can formulate hybrid models that incorporate a second transmission route through uniform/random contacts.

In network models, populations are represented as *graphs* (static or dynamic) and one can use a color coding to denote the status of individuals with respect to some infectious disease (e.g., an infective individual could be denoted by a red pixel in a population that is represented by a two-dimensional square lattice). Thus the results of a computer implementation and subsequent simulation can be visualized, and one can literally watch the agent spread. Such experiments catalyze the sharpening of our intuition. We will use the words 'network' and 'graph' interchangeably.

We shall briefly indicate some characteristic features, while providing pointers to the literature (i.e., we try to give you a taste of the flavor, without serving the dish, but do give directions to the restaurant where the dish can be ordered). Much attention has been paid in ecology[12] and evolution, and increasingly in epidemiology, to the *interacting particle systems*. As described above, the basic idea is to represent space as a lattice of cells, to allow each cell to be in a number of different states, and to specify rules according to which the state of a cell at time $t + 1$ is

---

[12]R. Durrett and S. Levin: Stochastic spatial models: a user's guide to ecological applications. *Phil. Trans. R. Soc. B*, **343** (1993), 329–350.

determined (we consider only the discrete-time version, but continuous-time formulations are possible). If we think of an infectious agent of plants, the cells could, for example, be individual plants in a field, or fields in a region. Suppose each cell can be in either of three different states: susceptible (0), infected (1) or immune/dead (2). Part of the specification of the reconfiguration rules involves making a list of the cells that can, with their state at time t, influence the possible change in state of a given cell x between t and $t + 1$. Those cells are often called the 'neighborhood' of x. One possible choice, in a square lattice, is to take only the four cells that share a boundary of positive measure with x, i.e. its nearest neighbors. This is called the von Neumann neighborhood, after the mathematician who originally introduced the idea of a cellular automaton. (Different choices of neighborhood can of course have a large effect on the dynamics.) One could specify that the infection probability for cell x is β times the number of infected cells in the neighborhood of x and that the removal probability is α for each infected cell. In this situation one can show[13] that, starting from one infected cell, under certain conditions on β and α, we obtain a wave of infection traveling outward over the lattice with constant velocity and asymptotically a fixed wave front (roughly circular). This is similar to results we discussed in Chapter 10.

The conditions on β and α will be shaped as a condition such as $R_0 > 1$ in the situation that contacts can be made with all individuals in the population (that we focussed on for most of the book). Here, however, one should realize that the arrangement of the cells in the lattice determines what the threshold will be above which an infection will take off (think of loops, recall Section 10.7). To determine analytically (as opposed to experimentally/numerically) how this threshold depends on the spatial arrangements in the lattice, neighborhood choices and reconfiguration rules is a very difficult open mathematical problem.[14] In this chapter we shall always have networks or graphs in mind that describe existing contact structures, on which transmission is then superimposed. A somewhat different approach is to consider the graph as 'constructed' by the transmission events and to use results from random graph theory to deduce information about the spread of the agent.[15]

Regular spatial lattices are easy to describe, characterize and represent in a computer. Yet we may want to introduce other types of graphs when modeling, for example, networks of sexual interaction between individuals. The ideal is to be able to describe a graph in terms of recognizable/observable *local characteristics* such as

- *Degree distribution*: for each node (i.e., individual), count how many edges (i.e., connections) attach to it.

- *Loops*: starting in a node and following edges, what is the minimal length of a round trip? And how many alternative routes exist?

- *Clusters*: does the graph decompose into loosely coupled components?

---

[13]See e.g., R. Durrett: Spatial epidemic models. In: Mollison (1995), pp. 187–201; D. Mollison: Dependence of epidemic and population velocities on basic parameters. *Math. Biosci.*, **107** (1991), 255–287.

[14]See e.g. R. Durrett: Stochastic growth models: bounds on critical exponents. *J. Appl. Prob.*, **29** (1992), 11–20; S.A. Levin and R. Durrett: From individuals to populations. *Phil. Trans. R. Soc. B*, **351** (1996), 1615–1621.

[15]B. Bollobás: *Random Graphs*. Cambridge University Press, 2nd edition, 2001; R, Durrett: *Random Graph Dynamics*. Cambridge University Press, 2010.

One furthermore wants to add some simple rule for *reconfiguration*, which determines the *temporal pattern*, and finally from such a description infer certain consequences for the spread of an infectious disease. In other words, the ideal is to deduce relevant characteristics of epidemic spread from a *statistical description* of the spatio-temporal patterns in a network. Obviously there is a long way to go before this ideal can be achieved, but once we can make the right connection, we can — in order to derive conclusions about the spread of the agent — rather directly use *observations* about quantities such as the following:

- The frequency of forming new connections (e.g., the frequency of acquiring a new sexual partner);

- The length of time that a connection exists;

- The degree distribution (e.g., simultaneous partnerships);

- Types of connections (e.g., to distinguish casual from serious partnerships, wives from mistresses, ...).

An important class of graphs for epidemic application are the random graphs, where one can explicitly model birth and death of new vertices and forming and breaking up of edges.[16]

But, is there a good reason to expect any effect of incorporating spatio-temporal pattern? Yes, there is. The key point is that, from the point of view of the infective agent, contacts between two infected individuals are *wasted*. When contacts are repeated and/or when a graph has many short cycles, the probability of contacts between two infectives is higher than in a randomly mixing population.

To measure the effect, we should investigate how the following indicators depend on the spatio-temporal pattern:

- The basic reproduction ratio R_0,

- The intrinsic growth rate r,

- The size of a major outbreak,

- The probability of a minor outbreak,

or, when incorporating host demography,

- The probability of extinction after a first major outbreak,

- The critical community size,

- The endemic level,

- The effort needed for eradication.

---

[16]A.D. Barbour and D. Mollison: Epidemics and random graphs. In: Gabriel et al. (1990), pp. 86–89; Ph. Blanchard, G.F. Bolz and T. Krüger: Modelling AIDS-epidemics or any venereal disease on random graphs. In: Gabriel et al. (1990), pp. 104–117; Bollobas, 2001; Durrett, 2006.

A characteristic (but rather annoying) difficulty is that we want to gauge the models that we compare in such a way that the contact 'intensity' is the same, while the contact 'pattern' differs, but that there often is not a unique well-defined way to specify 'intensity.'

As an alternative to the direct statistical description of pattern, we can postulate *mechanisms* (or, at least, phenomenological rules) that govern the formation and reconfiguration of the network. Here we can think, for example, of the tendency for individuals to form a new partnership when single or when having already some given number of partners (in principle, one can include other characteristics, such as age). The advantage of *simulation* models is that one can specify rules directly in terms of a computer program (although an interaction between two symmetric individuals remains a subtle modeling challenge; recall Section 12.3 on consistency conditions). The disadvantage is, of course, that it is difficult to extract insight and understanding from the myriad of simulation runs. With *analytical* models it is usually the other way around: they are, as a rule, intractable, but *if* we manage to analyze them, we obtain qualitative insight into the relationship between pattern and the spread of the agent. This makes the quest for tractable analytical network models a valuable endeavor.

12.6.2   Network jargon

Network models and their properties are very actively studied in recent years.[17] One reason for this is that more and more 'explicit' networks exist, for example on the internet, and another reason is perhaps the popular paper by Watts and Strogatz (1998)[18] and the by now famous saying that anyone is only six hand-shakes away from the American President. This feature, apparent in most empirical networks, that the distance between any two nodes in a network is small even though the graph/network is sparse (in the sense of each node being connected to only a few neighbors), is called the *small-world property*. The 'distance' here means the smallest number of edges one has to traverse in order to reach the other node. The small-world property was not present in many networks previously studied, for example spatial networks such as the integer lattice Z^2.

In this short presentation a network means a set of nodes, some of which are pairwise connected by undirected edges, see Figure 12.2, but imagine a much larger version of this picture.[19] There are many variants and generalizations of networks. For example, there may be different types of nodes and/or edges, the network may be time-dynamic, i.e., evolving over time, the edges may be directed rather than undirected, etc. In what follows we restrict ourselves to static networks with one type of nodes, often referred to as individuals, and one type of (undirected) edges. We will sometimes refer to individuals being directly connected by an edge as being 'friends,' or 'acquaintances,' hinting at a possible context. A network is said to be random if its construction is specified in terms of a stochastic process.

---

[17]See, for example, M.E.J. Newman: *Networks: An Introduction*. Oxford University Press, 2010; A collection of research papers that shaped the field can be found in M. Newman, A-L. Barabási and D.J. Watts: *The Structure and Dynamics of Networks*. Princeton University Press, 2006.

[18]D.J. Watts and S.H. Strogatz: Collective dynamics of 'small-world' networks. *Nature*, **393** (1998), 409–410.

[19]For a more thorough presentation one can consult for example M.E.J. Newman: The structure and function of complex networks. *SIAM Rev.*, **45** (2003), 167–256.

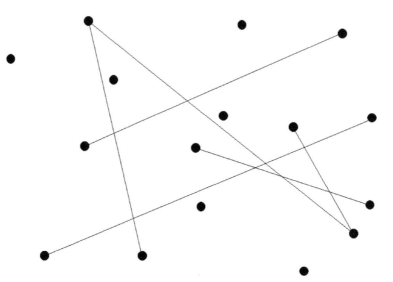

Figure 12.2: A network with nodes and edges.

Much research has focused on defining suitable stochastic models resulting in random networks having some desired 'local' properties, and to derive what global properties such networks have. One such important local property is the *degree distribution* $\{p_k; \ k \geq 0\}$ in the population (cf. the previous section), where the degree D of a randomly selected individual denotes the number of neighbors he or she has (i.e., how many individuals he or she is directly connected to), and where $P(D = k) = p_k$. There are different ways in which a network of nodes and edges can be called 'fully' connected. One extreme is to demand that each node is connected to each other node, i.e., there is an edge between any given pair of nodes. Less restrictive is a view where full connectivity means that each pair of nodes is connected by a path of edges, i.e., when each node can be reached from each other node by a path of edges.

Another local property is the *clustering coefficient*, often denoted by c. There are a few, asymptotically equivalent, definitions but they all measuring to what extent (i.e., with what probability) two friends of a given individual are also friends themselves. As a consequence, if c is close to 1 there are many triangles present in the network, whereas if c is close to 0 there are hardly any. Figure 12.3 shows two networks, one having low clustering ($c = 0$) and the other quite high clustering ($c = 9/21 = 3/7$). The clustering coefficient for the network on the right is computed as follows: there are in total 21 *pairs* of edges connected by a node, and out of these, nine form a triangle.

A third commonly studied local property measures how assortative or disassortative the network is with respect to some characteristic. A network is *assortative* (sometimes also referred to as homophilic) if individuals having the same characteristic are more likely to be connected than individuals having different characteristic, whereas the network is *disassortative* if individuals having different characteristics are more likely to be connected. To give a concrete example, consider a population of males and females (sex is hence the characteristic) and let edges correspond to

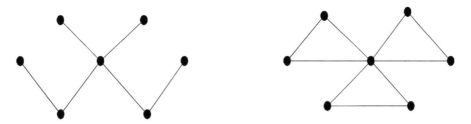

Figure 12.3: Two networks, each with seven nodes. The network on the left has $c = 0$ since there are no triangles. The network on the right has quite high clustering since there are 'many' triangles.

having a sexual relationship. Then a homosexual community (of both males and females) is highly assortative, a heterosexual community is highly disassortative, and a bisexual community lies somewhere in between. If the considered character-istic is numerical this aspect is often measured in terms of correlation: what is the correlation ρ between the numerical characteristic of two arbitrary neighbors? One particular example is the *degree correlation* ρ_D: what is the correlation between the degrees of neighbors? A positive degree correlation implies that individuals with high degree tend to have neighbors with high degree (and low-degree individuals are neighbors of other low-degree individuals), whereas a negative degree correlation corresponds to individuals with high degree typically being connected to individuals with low degree and vice versa. In Figure 12.4 we illustrate positive and negative degree correlation using two small (not fully connected) networks. The network on the left has positive degree correlation (in fact $\rho_D = 1$) since individuals with high (low) degree are neighbors with other individual having high (low) degree. The network on the right has negative degree correlation since it is the other way around: individuals with high degree are neighbors of individuals with low degree and vice versa.

Figure 12.4: Two (not fully connected) networks. The network on the left has eight individuals and positive degree correlation and the network on the right has 10 individuals and negative degree correlation.

Some other local features are more directly related to social structures, the main example being households (where all individuals of a household are pairwise connected) but sometimes also schools or workplaces are considered.

Beside local properties, networks also possess *global* properties. One such global property is referred to as *connectedness*; the relevant question is: are there only small connected components (sub-graphs in which any two vertices are connected by a path, i.e., sub-graphs that are fully connected in the weaker sense described before) or is there also a *giant* component, and in case of the latter: what fraction belongs to the giant? Another global property refers to the distance between individuals. This is often measured by the *diameter* of the network, usually defined as the distance between two randomly selected individuals. In terms of infectious disease dynamics, the size of the giant is of importance when studying the size of an epidemic outbreak, whereas the diameter is more related to the speed, or the growth rate, of the epidemic.

There are numerous different applications where random networks may be used: nodes may be web-sites and edges links between sites (in general a directed network), nodes may be some type of 'particle' and edges reflecting some type of connection between the particles, nodes may be jazz-musicians and edges reflecting if they have played together, etc. Here we focus on the situation where the network describes a population of individuals, and individuals being connected means that they have some relationship (friendship, sexual relationship, having shared the same room in a hospital, ...).

In real life the complete network of interest is quite often unknown. Then the best we can hope for is to know something about the local properties of the network. Empirical evidence indicates that most real world networks share some common properties in terms of these local features. For example, most networks tend to have very skew degree distributions, meaning that most individuals have few neighbors, but there are a few exceptional individuals having very high degree. The latter type of individual is often referred to as a 'hub' or, in our context, a super-spreader. This skewed distribution is often modeled using a so-called *power-law*, where it is assumed that for some $\alpha > 1$, $p_k \sim k^{-\alpha}$ for large k. The case where the power-law exponent α satisfies $2 < \alpha < 3$ has received special attention and is referred to as a *scale-free* degree distribution (or network). The reason why this case has received special attention is not only motivated by empirical support, but also by the fact that such networks exhibit quite different global characteristics, the reason being that the *variance* of the degree distribution is infinite whereas the mean is finite, see (12.32). Another empirical experience is that nearly all real-world networks have positive clustering ($c > 0$). With regard to how assortative networks are, experience is more mixed: some networks tend to be assortative whereas others are disassortative.

In light of the discussion in the previous paragraph a relevant question to pose is hence the following: given some specified local properties of a network, what global properties does it have? Unfortunately this is most often an ill-posed question: it is often possible to construct different networks, all having the 'correct' local properties, but having very different global properties. The common procedure is then to use probabilistic reasoning: pick a uniformly selected network among all networks having the desired local properties, and study the (random) global properties of this network.

The more local structure that is imposed, the harder it is to answer these type of questions. There are two sorts of complications: the first is to find a way to construct a graph having the desired local properties, and the second, having overcome the first, is to determine what global properties this graph has. There are at least two approaches to solving this type of problem. One, the probabilistic approach, is to define a (random) algorithmic construction of a network and to show that its local properties are the ones desired.[20] The second approach, the simulation approach, is to simulate a random network evolving in discrete time steps, where in the next time step a slightly different network is selected randomly in such a way that 'desirable' local properties are favored and undesirable properties are penalized. After having jumped around in the space of networks according to these rules, the network should be a 'typical' network having the desired properties. So called *exponential random graphs* are constructed by using this approach. An advantage with this approach, compared to a probabilistic one, is that it is easier to obtain complicated networks by penalizing/rewarding many different types of local properties, although it is not always easy to calibrate the penalizing parameters to obtain the desired local properties exactly.[21]

The networks discussed above often aim, in the epidemic context, at describing the social structure in the community. The spread of an infectious disease is then modeled 'on' the network. Starting with one *index case* (or several) this individual can infect some or all of its neighbors, according to some rule, and these individuals may in turn infect their other neighbors, and so on. The simplest rule is that an infectious individual infects each of its susceptible neighbors independently with probability p. This epidemic model can be 'realized' on the network by erasing each edge, independently, with probability $1 - p$ (and keeping the edge with probability p), a procedure called *thinning* of edges. The interpretation of the edges remaining is then that if one of the adjacent individuals gets infected, so will the other. The individuals infected at the end of the epidemic will consist of all individuals connected to the index case(s) in the edge-thinned network. As a consequence, by studying local and global properties of the edge-thinned network, it is possible to draw conclusions about the situation after the outbreak.

It is also possible to consider vaccination prior to the introduction of the disease. Assuming a perfect vaccine this means that vaccinated individuals cannot get infected nor spread the disease further. In the network this means that vaccinated individuals, and all edges connected to them, should be deleted. So, if a randomly selected community fraction v is vaccinated, this corresponds to *thinning* of *nodes* (and their adjacent edges) with probability v in the network. Clearly, there are better ways to select individuals to be vaccinated than random allocation: trying to find, and vaccinate, individuals with high degree, the super-spreaders, is more efficient than selecting people at random.[22]

In the next subsection, we describe in more detail a network model, concentrating on the degree distribution but ignoring other local structure in the network.

---

[20] An example of the probabilistic approach for constructing a random network having an arbitrary but pre-defined degree distribution is given by T. Britton, M. Deijfen and A. Martin-Löf: Generating random graphs with prescribed degree distribution. *J. Stat. Phys.*, **124** (2006), 1377–1397. A much more general (and mathematically involved) treatment is given in the two long papers B. Bollobás, S. Janson and O. Riordan: The phase transition in inhomogeneous random graphs. *Rand. Str. Alg.*, **31** (2007), 3–122; and B. Bollobás, S. Janson and O. Riordan: Sparse random graphs with clustering. *Rand. Str. Alg.*, **38** (2011), 269–323.

[21] For more information about exponential random graphs we refer for example to M. Handcock, G. Robins, T. Snijders, P. Wang and P. Pattison: Recent developments in exponential random graph ($p*$) models for social networks. *Social Networks*, **29** (2006),192–215.

[22] See T. Britton, S. Janson and A. Martin-Löf: Graphs with specified degree distributions, simple epidemics and local vaccination strategies. *Adv. Appl. Prob.*, **39** (2007), 922–948, and the references given there, for one network model considering both epidemics and various vaccination strategies.

12.6.3   A network with hardly any structure

We saw that, apart from demographic stochasticity, there are two key features that distinguish network models from standard deterministic models, namely:

- Repeated contacts between the same individuals;

- Local correlations due to short cycles in the graph.

In this subsection, we focus on a caricature of a network that displays the first of these features but hardly anything of the second. One could say that we sacrifice most of the second in order to achieve tractability of the first.

We assume that a connection is symmetric, and we shall call individuals that are connected *acquaintances* of each other. The number of acquaintances of an individual is a stochastic variable D with probability distribution $\{\mu_k\}$. That is, with probability μ_k, an individual has k acquaintances (degree k). We assume that the mean

$$E(D) := \sum_{k=1}^{\infty} k\mu_k \tag{12.29}$$

is finite.

The crucial assumption of the model is that acquaintances are a random sample of the population. By this we mean two things:

- An acquaintance of some individual x is, with probability

$$\nu_k := \frac{k\mu_k}{E(D)}, \tag{12.30}$$

   of degree k, irrespective of the degree of x itself (so the probability that an arbitrary acquaintance is of degree k is proportional to the occurrence of individuals of this type, with constant of proportionality k; the $E(D)$ in the denominator serves to make ν into a probability distribution, i.e., it serves to guarantee that the ν_k sum to one).

- The probability that acquaintances have other acquaintances in common equals zero in the limit of an infinite population.

As we shall indeed restrict our considerations to the infinite-population limit, it is this second assumption that eliminates cycles.

Having thus specified the network structure (or rather, the lack of it), we now turn to transmission. We assume that the contact intensity between two acquaintances is independent of the degrees k_1 and k_2 of these two individuals (allowing dependence would create either asymmetry or inconsistency or both). So, individuals with high k have a high overall contact intensity.

Infected individuals may differ in the amount of infectious material that they disseminate to their acquaintances. Depending on what we want to compute, we have to describe this variability in more or less detail. We will use this as a guiding principle for the order in which we present the results, the key idea being that we add detail only when needed to proceed.

Let $\bar{q}$ denote the probability that an individual picked at random will transmit the infective agent to any one of its acquaintances, when infected itself. Then

$$R_0 = \bar{q} \sum_{k=1}^{\infty} (k-1)\nu_k. \tag{12.31}$$

Note that $R_0 = \infty$ if $\sum k^2 \mu_k$ diverges.

Exercise 12.31 Explain the rationale behind the expression (12.31).

Exercise 12.32 Rewrite (12.31) to obtain

$$R_0 = \bar{q} \left(E(D) - 1 + \frac{V(D)}{E(D)} \right). \tag{12.32}$$

This should remind you of an earlier exercise. Which one?

In line with earlier conclusions (Section 2.2. Exercise 7.18, Section 12.4.2), we see that heterogeneity increases R_0. But what is the effect of the local finiteness of the network?

To focus on that, let us eliminate the heterogeneity by assuming that every individual has the same number of acquaintances $\bar{k}$ (so $\mu_k = 0$ for $k \neq \bar{k}$, and $\mu_{\bar{k}} = 1$, for some integer $\bar{k}$, and hence $E(D) = \bar{k}$). Then (12.18) reduces to

$$R_0 = \bar{q}(\bar{k} - 1). \tag{12.33}$$

The effect of repeated contacts between the same individuals is hidden in $\bar{q}$ (it will be exposed if we calculate $\bar{q}$ from a sub-model for infectivity; see below). The effect of the loop between two acquaintances (recall that we assumed the acquaintance relation to be symmetric!) is in the -1. Indeed, a derivation ignoring the network structure would argue that R_0 equals the product of the number of acquaintances $\bar{k}$ (during the infectious period, say) and the probability $\bar{q}$ of transmission, and would therefore lead to

$$R_0 = \bar{q}\bar{k}. \tag{12.34}$$

The difference is that in this second way of reasoning it is ignored that the individual that caused the infection of the individual we consider is still among the $\bar{k}$ acquaintances.

To illustrate the effect of the -1, we note that in order to have $R_0 = \bar{q}(\bar{k}-1) > 1$, we require

$$\bar{k} > 1 + \frac{1}{\bar{q}} > 2. \tag{12.35}$$

And indeed, when every individual has exactly two acquaintances (so for $\bar{k} = 2$), the 'network' is a chain. As we allow for nearest neighbor transmission only, spread in any of the two directions will stop as soon as one link fails (by chance) to transmit the infection, recall Section 10.7. Thus major outbreaks are precluded.

We now return to the case with arbitrary degree distribution $\{\mu_k\}$. Our next aim is to justify the expression $1 - s_\infty$ for the final size, where

$$s_\infty = \sum_{k=1}^{\infty} \mu_k (1 - \bar{q} + \pi_b \bar{q})^k \tag{12.36}$$

and where π_b is the unique root (which exists when we assume $R_0 > 1$) in the interval $(0,1)$ of the equation

$$\pi = g_b(\pi), \tag{12.37}$$

where

$$g_b(z) := \sum_{k=1}^{\infty} \nu_k (1 - \bar{q} + z\bar{q})^{k-1}. \tag{12.38}$$

This smells like a branching process, and that is indeed what is lurking behind it: the 'b' refers to 'backward' and we shall motivate the expression (12.36) by interpreting s_∞ as the probability of extinction of the so-called backward (or dual) branching process when starting (or, rather, ending) with a randomly chosen individual.

The idea is as follows. For each individual x we ask for each of its acquaintances y: will x be infected by y, should y itself become infected? Here we ignore the fact that x may already have been infected earlier by one of its other acquaintances. If the answer to the question is 'yes' for a given y, we call y a potential child of x. Note that this explains the name 'backward,' since we turn the parent-offspring relation, defined by 'infection,' upside down.

For this process of producing children, the distribution of type at birth is given by ν as well. Each individual begets offspring according to a binomial distribution with probability parameter $\bar{q}$ and a type-dependent parameter for the maximum number of offspring. In fact, this maximum number equals $k - 1$ for a k-degree individual, once the process is started, since one of its k acquaintances is the 'mother' of x, and therefore cannot be a 'daughter' of x as well. In other words, a potential child becomes a true child unless it actually is the parent of the individual under consideration.

Exercise 12.33 Now check that $g_b(z)$ defined in (12.38) is the generating function. Verify that $g_b'(1) = R_0$.

The root π_b of equation (12.37) therefore equals the probability of extinction for this backward process if we start with one 'typical' individual. However, our starting point is somewhat different. To begin our thought experiment, we pick one individual at random, which gives it degree k with probability μ_k (and not ν_k!). This first individual, if it is of degree k, can beget k offspring (rather than $k - 1$!). In other words, *all* of its potential children become its true children.

Exercise 12.34 Our aim is to interpret the right-hand side of (12.36) as the probability of extinction of the backward process when starting from (read: ending with) a randomly chosen individual. Finish the argument.

It remains to relate this probability of extinction to the final size s_∞. To do so, we imagine that a large outbreak has occurred. The collection of victims then forms a giant component of the graph.[23] Pick an arbitrary individual and ask what the probability is that it belongs to the giant component. By definition, this probability is $1 - s_\infty$. The key observation now is that the individual belongs to the giant component if and only if the backward branching process starting from this individual does *not* go extinct. Or, equivalently, the individual does not belong to the giant component if and only if the backward branching process starting from this individual does go extinct. The point is that only members of the giant component have an infinite 'family' tree. Therefore s_∞ equals the probability of extinction and we have arrived at (12.36). We have deliberately kept this derivation somewhat heuristic, but it can be made precise.[24]

---

[23]B. Bollobás: *Random Graphs*, 2nd edition. Cambridge University Press, 2001.

[24]O. Diekmann, M.C.M. de Jong and J.A.J. Metz: A deterministic epidemic model taking account of repeated contacts between the same individuals.*J. Appl. Prob.*, **35** (1998), 448–462; H. Andersson: Epidemic in a population with social structures. *Math. Biosci.*, **140** (1997), 79–84; H. Andersson: Limit theorems for a random graph epidemic model. *Ann. Appl. Prob.*, **8** (1998), 1331–1349.

We have characterized R_0 by (12.31) and s_∞ by (12.36). Implicitly this defines how these two quantities are related to each other. We next make one aspect of this relationship explicit.

Exercise 12.35 Show that for R_0 only slightly bigger than 1, we have

$$1 - s_\infty = \frac{2E(D)}{\bar{q} \sum_{k=3}^{\infty}(k-1)(k-2)\nu_k}(R_0 - 1) + \text{h.o.t.} \tag{12.39}$$

where 'h.o.t.' again denotes 'higher order terms.' Hint: Assume $\pi_b = 1 - \eta$ with η small. First expand the identity $\pi_b = g_b(\pi_b)$ in powers of η; next do the same with the expression (12.36) for s_∞ and, finally, combine these results to eliminate η from the lowest-order terms.

Now recall, from Exercise 1.20-ii, that for the standard final-size equation we have

$$1 - s_\infty = 2(R_0 - 1) + \text{h.o.t}$$

So the difference resides in the factor

$$\frac{E(D)}{\bar{q} \sum_{k=3}^{\infty}(k-1)(k-2)\nu_k}.$$

To investigate this factor, we employ a (rough) approximation: we replace $k - 2$ in the denominator by $E(D) - 2$ (note that this is exact when μ is concentrated in one point, cf. the text preceding (12.33), while extending the sum to include $k = 2$. Using the expression (12.31) for R_0 and our assumption that R_0 is only slightly bigger than one, we find that the factor is approximately equal to

$$\frac{E(D)}{E(D) - 2}.$$

For relatively small values of $E(D)$, this is substantially bigger than one. Our conclusion is therefore that the 'steepness' of the relation between $1 - s_\infty$ and $R_0 - 1$ is much enhanced by the network structure. In other words, the all $(R_0 > 1)$ or nothing $(R_0 < 1)$ dichotomy, which in general is an exaggeration, is closer to the truth in a network setting than in a situation with random mixing.

It remains to characterize the real-time growth rate r and the probability of a minor outbreak. To do so, however, we need to be more specific about the sub-model for infectivity. Assume that the dissemination of infectious material is determined by a stochastic variable ξ, the distribution of which is given by a probability measure m on a set Ω. If we restrict attention to a generation perspective, as we do when calculating the probability of a minor outbreak, we only need to specify the overall probability of transmission $q(\xi)$. The quantity $\bar{q}$, introduced earlier, is related to $q(\xi)$ through the formula

$$\bar{q} = \int_\Omega q(\xi)m(d\xi). \tag{12.40}$$

Again we shall consider a branching process, but now the *forward* process (hence the index f for π and g in Exercise 12.36) in which the parent-offspring relation corresponds to true transmission of infection. Hence, if we pick an individual whose infectivity is characterized by ξ, all its susceptible acquaintances experience that infectivity. (Note the difference with the backward branching process, where it was the infectivity 'at the other end of the edge' that mattered; in that case we have independence, whereas now, when looking forward, we have dependence.)

Exercise 12.36 Show that the probability of a minor outbreak equals

$$\sum_{k=1}^{\infty} \mu_k \int_{\Omega} \left(1 - q(\xi) + \pi_f q(\xi)\right)^k m(d\xi),$$

where, for $R_0 > 1$, π_f is the unique root in the interval $(0, 1)$ of the equation

$$\pi = g_f(\pi),$$

with

$$g_f(z) := \sum_{k=1}^{\infty} \nu_k \int_{\Omega} \left(1 - q(\xi) + z q(\xi)\right)^{k-1} m(d\xi).$$

Exercise 12.37 Check that $g_f'(1) = R_0$.

Before turning to the equation for r, we focus on the sub-model for infectivity. As an example, consider the case of an infectious period that is exponentially distributed with parameter α. Assume that within this period each susceptible acquaintance has probability β per unit of time of becoming infected. This implies that

$$q(\xi) = 1 - e^{-\beta\xi}. \tag{12.41}$$

To calculate $\bar{q}$, we average $q(\xi)$ with respect to the measure m, which has density $\alpha e^{-\alpha\xi}$ (compare Exercise 1.11):

$$\bar{q} = \int_0^{\infty} (1 - e^{-\beta\xi}) \alpha e^{-\alpha\xi} \, d\xi = 1 - \frac{\alpha}{\alpha + \beta} = \frac{\beta}{\alpha + \beta}. \tag{12.42}$$

It is now possible to determine the effect of the repetition of contacts. If we disregard the reduction in candidate victims due to the local network structure and focus once more on the special case that every individual has the same number $\bar{k}$ acquaintances, our calculation yields (see (12.34))

$$R_0 = \bar{q}\bar{k} = \frac{\beta\bar{k}}{\alpha + \beta}, \tag{12.43}$$

whereas a calculation that, in addition, disregards repetitions of contacts would give R_0 as the product of the rate $\beta\bar{k}$ of making effective contacts and the expected length $1/\alpha$ of the infectious period; i.e., it would yield

$$R_0 = \frac{\beta\bar{k}}{\alpha}. \tag{12.44}$$

More generally one should replace $\bar{k}$ in (12.44) and (12.45) by $\sum k\nu_k$. We conclude that allowing for repetition reduces R_0 by a factor $\alpha/(\alpha + \beta)$.

For general considerations, it is helpful to introduce $\mathcal{F}(\tau, \xi)$, the probability that a susceptible acquaintance of a type-ξ infectious individual is still uninfected at time τ after the start of the infectious period.

Exercise 12.38 As an alternative sub-model for infectivity, assume that the infectious period lasts from $\tau = 1$ to $\tau = \xi$, where ξ is uniformly distributed over the interval $[2, 3]$. Calculate $q(\xi)$ and $\bar{q}$. Determine the reduction factor of R_0 due to the repetition of contacts.

Exercise 12.39 Returning to the exponentially distributed infectious period, calculate the mean $\overline{\mathcal{F}}(\tau)$ of $\mathcal{F}(\tau, \xi)$ with respect to ξ.

Exercise 12.40 Explain that the initial real-time growth rate r is determined by the equation

$$1 = \sum_{k=1}^{\infty} \nu_k (k-1) \int_0^\infty e^{-r\tau} \frac{d}{d\tau} \big(1 - \overline{\mathcal{F}}(\tau) d\tau\big). \qquad (12.45)$$

We conclude that, for this particular network with hardly any structure, one can calculate (or at least characterize) all indicators of infectivity for a demographically closed host population. In particular, one can disentangle and quantify the effects of repeated contacts on the one hand, and of reduction of candidate victims on the other hand. Reluctantly, we admit that this is an exceptional situation and that, as a rule, network models are (as yet) not very amenable to mathematical analysis.

12.6.4   An STD in a network with hardly any structure

The aim of this set of exercises is to model the spread of a sexually transmitted infection within a heterosexual community. So the main difference with the preceding subsection is that now we consider individuals that are not only characterized by their degree, but also by their sex. We will focus on modeling the initial phase of an outbreak by means of a branching process approximation of an epidemic on a network. For further reading we refer to Britton et al. (2007).[25]

Consider an adult population consisting of N_f females and N_m males and let $N = N_f + N_m$ denote the total population size. Suppose further that the number of sex-partners that a male has, during a period corresponding to the length of a typical infectious period, is distributed according to $\{\pi_i^{(m)}; i \geq 0\}$, and let $\{\pi_i^{(f)}; i \geq 0\}$ denote the corresponding distribution for females (see Figure 12.5 for an illustration). The choice of partners is completely random in that males and females choose partners randomly on 'the market' of available potential partners. As a consequence, a male chooses a female having i partners with a probability proportional to $i\pi_i^{(f)}$, and a female chooses a male having j partners with probability proportional to $j\pi_j^{(m)}$. We denote these *size-biased* distributions by $\{\tilde{\pi}_i^{(f)}\}$ and $\{\tilde{\pi}_j^{(m)}\}$. These are given by

$$\tilde{\pi}_i^{(f)} = \frac{i\pi_i^{(f)}}{\sum_r r\pi_r^{(f)}} \quad \text{and} \quad \tilde{\pi}_j^{(m)} = \frac{j\pi_j^{(m)}}{\sum_r r\pi_r^{(m)}}.$$

Exercise 12.41 The total number of female partners that males have must equal the total number of male partners that females have. What relation must therefore hold if there are equally many males and females, $N_f = N_m$? (See Section 12.3 for discussion of the consistency issue.)

---

[25]T. Britton, M.K. Nordvik and F. Liljeros: Modelling sexually transmitted infections: the effect of partnership activity and number of partners on R_0. *Theor. Pop. Biol.*, **72** (2007), 389–399.

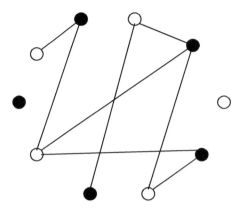

Figure 12.5: An illustration of a heterosexual network. Men are represented by black and women by white circles, and edges signify a sexual relationship. The empirical distribution of sexual partners for females is: $\pi_0^{(f)} = 0.2$, $\pi_1^{(f)} = 0.2$, $\pi_2^{(f)} = 0.4$, $\pi_3^{(f)} = 0.2$, and the same for men.

We now define a model for the spread of a sexually transmitted disease (STD) on this heterosexual network. A male that gets infected spreads the disease to each of his not yet infected (female) partners independently with probability $p^{(fm)}$. Similarly, a female who gets infected spreads the infection to each of her not yet infected (male) partners independently with probability $p^{(mf)}$. Assume that the epidemic is initiated by one randomly selected male individual being infected, the index case.

The male index case will randomly infect a fraction of his partners. Those females who get infected will in turn infect random numbers of their partners, however they will of course not infect the index case since he was already infected. In a large community, which is what we are considering, it is very unlikely that some of these females, by chance, should share other partners. Because of this, females infect new individuals more or less independently of each other, and the same holds for men in the generation thereafter, and so on. This motivates that the initial phase of the epidemic can be approximated by a suitable branching process; the question is which branching process.

Exercise 12.42 i) Each female that the index case infects has a random number $\tilde{X}_f$ of male partners. What distribution does such a random variable have?

ii) Let X_f denote the number of male partners a randomly selected female has. Then X_f has distribution $P(X_f = i) = \pi_i^{(f)}$, which has mean $E(X_f) = \sum_i i\pi_i^{(f)}$, second moment $E(X_f^2) = \sum_i i^2\pi_i^{(f)}$, and variance $V(X_f) = E(X_f^2) - (E(X_f))^2$. Express the mean $E(\tilde{X}_f)$ using these quantities.

iii) Each male that is infected by one of these females has a random number $\tilde{X}_m$ of female partners. What distribution does this random variable have? Express its mean using $E(X_m)$ and $V(X_m)$.

Exercise 12.43 i) Given that a female has i partners, she infects each of them independently with probability $p^{(mf)}$, except for the male she was infected by herself. Conditional on that $\tilde{X}_f = i$, what is the distribution of $Y_f =$ the number of males she will infect? What is the expected number she will infect, still conditioning on $\tilde{X}_f = i$?

ii) Given that a male who was infected by one of these females has i partners, he infects each of them independently with probability $p^{(fm)}$, except for the female he was infected by himself. So, conditional on that $\tilde{X}_m = i$, what is the distribution of $Y_m =$ the number of females he will infect? What is the expected number he will infect, still conditioning on $\tilde{X}_m = i$?

Exercise 12.44 In the two previous exercises the expected number of partners was derived, as was the expected number an individual will infect, given the number of partners. Combine these two results to compute:

i) The (unconditional) expected number of males a female will infect, $E(Y_f)$.

ii) The (unconditional) expected number of females a male will infect, $E(Y_m)$.

There are two types of individuals, males and females, each 'giving birth to' (i.e., infecting) new individuals more or less independently. This motivates that the network epidemic model just defined can be approximated by a two-type branching process when the population is large. The approximation relies on that, during the early stages, it is very unlikely that an infected is partner with someone already infected beside the person he/she was infected by, an approximation which can be made precise for the present model.

The next-generation matrix (here the matrix giving the mean offspring in the multi-type branching process) is given by

$$K = \begin{pmatrix} 0 & E(Y_m) \\ E(Y_f) & 0 \end{pmatrix}.$$

As before, the largest eigenvalue R_0 of K determines if a major outbreak is possible.

Exercise 12.45 Compute the largest eigenvalue R_0 of K using the expressions derived for $E(Y_m)$ and $E(Y_f)$ in the previous exercise.

If we study the basic reproduction number and treat the average number of male and female partners, as well as the transmission parameters, as fixed, it is seen that the basic reproduction number increases with the variances of the number of partners. This means that the more skewed the partnership distributions are, the more vulnerable the community is for invasion by sexually transmitted infections. In particular, if the partnerships follow a scale-free distribution with tail probabilities of the order $\pi_i^{(f)} \sim i^{-\alpha}$ and/or $\pi_i^{(m)} \sim i^{-\alpha}$, where $2 < \alpha < 3$, then the partnership distribution has finite mean but infinite variance, implying that $R_0 = \infty$. The fact that $R_0 = \infty$ has drastic consequences in terms of vaccination, as we shall see in the next exercise.

Exercise 12.46 Suppose that, prior to the arrival of the disease, a proportion v of randomly selected males and females are vaccinated with a vaccine giving 100% immunity. Similar to before, the initial phase of the epidemic can be approximated by a branching process. The difference is that, given that an infected male has i partners he will infect $Bin(i-1, p^{(fm)}(1-v))$ of them.

The probability to infect a partner (beside the partner he was infected by) is now $p^{(fm)}(1-v)$ since an infectious contact has to occur *and* the partner must be unvaccinated. The same change happens with the number of males a female infects.

i) Compute R_v, the reproduction number after vaccinating a fraction v uniformly in the community.

ii) The critical vaccination coverage v_c is defined as the proportion v necessary to vaccinate in order to reduce R_v to below the critical value of 1. Determine v_c for the case where the average number of partners for both males and females is 3 and the transmission parameters are both equal to 0.8, but for three different assumptions about the variance of the partnership distribution. First assume that both partnership distributions have variance 0, then variance 3, and finally infinite variance.

Of course, a better vaccination strategy would be to vaccinate individuals having many partners, i.e., promiscuous individuals. It is not likely that authorities have information about the number of partners individuals have. Trying to reach core groups when aiming at reducing the spread of STDs is, nevertheless, common. If no such 'global' information is available, vaccination can still be improved as the following toy model illustrates. The model is inspired by contact tracing and goes under the name 'acquaintance vaccination.' The model suggests that individuals should be picked at random. But, instead of vaccinating them, the strategy suggests that a randomly selected partner of the chosen individual should be vaccinated. By doing this it is possible to show that the distribution of the number of partners of vaccinated individuals follows the size biased distribution $\tilde{X}_m$ and $\tilde{X}_f$ rather than the community distributions of partnerships X_m and X_f. By performing this type of vaccination the critical vaccination coverage v_c is reduced, in particular when the original partnership distributions have heavy tails. For a detailed study of this vaccination strategy in a one-sex community, also studying outbreak probabilities and outbreak sizes, we refer to Britton et al. (2007).[26]

12.7   A PRIMER ON PAIR APPROXIMATION

Except in the previous section, we have often assumed that individuals have contacts according to the law of mass action, adapted from chemistry. This implies that we assume that individuals 'bump into' contact individuals randomly and that, if we neglect spatial structure, they have equal probability to bump into any specific other individual. There has been substantial advance in approaches that allow one to relax this assumption and to take into account that not all pairs of contact individuals are equally likely,[27] mainly concentrating on approaches involving spatial or social networks of individuals that prescribe which contact pairs are possible. Physical space or social space (e.g., a set of acquaintance relationships) make that some contact pairs are more likely than others. Rather than providing a detailed account of possibilities, we here concentrate on an extension of mass action that

[26]T. Britton, S. Janson and A. Martin-Löf: Graphs with specified degree distributions, simple epidemics and local vaccination strategies. *Adv. Appl. Prob.*, **39** (2007), 922–948.

[27]For a recent overview see M.J. Keeling: Extensions to mass action mixing. In: *Ecological Paradigms Lost*, K. Cuddington and B. Beisner (eds.), Elsevier, 2005, pp. 107–142, and the book by Keeling and Rohani, 2008.

takes some of the spatial or social correlation between individuals into account: looking at pairs of individuals, as an intermediate between regarding only single individuals (and looking at the 'mean field') and explicitly modeling the entire network structure (as we did in Section 12.6 for some simple networks). We refer to the exposition in the book by Keeling and Rohani (2008, Section 7.8) for more details and references to the recent literature. The idea of taking low-order correlations between individuals into account was introduced and formulated as a theory for use in biology independently by Matsuda et al. (1992) and by David Rand (1999).[28]

For clarity of notation we denote by $[S]$ and $[I]$, the number of susceptible and infectious individuals, respectively, and by N the total number of individuals. In Chapter 1, under the mass action assumption, we argued that the incidence (the appearance of new infecteds) at a given point in time, is proportional to the number of contacts between a susceptible and an infectious individual, i.e., to the number of contact pairs consisting of an S and an I individual at that time. Previously we modeled this contact rate as simply being proportional to $[S][I]$. Now, however, we recognize SI-pairs as a separate variable, denoted by $[SI]$, and say that the transmission rate is proportional to $[SI]$. The basic SIR-type system, such as for example the system in Exercise 1.28, now reads

$$\frac{d[S]}{dt} = -\theta[SI],$$
$$\frac{d[I]}{dt} = \theta[SI] - \alpha[I].$$

Because $[SI]$ is a variable we have to specify a differential equation that governs its change. First let us reflect on the differences between these pairs and the pair-formation process we discussed in Section 12.2. In both cases pairs are formed and dissolved at certain rates and pairs exist for definite amounts of time, and while being in a pair potentially infectious contacts take place between the individuals involved at a certain rate θ. In the case of Section 12.2 the network structure is completely ignored: there are only single individuals and pairs. While an individual is part of a pair it has no contact with other individuals, but when it is single pairing up with any given other single in the population is equally likely for all singles (mass action contacts). In contrast: in the present section we view individuals as being part of some fixed network. As before, individuals are the nodes of the network and an edge linking two individuals signifies that these individuals are acquaintances of each other and have the type of contacts relevant for transmission. An individual can therefore be part of numerous pairs, for example in a star-like arrangement where the individual has exactly k, say, acquaintances. The rate of transmission, written as θ above, is a function of the number of contacts per unit of time across an edge (and the success probability per contact). We use a different symbol in this section than the β and γ used earlier to highlight the different interpretation. If each individual is part of exactly k edges, the total number of infectious contacts per unit of time for an individual is $k\theta$.

The network is considered to be fixed, it is only the infection status of the individuals that can change. Instead of describing the network in detail, i.e., specifying precisely which individual has contact to which other individuals, we take the structure as given and only give a coarse summary of network statistics: total number

---

[28]H. Matsuda, N. Ogita, A. Sasaki and K. Sato: Statistical mechanics of populations — the lattice Lotka-Volterra model. *Prog. Theor. Phys.*, **88** (1992), 1035–1049; D. Rand in J. McGlade (ed.) *Advanced Ecological Theory*. Blackwell, 1999, pp. 99–143.

of individuals, number of pairs, number of triplets, quadruplets, etc. We note that
in the literature the variables $[S]$ and $[I]$ are sometimes referred to as 'singles,' but
this is incorrect and confusing: single individuals are not relevant in the static network. What is meant is that there are $[S]$ S-individuals in the network, involved in
various pairs, triplets, quadruplets etc.

Exercise 12.47 Given that the labels S and I are distributed randomly over individuals and that each individual is connected to exactly k acquaintances,
argue that the number of $[SI]$ pairs is

$$[SI] = k\frac{[S][I]}{N}.$$

To specify an equation for the change in $[SI]$ we concentrate on the case of SIR-
type dynamics. We envisage several different potential changes: i) in an existing
$[SS]$ pair, one of the two individuals involved can become infected (through a
contact by one of the S-individuals with an infectious individual that it is also paired
with); ii) in an existing $[SI]$ pair the I-individual recovers and becomes immune;
iii) in an existing $[SI]$ pair the S-individual becomes infected by the particular
I-individual of the pair we are looking at; iv) as in iii) but now the S-individual
becomes infected by another I-individual that this S-individual happens to be paired
with. Note that two of these four changes are brought about by individuals outside
the pair. This means that in order to describe this change we need to know how
many triples there are of the form $[SSI]$ and $[ISI]$ and how these numbers change.
This sets in motion a cascade of dependencies: the equations for triples will depend
on the number of quadruples, etc.

Our aim is to take one step, not to climb an infinite staircase. We therefore limit
the description to the pairs only. Clearly, we then have to make approximations.
The so-called moment-closure approximation (or pair approximation, terminology
introduced by David Rand; a technique adapted from statistical physics) allows one
to make this precise. For this we ignore that the network has triangles or higher-
order clusters. In other words: in a triple it is considered extremely unlikely that the
two partners of the common central individual are also each other's acquaintance.
This may of course be unrealistic for many types of social contact, but is an attempt
to improve upon the mean-field approximation, in which every aspect of the network
structure is ignored.

Exercise 12.48 Assume that edges between individuals have been formed randomly. Argue that, in the absence of clusters and in the case where everyone
has exactly k acquaintances, the number of triples $[XYZ]$ is

$$[XYZ] = \frac{k-1}{k}\frac{[XY][YZ]}{[Y]}.$$

If we do not assume random choice of partners, the same formula for the number
of triples holds as a first approximation. The idea is to explicitly take correlation
between both X and Y into account by writing the number of Y acquaintances of
a given X-individual i as the value expected under random partnering $[XY]/[X]$
plus an error-term. For the latter one then specifies a distribution and one can approximate $[XYZ]$ as above. As an extension it is possible to soften the assumption

that the network is 'triangle free' (see Keeling and Rand): when ϕ is defined as the number of triangles in the network divided by the number of triplets, then

$$[XYZ] \approx \frac{k-1}{k} \frac{[XY][YZ]}{[Y]} ((1-\phi) + \phi \frac{N}{k} \frac{[XZ]}{[X][Z]}).$$

Note that for the 'mean-field case' of mass action contacts, N is large, and $k \to N$, and $\phi = 1$ (because every individual can have contact to each other individual and every triplet is therefore a triangle). If we do not allow any clusters in our hypothetical network, one has that $\phi = 0$ because triangles are absent, and we recover the formula given in Exercise 12.48.

Exercise 12.49 Show that the system of differential equations for the simple SIR model in the case of a cluster-free network, taking into account pairs of individuals and using the pair approximation from Exercise 12.48, is given by

$$\frac{d[S]}{dt} = -\theta[SI],$$

$$\frac{d[I]}{dt} = \theta[SI] - \alpha[I],$$

$$\frac{d[SI]}{dt} = \frac{\theta[SI]}{k[S]}((k-1)[SS] - (k-1)[SI] - k[S]) - \alpha[SI],$$

$$\frac{d[SS]}{dt} = -2\frac{k-1}{k} \frac{\theta[SI][SS]}{[S]},$$

$$\frac{d[II]}{dt} = 2\theta(\frac{k-1}{k} \frac{[SI]^2}{[S]} + [SI]) - 2\alpha[II].$$

Note that we can dispense with providing equations for the remaining variables $[SR], [IR]$ and $[RR]$ by making use of the bookkeeping equalities such as $[SS] + [SI] + [SR] = k[S]$.

The nice thing about the system of differential equations in Exercise 12.49 — as compared to the standard SIR-system on the one hand, and actual network approaches such as described in Section 12.6 on the other — is that this system lends itself to analytic exploration, while at the same time incorporating more structure than the mean-field approach (in the sense that some of the correlations between individuals are taken into account). Numerical comparison, see the Keeling references in this Section, between the mean-field description, the full random network model (through stochastic simulation) and the 'paired' model in Exercise 12.49, shows that the 'paired' model describes the mean of the stochastic simulations well.

Exercise 12.50 What do you expect, broadly, to be the differences in epidemic dynamics between outbreaks generated by the standard SIR model and the 'paired' counterpart given in Exercise 12.49?

Part III

Case studies on inference

Chapter Thirteen

Estimators of R_0 derived from mechanistic models

13.1 INTRODUCTION

Up to now we derived, occasionally and non-systematically, formulas expressing reproduction numbers in terms of observable/measurable quantities, on the basis of various assumptions concerning demography, contact and transmission. In the present chapter we want to provide a modest selection of tools for the estimation of the basic reproduction number from data. We shall do so by providing (slightly systematically) a, by no means exhaustive, review of the literature. The estimation of reproduction numbers is important, as these play a major role, for example, in gauging outbreak potential, and in public health decisions on prevention and control effort.

There are, at least, two steps involved in the estimation: the first is to derive an *estimator*, usually from mechanistic considerations; the second is to obtain an *estimate* (i.e., a numerical value for the estimator) from the available data. We concentrate on the former, but will provide a few examples of the latter. We will denote estimators for R_0 by $\hat{R}_0$. Note that there is a third step: to determine how reliable the estimate is. We will not address this issue; there is no independent 'gold standard' to determine the 'true value' of R_0 in any given situation. The reliability does not only depend on the quality (and quantity) of the data, but also on the degree of compatibility between the model and reality. How important the discrepancy between the estimated and the 'real' or 'true' quantity is, depends on the goal. For example, in transmission experiments for the evaluation of the ability of a vaccine to significantly reduce transmission of an agent (on top of protecting the individual host against clinical manifestation of the infection), one generally is concerned only with establishing whether or not $R_v < 1$ in the vaccinated population.

Trying to estimate R_0 for an infectious agent during its very first (major) outbreak, is very different from estimating R_0 for an infectious agent for which one has a large body of knowledge about infection and transmission mechanisms, as well as good estimates of basic parameters such as incubation period, latency period, infectious period, and their variability. This is different again from estimating R_0 for an infectious agent for which many detailed records of previous outbreaks are available.

The method one can use for inference, depends on the available data. If one has data from an endemic infection, it is probable that stochastic effects can be ignored (owing to the law of large numbers), and that therefore methods based on deterministic models can be used (see for example the section on age-structure below). This is in contrast to a situation where one only has data from the initial phase of an outbreak.

In this chapter we concentrate on three topics: i) generalizing estimators presented earlier in the book based on final-size epidemic data or age-structured endemic data; ii) controlled (transmission) experiments performed to estimate transmission potential and duration of infectivity; iii) the infectious agent is emerging, either for the first time or for the first time in a particular host population, so very little is known and outbreak data are becoming available 'in real time,' as the epidemic progresses.

We will not concern ourselves with situations where detailed biological and epidemiological knowledge is available, allowing for both the formulation and the parametrization of appropriate mechanistic models. In the earlier chapters of the book we have described in detail how one should derive expressions for R_0 in such cases, and that this will lead to numerical estimates of the dominant eigenvalue of the next-generation matrix (in the case of a finite number of states-at-infection) if parameter values are prescribed. Of course one never knows the exact value of a parameter (note the philosophical difficulty that a parameter obtains its meaning by way of a model, which is in turn an abstraction of reality; so in order to speak about the 'exact value,' we need to *assume* that the model yields an exact description of the real system). Also, values of which one is reasonably certain in a given population may deviate substantially from this 'exact' value in another population where the same infectious agent is studied. In such cases it seems sensible to postulate a suitable range for the values of each parameter and prior distributions across those ranges. Of course additional information should help in choosing a suitable distribution: for example, a lower limit, a (finite) upper limit, and a mean (or a most likely) value. One can next do so-called Latin hypercube sampling to arrive at a range (distribution) for R_0, factoring in the uncertainties in the values of parameters.[1] This uncertainty analysis is straightforward in that one repeatedly computes the dominant eigenvalue of the matrix K, each time choosing the value for each parameter at random from the prescribed range (according to the specified probability distribution). It is akin to a Bayesian approach to estimation where prior information about the distribution of the parameters is postulated to derive a posterior distribution of a quantity of interest (for example by using Markov chain Monte Carlo simulation, see Chapter 15 for a brief introduction).

We end this introductory section with a number of a priori difficulties, that are good to be aware of and keep in mind when estimating R_0 from data. Several of these difficulties are a direct consequence of the definition of the basic reproduction number.

Exercise 13.1 Give a number of reasons, linked to the way R_0 is defined, why deriving an accurate estimate is problematic.

In addition to the above, there are other difficulties that need to be overcome related to what data can be obtained. Data concerning time of onset of symptoms, number of contacts, infectiousness, etc, relate to individuals and can be subject to bias on several levels.

Exercise 13.2 Give a number of ways in which variability in epidemiological characteristics of individuals can cause difficulties with estimates.

We conclude that, for general reasons, there is not necessarily close agreement between the 'real' value of R_0 for a particular host-agent-environment system and its

---

[1]See M.A. Sanchez and S.M. Blower: Uncertainty and sensitivity analysis of the basic reproductive rate: Tuberculosis as an example. *Am. J. of Epid.*, **145** (1997), 1127–1137, and the references given there.

estimate, obtained by whatever method. Specific estimation methods are, moreover, often derived from simplifying assumptions. For example, many methods assume that contact between any two individuals in a population is equally likely (and in fact the individuals are assumed to be epidemiologically identical as well), i.e., contacts are according to mass action. We have encountered a number of simple estimation formulas earlier in the book — all based on the assumption of mass-action contacts. The simplest case, encountered in Exercise 4.2, uses only the fraction $s^* = \overline{S}/N$ of the population that is susceptible in a situation where the infectious agent is endemic (i.e., is present in a constant, positive, fraction of the population, and the force of infection is constant):

$$\hat{R}_0 = \frac{1}{s^*}.$$

13.2 FINAL SIZE AND AGE-STRUCTURED DATA

13.2.1 Introduction

Earlier in the book we described estimators of R_0 that use primarily epidemic information, notably the number (or fraction) of susceptibles at the end of an outbreak. First of all we encountered the final-size equation in Section 1.3, in its simplest form given by equation (1.11), assuming that all individuals are susceptible when the outbreak starts:

$$\hat{R}_0 = \frac{-\ln s(\infty)}{1 - s(\infty)}$$

where $s(\infty) = S(\infty)/N$, and where '∞' indicates measuring S at the end of the outbreak. We can equally easily relate $\hat{R}_0$ to an estimate of both the fraction of a population that is susceptible to a given infectious agent at the start of an outbreak and the same fraction after the outbreak has run its course (such estimates are often obtained through serological survey)

$$\hat{R}_0 = \frac{\ln s(0) - \ln s(\infty)}{1 - s(\infty)},$$

which we also encountered in Section 1.3, equation (1.13). (To emphasize that such estimators are based on various assumptions, we recall Exercise 1.26 about a fatal infection that spreads among hosts living in herds.)

We have seen in Section 8.3 that the corresponding final-size relation in structured populations involves the integral kernel $\int_0^\infty A(\tau, x, \eta)d\tau$ rather than just the scalar R_0. For the purpose of estimating R_0 these structured relations are not very helpful because they require substantially more detailed information, for example where the stratified fraction susceptibles and details about the infectivity and susceptibility are concerned. Even in the case of separable mixing (Section 8.4) there does not seem to be a way to express the unknown R_0, as characterized in Exercise 8.24 and 8.26, in terms of an observed $w(\infty)$ that satisfies equation (8.33).

The simplest case of an estimator that uses endemic 'age-structured' data was encountered in Exercise 4.6, and the discussion following from it in Section 4.2. In that case the force of infection is assumed to be constant and one expresses $\hat{R}_0$ as the ratio between the average life expectancy L and the mean age at infection $\bar{a}$:

$$\hat{R}_0 \approx \frac{L}{\bar{a}},$$

It will be shown in Exercise 13.7 that this relation holds in the case of an exponentially distributed life expectancy (i.e., a constant per capita rate of death, often referred to as Type II mortality), and in the case of a fixed life-length (i.e., with survival described by a Heaviside function, often referred to as Type I mortality).

In the following subsections we will first derive additional estimators based on final-size data, and then focus on estimators that use age-structured data as input.

13.2.2   Final-size data from small host populations

For small populations there are a number of methods to estimate R_0. The most well-known method uses only the final-size information. This method was introduced by Becker (1989). We have shown in Section 5.4.3 how to derive, by using martingale methods, the estimator

$$\hat{R}_0^{mrt} = \frac{N}{Y+1} \sum_{k=N-Y+1}^{N} \frac{1}{k}$$

where N and Y are the initial number of susceptibles and the final number that were infected during the outbreak (plus the initial infected individual), respectively (in Section 5.4.3 also a standard error is derived for this estimator).

A frequently encountered situation is that one deals with many different small populations that are somewhat dependent. Think, for example, of situations where individuals form many, but small, households within which contacts are more intensive than contacts with members from other households (see Andersson and Britton, 2000).[2] We have treated household models in Section 12.5.

Another situation where estimates need to be obtained from small populations is that of transmission experiments for animal infections. Typically the set up is that a small number of inoculated animals, five say, are housed as one group with a small number of susceptible contact animals. These animals are then monitored for signs of infection and infectivity for a number of days and at some specified end point the final size of the 'outbreak' is determined. Information then includes not only the final number of contact-infected animals, but also details of the time evolution of infection of all individuals. We will return to this setting in Section 13.3 below. A method that uses only the final-size data of such a transmission experiment is based on maximum likelihood estimation (see Section 5.2) and was introduced by Kroese and de Jong.[3]

13.2.3   Estimate from multi-type final-size distribution

In Exercise 7.7 we derived, for an example with two types of individuals, estimates for the minimum and maximum value of R_0 as a result of variation in the values of two parameters. The parameters governed the partitioning of contacts by the two types over both types. We did this by directly analyzing the explicit expression of

---

[2]See also the papers F. Ball: A unified approach to the distribution of total size and total area under the trajectory of infectives in epidemic models. *Adv. App. Prob.*, **18** (1986), 289–310; and F. Ball and P.D. O'Neill: The distribution of general final state random variables for stochastic epidemic models. *J. Appl. Prob.*, **36** (1999), 473–491.

[3]A.H. Kroese and M.C.M. de Jong: Design and analysis of transmission experiments. In: *Proceedings Society for Veterinary Epidemiology and Preventive Medicine Conference.* F. Menzies and S. Reid (eds.)Noordwijkerhout 2001, pp. 21–37. For this and other methods, for example using MCMC, see also the paper by M. Höhle, E. Jørgenson and P.D. O'Neill: Inference in disease transmission experiments by using stochastic models. *J. R. Stat. Soc. C*, **54** (2005), 349–366.

the dominant eigenvalue of the next-generation matrix. In this subsection[4] we aim to provide minimum and maximum values of R_0 derived from the next-generation matrix and estimated from type-stratified final-size data, which allow study of these partitioning questions for an arbitrary number of types. Similar questions were studied, but not from the point of view of estimating R_0, in Section 12.5.

Let there be n types of individuals in a population of known size N, and let π_i be the fraction of the population that is of type i. A fraction s_i of the subpopulation of type i is assumed to be susceptible at the time of introduction of the infectious agent (the individuals in the remaining fraction $1 - s_i$ having prior immunity). Neglect of information on prior immunity will lead to underestimates of R_0.

Let $p_1, \ldots, p_n$ be the final proportions infected after the outbreak, as a fraction of the originally susceptible individuals in the group. More precisely, let there be $(1 - p_i)s_i\pi_i N$ susceptibles of type i at the end of the outbreak, $i = 1, \ldots, n$.

Exercise 13.3 Regard the equation for the final size in a structured population, given in Exercise 8.24, equation (8.24). Take as state space (in which both x and η assume their values) the finite set $\{1, \ldots, n\}$. Translate the final-size relation (8.24) into the symbolic language used above, letting x correspond to i, and η to j, to derive

$$1 - p_i = \exp(-\sum_{j=1}^{n} \pi_j s_j p_j k_{ij}). \tag{13.1}$$

As in Subsection 12.5.4 one can show that the random final proportions of infecteds $\{\widetilde{p}_i\}$ converge in probability to $\{p_i\}$ as population size $N \to \infty$, where $\{p_i\}$ is the unique positive solution of (13.1). The system of equations expresses that, in the limit, the probability to escape infection equals the probability of escaping infection from the total infection pressure generated in the outbreak. For a single type $\pi_1 = 1$ and $k_{11} = R_0$, so we obtain the relation $1 - p = \exp(-spR_0)$, which we have encountered several times in the book (see Section 1.2.2, Section 1.3.1, Section 3.4.1 and Section 5.3.1).

In the multi-type case, one cannot hope to calculate R_0 when there are only n observations, because the next-generation matrix has n^2 entries. We will encounter this problem again in the next subsection where we discuss estimation from age-dependent data. One can, however, derive an upper and lower bound for R_0 and prove the following result:

Theorem 13.4 *Let the vectors* $\{p_j\}, \{s_j\},$ *and* $\{\pi_j\}$ *be defined as above, then* R_0 *lies between* $\hat{R}_0^{\min}$ *and* $\hat{R}_0^{\max}$ *given by*

$$\hat{R}_0^{\min} = \min_i\{-\frac{\log(1 - p_i)}{s_i p_i}\}$$

$$\hat{R}_0^{\max} = \max_i\{-\frac{\log(1 - p_i)}{s_i p_i}\}.$$

---

[4]The analysis in this subsection is based on T. Britton: Epidemics in heterogeneous communities: estimation of R_0 and secure vaccination. *J. R. Stat. Soc. B*, **63** (2001), 705-715. See there for more details and the proof of the theorem in the main text.

13.2.4   Age-dependent data

For infectious agents in human populations there are often serological data available that are stratified by age (for endemic infections these are derived from serological surveys, while for epidemic outbreaks they arise from case notification). Such data are very useful for the estimation of the age-dependent force of infection and R_0. One well-known difficulty with such estimates is, however, that there is indeterminacy: one can have observations of numbers of seropositive individuals, or of newly infected individuals, for each age class that is specified, but the matrix describing age-specific contact frequencies for n age classes contains n^2 elements that need to be estimated (age-specific contact frequencies between individuals are not likely to be observed). Hence one always needs additional assumptions when detailed contact data are lacking.

Assume that the infectious agent is in a steady endemic state in the population, implying that the age-specific force of infection is constant in time. We start from the next-generation operator as given in Section 9.3

$$(K\phi)(a) = \int_0^\infty k(a,\alpha)\phi(\alpha)d\alpha \tag{13.2}$$

with

$$k(a,\alpha) = \int_0^\infty h(\tau,\alpha)c(a,\alpha+\tau)N(a)\frac{\mathcal{F}_d(\alpha+\tau)}{\mathcal{F}_d(\alpha)}d\tau \tag{13.3}$$

being the expected number of new cases among susceptibles of age a that an individual that itself becomes infected at age α will produce during its entire infectious period (h measures infectiousness, c quantifies the age-dependent contact structure, N is the steady state population size, and the ratio of survival functions gives the probability for the infected to survive from age at infection α to at least age $\alpha+\tau$). One convenient combination of assumptions, for infections where the length of the infectious period is much shorter than the average individual lifespan, was referred to earlier in Section 9.3 as the 'short-disease approximation': $\mathcal{F}_d(\alpha+\tau)/\mathcal{F}_d(\alpha) \approx 1$ and $c(a,\alpha+\tau) \approx c(a,\alpha)$. We also put: $H(\alpha) := \int h(\tau,\alpha)d\tau$. We will use this approximation below.

When the age-specific force of infection $\Lambda(a,t)$ is assumed to be constant in time we have derived in Section 9.5, equation (9.7), that $\Lambda(a)$ must necessarily satisfy

$$\Lambda(a) = \int_0^\infty \int_0^\infty h(\tau,\alpha)c(a,\alpha+\tau)N(\alpha)\Lambda(\alpha)\mathcal{F}_i(\alpha)\frac{\mathcal{F}_d(\alpha+\tau)}{\mathcal{F}_d(\alpha)}d\tau d\alpha \tag{13.4}$$

where $\mathcal{F}_i(\alpha)$ is the susceptibility survival function, i.e., the probability for a susceptible to escape from becoming infected at least until age α:

$$\mathcal{F}_i(\alpha) = \exp(-\int_0^\alpha \Lambda(\sigma)d\sigma).$$

Exercise 13.5 How should one compare the risk of dying at an early age due to inoculation with the risk of dying at a later age due to infection? The current exercise relates to a historic treatment of such questions. Assume that individuals survive a given disease with probability θ. Assume further that the disease has negligible duration compared with the expected life-time of individuals.

   i) Express the probability to be alive at age a in terms of $\mathcal{F}_d, \mathcal{F}_i$ and θ.

ii) Next, in order to decompose mortality into causes related to the disease and other causes, compute the probability $P(a)$ to be alive at age a, given that one does not die from other causes.

iii) Argue that elimination of the infection leads to a survival function that is obtained from $\mathcal{F}_d(a)$ by multiplication with $1/P(a)$.

iv) Show, for a constant force of infection Q, that $1/P$ describes a logistic curve, and compute the limit of $1/P(a)$ for a tending to infinity. This result was obtained by Daniel Bernoulli in 1760 concerning smallpox, in what is probably the first paper (published much later in 1766) dealing with mathematical epidemiology.[5]

Exercise 13.6 (cf. Exercise 4.6) Show that the *mean age at infection* $\bar{a}$ is given by

$$\bar{a} = \frac{\int_0^\infty a\Lambda(a)\mathcal{F}_i(a)\mathcal{F}_d(a)\,da}{\int_0^\infty \Lambda(a)\mathcal{F}_i(a)\mathcal{F}_d(a)\,da}. \tag{13.5}$$

Consider the special case in which every individual lives exactly M units of time. Use partial integration to rewrite $\bar{a}$ in the form

$$\bar{a} = \frac{\int_0^M \mathcal{F}_i(a)\,da - M\mathcal{F}_i(M)}{1 - \mathcal{F}_i(M)}.$$

Exercise 13.7 i) Assume $c(a,\alpha) = \sum_{k=1}^n f_k(a)g_k(\alpha)$. Reduce the equation for Λ to a finite-dimensional problem. Pay particular attention to the case $n = 1$.

ii) Elaborate for the special case of interval decomposition and the short-disease approximation.

iii) Suppose that the c_{ij} are unknown, but that we can measure in some way the force of infection in the various age intervals. Suppose that we want to determine the c_{ij} from these data. How many equations do we have? And how many unknowns? So, can we realise our objective?

iv) In the literature one finds reference to the WAIFW matrix (Who Acquires Infection From Whom; see e.g., Anderson and May 1991). Can you guess what matrix is meant? Hint: Do not take the name too literally.

v) Design additional assumptions on the c_{ij} that reduce the number of unknowns to the number of equations. Hint: Consult Anderson and May (1991) and Greenhalgh and Dietz (1994).[6]

Exercise 13.8 i) Consider separable mixing, i.e., the special case of Exercise 13.7 with $n = 1$, and let additionally $f(a) \equiv 1$ (in other words, suppose that data about $\mathcal{F}_i(a)$ suggest that $\mathcal{F}_i(a) = e^{-Qa}$, for some Q, is not too bad an approximation). How would you estimate the force of infection Q?

---

[5] D. Bernoulli: Essai d'une nouvelle analyse de la mortalité causée par la petite vérole et des avantages de l'inoculation pour la prévenir. *Mém. Math. Phys. Acad. Roy. Sci., Paris*, (1766) 1–45; K. Dietz and J.A.P. Heesterbeek: Daniel Bernoulli's epidemiological model revisited. *Math. Biosci.*, **180** (2002), 1–21.

[6] D. Greenhalgh and K. Dietz: Some bounds on estimates for reproductive numbers derived from the age-specific force of infection. *Math. Biosci.*, **124** (1994), 9–57.

ii) Additionally, assume that g is constant and that h is independent of α, and, finally that $\mathcal{F}_d(a)$ is an exponential function, say $e^{-\mu a}$. Show that

$$\hat{R}_0 = 1 + \frac{Q}{\mu + r}.$$

Show that, under the same conditions, the mean age of infection is

$$\bar{a} = \frac{1}{\mu + Q},$$

so that for $r \approx 0$ we have

$$\hat{R}_0 = \frac{L}{\bar{a}},$$

where L is the life expectancy.

iii) Assume again that f is identically 1 and that $\mathcal{F}_i(a) = e^{-Qa}$. Assume in addition that ψ (defined by (9.3), reiterated as (13.7) below) is approximately constant. Express R_0 in terms of the observable quantities $N(a)$ and Q as

$$\hat{R}_0 = \frac{\int_0^\infty N(a)\,da}{\int_0^\infty N(a)e^{-Qa}\,da}.$$

iv) Now repeat ii) for the situation where individuals live until a fixed age L and then die, while retaining the other assumptions. Show that in that case $\bar{a} = 1/Q$ and that

$$R_0 \approx QL = \frac{L}{\bar{a}},$$

when $r \approx 0$ and $QL \gg 1$.

We have seen in Chapter 9 that if we assume separable mixing, i.e., if we let $c(a, \alpha) = f(a)g(\alpha)$ for certain functions f and g, the eigenvalue problem for K becomes one-dimensional and we can write down an explicit expression for R_0. The result is given by equations (9.2)–(9.3):

$$R_0 = \int_0^\infty \psi(\alpha)f(\alpha)N(\alpha)d\alpha \tag{13.6}$$

with

$$\psi(\alpha) = \int_0^\infty h(\tau, \alpha)g(\alpha + \tau)\frac{\mathcal{F}_d(\alpha + \tau)}{\mathcal{F}_d(\alpha)}d\tau. \tag{13.7}$$

In order to derive the estimating formulas for R_0 in Exercise 13.8 we needed to make severe additional assumptions. Basically this approach amounts to ignoring much of the heterogeneity that age structure creates, but taking this heterogeneity into account was the main reason for looking at the age-structured models in the first place, so subsequently ignoring most of this structure for estimation is undesirable. The same would hold if we would like to use equation (13.6) for estimating purposes, as the right-hand side does not correspond to routinely available data, such as results from serological surveys, but instead involves ingredients for which no information is available. We saw that an important 'summary ingredient' appearing in the estimators is the force of infection, combining several other ingredients, the values of which cannot be separately estimated. However, that force of infection is assumed to be both constant in time and independent of age in Exercise 13.8. A formulation that uses age-structured data from serological surveys would be helpful as one can use such data to estimate an age-dependent force of infection. Indeed, assuming that all individuals are susceptible at birth and that natural mortality is independent of the force of infection, we have that the fraction of individuals of age a who will have antibodies to infection is given by

$$G(a) = 1 - \mathcal{F}_i(a) = 1 - e^{-\int_0^a \Lambda(\alpha)d\alpha}. \tag{13.8}$$

We therefore look for an expression for R_0 in terms of the force of infection. Of course there is a price to pay, as we will see below, but then additional assumptions provide concrete useful estimators that still incorporate some of the heterogeneity relating to age. We base our exposition on the work by Farrington, Kanaan and Gay.[7] The formulation makes use of the left eigenvector of K associated with R_0.

Let now $l(a)$ be the dominant (left) eigenvector (more precisely: eigenfunction) of K corresponding to R_0, i.e.

$$R_0 l(\alpha) = \int_0^\infty l(a)k(a,\alpha)da. \tag{13.9}$$

If we multiply the left and right-hand sides by $\Lambda(\alpha)\mathcal{F}_i(\alpha)\mathcal{F}_d(\alpha)$ and integrate the result over α we obtain

$$R_0 \int_0^\infty \Lambda(\alpha)\mathcal{F}_i(\alpha)\mathcal{F}_d(\alpha)l(\alpha)d\alpha = \int_0^\infty \Lambda(\alpha)\mathcal{F}_i(\alpha)\mathcal{F}_d(\alpha) \int_0^\infty l(a)k(a,\alpha)da\,d\alpha.$$

We now switch the order of integration on the right-hand side, and insert the term $\frac{N(\alpha)N(a)}{N(\alpha)N(a)}$. We then use the relation $\Lambda(a) = \int \Lambda(\alpha)\mathcal{F}_i(\alpha)\frac{N(\alpha)}{N(a)}k(a,\alpha)d\alpha$, which is shorthand for (13.4) combined with (13.3). In a population of constant size, we have from (9.1) that $N(\alpha)/N(a) = \mathcal{F}_d(\alpha)/\mathcal{F}_d(a)$. The right-hand side then simplifies to $\int_0^\infty l(a)\Lambda(a)\mathcal{F}_d(a)da$. Finally, we can substitute a for α on the left-hand side, and obtain the following general expression for an estimator of R_0

$$\hat{R}_0 = \frac{\int_0^\infty l(a)\Lambda(a)\mathcal{F}_d(a)da}{\int_0^\infty l(a)\Lambda(a)\mathcal{F}_i(a)\mathcal{F}_d(a)da}. \tag{13.10}$$

Without additional assumptions, even when $\Lambda(a)$ can be estimated from sero-prevalence data, we cannot use this formula for R_0 because $l(a)$ is not known. The eigenfunction depends on the kernel $k(a,\alpha)$, and therefore on its ingredients, where notably the infectivity h during the infectious period and the contact structure c cannot be obtained from the serological data. One of the most attractive ways to proceed is to impose separable mixing on the contact structure as above.

Exercise 13.9 Assume the short-disease approximation with separable mixing, i.e., $c(a, \alpha + \tau) \approx c(a, \alpha) = f(a)g(\alpha)$.

i) Show that the force of infection is proportional to the function f.

---

[7]C.P. Farrington, M.N. Kanaan and N.J. Gay: Estimation of the basic reproduction number for infectious diseases from age-stratified serological survey data. *J. R. Stat. Soc. C*, **50** (2001), 251–292. We present only the most elementary part of that work, however, and refer to the paper for more sophisticated approaches.

ii) Now assume additionally that $h(\tau,\alpha) \equiv h(\tau)$ and that the per-capita death rate μ is age-independent. Show that the left eigenfunction l becomes proportional to the function g.

iii) Show that under the more severe assumption of proportional mixing, i.e., when $f \propto g$, one has that

$$\hat{R}_0 = \frac{\int_0^\infty \Lambda^2(a)e^{-\mu a}da}{\int_0^\infty \Lambda^2(a)\mathcal{F}_i(a)e^{-\mu a}da}. \tag{13.11}$$

This expression was first derived by Dietz & Schenzle.[8]

Now that we have an expression for R_0 in terms of the age-dependent force of infection, we need to estimate the latter from the seroprevalence data in order to obtain a concrete estimator based on, for example, approximation (13.11) above. Suppose we have data from n non-overlapping age intervals $(a_i, a_{i+1}]$, $i = 0, \ldots, n-1$. A possibility then is to assume that the force of infection is piecewise constant over these intervals, i.e., $\Lambda(a) = \Lambda_i$, for $a \in [a_i, a_{i+1})$, and subsequently estimate values for the Λ_i directly from the data. Suppose we have the following data: we have sampled n_i individuals that were seronegative upon entering age interval i and m_i of those become seropositive in that interval. The binomial likelihood of observing this outcome is

$$\binom{n_i}{m_i}G_i^{m_i}(1 - G_i)^{n_i - m_i}$$

where $G_i = 1 - \exp(-\int_{a_i}^{a_{i+1}}\Lambda(a)da) = 1 - \exp(-\Lambda_i(a_{i+1} - a_i))$. We can then estimate values for the Λ_i by maximizing the likelihood function

$$L = \prod_{i=0}^{n-1}\binom{n_i}{m_i}G_i^{m_i}(1 - G_i)^{n_i - m_i}. \tag{13.12}$$

Inserting the resulting estimated force of infection function into (13.11) then leads to a concrete estimator for R_0.

Exercise 13.10 Show that the maximum likelihood estimate from (13.12) is given by

$$\hat{\Lambda}_i = \frac{-\ln(\frac{n_i - m_i}{n_i})}{a_{i+1} - a_i}.$$

An alternative to assuming that the force of infection is piecewise linear, is to impose that the force of infection can be described by a specific flexible function of age, containing one or more parameters. A flexible choice for this is a gamma-type distribution:

$$\Lambda(a) = (ua - w)e^{-va} + w, \tag{13.13}$$

where u, v, and w are parameters, with values assumed to be greater than or equal to 0. This can again be combined with a binomial likelihood in a similar analysis to the preceding one, where maximizing the binomial likelihood function now leads to estimates for the values of u, v and w.

---

[8]K. Dietz and D. Schenzle: Proportionate mixing models for age-dependent infection transmission. *J. Math. Biol.*, **22** (1985), 117–120.

13.3 ESTIMATING R_0 FROM A TRANSMISSION EXPERIMENT

When in addition to data on the final outcome, also data on the course of the 'outbreak' are available, one can use a generalized linear model (GLM) approach based on the SIR model (1.12). In Becker (1989) this is applied to outbreaks in small populations. Such detailed data are usually not available, but in the controlled setting of a transmission experiment they are in principle obtainable. The idea is to estimate the transmission rate constant β and the mean length of the infectious period separately. As a rule such experiments are not intended to estimate R_0, but to compare the effect of different interventions on transmissibility. They are used, for example, to show that a certain vaccine is able to substantially reduce transmission, compared to the situation in an unvaccinated population. Effectively therefore, one tests a null-hypothesis such as: $R_v \geq 1$, or that the value of β in the vaccinated and unvaccinated population is the same.

In a typical experiment $N = 10$ individuals are housed together, of which five have been inoculated with a particular infectious agent, while the other five individuals are susceptible.[9] The idea is to follow the individuals through time, for a specified period, and assess the infection status and infectivity of each animal in subsequent time intervals of some specified length $\triangle t$. These are usually assessed by determining whether, and in which amounts, individual animals are shedding the infectious agent (i.e., their infectious output is measured). The data consist, for each interval, of the number of susceptibles (s) at the beginning of the interval, the number of infectious individuals during the interval multiplied by the length of the interval ($i\triangle t$), and the number of individuals (c) that start being infectious during the interval $\triangle t$. The latter individuals are counted as the new cases arising in interval $\triangle t$, which is consistent with the assumption underlying the standard SIR model, namely, that there is no latency period.

For the particular GLM we consider, we assume a binomial distribution for the number of new cases in each interval. First consider the probability that a susceptible escapes infection in interval $\triangle t$: $\exp[-\beta i\triangle t/N]$. From this follows the probability that the susceptible does become infected in the interval $\triangle t$: $1 - \exp[-\beta i\triangle t/N]$. Finally, the expected number of new cases in interval $\triangle t$ is

$$E(c) = s(1 - e^{-\beta si\triangle t/N}).$$

We can rewrite this expression again as a linear regression to estimate $\log \beta$ but we need a more complicated transformation for this. One can use, instead of the function $\log[x]$, the function $\log(-\log[1-x])$ (also written as $C\log\log[x]$, 'complementary loglog'). This transformation leads to the regression formula (again with slope 1):

$$C\log\log[E(c)/s] = \log\beta + \log[i\triangle t/N].$$

The above gives an estimate of β once we replace $E(c)$ by its estimate c. To estimate the mean length of the infectious period one takes the mean of all periods in which the individuals in the experiment had detectable infectivity, as observed in the daily testing, restricted to those individuals that stopped being detectably infectious before the end of the experiment, but were still alive at that time (thus assuming that such individuals had recovered). Together these two estimates provide an estimate for R_0.

---

[9]For an example, with references to other work, see J.A. van der Goot, G. Koch, M.C.M. de Jong and M. van Boven: Quantification of the effect of vaccination on transmission of avian influenza (H7N7) in chickens. *PNAS* **102** (2005), 18141–18146.

13.4   ESTIMATORS BASED ON THE INTRINSIC GROWTH RATE

One of the major challenges is to come up with reliable methods for estimating R_0 in the initial stages of a developing outbreak, especially when the infectious agent has never been seen before. Indeed, in that case epidemiological information about modes of transmission, latency, incubation and infectious periods and their distributions, estimates of the generation interval (i.e., the average time between the start of the infectious period for successive cases) or the serial interval (i.e., the average time between the onset of clinical symptoms for successive cases), is not available, at least not immediately and reliably. For the planning of intervention measures, and for 'guestimating' their effect on the spreading epidemic, estimates of both the basic and the effective reproduction number for the situation at hand are required. It is of course difficult to specify what 'reliable' means, as there is usually no frame of reference for an epidemic of an emerging infection. One aims for ballpark estimates that are 'model free' in that they do not presuppose a specific transmission model, with precision to be determined afterwards. Note in particular that the data that come available as the epidemic unfolds are often unreliable, even case data, as they may suffer from both misdiagnosis and underreporting, and notably suffer from influences of chance events when numbers are still low. One of the major challenges is to estimate the time-dependent effective reproduction number $R(t)$ as the epidemic unfolds in order to judge the impact that the implemented control and intervention measures have on the outbreak. We concentrate here on R_0 only.[10]

An approach is to use the exponential or intrinsic growth rate r, that we have seen before, and to estimate R_0 from the exponentially growing cumulative incidence of cases in the early phase of an outbreak (see for example Section 1.2.3). Suppose we regard the simple SIR model with $dI/dt = \beta SI/N - \gamma I$.

Exercise 13.11 i) Derive from this the estimator

$$\hat{R}_0 = 1 + \frac{r}{\gamma} \approx 1 + \frac{\ln 2}{\gamma t_d}$$

where t_d is the doubling time of the incidence and r is the exponential growth rate of the observed incidence.

ii) How would you estimate r or t_d from the cumulative incidence curve, say denoted by $g(t)$?

We now discuss more sophisticated methods based on the same principle, but go beyond the simple model and use the generation time, rather than the doubling time.

A frequently used approach is to fit an exponential function to the (cumulative) incidence and to use the estimator $\hat{R}_0 = e^{rT_G}$ to estimate R_0, where T_G is the generation interval of the epidemic. For rT_G small the simpler estimator $\hat{R}_0 = 1 + rT_G$ is sometimes used. Estimating R_0 then comes down to estimating both r and T_G from the data. This method is easy and intuitive and one can wonder in which circumstances it would yield a 'good' approximation, and how large discrepancies

---

[10]This subsection is based on M.G. Roberts and J.A.P. Heesterbeek: Model-consistent estimation of the basic reproduction number from the incidence of an emerging infection. *J. Math. Biol.*, **55** (2007), 803–816. Very similar results can be found in J. Wallinga and M. Lipsitch: How generation intervals shape the relationship between growth rates and reproductive numbers. *Proc. R. Soc. B*, **274** (2007), 599–604.

can be when these circumstances are not met. In this section we show how one can investigate these questions.

To do this analysis, we derive estimators of R_0 based on specific models and compare these with the simple estimators given above. To distinguish the various estimators, we introduce some notation and write $\hat{R}_0^+ = e^{rT_G}$ and $\hat{R}_0^- = 1 + rT_G$. We need to emphasize that R_0 is independent of timescale, whereas T_G has dimension time and r has dimension time^{-1}. We also emphasize that we do not know r or T_G, we assume that they have been estimated from data in some way. We do not write $\hat{r}$ or $\hat{T}_G$ for these estimates, as this would result in far too many hats in one section.

Assume for simplicity that the incidence $i(t)$ of an emerging infection may be characterized by a slight modification of equation (1.25):

$$i(t) = \delta(t) + \frac{S(t)}{N} \int_0^\infty A(\tau) i(t - \tau) \, d\tau \qquad (13.14)$$

for $t \geq 0$, where $\delta(t)$, the Dirac delta 'function' at $t = 0$, describes the introduction of the infection at time zero, and where the contact rate c has been absorbed into the kernel $A(\tau)$. The number of susceptibles in the population at time t is

$$S(t) = N - \int_{-\infty}^t i(u) \, du$$

with $i(t) := 0$ for $t < 0$. For an emerging infection we assume the entire population to be susceptible at time zero. If this is not the case, we take N to be the size of the susceptible population prior to infection.

We write $A(\tau) = R_0 f(\tau)$. If we regard the generation interval as a random variable for which we can assume a distribution, then f is the density function of that distribution, and the mean generation interval may be determined from the formula

$$T_G = \int_0^\infty t f(t) \, dt. \qquad (13.15)$$

For an emerging infection we have little information about the density function f. We may have observations of the latency period T_E; the incubation period (the time from infection to the onset of symptoms) that we may in some cases assume to equal T_E; or the infectious period T_I. Given these we may wish to impose a particular form on f, and use our limited knowledge to estimate parameter values for the distribution. These estimates may be revised as more information becomes available.

Given a probability distribution for the generation interval, described by the density f, and an estimated initial rate of exponential increase for the epidemic, r, we approximate the initial stages of the epidemic by $i(t) = e^{rt}$ with $S(t) \approx N$. Equation (13.14) then leads to a model-consistent estimate of the basic reproduction number via the formula

$$R_0 \int_0^\infty e^{-rt} f(t) \, dt = 1. \qquad (13.16)$$

If $f(t)$ were a Dirac delta 'function,' then equations (13.15) and (13.16) would lead to the estimate $\hat{R}_0 = \hat{R}_0^+$. For the standard SIR model, where we have $f(t) = \gamma e^{-\gamma t}$, equations (13.15) and (13.16) lead to the estimate $\hat{R}_0 = \hat{R}_0^-$.

We now compute R_0 for three possible choices for f: a density with a latency and infectious period of fixed length, a density with exponentially distributed latency and infectious periods, and a density with a gamma distribution. We refer to the respective estimators as $\hat{R}_0^{\text{fix}}$, $\hat{R}_0^{\text{exp}}$ and $\hat{R}_0^{(\text{m,n})}$.

Exercise 13.12 First regard the case of an exponentially distributed latency period and infectious period, with parameter θ and α, respectively. Give an expression for the density of the generation-time distribution $f(\tau)$, and derive from that the estimator

$$\hat{R}_0^{\text{exp}} = 1 + r(\frac{1}{\theta} + \frac{1}{\alpha}) + \frac{r^2}{\theta\alpha} = 1 + rT_G + r^2 T_E (T_G - T_E).$$

Note that the estimator $\hat{R}_0^{\text{exp}}$ gives a correction to the estimator $R_0^- = 1 + rT_G$ for the standard SIR model (as this is obtained by letting $\theta \to \infty$). Furthermore, we see that $\hat{R}_0^- \leq \hat{R}_0^{\text{exp}}$ because the generation time T_G is by definition at least as long as the latency period T_E.

One can do the same analysis as in Exercise 13.12 when we allow the latency period to be divided into m consecutive classes E_i, and the infectious period to be divided into n consecutive classes I_j. We will denote this distribution by $\hat{R}_0^{(m,n)}$. The mean total time spent in the latency class and the infectious class has a gamma distribution with mean $T_E = 1/\theta$, and $T_I = 1/\alpha$, respectively, identical to those for the exponential case. Moreover, we have $\hat{R}_0^{\text{exp}} = \hat{R}_0^{(1,1)}$ and $\hat{R}_0^{\text{fix}} = \lim_{m,n \to \infty} \hat{R}_0^{(m,n)}$. In Wearing et al. (2005)[11] one can find that

$$\hat{R}_0^{(m,n)} = \frac{\frac{r}{\alpha}(1 + \frac{r}{m\theta})^m}{1 - (1 + \frac{r}{n\alpha})^{-n}} = \frac{\frac{2nr}{n+1}(T_G - T_E)(1 + \frac{r}{m}T_E)^m}{1 - (1 + \frac{2r}{n+1}(T_G - T_E))^{-n}}.$$

Suppose finally that the latency and infectious periods have fixed lengths, T_E and T_I, respectively. In that case $T_G = T_E + T_I/2$, and $f(t) = 1/T_I$ for $T_E < t < T_E + T_I$, and $f(t) = 0$, otherwise. If one then works out the relation (13.16) for this choice of kernel, one obtains

$$\hat{R}_0^{\text{fix}} = \frac{r(T_G - T_E)}{\sinh r(T_G - T_E)} e^{rT_G}.$$

From this we have that $\hat{R}_0^{\text{fix}} \leq \hat{R}_0^+$. Actually one can show that $\hat{R}_0^+$ is an upper bound for any density function for the generation time (see Wallinga and Lipsitch 2007, in footnote 10).

Using all estimators defined above one can show:

$$\hat{R}_0^- \leq \hat{R}_0^{\text{exp}} \leq \hat{R}_0^{(m,n)} < \hat{R}_0^{\text{fix}} \leq \hat{R}_0^+ \tag{13.17}$$

for any fixed positive numbers m and n, and fixed values of r, θ, and α. The proof for the middle inequalities is based on the property that $\hat{R}_0^{(m,n)}$ is increasing in both m and n. This is shown graphically in Keeling and Rohani (2008, Figure 3.16 on p. 97) and proved in Roberts and Heesterbeek (2007) (referred to in footnote 10). This property is an interesting observation in its own right. One sees that, simply by increasing the number of classes in the latency and/or infectious period, one obtains a higher estimate for R_0, for the same value of r.

---

[11]H.J. Wearing, P. Rohani and M.J. Keeling: Appropriate models for the management of infectious diseases. *PLoS Med.*, **2(7)** (2005), e174; see also the discussion in Keeling and Rohani, 2008, pp 96–98.

The relation (13.17) gives a broad picture. In fact, in specific applications there can be marked differences between the different estimates for the same basic ingredients. These differences are purely due to the different choices of density functions for which only the mean values are known from observations. This can better be explored graphically.

At the risk of sounding patronizing, we end this chapter with the advice to always think long and hard about the assumptions underlying the estimator one is using, and contemplate how these assumptions compare to what is known about the reality that yielded the data, and what the potential effect might be of the unavoidable discrepancies.

Chapter Fourteen

Data-driven modeling of hospital infections

14.1 INTRODUCTION

As already explained in Section 4.6, the treatment of hospital patients suffering from bacterial infections is increasingly hampered by antibiotic resistance. When the means of curing infections diminish, the prevention of infection gains importance. Thus, for example, Scandinavian countries and the Netherlands have implemented already in the 1980s a 'search and destroy' policy in order to prevent the rise of MRSA prevalence.[1]

Before we proceed, we first have to clarify the terminology. We shall say that a patient is 'colonized,' when this person carries the antibiotic resistant bacteria at a detectable level, meaning that when a swab/sample is taken and cultured, the outcome is 'positive.' There are two points to note: 1) Only a fraction of the colonized patients will develop clinically apparent infections (when, for instance, the bacteria start multiplying in the lungs or in the bloodstream). So those patients that are diagnosed as 'infected' by medical doctors possibly form the figurative tip of an iceberg of prevalence; 2) A patient may carry the antibiotic resistant bacteria at an *un*detectable level and is then classified as 'susceptible,' even though, strictly speaking, this is a misclassification. If such a patient is treated with antibiotics, the colony of resistant bacteria may profit from the extermination of its competitors and grow to a detectable level. Unjustly we shall say that in such a case 'colonization' took place.

Now recall from Section 4.6 that in an ICU the prevalence of colonization changes through various processes: discharge and admission, cross-transmission (mediated by health-care workers), opportunistic growth. By the latter we mean the scenario that a small colony of resistant bacteria can grow to an appreciable size due to the elimination of vulnerable competitors (point 2). Note that the 'small colony' may just be a single individual that arose by mutation.

If we want to ascertain the effectiveness of various potential control measures, we need to know whether cross-transmission or opportunistic growth is the predominant colonization route. For the former, advocating hand washing and other traditional methods of infection prevention makes sense, whereas in the latter case a more prudent use of the arsenal of antibiotics should be pursued. But how to assess the relative importance of these two routes to colonization?

The secure way is laborious and costly; it combines epidemiological surveillance with genotyping. It is of course very convenient that such a secure way exists, as it provides a gold standard against which to judge how well other, less costly and

---

[1]This is a compound strategy where the contribution of the various components can be evaluated quantitatively; see M.C.J. Bootsma, O. Diekmann and M.J.M. Bonten: Controlling of methicillin-resistant *Staphylococcus aureus*: quantifying the effects of interventions and rapid diagnostic testing. *PNAS*, **103** (2006), 5620–5625.

less laborious, methods perform. The aim of this chapter is to explain in detail one such alternative method, suitable for application to data that are readily available.

In terms of the Markov chain model introduced in Section 4.6, our aim is to determine the values of the parameters α, corresponding to opportunistic growth, and β, corresponding to cross transmission. The key difference is that, with respect to the processes captured by α, the patients are independent from one another, whereas cross transmission clearly creates dependence. In deterministic jargon this is formulated by saying that, at the population level, opportunistic growth is a linear process, while cross transmission is nonlinear. If we call the probability per unit of time for a susceptible patient to become colonized the 'force' of colonization, then this force increases linearly with the prevalence due to the possibility of cross transmission. This reinforces the natural autocorrelation of the day by day prevalence (due to the fact that patients stay in the ICU for a number of days in a row). So the hope is that the prevalence fluctuations contain enough information to disentangle the relative importance of opportunistic growth and cross transmission.

In the present chapter we show that maximum likelihood estimation is feasible when bookkeeping is done in an efficient way. An alternative approach would be to iteratively improve guesses about missing data in a Bayesian setting, using the Markov chain Monte Carlo method described in the next chapter (in the present situation, 'missing data' refers to the vector v introduced below; the key point is that for known v one has a simple explicit formula for the likelihood).

14.2 THE LONGITUDINAL SURVEILLANCE DATA

We assume that for each patient the following data are available:

- The day of admission t_0;

- The colonization status on admission;

- The day of discharge t_e;

- The days $t_1, t_2, \ldots$ at which samples are taken;

- The result, either a '+' or a '−', of culturing such a sample.

We ignore both the possibility of a false positive and of a false negative, i.e., we assume that the culture technique has 100% sensitivity and specificity. We assume that colonization is never lost during the stay in the ICU (so that if a patient is tested positive on a given day t, then this patient remains positive until t_e). We label a patient C, for colonized, each day where we are sure that this patient is colonized. We label a patient Q, for questionable, each day for which: the patient was not tested; and there is no later '−' test (so either there is no later test or, if there is one, it is positive); and there is no earlier '+' test (so either there was no earlier test or, if there was one, it was negative). We label a patient U, for uncolonized, each day for which we are sure that this patient is not colonized (and is hence not labeled either C or Q). Note that all uncertainty lies in the Q category.

As an idealization we assume that admission, discharge, and the taking of samples all occur at noon, 12.00 h. We call 12.01 h the start of the day, and 11.59 h on the next day the end of the day.

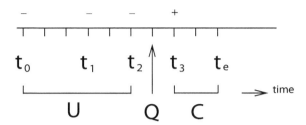

Figure 14.1: Timeline of information for a patient from day of admission (t_0) to discharge (t_e), with colonization test results on particular days, labeling the patient as either Uncolonized, Questionable, or Colonized.

Suppose that our (partial) observations cover $T + 1$ days. We number these days $0, 1, 2, \ldots, T$. We denote the total of these observations by $Obs(0, T)$. The parameter estimation will be based on maximizing $P(Obs(0, T))$.

By $Obs(0, t)$ we shall denote: information about admission and discharge before or at t, and the U, Q, C distinction at t and before t, where $0 \leq t < T$.

Note that although the C label is indeed based on information obtained before or at t, the Q-U distinction is based on information obtained at *later* times. So the notation seems a bit misleading. And indeed, if we would carry out calculations in real time, we would restrict the U label to patients tested negatively the very same day. Such patients would hence return to Q the next day, unless they are once more tested and found negative that next day. This leads to many more 'possibilities' and hence to a much larger computational effort. As we actually do the computations (long) after day T, we incorporate everything we are certain about in $Obs(0, t)$, so that we can focus our attention on the things we do not know for sure.

We denote by $Obs(t + 1, T)$ all the information about admission and discharge at time $t + 1$ and later, and the U, Q, C distinction at $t + 1$ and later.

Note that $Obs(0, T) = Obs(0, t)$ & $Obs(t + 1, T)$.

14.3 THE MARKOV CHAIN BOOKKEEPING FRAMEWORK

Suppose that at the start of the day

$$u \text{ patients are labeled U,} \tag{14.1}$$
$$q \text{ patients are labeled Q,}$$
$$c \text{ patients are labeled C.}$$

These known numbers partially characterize the 'state' of the ICU. A full specification of the state involves information we do not have at that time: we need to tell for each of the q patients with questionable status whether they are colonized or not. So, for bookkeeping purposes, we introduce the vector

$$v = (v_1, v_2, \ldots, v_q) \tag{14.2}$$

with v_i being 1 if individual i with questionable status is actually colonized, and 0 otherwise. Here we order the q individuals, for instance according to the time of entering the Q category, the patient numbered 1 being the last to have acquired the Q label. As we do not know v, we shall work with the *probability* that the ICU state is described by the vector v (and the numbers u, q, c). There are 2^q possible vectors v, and we *number* these by

$$Nv := \sum_{i=1}^{q} v_i 2^{q-i}. \tag{14.3}$$

We note three things: i) Nv takes the values $0, 1, \ldots, 2^q - 1$; ii) we have chosen to let low values of the index correspond to high powers of 2; and iii) N is invertible. For example, if $q = 1$ we have $N(0) = 0$ and $N(1) = 1$, while for $q = 2$ we have $N(00) = 0, N(01) = 1, N(10) = 2$ and $N(11) = 3$. The inverse N^{-1} reconstructs the vector v from the binary representation of the number.

The state of the ICU at a particular time t is now fully described by the numbers u, q, c and the number j such that $v = N^{-1}(j)$. Since only j is unknown, we focus our attention on this part of the state and, with some abuse of language, call it from now on 'the state of the ICU' (and use $s(t)$ to denote it).

Apart from estimating the values of the parameters, we want to estimate other quantities, such as the expected number of acquisitions (both in total and stratified according to the route: cross transmission or opportunistic growth). In order to do so, we need to know the probability that the ICU state on day t is indeed described by the vector v that carries the number j, for all relevant t and j. Let $s(t)$ denote the state of the ICU on day t. We define

$$p_j(t) := P(s(t) = j \mid Obs(0, T)). \tag{14.4}$$

We then aim to compute the vector $p(t)$ of length 2^q (with q depending on t).

For any time t with $0 \le t < T - 1$ we can quite naturally split the observations in those that are, in principle, available at that time, i.e., $Obs(0, t)$, and those made at later times, i.e., $Obs(t + 1, T)$. When it comes to iterative updating, the vector $f(t)$ of length 2^q defined by

$$f_j(t) := P(s(t) = j \ \& \ Obs(0, t)) \tag{14.5}$$

is easier to handle than $p(t)$. The vectors f and p are related to each other by a further building block, the vector b defined by

$$b_j(t) := P(Obs(t + 1, T) \mid s(t) = j). \tag{14.6}$$

Indeed, we claim that

$$p_j(t) = \frac{b_j(t) f_j(t)}{\sum_{i=0}^{2^q - 1} b_i(t) f_i(t)}. \tag{14.7}$$

To derive this relationship we argue as follows:

$$
\begin{aligned}
p_j(t) P(Obs(0, T)) &= & P(s(t) = j \ \& \ Obs(0, T)) \\
&= & P(s(t) = j \ \& \ Obs(0, t) \ \& \ Obs(t + 1, T)) \\
&= & P(Obs(t + 1, T) \mid s(t) = j \ \& \ Obs(0, t)) \\
& & \times P(s(t) = j \ \& \ Obs(0, t)).
\end{aligned}
$$

According to the Markov property, information about the past or the present is irrelevant when it is given that $s(t) = j$. So the first factor is exactly $b_j(t)$. The second factor equals $f_j(t)$. Because p is a probability vector, (14.7) follows. Note in particular that the denominator in (14.7) does not depend on t and is equal to $P(Obs(0, T))$. This should be no surprise as the relation

$$P(Obs(0, T)) = \sum_{i=0}^{2^q-1} b_i(t) f_i(t) \qquad (14.8)$$

follows directly from the definitions of $b(t)$ and $f(t)$.

In the next two sections we explain how $f(t)$ and $b(t)$ can be computed recursively on the basis of a mechanistic acquisition model and the observations. In conclusion of this section we specify the initial condition, in the case of f, and the end condition, in the case of b, for these recursions. The definition (14.6) of the vector b does not work for $t = T$, simply because there are no further observations. We replace it by

$$b_j(T) = 1 \qquad (14.9)$$

in order to express that, for all we know, the patients in the Q-category can be either colonized or uncolonized.

The study period starts at $t = 0$ and the observations at this time point are usually of a somewhat different character. The ideal situation is that all patients are cultured at $t = 0$, so that we know the ICU state at that time with certainty. The component of $f(0)$ corresponding to this state will then be one, while all other components are zero. If some information is missing, an (educated) guess must be made (and one may then wish to determine the possible influence on the final conclusions by way of a sensitivity analysis). In any case, one needs to specify $f(0)$ with $\sum_j f_j(0) = 1$. Note that for $t > 0$ we have $\sum_j f_j(t) = P(Obs(0, t)) < 1$.

14.4   THE FORWARD PROCESS

The day-to-day updating of f involves several steps:

- *Step 1*: Transform f to account for the *probabilities* with which $0 \to 1$ transitions have taken place, while incorporating the *fact* that u susceptible patients did *not* become colonized. So this step concerns what happens between 12.01 h and 11.59 h the next day and it deals with both unobserved and observed events.

- *Step 2*: Account for those patients that leave Q to move on to C. In particular, update the probabilities of unobserved aspects of the state on the basis of the partial observation. While doing so, also compute the *likelihood* of the observed events.

- *Step 3*: Account for discharge.

- *Step 4*: Account for admission.

Note that the value of q may change during the last three steps so that, when we iterate the four steps, the length of the vector f is variable. Also note that the last three steps deal with the information obtained in the 'infinitesimal' time interval between 11.59 h and 12.01 h.

Before describing the steps in technical detail, we introduce some more notation. When v and w are vectors of the same length, say q, we write $v \geq w$ when this inequality holds component-wise, i.e.,

$$v \geq w \iff v_i \geq w_i \text{ for } i = 1, 2, \ldots, q, \qquad (14.10)$$

and write $v \not\geq w$ if this inequality does *not* hold, i.e.,

$$v \not\geq w \iff \exists j \in \{1, 2, \ldots, q\} \text{ such that } v_j < w_j. \qquad (14.11)$$

Finally, we use $|v|$ to denote the l_1-norm of the vector v, i.e.,

$$|v| := \sum_{i=1}^{q} |v_i|. \qquad (14.12)$$

Note that we can omit taking the absolute values when, as in our case, v_i is always non-negative. In fact, if v_i is either 0 or 1, then $|v|$ is simply the number of components that are equal to 1 or, in our case, the number of Q-patients who are actually colonized. When $v \geq w$ then $|v - w| = |v| - |w|$ counts how many additional patients are colonized if we change from state w to state v.

As in Exercise 4.32 we assume that uncolonized patients can become colonized on a given day only from those patients who were already colonized at the start of that day and not through patients who themselves got colonized earlier during that very same day. Consider an ICU with $u + q + c$ patients in total and assume that k of these are colonized at the start of the day (so $k \geq c$). We denote the probability that an uncolonized patient gets colonized during that day by $\pi(k)$. When we recall Exercises 4.30–32 we feel motivated to postulate that

$$\pi(k) = 1 - e^{-(\alpha + \beta \frac{k}{u+q+c})}. \qquad (14.13)$$

Note that $\pi(k)$ depends on the state information $u + q + c$ and on the parameters α and β, but that this is not expressed in the notation (only in order to avoid long expressions). The dependence on α and β forms the basis for the estimation of these parameters. Our method works equally well if one replaces the explicit formula (14.13) by any reasonable alternative.

Technical specification of Step 1 In Step 1 we act on the vector f with a $2^q \times 2^q$-matrix A defined by

$$A_{mn} := \begin{cases} 0 & \text{if } N^{-1}(m) \not\geq N^{-1}(n), \\ (1 - \pi(k))^{u+q-l}\pi(k)^l & \text{if } N^{-1}(m) \geq N^{-1}(n), \end{cases} \qquad (14.14)$$

with $k = c + |N^{-1}(n)|$ and $l = |N^{-1}(m)| - |N^{-1}(n)|$. The idea behind this definition is first of all that during the day the only possible changes are that components of v change from 0 into 1. Hence a non-zero transition probability requires that the new state has 1 at all positions where the old state had 1. If we restrict our attention to the Questionable patients, a transition in which l *specified* patients get colonized occurs with probability

$$(1 - \pi(k))^{q-l}\pi(k)^l.$$

Note that we should not multiply with the familiar binomial coefficient, simply because our encoding reveals not only the number of new colonizations but also their 'identity.' The extra factor $(1 - \pi(k))^u$ in (14.14) takes into account that there are u more patients that could in principle become colonized during the day that we consider, but for which we know with certainty that they actually were not. As a result of this factor, Af is not a probability vector.

As a concrete example, consider $q = 2$. The possible states are $(0\ 0), (0\ 1), (1\ 0)$, and $(1\ 1)$, which are numbered 0,1,2 and 3, respectively. Then A is the 4 x 4-matrix

$$
\begin{pmatrix}
(1 - \pi(c))^{2+u} & 0 & 0 & 0 \\
\pi(c)(1 - \pi(c))^{1+u} & (1 - \pi(c+1))^{1+u} & 0 & 0 \\
\pi(c)(1 - \pi(c))^{1+u} & 0 & (1 - \pi(c+1))^{1+u} & 0 \\
(\pi(c))^2(1 - \pi(c))^u & \pi(c+1)(1 - \pi(c+1))^u & \pi(c+1)(1 - \pi(c+1))^u & A_{44}
\end{pmatrix}
$$

with $A_{44} = (1 - \pi(c+2))^u$.

On average a fraction $\alpha/(\alpha + \beta\frac{k}{u+q+c})$ of the new cases is due to opportunistic growth and the complementary fraction is due to transmission. So if we weight these fractions in the appropriate way, we can estimate the proportion of new cases that we should attribute to either mechanism. But what is the 'appropriate' way? We shall return to this significant question.

Technical specification of Step 2 In Step 2 we act upon the vector Af with a $2^q \times 2^q$-matrix O defined by

$$
O_{mn} := \begin{cases} 1 & \text{if } n = m \text{ and } N^{-1}(n)_j = 1 \text{ for all cultured } j; \\ 0 & \text{otherwise.} \end{cases} \tag{14.15}
$$

(The symbol O refers to incorporation of the Observations.) The idea behind this definition is first of all that if patient j is cultured, this patient is bound to be positive because otherwise the patient would not belong to Q. The culturing results thus reveal with certainty that certain states are impossible and we therefore put the corresponding component of f equal to zero, while not altering any components that correspond to ICU states that are compatible with the data. Note that we need to renormalize if we want to obtain a probability vector in the end; without normalization the components of OAf are just relative probabilities.

According to the model, how likely are the actual observations? If the initial vector f had l_1-norm 1, then the l_1-norm of OAf tells us how much of the probability distribution that the model predicts is concentrated in the states that are compatible with the observations. Hence, given the probabilistic description at the start of the day as embodied in the vector f, the *likelihood* L of the observed events is precisely $|OAf|$. If f is not normalized we set

$$
L = \frac{|OAf|}{|f|}. \tag{14.16}
$$

Because A depends on the parameters α and β, so does L, and this forms the basis for the maximum likelihood estimation of these parameters. Of course we use for this estimation the observations of all the available days and we postpone for a little while the elaboration, which is based on (14.8).

Technical specification of Step 3 If the jth patient is removed from Q, because this patient was either discharged or tested, we decrease q by 1 and form a new vector f as follows. With a vector $v = (v_1, v_2, \dots, v_{q-1})$ of length $q-1$ we associate two vectors v^0 and v^1 of length q by inserting between v_{j-1} and v_j either a 0 or a 1 (if $j = 1$ we do this in front of v_1). Next we set

$$f^{\text{new}}_{N(v)} = f^{\text{old}}_{N(v^0)} + f^{\text{old}}_{N(v^1)}. \tag{14.17}$$

Effectively this amounts to switching to the marginal distribution for the remaining $q - 1$ patients. As a short-hand notation we write

$$f^{\text{new}} = R f^{\text{old}} \tag{14.18}$$

where R refers to Removal and incorporates all patients who leave Q.

As a concrete example, consider the situation that one of a total of two patients is discharged. Suppose they are numbered such that patient 1 is discharged, while patient 2 remains. With $v = (0)$ we associate $v^0 = (0\ 0)$ and $v^1 = (1\ 0)$, while with $\tilde{v} = (1)$ we associate $\tilde{v}^0 = (0\ 1)$ and $\tilde{v}^1 = (1\ 1)$. According to (14.3) $Nv^0 = 0$, $N\tilde{v}^0 = 1$, $Nv^1 = 2$, and $N\tilde{v}^1 = 3$, while $Nv = 0$ and $N\tilde{v} = 1$. Hence

$$R = \begin{pmatrix} 1 & 0 & 1 & 0 \\ 0 & 1 & 0 & 1 \end{pmatrix}. \tag{14.19}$$

Also in general, each column of R has exactly one component equal to 1 and all other components equal to 0. If $\mathbf{1}$ denotes the vector with all components equal to 1, then it follows that $\mathbf{1}R = \mathbf{1}$ (but note that the length of the vector $\mathbf{1}$ on the left is less than the length of the vector $\mathbf{1}$ on the right). If r denotes the number of removed patients, then the number of rows of R equals the number of columns minus $2r$.

Technical specification of Step 4 For each patient who enters Q we form a new vector f according to

$$f^{\text{new}}_{N(1,v_1,\ldots,v_q)} = 0,$$
$$f^{\text{new}}_{N(0,v_1,\ldots,v_q)} = f^{\text{old}}_{N(v_1,\ldots,v_q)},$$

and we raise q by 1. This reflects that the new patient is the first in the row and that we know for sure that this patient is not colonized (because the patient enters Q on the day of the last negative test). We write

$$f^{\text{new}} = I f^{\text{old}} \tag{14.20}$$

where the symbol I refers to input, and where I now incorporates all patients that enter Q. As a concrete example, consider the situation that one new patient enters an ICU with just one existing patient. Then

$$I = \begin{pmatrix} 1 & 0 \\ 0 & 1 \\ 0 & 0 \\ 0 & 0 \end{pmatrix}.$$

In general, each column of I has exactly one component equal to 1, and 0's otherwise, so $\mathbf{1}I = \mathbf{1}$. If n denotes the number of new patients, then the number of rows of I equals the number of columns plus $2n$.

The recurrence relation

$$f(t) = I(t)R(t)O(t)A(t-1)f(t-1) \tag{14.21}$$

summarizes how the vector f is updated from day to day, on the basis of parameters α and β, the values of which we do not know, and the known information about admission, discharge and testing.

14.5 THE BACKWARD PROCESS

Our next aim is to derive the recurrence relation

$$b(t) = b(t+1)I(t+1)R(t+1)O(t+1)A(t) \tag{14.22}$$

for back-dating the vector b (the easiest way to read this relation is to think of b as a row-vector that is multiplied from the right by a product of four matrices).

We first check (14.22) for $t = T - 1$. Because no new patients are added to the category Q at time T, the matrix $I(T)$ is simply the identity. The matrix $R(T)$ brings about an adjustment of the bookkeeping that preserves the l_1-norm: $\mathbf{1}R(T) = \mathbf{1}$. According to the definition (14.6) we have

$$b_i(T-1) = P(Obs(T,T) \mid s(T-1) = i),$$

which we can rewrite as

$$b_i(T-1) = \sum_j P(\tilde{s}(T) = j \ \& \ Obs(T,T) \mid s(T-1) = i)$$

where $\tilde{s}(t)$ denotes the state of the ICU at day t *before* we account for discharge and admission. Hence

$$P(\tilde{s}(T) = j \ \& \ Obs(T,T) \mid s(T-1) = i) = O_{jj}(T)A_{ji}(T-1)$$

Because $O(T)$ is a diagonal matrix, we have

$$O_{jj}(T)A_{ji}(T-1) = \sum_k O_{jk}(T)A_{ki}(T-1).$$

We conclude that $b_i(T-1)$ is the ith component of the vector $\mathbf{1}O(T)A(T-1)$, which is also the ith component of $\mathbf{1}I(T)R(T)O(T)A(T-1)$. Thus we have verified (14.22) for $t = T$.

In the same spirit we have

$$P(Obs(t+1,T) \mid s(t) = i) = \sum_j P(Obs(t+2,T) \mid \tilde{s}(t+1) = j)$$

$$\times P(\tilde{s}(t+1) = j, Obs(t+1,t+1) \mid s(t) = i)$$

$$= \sum_j P(Obs(t+2,T) \mid \tilde{s}(t+1) = j)O_{jj}(t+1)A_{ji}(t),$$

which we may rewrite as

$$b_i(t) = (\tilde{b}(t+1)O(t+1)A(t))_i$$

where $\tilde{b}$ is defined as b in (14.6), but with s replaced by $\tilde{s}$. So (14.22) follows if we verify that

$$\tilde{b}(t+1) = b(t+1)T(t+1)R(t+1)$$

or, written out in components,

$$\tilde{b}_j(t+1) = \sum_l \sum_k b_k(t+1)I_{kl}(t+1)R_{lj}(t+1).$$

If we denote by $\bar{s}(t)$ the state of the ICU at day t *after* we account for discharge, but *before* we account for admission, and by $\bar{b}$ the quantity corresponding to b, but with s replaced by $\bar{s}$, then we should verify that

$$\bar{b}_l(t+1) = \sum_k b_k(t+1)I_{kl}(t+1)$$

and

$$\tilde{b}_j(t+1) = \sum_l \bar{b}_l(t+1)R_{lj}(t+1).$$

The first amounts to

$$\bar{b}_l(t+1) = b_{k(l)}(t+1)$$

where $k(l)$ is the position of the 1 (and the only 1) in the lth column of $I(t+1)$. Likewise, the second amounts to

$$\tilde{b}_j(t+1) = \bar{b}_{l(j)}(t+1)$$

where $l(j)$ is the position of the 1 (and the only 1) in the jth column of $R(t+1)$. Both are true by definition, as they only reflect aspects of the bookkeeping, with neither information loss, nor information gain. A peculiar feature of the second is that those components of $\tilde{b}$ that relate to the same component of $\bar{b}$ should take the same value (for instance, with R given by (14.19), the first and third component of $\tilde{b}$ take the same value and so do the second and fourth component). This simply reflects that the likelihood of future observations does not depend, according to the model, on the unknown status of patients that are discharged.

We conclude that $b(t)$ can be computed recursively from (14.22) provided with the 'end'-condition $b(T) = \mathbf{1}$.

14.6   LOOKING BOTH WAYS

Recall that in Section 14.3 we defined

$$p_j(t) = P(s(t) = j \mid Obs(0,T))$$

and then showed that

$$p_j(t) = \frac{b_j(t)f_j(t)}{\sum_{i=0}^{2^q-1} b_i(t)f_i(t)} = \frac{b_j(t)f_j(t)}{b(t) \cdot f(t)} \tag{14.23}$$

where $b \cdot f$ is the standard Euclidean inner product. Now that we know that f is updated according to (14.21) and b is back-dated according to (14.22), it follows once more that the denominator at the right-hand side of (14.7) is independent of t. Hence it is equal to the value for $t = T$, which is $(\mathbf{1}, f(T)) = |f(T)|$, the overall likelihood of the observations, and likewise equal to the value for $t = 0$, which is $b(0) \cdot f(0)$.

So for maximum likelihood estimation we can restrict our attention to (14.21) to compute $f(T)$ (or, alternatively, to (14.22) to compute $b(0)$). But there is more detailed relevant information that we can reveal by computing both $f(t)$ and $b(t)$. For instance

$$P(s(t) = j \ \& \ \tilde{s}(t+1) = k \ \& \ Obs(0,T)) = \tilde{b}_k(t+1)O_{kk}(t+1)A_{kj}(t+1)f_j(t) \tag{14.24}$$

as one can verify by first writing the left-hand side in the form

$$P(Obs(t+2,T) \mid s(t) = j \ \& \ \tilde{s}(t+1) = k \ \& \ Obs(0,t+1))$$
$$\times \, P(s(t) = j \ \& \ \tilde{s}(t+1) = k \mid Obs(0,t+1))$$

and next noting that, by the Markov property, the first factor equals $\tilde{b}_k(t+1)$. The second factor equals

$$P(\tilde{s}(t+1) = k \ \& \ Obs(t+1,t+1) \mid s(t) = j \ \& \ Obs(0,t)) P(s(t) = j \mid Obs(0,t)).$$

The last factor of this expression is precisely $f_j(t)$ while in the first factor the conditioning on $Obs(0,t)$ is irrelevant. Hence that first factor is equal to

$$O_{kk}(t+1) A_{kj}(t+1).$$

Thus we arrive at (14.24).

If a transition from $s(t) = j$ to $\tilde{s}(t+1) = k$ occurs, then this involves

$$\mid N^{-1}(k) \mid - \mid N^{-1}(j) \mid$$

acquisitions of colonization. Any of these is with probability $\alpha/g(t)$, with

$$g(t) := \alpha + \beta \frac{c(t) + \mid N^{-1}(j) \mid}{u(t) + \mid N^{-1}(j) \mid + c(t)},$$

by the endogenous route, and with $1 - \alpha/g(t)$ due to transmission. So, given $s(t) = j$ and $\tilde{s}(t+1) = k$, the expected number of acquisitions by the endogenous route equals

$$(\mid N^{-1}(k) \mid - \mid N^{-1}(j) \mid) \frac{\alpha}{g(t)}.$$

We denote such an expected number of acquisitions by a specified route in general by $q_{kj}(t)$ and define

$$q(t) = \sum_{j,k} q_{kj}(t) P(s(t) = j \ \& \ \tilde{s}(t+1) = k \mid Obs(0,T)).$$

It follows that

$$q(t) = \sum_{j,k} q_{kj}(t) \frac{P(s(t) = j \ \& \ \tilde{s}(t+1) = k \ \& \ Obs(0,T))}{\mid f(T) \mid} \tag{14.25}$$

$$= \sum_{j,k} \tilde{b}_k(t+1) O_{kk}(t+1) q_{kj}(t) A_{kj}(t) f_j(t), \tag{14.26}$$

which we can rewrite as

$$q(t) = b(t+1) I(t+1) R(t+1) O(t+1) M(t) f(t)$$

where $M(t)$ is, by definition, the matrix with entries

$$M_{kj}(t) = q_{kj}(t) A_{kj}(t).$$

By summing over t we finally obtain the expected total number of acquisitions during the study period for the particular route used in the specification of $q_{kj}(t)$.

These expressions are then used to calculate the relative importance of each route. We do this at first for the parameter values obtained by maximizing the likelihood, but subsequently we can also determine confidence intervals for the relative importance of the routes.

After all this detailed modeling of intensive-care unit infection in hospitals, the reader may wonder about the link to actual data, and whether the approach yields useful medical information and insight. In Bootsma et al. (2007),[2] the method outlined in this chapter is applied to data about colonization with cephalosporin-resistant Enterobacteriaceae in two ICU of the Utrecht University Hospital during the period September 2001–May 2002. The conclusion is that the endogenous route predominates. Genotyping and epidemiological linkage, used to provide a reference standard, led to the same conclusion.

So the Markov methodology explained in this chapter provides a promising tool for disentangling the contributions of various acquisition routes on the basis of longitudinal data, without requiring labor intensive and costly genotyping procedures.

---

[2]M.C.J. Bootsma, M.J.M. Bonten, S. Nijssen, A.C. Fluit and O. Diekmann: An algorithm to estimate the importance of bacterial acquisition routes in hospital settings. *Am. J. Epid.*, **166** (2007), 841–851.

Chapter Fifteen

A brief guide to computer intensive statistics

In Chapters 5, 13 and 14, we have presented methods for making inference about infectious diseases from available data. This is of course one of the main motivations for modeling: learning about important features, such as R_0, the initial growth rate, potential outbreak sizes and what effect different control measures might have in the context of specific infections. As we have seen, quite a lot can be learned from such modeling and inference. The models considered in these chapters have all been simple enough to obtain more or less explicit estimates of just a few relevant parameters. In more complicated and parameter rich models, and/or when analyzing large data sets, it is usually impossible to estimate key model parameters explicitly. In such situations there are (at least) two ways to proceed. One uses Bayesian statistical inference by means of Markov chain Monte Carlo methods (MCMC), and the other uses large scale simulations, together with numerical optimization to fit parameters to data. In the present chapter we mainly describe Bayesian inference using MCMC and only briefly some large simulation methods. First we return to inference using simple models and show some of its shortcomings.

15.1 INFERENCE USING SIMPLE EPIDEMIC MODELS

In this book many different epidemic models have been presented, both deterministic and stochastic. In general, deterministic models are easier to analyze than the 'corresponding' stochastic models. However, a big advantage of stochastic models when making inference is that they enable quantification of uncertainty of parameter estimates — usually expressed in terms of standard errors. Since this part of the book is devoted to inference, we restrict ourselves to stochastic models from now on.

In Section 5.4 we presented methods for making inference for the standard stochastic epidemic model, in which all individuals mix uniformly (i.e., there are no social preferences) and all individuals are equally susceptible to the infection. Even for this simple (and unrealistic) model, inference methods are non-trivial. The reason for this is two-fold. First, a generic feature of infectious disease spread is that individuals infect each other, thus causing dependencies between individuals, so standard statistical theory based on independent events is not applicable. Secondly, as was illustrated in Section 5.4 where inference was easier when assuming time continuous data as opposed to final size data, the epidemic process is nearly always only partly observed. For instance, even if temporal data is available, it is very unusual that the actual time of infection is known (usually the time of first show of symptoms is the available information) and it is even more rare to have information about actual contacts and which of them caused transmission. This fact, that the epidemic process is only partly observed, complicates inference in that the likelihood for the available data usually becomes rather involved (see for example Chapter 14). Additional to this, partial observation of the epidemic process may also make it impossible to estimate certain parameters separately. For example, in the standard stochastic epidemic model individuals have contact with other individuals at rate c, and if one of the persons in such a contact was infectious and the other susceptible then transmission takes place with probability p. If there are no data about contacts, but only infections, it is impossible to estimate c and p separately — all we can hope for is to make inference about their product cp, the rate of having *infectious* contacts (cf. Section 5.6 for comments on such *identifiability* problems).

One advantage of simple models, compared to more complicated models, is that they are often more simple to analyze and this makes deductions easier. In particular, many important model 'features' (such as R_0 and final outbreak sizes) as well as parameter estimates and their standard errors are 'explicit' in terms of model parameters and data statistics respectively. As a consequence, such analyses increase the general understanding of model features and estimates, e.g. by showing which parameters have biggest impact on R_0, or, how much we gain in precision by collecting temporal data as opposed to final-size data.

A disadvantage of simpler models is of course that the model and its parameter estimates have a reduced bearing on the outbreak under study, i.e., on reality. If, for example, a model assuming identical individuals that have contact uniformly is used for analyzing a real epidemic outbreak in which individuals are very diverse and contacts are far from uniform, then the conclusions drawn are not very reliable, and they should not be used without precautions.

If one is to make a general conclusion for when simple models may be applicable (and when they are not), the level of infectiousness is of importance. The more highly infectious a disease agent is, and the less close the contact needs to be for transmission to occur, the less do heterogeneities and social structure tend to matter. A prime example for such a disease agent is the measles virus, which indeed infects large fractions of the susceptible community thus allowing for deterministic modeling (e.g., Anderson and May, 1991), but most other childhood diseases also fall in this category. The basic reproduction number of measles is often estimated to be around $R_0 = 15$ indicating very high infectiousness. Of course, the level of complexity of the model also depends on the question being studied — more complicated models have also been used when studying measles, for example when trying to understand the relation between community size and the ability for the disease to become endemic, or when trying to understand the predominance of two year cycles (before mass vaccination).

15.2   INFERENCE USING 'COMPLICATED' EPIDEMIC MODELS

Epidemic models can be made more realistic (as well as complicated) in many ways. Indeed several such more complicated models have been discussed in the book. For example, the model can be made stochastic rather than deterministic, individuals can be considered as differing in characteristics that influence susceptibility and/or infectivity, mixing may not be uniform but instead having some random (e.g., network) and/or given (e.g., households, schools) structure. Introduction of a spatial component, and/or allowing for temporal changes due to seasonality and/or changing behavior due to the ongoing outbreak, are other such extensions.

The complications serve to make the model more flexible and, hopefully, to capture reality better. The down-side is that these models are more difficult to analyze and, when making inference, more parameters have to be estimated, thus potentially introducing more uncertainty. Also, certain model parameters may not be identifiable at all making it impossible to estimate them from the available data (see for example the previous section). In general, the more data that is available, both concerning the community, transmission dynamics and infectious status of individuals, the more complicated models can be used when making inference.

If both the size of the community in which the disease is spreading and the observed number that get infected are very large, say tens of thousands, then stochastic models might not even be needed. Parameters may simply be estimated by fitting (using e.g., the method of least squares) the deterministic model curve to the corresponding observed curve. The reason for this is that uncertainty in the parameter estimates is usually negligible when the community is very large and can hence be neglected. See for example Anderson and May (1991) and Keeling and Rohani (2007) for many important applications of deterministic models fitted to data of many diseases.

If the community size is not very large, or data only covers few infections, for example if only the beginning of an outbreak is observed or if only a small number of individuals are affected by the disease, then stochasticity does matter. When using more complicated stochastic epidemic models it may in some situations still be possible to obtain parameter estimates, also equipped with standard errors, for specific diseases. For example, Addy et al. (1991)[1] use likelihood methods for making inference about influenza using data from two outbreaks in which both households and age structure are taken into account in the modeling.

As mentioned previously it is not possible for many models to yield explicit parameter estimates using likelihood methods. An alternative approach that has been effective in analyzing epidemic outbreak data having more complicated structure, as well as in numerous other applications, is to use Markov chain Monte Carlo (MCMC) methods, which are based on the statistical approach known as *Bayesian statistics*.[2] In the next subsections we present the principle of Bayesian statistics and thereafter the ingenious idea which makes MCMC work.

15.3 BAYESIAN STATISTICS

We present the idea behind Bayesian statistics in general, rather than for epidemic models. Suppose that we have a random vector $X = (X_1, \ldots, X_n)$ and that its distribution depends on a vector $\theta = (\theta_1, \ldots, \theta_k)$ of parameters (in many applications both the data vector as well as the parameter vector are high dimensional). The model may be expressed by the probability distribution $p(x|\theta)$, $x \in \Omega$, the space of possible outcomes. If X is discrete $p(x|\theta) = P(X = x|\theta)$ is the probability function, and if X is continuous it denotes the probability density function.

Suppose data is collected by observing the random vector X, i.e., we observe $X = x_{obs}$. If, rather than viewing $p(x|\theta)$ as a function of x for fixed θ, we regard it as a function of θ for our observed data point x_{obs}, we have the likelihood function

---

[1]C.L. Addy, I.M. Longini and M. Haber: A generalized stochastic model for the analysis of infectious disease final size data. *Biometrics*, **47** (1991), 961–974.

[2]e.g., P.D. O'Neill: A tutorial introduction to Bayesian inference for stochastic epidemic models using Markov chain Monte Carlo methods. *Math. Biosci.*, **180** (2002), 103–114; and I. Chis Ster and N.M. Ferguson: Transmission parameters of the 2001 foot and mouth epidemic in Great Britain. *PLoS ONE*, **2(6)** (2007), e502.

$L(\theta; x_{obs}) = p(x_{obs}|\theta)$ (quite often the likelihood is simply written as $L(\theta)$ since x_{obs} is fixed). Up until now this has nothing to do with Bayesian statistics.

The Bayesian idea is to treat the parameter θ as a random vector rather than a fixed unknown constant. This means that both θ and X are random vectors, the difference is only that we have an observation of X. The knowledge of a random variable, or the lack of knowledge, is expressed by a probability distribution. Before we make our observation we express our knowledge about the parameter vector θ by the *prior distribution* $f_{pri}(\theta)$ (in what follows $f(\cdot)$ will denote a generic probability/density function). In many situations little is known about θ before data are collected — this is reflected by a so-called *uninformative* or *flat* prior. In other situations there might be prior information about the parameter, thus urging us to choose a prior distribution being more concentrated around some probable values.

Once we have collected data ($X = x_{obs}$) we obtain more information about the distribution of θ. This is reflected in that the prior distribution is updated to the *posterior* distribution $f_{post}(\theta|x_{obs})$, which combines the prior distribution with the observed data x_{obs}. The posterior distribution is defined as the distribution of θ conditional on having observed x_{obs}, and it can be expressed in terms of the likelihood and the prior distribution using Bayes' theorem (which explains why the method is called Bayesian inference):

$$\begin{aligned} f_{post}(\theta|x_{obs}) &= \frac{f(\theta, x_{obs})}{f(x_{obs})} = \frac{f_{pri}(\theta) f(x_{obs}|\theta)}{f(x_{obs})} \\ &= \frac{f_{pri}(\theta) L(\theta)}{\int f_{pri}(\theta') L(\theta') d\theta'} \\ &\propto f_{pri}(\theta) L(\theta). \end{aligned} \tag{15.1}$$

The posterior distribution $f_{post}(\theta|x_{obs})$, our quantity of interest, is a function of θ, and since the denominator in the second row does not contain θ it is disregarded in the last row. The conclusion is hence that the posterior distribution is proportional to the product of the prior distribution and the likelihood $L(\theta) = L(\theta; x_{obs})$. Our knowledge about θ after having observed x_{obs} is expressed in this posterior distribution. Sometimes one wants to summarize the key feature of our knowledge by one numeric value or an interval, just like when estimating parameters using maximum likelihood methods. It is then common to use the posterior mean

$$\hat{\theta}_{post} = E_{post}(\theta) = \int \theta f_{post}(\theta|x_{obs}) d\theta$$

as point estimate, and so-called credibility intervals defined from quantiles of the posterior distribution to express uncertainty. A 95% credibility interval is for example given by the interval between the 2.5% quantile and the 97.5% quantile of the posterior distribution.

Exercise 15.1 Suppose that an experiment can end in two ways, 'success' or 'failure,' and that $P(\text{success}) = \pi$. Suppose that the experiment is repeated $n = 3$ times, and the random variable X denotes the number of successes. Then X follows the binomial distribution $X \sim Bin(n = 3, \pi)$. In the Bayesian perspective π plays the roll of θ and is hence treated as a random variable. Suppose now that we observe $X = 2$ successes and $3 - X = 1$ failures.

i) Estimate π using maximum likelihood.

ii) Derive the posterior distribution for π if no prior information about π is available, thus suggesting a uniform distribution between 0 and 1 for π ($f_{pri}(\pi) = 1$, $0 \leq \pi \leq 1$). Compute also the posterior mean $\hat{\pi}_{post}$.

iii) Repeat i) and ii) but now assuming that we observe $n = 30$ experiments out of which $X = 20$ were successes (so we observe the same *proportion* of successes as before).

In the history of statistics there have been many controversies between 'frequentists' (relying on likelihood theory) and 'Bayesians.' The main criticism of Bayesian statistics concerns its subjectivity: two scientists having different prior beliefs will also have different posterior beliefs; a fact that has been said to be contradictory to the nature of empirical science. Bayesians have defended themselves by saying: why should we not also include what is known about the situation before data is collected. Nowadays, there is less controversy — most statisticians find Bayesian statistics to be useful sometimes and the frequentist view point other times — most often they also lead to similar parameter estimates, which of course is reassuring.

The use of Bayesian statistics has more or less exploded over the last few decades. The main reason for this is not that many 'frequentists' have converted to become 'Bayesians,' but rather that improved technology allowing for more or less 'automatic' measurements has made many data sets very large. It is then much harder to numerically compute parameter estimates using any method/principle. However, in Bayesian inference there exists the ingenious method of Markov chain Monte Carlo, which we present in the next section.

15.4 MARKOV CHAIN MONTE CARLO METHODOLOGY

We now present the general idea underlying Markov chain Monte Carlo (MCMC), for details we refer to other sources, e.g., Gilks et al. (1996). MCMC is a numerical method that can be used for obtaining posterior distributions of model parameters. The method is used in more or less all areas of applied statistics and there is also much theoretical research devoted to its performance.

Consider the situation where we have a statistical model containing a (typically high-dimensional) parameter θ, and that the data d, also often containing many observations, each having high-dimension, is collected in order to make inference about the model parameters contained in θ. Suppose furthermore that we can compute the likelihood $L_d(\theta)$, or at least a quantity proportional to it, and that we have a prior distribution $f_{pri}(\theta)$ that is also computable. From the previous section we then know that the posterior distribution $f_{post}(\theta|d)$ is proportional to the product of the prior and the likelihood:

$$f_{post}(\theta|d) \propto f_{pri}(\theta)L_d(\theta).$$

Estimating θ, or some of its components, then consists of maximizing $L_d(\theta)$ with respect to θ (for ML-estimates) or identifying θ-regions where the posterior $f_{post}(\theta|d)$ is large (for Bayesian estimates). In the situation where both data and the parameter vector have high dimension, this is often very numerically complicated even with efficient software and computer. In this situation MCMC is useful.

Another frequently encountered situation is where the likelihood of the observed data d is very complicated, but where the likelihood for some more detailed data is much simpler to compute. Then one can consider the *latent* (unobserved) variables ξ as unobserved random variables and include them in the parameters: $\theta' = (\theta, \xi)$. The posterior distribution is then a distribution of the parameters and latent variables, given the observed data d. By analyzing the posterior distribution one obtains information about the parameters as well as the latent variables (see Exercise 15.2 for an example).

What MCMC does is to generate a sample $\theta^{(1)}, \ldots, \theta^{(N)}$, of arbitrary length N, from the posterior distribution $f_{post}(\theta|d)$ (or a distribution very close to it). By studying the empirical distribution of $\theta^{(1)}, \ldots, \theta^{(N)}$ one can then make statements about the parameters. For example, a 95% credibility interval for one of the components θ_k of θ is simply estimated by the interval between the 2.5 percentile and the 97.5 percentile (of that component) in the sample. We now describe how this is done. We present the method in a general form since it may also be used in situations other than Bayesian inference.

Suppose we want to find an approximation to the distribution $\pi(x)$, where we know and can compute $g(x) = c\pi(x)$ but don't know the constant c. (Note that the situation just described fits into this with θ, and possibly some latent variables added as discussed, in the role of x, and $f_{pri}(\theta)L_d(\theta)$ in the role of $g(x)$.) As a consequence, we are able to compute $\pi(x)/\pi(y) = g(x)/g(y)$ exactly, a fact that we will use.

The idea is to construct a Markov chain $X_1, X_2, \ldots$ having $\pi(\cdot)$ as its stationary distribution; this means that the components in the Markov chain are dependent but, as n tends to infinity, the distribution of X_n will converge to the stationary distribution. Having done this, which is the tricky part, we can simply estimate $\pi(\cdot)$ by simulating the Markov chain as long as desired and estimating $\pi(\cdot)$ from the empirical distribution. The Markov chain will approach the stationary distribution the longer the chain is simulated; the sample will in general be dependent. For this reason one usually rejects the first 10,000 or 100,000 simulations (called the burn-in period) such that the chain is close to stationarity, and from this step one usually only keeps every 100th or every 1000th step in the chain (called thinning), thus making the sample close to independent.

It remains to construct the Markov chain having $\pi(\cdot)$ as its stationary distribution. We recall that a Markov chain $X_1, X_2, \ldots$, is a sequence of random variables/vectors, where the distribution of X_{n+1}, conditional on X_n, is independent of $X_1, \ldots X_{n-1}$ (given the present, the future is independent of the past). The random properties of the Markov chain are defined by the transition matrix (or kernel) $p(y|x) := P(X_{k+1} = y | X_k = x)$, together with the starting point $X_0 = x_0$. The stationary distribution $\pi(\cdot)$ of the Markov chain is defined as the solution to

$$\pi(y) = \sum_x \pi(x) p(y|x)$$

(or the corresponding integral in the case of a continuous variable x). In words, if the state of the chain at some point in time is described by the stationary distribution, then the same is true one time step later. It is known that the stationary distribution is unique under certain regularity conditions, and also that the distribution of X_n converges to $\pi(\cdot)$ irrespective of the starting value x_0. This triggers the hope that our Markov chain (yet to be constructed!) will approach the desired distribution $\pi(\cdot)$.

Assume now that $X_k = x$. We define our transition matrix in two steps:

- First we *propose* a *candidate* y by drawing from a proposal distribution $q(y|x)$, that is, we generate a value y from the conditional distribution $q(\cdot|x)$.

- Then we *accept* our candidate y with probability $\alpha(x, y)$, defined by

$$\alpha(x, y) := \min\left(1, \frac{\pi(y)q(x|y)}{\pi(x)q(y|x)}\right).$$

By this is meant that we set $X_{k+1} = y$ with probability $\alpha(x, y)$ (i.e., if a uniform $[0, 1]$ random number u is smaller than $\alpha(x, y)$), otherwise we let the chain stay where it is: $X_{k+1} = x$.

First we note that we are able to compute $\alpha(x, y)$ even though it involves $\pi(\cdot)$. This is true since it was assumed that we can compute the ratio $\pi(y)/\pi(x)$ as pointed out earlier.

The remarkable thing is that this Markov chain will have the desired distribution $\pi(\cdot)$ as its stationary distribution. What is even more remarkable is that this is true *whatever* proposed distribution $q(\cdot|\cdot)$ is chosen! This does not mean that the choice of $q(\cdot|\cdot)$ is irrelevant. The distribution should ideally have two characteristics: it should be simple to generate random numbers from it and it should give a large, or at least not too small, acceptance probability. The latter is true if $q(y|x) \approx \pi(y)$. To have exact equality is impossible: then we can simulate from $\pi(\cdot)$, but not knowing $\pi(\cdot)$ was how the problem originated. These two features of $q(\cdot|\cdot)$ will affect respectively how fast we can simulate the Markov chain and its rate of convergence towards stationarity.

To show that our Markov chain has $\pi(\cdot)$ as its stationary distribution is in fact very easy. Pick arbitrary x and y ($x \neq y$) and assume that $\alpha(x, y) < 1$ (otherwise we study $p(x|y)$ instead). Then, by construction we have

$$p(y|x) = q(y|x)\alpha(x, y) = q(y|x)\frac{\pi(y)q(x|y)}{\pi(x)q(y|x)} = \frac{\pi(y)}{\pi(x)}q(x|y) = \frac{\pi(y)}{\pi(x)}p(x|y).$$

The last equality is true because if $\alpha(x, y) < 1$, then $\alpha(y, x) = 1$ implying that $q(x|y) = p(x|y)\alpha(y, x) = p(x|y)$. We hence have that $\pi(x)p(y|x) = \pi(y)p(x|y)$, called 'detailed balance,' which clearly also holds if $x = y$. This implies that

$$\sum_x \pi(x)p(y|x) = \sum_x \pi(y)p(x|y) = \pi(y)\sum_x P(X_{k+1} = x|X_k = y) = \pi(y).$$

The last identity follows trivially since we sum a probability function over all possible outcomes. We have thus shown that our Markov chain satisfies $\sum_x \pi(x)p(y|x) = \pi(y)$, but then, as discussed previously, it will have $\pi(\cdot)$ as its stationary distribution.

The algorithm described above is called the Metropolis-Hastings algorithm. For more about this algorithm and many special situations and choices of proposal distributions we refer to Gilks, et al. (1996).

Exercise 15.2 Consider the standard SEIR epidemic model but assume that there is no latent period. Assume further that the infectious period I follows a Gamma-distribution, and assume a controlled setting, for example when studying disease propagation in domestic animals, where there are $N = 10$

susceptible individuals and one initially infected individual at time $t = 0$ (the index case). Assume that five of the remaining 10 individuals become infected, and that these animals showed symptoms (were diagnosed), respectively, on days 6, 8, 9, 13 and 15, and that the index case showed symptoms on day 7. Diagnosed individuals were removed immediately and isolated. These times can hence be used as time of removal, but there are no measurements for times of infection (except the index case). The remaining five individuals escaped infection during the course of the outbreak.

The parameters of the model are: $\lambda(= cp)$: the rate of infectious contacts, and α and β in the Gamma distributed infectious period. We also treat the unobserved infection times $i_1, \ldots, i_5$ as parameters. The data consists of $i_0 = 0$ and the removal times $r_0 = 7, r_1 = 6, \ldots, r_5 = 15$ together with the knowledge that there were no further infections during the outbreak. The individuals are labeled as '0' for the index case, and then in the order of observation of symptoms (removal times). The infection times need not necessarily be monotone in this ordering — for example the removal time of the index case was not the first of all individuals, whereas its infection time was.

Assume for simplicity that all three parameters (λ, α and β) have $Exp(1)$ priors.

i) What is, modulo a normalization factor, the posterior distribution

$p(\lambda, \alpha, \beta, i_1, \ldots, i_5 | i_0 = 0, r_0 = 7, r_1 = 6, \ldots, r_5 = 15)$?

ii) Define suitable proposal distributions for the parameters and the unobserved infection times $i_1, \ldots, i_5$.

iii) Simulate an MCMC chain and remove the first 10,000 observations and keep every 100th after this burn in, until there are 1000 observations kept. Give the posterior means and 95% credibility intervals of λ, α and β, and also of the mean duration of the infectious period: $\alpha/(\alpha + \beta)$.

15.5 LARGE SIMULATION STUDIES

An alternative, more explorative, approach for making inference from data is to use large scale simulation studies. More often these type of methods are used when making predictions of future outbreaks, but they may also be used for fitting parameters to data.

When a specific disease and a specific country is of interest, there is usually much information available about the community: age structure, school structure, spatial structure in the community, information about travel patterns, etc. On top of this, assumptions are often made about the disease agent: transmissibility, latency periods, etc. With this knowledge as input it is possible to build up a complicated model that is intractable for analytical analysis and perhaps also for MCMC. Then an alternative approach is to simulate a disease outbreak given all the described features, and to study the resulting outbreak. Preferably this is of course done many times. It is also recommended to slightly perturb different parameters to see if this has a dramatic effect on the outbreak — if it does, a closer analysis is necessary.

From an inferential point of view one might first 'guess' some initial parameter values and simulate outbreaks to see if the simulated outbreaks appear similar to the observed outbreak. If not, one can modify the parameter guesses and repeat the procedure, up until the point where the fit of the simulations to the observed outbreak is adequate. The parameter values used for the simulations that are in agreement with data can then serve as 'parameter estimates.' Of course, there could very well exist other parameter sets that generate outbreaks even more similar to the observed outbreak, so this method might lead to erroneous conclusions. However, in certain situations there is not much else one can do, and in other situations one may be convinced that the explored parameter region is really where the 'truth' lies.

The epidemic models used in these situations most often involve some random components (numerical evaluations of deterministic models are however also frequent). Most often individuals are put into various groups according to age, school, geographic location and so on, and then transmission between individuals of different groups are modeled as random events according to some predefined random rules. An alternative, more computer intensive, method is to actually keep track of each individual separately, and to model how the epidemic progresses in the population of individuals. Sometimes each individual is followed through time in close detail: during day the individual goes to school/work, in the evening the individual is home with the family and sometimes out in the community at large, all such events affecting transmission probabilities. Such detailed models/simulations are called agent-based, individual-based or micro-simulations. Their advantage is of course that very detailed descriptions of the behavior of individuals can be included. Their disadvantage is that such detailed description also gives rise to many parameters that have to be estimated or assigned values, and it may sometimes be difficult to understand which parameters are most influential.

Part IV

Elaborations

Chapter Sixteen

Elaborations for Part I

16.1 ELABORATIONS FOR CHAPTER 1

Exercise 1.1 Let c denote the number of blood meals a mosquito takes per unit of time. Suppose a human receives k bites per unit of time. Consistency requires that $kD_{\text{human}} = cD_{\text{mosquito}}$. Our assumption is that c is a given constant. Hence necessarily

$$k = c\frac{D_{\text{mosquito}}}{D_{\text{human}}}.$$

Exercise 1.2 $cT_m p_m$.

Exercise 1.3 $kT_h p_h$.

Exercise 1.4 Consider one infected mosquito. It is expected to infect $cT_m p_m$ humans, each of which is expected to infect $kT_h p_h$ mosquitoes. So, going full circle, we have a multiplication factor

$$cT_m p_m k T_h p_h = c^2 T_m T_h p_m p_h \frac{D_{\text{mosquito}}}{D_{\text{human}}}.$$

When this multiplication factor is below one, an initial infection will die out in a small number of 'generations.' If, however, it is above one then most likely (see Section 1.2.2 for a qualitative and quantitative elaboration of how likely this actually is) an avalanche will result.

Exercise 1.5 i) $R_0 = q_1 + 2q_2$. Using $q_0 + q_1 + q_2 = 1$ we can rewrite this as $R_0 = 1 + q_2 - q_0$. If follows that $R_0 > 1$ if $q_2 > q_0$, and that $R_0 \leq 1$ if $q_2 \leq q_0$.

ii) Individuals are indistinguishable, so for each the probability that its line of descent will stop equals z. By independence, the probability that both lines stop then equals z^2.

iii) Consider one individual. If this individual has no offspring, its line of descent will stop with certainty. If it produces one offspring, its line of descent will stop with probability z. If it produces two offspring, its line of descent will stop with probability z^2. Hence consistency requires that

$$z = q_0 1 + q_1 z + q_2 z^2.$$

iv) If we substitute $q_1 = 1 - q_0 - q_2$ we obtain $z = q_0 + z - q_0 z - q_2 z + q_2 z^2$ and next $q_0(z - 1) = q_2 z(z - 1)$. From this we see that either $z = 1$ or that $q_0 = q_2 z$.

v) We recall from i) that $R_0 \leq 1$ corresponds to $q_2 \leq q_0$ and hence to $q_0/q_2 \geq 1$. As z should be a probability, we conclude that only the possibility $z = 1$ remains.

vi) If $R_0 > 1$ then $q_2 > q_0$ so $q_0 q_2 < 1$. We expect that in this case $z = q_0 q_2$ and that with complementary probability $1 - z > 0$ the line of descent of the individual will grow exponentially.

Exercise 1.6 i) Direct substitution.

ii) $g(1) = \sum_{k=0}^{\infty} q_k$ and, since $\{q_k\}$ should be a probability distribution, we require that $\sum_{k=0}^{\infty} q_k = 1$.

iii) By term-by-term differentiation, $g'(z) = \sum_{k=0}^{\infty} k q_k z^{k-1} = \sum_{k=1}^{\infty} k q_k z^{k-1}$ and therefore $g'(1) = \sum_{k=1}^{\infty} k q_k = R_0$.

iv) All $q_k \geq 0$ (by their interpretation) and $\sum_{k=0}^{\infty} q_k = 1$ guarantee that at least one of the q_k is strictly positive; the expression for $g'(z)$ derived in 1.6-iii then implies $g'(z) > 0$, provided that $q_0 < 1$.

v) $g''(z) = \sum_{k=0}^{\infty} k(k-1) q_k z^{k-2}$, and the argument is now identical to that used in proving 1.6-iv. Note that if $q_0 + q_1 = 1$ then $g''(z) = 0$. The strict inequality $g''(z) > 0$ holds whenever $q_0 + q_1 < 1$.

Exercise 1.7 Regard Figure 1.1. Recall the assumption that $q_0 > 0$ and note that consequently $g(0) = q_0 > 0$. Graphically, we can construct the sequence z_n as in the picture, that is, by going to the graph of g, then going to the 45°-line, then going to the graph of g, etc. We then observe that z_n converges to the intersection of the two graphs that is nearest to the origin, or, in other words, to the smallest root of $z = g(z)$.

This graphical argument can be elaborated analytically. Since g is increasing, $g(q_0) > g(0) = q_0$ or $z_1 > z_0$. This in turn implies $z_2 = g(z_1) > g(z_0) = z_1$, and inductively we find that the sequence z_n is increasing. Since $g(z) \leq g(1) = 1$, the sequence is also bounded. A bounded increasing sequence has a limit, which we shall call z_∞. By taking the limit $n \to \infty$ in the recurrence relation $z_n = g(z_{n-1})$, we obtain $z_\infty = g(z_\infty)$ (here we use that g is continuous). Now let w denote the smallest root of the equation $z = g(z)$; then $w > 0$ and hence $w = g(w) > g(0) = q_0 = z_0$. Applying g repeatedly, we find $w > z_n$ for all n. By taking the limit $n \to \infty$ we deduce that $w \geq z_\infty$. But since w is defined to be the smallest root, we must have equality.

The interpretation as a consistency condition is as follows. The probability to go extinct equals the probability to go extinct immediately plus, summed over k, the probability to go extinct conditional on infecting k other individuals. In mathematical terms this sentence reads

$$z_\infty = q_0 + \sum_{k=1}^{\infty} q_k z_\infty^k = g(z_\infty),$$

where we have used the independence to write the probability that all k 'branches,' or 'lines,' go extinct as z_∞^k.

Exercise 1.8 The function $z - g(z)$ can have at most two zeros, since its second derivative equals $-g''(z) < 0$ (and if a function has three or more zeros, its derivative has two (or more) zeros and consequently its second derivative has at least one zero). If $R_0 < 1$, this function is negative for z equal to zero and for z slightly less than one (recall Exercise 1.6-iii: $g'(1) = R_0$), so either $z - g(z)$ has no zero at all on the interval $[0, 1)$ or at least two. Since $z = 1$ is a root, the last case would imply that altogether three roots exist, which is impossible. Hence $z = 1$ is the smallest root. If, on the other hand, $R_0 > 1$ then, since $g'(1) = R_0$, the function $z - g(z)$ is positive for z slightly less than one, so a zero less than one must exist.

Exercise 1.9 For $R_0 = 1$ we have $z_\infty = 1$. Indeed, since $g''(1) > 0$, the graph of g lies above the 45°-line for z slightly less than one, and the same argument as used in Exercise 1.8 for the case $R_0 < 1$ applies.

Exercise 1.10 If k individuals are contacted, the probability that m of them are actually infected equals

$$\binom{k}{m} p^m (1-p)^{k-m}$$

(the first factor $\binom{k}{m} = \frac{k!}{m!(k-m)!}$ equals the number of ways one can choose m individuals out of k individuals). Hence the probability that m individuals are infected equals

$$\sum_{k=m}^{\infty} \binom{k}{m} p^m (1-p)^{k-m} \frac{(c\Delta T)^k}{k!} e^{-c\Delta T}$$

$$= \frac{p^m}{m!} (c\Delta T)^m e^{-c\Delta T} \sum_{k=m}^{\infty} \frac{(1-p)^{k-m}}{(k-m)!} (c\Delta T)^{k-m}$$

$$= \frac{p^m}{m!} (c\Delta T)^m e^{-c\Delta T} \sum_{l=0}^{\infty} \frac{(1-p)^l}{l!} (c\Delta T)^l$$

$$= \frac{p^m}{m!} (c\Delta T)^m e^{-c\Delta T} e^{(1-p)c\Delta T}$$

$$= \frac{(pc\Delta T)^m}{m!} e^{-pc\Delta T}.$$

Exercise 1.11 For given ΔT, the probability that m individuals are infected equals (Exercise 1.10)

$$\frac{(pc\Delta T)^m}{m!} e^{-pc\Delta T}.$$

The probability density function for ΔT is, by assumption, $\alpha e^{-\alpha \Delta T}$. Hence

$$q_m = \alpha \int_0^{\infty} e^{-\alpha \Delta T} \frac{(pc\Delta T)^m}{m!} e^{-pc\Delta T} \, d(\Delta T)$$

and

$$g(z) = \sum_{m=0}^{\infty} q_m z^m = \alpha \int_0^{\infty} e^{-\alpha \Delta T} e^{pc\Delta T(z-1)} \, d(\Delta T)$$

(here we have interchanged the summation and the integration and subsequently evaluated the sum, i.e., we have applied (1.6) with $\lambda = pc\Delta T$). Evaluation of the integral yields the expression

$$g(z) = \frac{\alpha}{\alpha - pc(z-1)}.$$

Exercise 1.12 i) If $g(z) = \frac{\alpha}{\alpha - pc(z-1)}$ then $g'(z) = \frac{\alpha pc}{(\alpha - pc(z-1))^2}$ and hence $R_0 = g'(1) = \alpha pc / \alpha^2 = pc/\alpha$.

ii) If ΔT is exponentially distributed with parameter α then its expected value equals $1/\alpha$ (if you don't know this by heart, compute $\alpha \int_0^{\infty} \Delta T e^{-\alpha \Delta T} d(\Delta T)$ and find, by using partial integration, that this integral equals $1/\alpha$). During this period contacts are made with rate (= probability per unit of time) c; hence the expected number of contacts equals c/α. Finally, a contact results in transmission with probability p, so the expected number of secondary cases R_0 equals pc/α.

Exercise 1.13 The equation is $z = \frac{\alpha}{\alpha - pc(z-1)}$, which, as can be verified by direct substitution, has roots $z = 1$ and $z = \alpha/pc$. As our earlier geometric argument in Exercise 1.8 showed, there are at most two roots, so we know all of them. Alternatively, one can multiply the equation by the denominator on the right-hand side and solve the resulting quadratic equation in z by using the explicit formula for the roots.

Exercise 1.14 A new case at time t results from a contact with an infective that was itself infected some time earlier, say at time $t - \tau$ for some positive τ. In order to be infective, τ should actually be restricted to $T_1 \leq \tau \leq T_2$. Infective individuals make contacts at rate c and, given a contact, transmit the agent with probability p. Hence, if we break down the incidence (i.e., the number of new cases per unit of time) according to the possibilities for the value of τ of the infector (i.e., the individual that is responsible for the transmission), we find (1.9).

Exercise 1.15 For $\lambda = 0$ the right-hand side of (1.10) equals $pc \int_{T_1}^{T_2} d\tau = pc(T_2 - T_1) = R_0$. Furthermore,

$$\frac{d}{d\lambda} \left(pc \int_{T_1}^{T_2} e^{-\lambda\tau} \, d\tau \right) = -pc \int_{T_1}^{T_2} \tau e^{-\lambda\tau} \, d\tau < 0.$$

For $\lambda \to \infty, e^{-\lambda\tau} \to 0$ uniformly for $T_1 \leq \tau \leq T_2$ and hence the limit of the integral equals zero. Likewise $e^{-\lambda\tau} \to +\infty$ for $\lambda \to -\infty$, uniformly for $T_1 \leq \tau \leq T_2$ (if $T_1 > 0$; if $T_1 = 0$, restrict to $\varepsilon \leq \tau \leq T_2$ for some fixed small ε and the same argument applies) and the limit of the integral equals $+\infty$.

Exercise 1.16 The key idea is that if you produce many 'children,' but only at a very advanced age, population growth may be relatively slow! If $T_2^* - T_1^* = T_2^{**} - T_1^{**}$ then $R_0^* = R_0^{**}$. But if, for instance, $T_i^* = T_i^{**} + S$, for some $S > 0$, then $r^* < r^{**}$, since, for given r, $e^{-r(\tau+S)} = e^{-rS}e^{-r\tau} < e^{-r\tau}$ and hence the right-hand side of (1.10) is, for the same r, smaller in the * case than in the ** case (now draw a picture and you see at once that $r^* < r^{**}$). If S is large then actually r^* is much smaller than r^{**}. If we subsequently make T_2^* a bit bigger, we make sure that $R_0^* > R_0^{**}$ while still $r^* < r^{**}$ (since both R_0 and r are continuous functions of T_1 and T_2).

Exercise 1.17 The question is for which time T_d we have $\exp(rT_d) = 2$. By taking logarithms, we find $T_d = \ln(2)/r$.

Exercise 1.19 The function $h(s) = \ln s - R_0(s-1)$ satisfies $h(1) = 0$, $h'(1) = 1 - R_0$, $h''(s) = -1/s^2 < 0$ (for $0 < s \leq 1$). Moreover, $h(s) \downarrow -\infty$ for $s \downarrow 0$ and $h(s) \downarrow -\infty$ for $s \to \infty$. The property $h''(s) < 0$ guarantees that there are at most two roots. Since $h(1) = 0$ and $h'(1) = 1 - R_0$, we know that $h(s) > 0$ for s slightly larger than one when $R_0 < 1$, and for s slightly less than one when $R_0 > 1$. The continuity of h and the fact that $h(s) \downarrow -\infty$ for both $s \downarrow 0$ and $s \to \infty$ then imply that for $R_0 \neq 1$ there is always a second zero, which is larger than one (and hence irrelevant) when $R_0 < 1$, but smaller than one (and then called $s(\infty)$) for $R_0 > 1$. For the exceptional case $R_0 = 1$ we have that $s = 1$ is a double zero.

Exercise 1.20 For notational convenience we omit the argument ∞ of s in the following. We are studying roots of the equation (1.11) as a function of R_0, in particular the root that is different from one (and the preceding exercise already showed that there exists exactly one such root).

For i) it is easiest to take the exponential on both sides of (1.11) to obtain

$$s = e^{R_0 s} e^{-R_0}.$$

Assuming that $R_0 s \to 0$ for $R_0 \to \infty$, we come to the consistent conclusion that

$$\lim_{R_0 \to \infty} se^{R_0} = 1,$$

which is what we mean when we write $s \sim e^{-R_0}$ for $R_0 \to \infty$. To justify the assumption, we multiply the equation by $R_0 e^{-R_0 s}$, give $R_0 s$ a new name, say ξ, and write the equation in the form

$$h(\xi) = h(R_0),$$

where we now define $h(x) := xe^{-x}$. The graph of h is depicted in Figure 16.1, the relevant properties being that h is strictly increasing on $[0, 1)$, attaining its maximum at one, and is strictly decreasing towards zero on $(1, \infty)$.

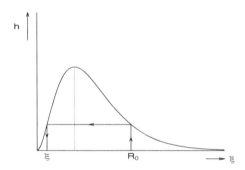

Figure 16.1: Graph of the function $h(x) = xe^{-x}$.

Graphically, we find ξ from R_0 by going from R_0 on the horizontal axis vertically up to the graph of h, then horizontally to find the second point of the graph with the same value of h, and finally down to find the corresponding value ξ on the horizontal axis (see Figure 16.1). The construction immediately shows that ξ is a decreasing function of R_0 with limit zero for $R_0 \to \infty$. Recalling that ξ is just a name for $R_0 s$, we conclude that our assumption is valid.

ii) The assertion is obtained by formal Taylor expansion near $R_0 = 1$ and is justified by the implicit function theorem. Introducing $\sigma = s - 1$, we can rewrite equation (1.11) as

$$\frac{\ln(\sigma + 1)}{\sigma} = R_0.$$

For small σ we have $\ln(\sigma + 1) = \sigma - \sigma^2/2 + O(\sigma^3)$. (For the sake of completeness, we recall that $O(\sigma^3)$ denotes a quantity that after division by σ^3 remains bounded when we let σ approach zero.) So the left-hand side equals $1 - \sigma/2 + O(\sigma^2)$. Bringing the 1 to the other side and multiplying by -2 we obtain

$$\sigma + O(\sigma^2) = -2(R_0 - 1),$$

which is the required result (recall that $\sigma = s - 1$ and that we want to consider σ as a function of $R_0 - 1$, rather than the other way around; but this relation shows that σ and $R_0 - 1$ are of the same order, so, when changing perspective, we may replace $O(\sigma^2)$ by $O((R_0 - 1)^2)$. For the important implicit function theorem, which justifies these formal calculations, we refer to any introductory analysis book (e.g., Dieudonné 1969) or to advanced calculus books.

Exercise 1.22 i) The expected duration of the infectious period is $1/\alpha$. During the infectious period, a force of infection β is exerted on all members of the susceptible population, which, in the initial phase, has size N. Hence $R_0 = \beta N/\alpha$. To determine r, the initial rate of increase of the I sub-population, we replace in the equation for dI/dt the variable S by the constant N (this amounts to linearization in the steady state $(S, I) = (N, 0)$). From $dI/dt = (\beta N - \alpha)I$, we conclude that I initially grows with rate $\beta N - \alpha$. Hence $r = \beta N - \alpha$. Clearly R_0 passes the value one exactly when r passes the value zero, whatever parameter α, β or N we vary. Concentrating on the dependence on N, we see that below the threshold value $N_{\text{th}} := \alpha/\beta$ we have $R_0 < 1$ while above it we have $R_0 > 1$.

ii) We have

$$\frac{dI}{dS} = \frac{\beta SI - \alpha I}{-\beta SI} = -1 + \frac{\alpha}{\beta}\frac{1}{S}$$

$$\Rightarrow \quad dI = \left(-1 + \frac{\alpha}{\beta}\frac{1}{S}\right) dS$$

$$\Rightarrow \quad \int^{I(t)} d\xi = \int^{S(t)} \left(-1 + \frac{\alpha}{\beta}\frac{1}{\sigma}\right) d\sigma$$

$$\Rightarrow \quad I(t) = -S(t) + \frac{\alpha}{\beta}\ln S(t) + C$$

$$\Rightarrow \quad \frac{\alpha}{\beta}\ln S(t) - S(t) - I(t) \text{ is independent of } t.$$

iii) In the context of models (and perhaps in reality sometimes as well) the first case of an infection is always a bit mysterious. In deterministic models we have lost the concept of 'one individual.' What we have in mind here is that when we go backwards in time we approach (asymptotically for $t \to -\infty$) the point $(S, I) = (N, 0)$, which describes the situation before the infectious agent appeared. What we leave completely open and unspecified is how and when the agent appeared, simply because such a discrete event affecting one individual cannot be incorporated in our deterministic description.

Because $R_0 > 1$, I at first increases. In general, S can only decrease. Once $\beta S - \alpha < 0$, I decreases too. So $I(t)$ has a limit for $t \to \infty$ and so has $S(t)$. Limits must be steady states. Looking at (1.12), ignoring the R variable, which is redundant, we see that all points on the S axis are steady states and these are the only ones. So, $I(\infty) = 0$. Since $\frac{\alpha}{\beta}\ln S(t) - S(t) - I(t)$ is independent of t, its values at $t = \pm\infty$ must be equal, that is

$$\frac{\alpha}{\beta}\ln S(\infty) - S(\infty) = \frac{\alpha}{\beta}\ln N - N.$$

Rewriting this identity as

$$\ln \frac{S(\infty)}{N} = \frac{\beta N}{\alpha}\left(\frac{S(\infty)}{N} - 1\right),$$

we only have to realize that the fraction $s(\infty)$ introduced earlier is precisely $S(\infty)/N$, to conclude that we have derived (1.11).

iv) A brute force solution to the difficulty of modeling the start of the epidemic is to specify at some particular time instant (which without loss of generality we take to be zero) values for S, I and R that add up to N. And if we really intend the time instant to be the 'start,' it makes sense to take $R(0) = 0$ and $I(0)$ small. But anyhow we know from ii) that

$$\frac{\alpha}{\beta} \ln S(t) - S(t) - I(t) = \frac{\alpha}{\beta} \ln S(0) - S(0) - I(0).$$

Subtracting $\ln N$ from both sides and rearranging, we find, using $s(t) := S(t)/N$,

$$\ln s(t) = \ln s(0) + \frac{\beta N}{\alpha} \left(s(t) - \frac{S(0) + I(0)}{N} + \frac{I(t)}{N} \right).$$

Arguing exactly as in iii), we deduce that $I(t) \to 0$ for $t \to \infty$. So if we choose $R(0) = 0$ or, equivalently, $S(0) + I(0) = N$ then, by taking the limit $t \to \infty$, we obtain

$$\ln s(\infty) = \ln s(0) + R_0(s(\infty) - 1).$$

Note that $s(0) < 1$, so $\ln s(0) < 0$. The arguments of Exercise 1.19 carry over and show that the equation $\ln s = \ln s(0) + R_0(s - 1)$ has exactly two roots, one bigger than one and (the relevant) one in the interval (0,1). See Figures 16.2 and 16.3, where the functions $y = \ln s$ and $y = \ln s(0) + R_0(s - 1)$ are plotted for $R_0 < 1$ and $R_0 > 1$.

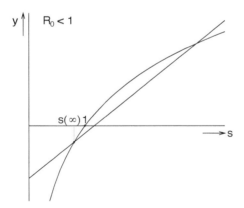

Figure 16.2: Representative graphs of the functions $y = \ln s$ and $y = \ln s(0) + R_0(s - 1)$ for a value of $R_0 < 1$.

For $R_0 < 1$ the relevant root is only slightly less than $s(0)$ when $s(0)$ itself is near one (and hence $\ln s(0)$ is almost zero). In fact it converges to 1 for $s(0) \uparrow 1$. This consolidates our conclusion that for $R_0 < 1$ the size of the epidemic is so small that the name 'epidemic' doesn't even apply.

In contrast, for $R_0 > 1$ and $s(0)$ close to 1, the relevant root is slightly smaller than the (relevant) root of (1.11), and for $s(0) \uparrow 1$ we have convergence towards

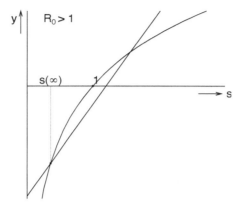

Figure 16.3: Representative graphs of the functions $y = \ln s$ and $y = \ln s(0) + R_0(s-1)$ for a value of $R_0 > 1$.

that root. So, indeed, for $R_0 > 1$ we observe a sizable epidemic, no matter how small we make the initial introduction of infectives.

v) See Figure 16.4.

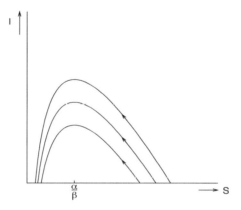

Figure 16.4: Schematic representation of typical trajectories of the ODE system given in Exercise 1.22-v.

vi) A necessary condition for I to be maximal is that $dI/dt = 0$. Since $dI/dt = (\beta S - \alpha)I$ and $I \neq 0$, this requires that $S = \alpha/\beta$. And, indeed, in the phase portrait of Figure 16.4 we see all curves reaching their maximum at this value of S.

If we 'freeze' S at a particular value, the argument presented in i) predicts that a case, on average, will produce $\beta S/\alpha$ new cases. As long as this number exceeds one, we will observe an increase in the infected sub-population, as soon as it falls below one, in contrast, we will observe a decrease. This is exactly what Figure 16.4 shows.

vii) In the elaboration of Exercise 1.19 we have already observed that $s(\infty)$ is a decreasing function of R_0. Since $R_0 = \beta N/\alpha$ is an increasing function of N, this observation suffices.

In vi) we noted that I starts to decrease as soon as S falls below the level α/β. Furthermore, we saw that, for S below this level, a case produces less than one new case, on average. The point, however, is that there may be very many cases around at the moment that S passes this level. So, even though one case is expected to produce less than one new case, all of these together may still produce many new cases.

How many cases there are when S reaches the critical level α/β depends on the population size. This is clearly demonstrated in the phase portrait in Figure 16.4. The further to the right on the S axis we start, the higher the peak of the I level will be, and the lower we 'end' on the S axis. So by 'overshoot phenomenon' we mean that there will be many new cases after the size of the susceptible sub-population has dropped below the critical level, simply because there are many infectives around.

viii) Let a small letter correspond to the quantity denoted by the corresponding capital letter divided by N (note that the symbol i then denotes the prevalence as a fraction and not, as in the main text following this exercise, the incidence). Dividing both sides of (1.12) by N, we get, in this notation, the system

$$\frac{ds}{dt} = -\beta N s i,$$

$$\frac{di}{dt} = \beta N s i - \alpha i,$$

$$\frac{dr}{dt} = \alpha i.$$

So the structure is preserved, but the effect is that β becomes multiplied by N.

The force of infection has dimension $(\text{time})^{-1}$ and equals, in this model, βI. The dimension of β thus depends on the dimension of I, which we have deliberately left unspecified. To make the connection with the quantities we introduced earlier, it helps to write $\beta I = \beta N I/N$ and to realize that I/N is the fraction of the population that is infective, i.e., in the deterministic setting and assuming random 'mixing,' the probability that a contact is with an infective. We may then compare βN with the product pc, i.e., with the number of contacts per unit of time multiplied by the probability that transmission occurs given a contact between a susceptible and an infectious individual. If we take β to be constant, this means that c is proportional to N.

Warning: In the text of Section 1.3.2 following Exercise 1.22 the symbol i has a different meaning.

Exercise 1.23 If we integrate both sides of

$$\frac{s'(t)}{s(t)} = c \int_0^\infty A(\tau) s'(t - \tau) \, d\tau$$

from $-\infty$ to $+\infty$, we find

$$
\begin{aligned}
\ln \frac{s(\infty)}{s(-\infty)} &= \int_{-\infty}^{\infty} \frac{s'(t)}{s(t)} \, dt = c \int_{-\infty}^{\infty} \int_{0}^{\infty} A(\tau) s'(t-\tau) \, d\tau \, dt \\
&= c \int_{0}^{\infty} A(\tau) \int_{-\infty}^{\infty} s'(t-\tau) \, dt \, d\tau \\
&= c \int_{0}^{\infty} A(\tau) \{ s(\infty) - s(-\infty) \} \, d\tau.
\end{aligned}
$$

(Here we have interchanged the order of the two integrations. The monotonicity of s, and hence the existence of $s(\pm\infty)$, is compatible with (1.16) and required by the interpretation, and can moreover be proved rigorously.[1] Since the proof would force us to discuss the way we specify the initial condition as well as the existence and uniqueness of a solution, we take the monotonicity of s for granted here.) Since $s(-\infty) = 1$ we can rewrite the identity as

$$
\ln s(\infty) = c \int_{0}^{\infty} A(\tau) d\tau \, (s(\infty) - 1)
$$

which is exactly (1.11) since $c \int_{0}^{\infty} A(\tau) d\tau = cp(T_2 - T_1) = R_0$.

Exercise 1.24 Whenever we have $c = \theta N(t)$ and mixing is at random, i.e., the probability that a contact is with a susceptible equals the fraction $S(t)/N(t)$ of susceptibles in the population, we have equation (1.18) and the dependence on $N(t)$ cancels, allowing us to derive (1.11). So the correct answer is: no, it does not!

Exercise 1.25 The fraction that got infected but survived is $n(\infty) - s(\infty)$. The fraction that got infected is $1 - s(\infty)$. So if f denotes the probability of surviving infection then (1.20) should hold.

Exercise 1.26 i) Taking logarithms of (1.21), we arrive at

$$
\ln n(\infty) = \frac{1-f}{R_0} \ln s(\infty).
$$

By (1.20), we have
$$
\ln n(\infty) = \ln(s(\infty) + f(1 - s(\infty))).
$$

The combination of these two identities implies (1.22).

ii) Consider the graphs in Figure 16.5 and Figure 16.6, where the functions are $y = \ln s$ and $y = \frac{R_0}{1-f} \ln(f + (1-f)s)$, respectively.

A first key element in the underpinning of these graphs is the computation of the derivatives in $s = 1$ of both curves:

$$
\frac{d}{ds} \ln s = \frac{1}{s}, \quad \text{so this derivative equals 1 for } s = 1;
$$

$$
\frac{d}{ds} \frac{R_0}{1-f} \ln(f + (1-f)s) = \frac{R_0}{1-f} \frac{1-f}{f + (1-f)s} = \frac{R_0}{f + (1-f)s},
$$

so this derivative equals R_0 for $s = 1$. The intersection of the two curves for the cases $R_0 > 1$ and $R_0 < 1$ is therefore correctly represented in the two graphs.

---

[1] O. Diekmann: Limiting behavior in an epidemic model. *Nonl. Anal. Theory, Meth. Appl.*, **1** (1977), 459–470.

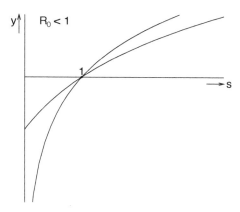

Figure 16.5: Typical graphs of the functions $y = \ln s$ and $y = \frac{R_0}{1-f} \ln(f + (1-f)s)$ for the case $R_0 < 1$, $f \in (0,1)$.

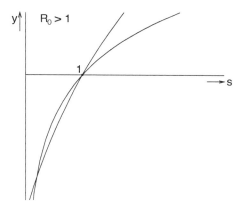

Figure 16.6: Typical graphs of the functions $y = \ln s$ and $y = \frac{R_0}{1-f} \ln(f + (1-f)s)$ for the case $R_0 > 1$, $f \in (0,1)$.

For $s = 0$ we have $\frac{R_0}{1-f} \ln(f + (1-f)s) = \frac{R_0}{1-f} \ln f < 0$ but finite (whereas $\ln s \downarrow -\infty$ for $s \downarrow 0$). The ordering of the values at $s = 0$ is therefore also correctly represented in the two graphs.

Next we look at the derivative of the difference:

$$\frac{d}{ds}\left(\ln s - \frac{R_0}{1-f} \ln(f + (1-f)s)\right) = \frac{1}{s} - \frac{R_0}{f + (1-f)s},$$

which tends to $+\infty$ for $s \downarrow 0$ and equals $1 - R_0$ for $s = 1$. Moreover, in order for the right-hand side to equal zero we should have $f + (1-f)s = R_0 s \Rightarrow s = \frac{f}{R_0 - 1 + f}$. The point is that this derivative has at most one zero in $(0,1)$. So for $R_0 < 1$, when the derivative is positive (at $s = 1$ and near $s = 0$), there can be no zero at all for this derivative: it is positive on the entire interval $(0,1)$. Consequently, the function

$\ln s - \frac{R_0}{1-f}\ln(f + (1-f)s)$ cannot have two or more zeros on $(0,1)$ and the picture for $R_0 < 1$ is confirmed. For $R_0 > 1$, on the other hand, the derivative is negative at $s = 1$ and positive near $s = 0$, so the derivative has exactly one zero. But then the function $\ln s - \frac{R_0}{1-f}\ln(f + (1-f)s)$ cannot have three or more zeros on $(0,1)$, which confirms our picture for the case $R_0 > 1$.

To summarize: 1 is always a root of (1.22) and for $R_0 < 1$ this is the only root in $[0,1]$, but for $R_0 > 1$ there is exactly one additional root $s(\infty) \in (0,1)$. For $R_0 = 1$ the root 1 is a double root.

iii) We use the fact that $\ln(1+x) = x + O(x^2)$ for $x \to 0$. Writing $f = 1 - \varepsilon$ and assuming $\varepsilon \downarrow 0$, we deduce $\ln(f + (1-f)s) = \ln(1 - \varepsilon(1-s)) = -\varepsilon(1-s) + O(\varepsilon^2)$. Inserting this into equation (1.22), we find

$$\ln s(\infty) = \frac{R_0}{\varepsilon}\{-\varepsilon(1 - s(\infty)) + O(\varepsilon^2)\} = R_0(s(\infty) - 1) + O(\varepsilon),$$

and in the limit $\varepsilon \downarrow 0$ we indeed arrive at (1.11).

iv) For $f \uparrow$ we have $(1-f) \downarrow$ and hence $1/(1-f) \uparrow$. Moreover, for $f \uparrow$ we have, for $0 < s < 1$, that $(s + f(1-s)) \uparrow$ and hence that $\ln(s + f(1-s)) = \ln(f + (1-f)s) \uparrow$. We conclude that the right-hand side of (1.22) is an increasing function of f. A graphical argument (see Figure 16.6 for $R_0 > 1$) now shows that $s(\infty)$ is a monotonically increasing function of f. An analytical elaboration can be based on the implicit function theorem and involves sign considerations about derivatives with respect to s, as presented in ii) above.

v) When $f \downarrow 0$ the right-hand side of (1.22) tends to $R_0 \ln s$ for any $s > 0$. So, since $\ln s < 0$ for $0 < s < 1$, the right-hand side is less than the left-hand side when $R_0 > 1$, for any $s \in (0,1)$, and f sufficiently small (depending on R_0 and s). Hence (see Figure 16.6), $s(\infty) \downarrow 0$ for $f \downarrow 0$. Next look at (1.20) and conclude that also $n(\infty) \downarrow 0$ for $f \downarrow 0$.

Exercise 1.27 The relation (1.20) implies that

$$f = \frac{n(\infty) - s(\infty)}{1 - s(\infty)}.$$

Taking logarithms of (1.21), we find, after a little rearrangement,

$$R_0 = (1-f)\frac{\ln s(\infty)}{\ln n(\infty)} = \frac{1}{1 - s(\infty)}\frac{s(\infty) - n(\infty) + s(\infty)}{1}\frac{\ln s(\infty)}{\ln n(\infty)},$$

giving the desired result.

Exercise 1.28 i) The parameter γ is the product of the expected number of contacts per unit of time and the probability of transmission given a contact between an infective and a susceptible. The product $\gamma\frac{SI}{N}$ may now be read as $\gamma\frac{S}{N}I$, in which case we consider the contacts made by the infectives and say that the probability that such a contact is with a susceptible equals the fraction S/N of susceptibles in the population. We can also read the product as $\gamma\frac{I}{N}S$, in which case we consider the contacts made by the susceptibles and say that the probability that such a contact is with an infective equals the fraction I/N of infectives in the population. Note that consistency is guaranteed. When we take γ to be a given constant (parameter), the number of contacts per unit of time is indeed independent of population size N.

A special feature of the model is the way in which we have incorporated mortality. The more usual way is to add a term $-\nu I$ to the equation for dI/dt, which is then interpreted as infectives being exposed to an additional force of mortality ν. A consequence is that the length of the infectious period depends on the parameter ν.

The present system expresses a different point of view: the length of the infectious period is exponentially distributed with parameter α, but whether or not the infectious period ends with fatality is determined by another chance process, which gives a probability f to survive.

ii) When $S = N$, the expected number of secondary cases per primary case equals γ/α and the real-time growth of the infected sub-population is governed by $dI/dt = (\gamma - \alpha)I$.

iii) We start from

$$\frac{dS}{dN} = \frac{\gamma \frac{SI}{N}}{(1-f)\alpha I} = \frac{\gamma S}{(1-f)\alpha N} = \frac{R_0 S}{(1-f)N}.$$

Now separate variables, $\frac{dS}{S} = \frac{R_0}{1-f}\frac{dN}{N}$, then integrate from σ to t,

$$\ln \frac{S(t)}{S(\sigma)} = \frac{R_0}{1-f}\ln\frac{N(t)}{N(\sigma)},$$

and finally take exponentials,

$$\frac{S(t)}{S(\sigma)} = \left(\frac{N(t)}{N(\sigma)}\right)^{\frac{R_0}{1-f}}.$$

For $\sigma \downarrow -\infty$ both $S(\sigma)$ and $N(\sigma)$ tend to the initial population size. So, by taking the limits $\sigma \downarrow -\infty$ and $t \uparrow \infty$, we arrive at (1.21).

iv) Following the hint, we have

$$N(\infty) - N(-\infty) = -(1-f)\alpha \int_{-\infty}^{+\infty} I(\tau)\,d\tau.$$

Since $I(t) \downarrow 0$ for both $t \downarrow -\infty$ and $t \uparrow \infty$, we find by integrating $dI/dt = -dS/dt - \alpha I$ that $0 = -S(\infty) + S(-\infty) - \alpha \int_{-\infty}^{+\infty} I(\tau)\,d\tau$. Next we eliminate $\int_{-\infty}^{+\infty} I(\tau)\,d\tau$ from these two identities and arrive at $N(\infty) - N(-\infty) = (1-f)(S(\infty) - S(-\infty))$. Now divide by $N(-\infty)$, realise that $S(-\infty) = N(-\infty)$ and obtain $n(\infty) - 1 = (1-f)(s(\infty) - 1)$, which is just another way of writing (1.20).

v) Start from

$$\left.\begin{array}{l} \frac{1}{S}\frac{dS}{dt} = -\gamma\frac{I}{N} \\[2ex] \frac{1}{N}\frac{dN}{dt} = -(1-f)\alpha\frac{I}{N} \end{array}\right\} \quad \frac{1}{N}\frac{dN}{dt} = \frac{(1-f)\alpha}{\gamma}\frac{1}{S}\frac{dS}{dt} = \frac{(1-f)}{R_0}\frac{1}{S}\frac{dS}{dt},$$

then integrate with respect to time from σ to t and take exponentials.

Exercise 1.29 i) With c equal to the number of contacts per unit time and $A(\tau)$ equal to the product of the probability to be alive τ units of time after infection and the probability that transmission takes place, given a contact with a susceptible, τ units of time after infection we have, by definition, $R_0 = c \int_0^\infty A(\tau)\,d\tau$.

ii) We derive a second identity involving $\int_0^\infty A(\tau)\,d\tau$ as follows: $f-1=\mathcal{F}(\infty)-1=\int_0^\infty \mathcal{F}'(\tau)\,d\tau=-\int_0^\infty \mu(\tau)\mathcal{F}(\tau)\,d\tau=-q\int_0^\infty a(\tau)\mathcal{F}(\tau)\,d\tau=-q\int_0^\infty A(\tau)\,d\tau$. If we eliminate $\int_0^\infty A(\tau)d\tau$ from this identity and the relation for R_0, we find

$$q=c\frac{1-f}{R_0}.$$

Exercise 1.30 Rewriting (1.26) as

$$N(t)=S(t)-\int_{-\infty}^t S'(\tau)\mathcal{F}(t-\tau)\,d\tau,$$

we obtain by differentiation

$$N'(t)=S'(t)-S'(t)\mathcal{F}(0)-\int_{-\infty}^t S'(\tau)\mathcal{F}'(t-\tau)\,d\tau.$$

The first two terms on the right-hand side cancel each other, and $\mathcal{F}'(t-\tau)=-\mu(t-\tau)\mathcal{F}(t-\tau)=-qa(t-\tau)\mathcal{F}(t-\tau)=-qA(t-\tau)$, so we have in fact derived that

$$N'(t)=q\int_{-\infty}^t S'(\tau)A(t-\tau)\,d\tau=q\int_0^\infty S'(t-\tau)A(\tau)\,d\tau,$$

and consequently

$$\frac{d}{dt}\ln N(t)=\frac{N'(t)}{N(t)}=\frac{q\int_0^\infty S'(t-\tau)A(\tau)\,d\tau}{N(t)}.$$

We can then rewrite (1.25) as

$$\frac{d}{dt}\ln S(t)=\frac{S'(t)}{S(t)}=\frac{c\int_0^\infty S'(t-\tau)A(\tau)\,d\tau}{N(t)}=\frac{c}{q}\frac{d}{dt}\ln N(t).$$

If we now integrate and use the expression for q derived in Exercise 1.29, we obtain (1.32).

Exercise 1.31 We rewrite, as above, (1.26) as

$$N(t)=S(t)-\int_{-\infty}^t S'(\tau)\mathcal{F}(t-\tau)\,d\tau.$$

For $t\to\infty$ we have $\mathcal{F}(t-\tau)\to f$ (uniformly on every bounded τ interval), so, by taking the limit $t\to\infty$, we find

$$N(\infty)=S(\infty)-f\int_{-\infty}^\infty S'(\tau)\,d\tau=S(\infty)-f(S(\infty)-N(-\infty)),$$

which, after division by $N(-\infty)$, is precisely (1.20).

Exercise 1.32 The differences are irrelevant if population size is fixed. They pertain to the dependence of contact intensity on population size/density. In the foregoing section we assumed that the combination of individual behavior and the way we measure population size is such that it is reasonable to assume that the number of contacts per unit time per individual is proportional to population size.

Even if infection may lead to fatalities, we can ignore the effect on population size in the very early phase of an epidemic, when there are just a few cases. This means that the determination of R_0 and r for an infection in a particular population is not influenced at all by the difference between the two models.

However, when comparing the effect of the introduction of the infection in two (or more) populations that differ in size, the difference between the two models will manifest itself.

In the model of the preceding section we have $R_0 = \beta N/\alpha$, $r = \beta N - \alpha$ and $s(\infty)$ is determined by (1.11). We see that both R_0 and r increase with N, while $s(\infty)$ decreases with N.

In the model of the present section we have $R_0 = \gamma/\alpha$, $r = \gamma - \alpha$ and $s(\infty)$ is determined by (1.22). All of these quantities are independent of population size.

In both cases the probability that the introduction of one initial case does not lead to an epidemic is given, according to (1.8), by $1/R_0$. So, again, this is independent of population size for the model in this section, but decreases with population size for the model of the preceding section.

Exercise 1.33 It is reasonable to take this as a constant, independent of population size. This clearly leaves out certain subtleties: people who want to be very sexually active may be attracted to big cities, and a small number of whales in a large ocean may have difficulty finding each other for mating.

Exercise 1.34 i) The rate of leaving state (n, m) equals $\beta nm + \alpha m$. State (n, m) is entered from state $(n, m + 1)$ at rate $\alpha(m + 1)$ and from state $(n + 1, m - 1)$ at rate $\beta(n + 1)(m - 1)$; see Figure 16.7 and Figure 16.8.

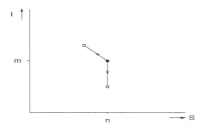

Figure 16.7: Schematic representation of state transitions from the starting state $(S, I) = (n, m)$.

ii) Initially, there are $N - 1$ susceptibles and one infective.

iii) An absorbing state is a state where, once the system attains it, the system will stay forever. In a diagram such as Figures 16.7 and 16.8, we can spot absorbing states by looking for points that have no outgoing arrows. If $m = 0$, there are no infectives and so, according to the interpretation, m should stay zero. We can verify this mathematically:

$$\frac{d}{dt}P_{(n,0)}(t) = \alpha P_{(n,1)}(t)$$

(recall the convention introduced at the beginning of the exercise), which tells us that there is only a positive contribution to the rate of change of $P_{(n,0)}$, i.e., one

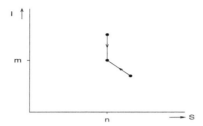

Figure 16.8: Schematic representation of state transitions leading to the state $(S, I) = (n, m)$.

can only enter the state $(n, 0)$ but not leave it. There are no other absorbing states because every other state has $m > 0$ and therefore the corresponding differential equation for $P_{(n,m)}(t)$ always contains a negative contribution $-\alpha m P_{(0,m)}(t)$ caused by an infective being removed.

iv) For $m > 0$, there is a positive probability per unit of time of moving 'down' due to removal, which cannot be compensated, in the long run, by the oblique movement due to infection (just look at the diagram, restricted to the feasible triangular region $n \geq 0, m \geq 0, n + m \leq N$). So $\lim_{t \to \infty} P_{(n,m)}(t) = 0$ for any $m > 0$. This intuitive argument is easily elaborated formally by analyzing the system of differential equations.

v) The probability that there are k victims in total is given by $P_{(N-k,0)}(\infty)$.

vi) For $m > 0$, the rate of leaving state (n, m) equals $\beta nm + \alpha m$, so the sojourn time is exponentially distributed with mean $(\beta nm + \alpha m)^{-1}$. The next state will be either $(n - 1, m + 1)$ or $(n, m - 1)$. The ratio of the probabilities equals the ratio of the rates, that is how one arrives at the expressions. More formally, if

$$\frac{d\mathcal{F}}{dt} = -(\lambda_A + \lambda_B)\mathcal{F}, \quad \mathcal{F}(0) = 1,$$

$$\frac{dP_A}{dt} = \lambda_A \mathcal{F},$$

$$\frac{dP_B}{dt} = \lambda_B \mathcal{F}$$

then $P_A(\infty) = \frac{\lambda_A}{\lambda_A + \lambda_B}$ and $P_B(\infty) = \frac{\lambda_B}{\lambda_A + \lambda_B}$.

vii) Again Figures 16.7 and 16.8 are helpful. But note carefully that for $m = 1$ there is no incoming arrow from the south-east, as there can be no new cases after all infectives have been removed. This observation, together with the results of vi), allow us to reformulate the diagrams in Figures 16.7 and 16.8 in the form of recurrence relations (note that a factor cancels in the expressions for the probabilities). The expression for the initial condition is self-evident.

viii) $Q_{(n,0)}(l)$ does not occur on the right-hand side of any of the recurrence relations.

ix) If everybody gets infected, we have the maximal number of events. These are $N - 1$ transmissions and N removals, so $2N - 1$ events in total.

x)

$$\sum_{l=1}^{2N-1} Q_{(N-k,0)}(l) = \Pr\{k \text{ victims in total}\}.$$

Note that this is easily evaluated numerically by programming the recurrence relations and the initial condition. In fact, however, all Q contributions in the sum are zero except for one. In order to go from state $(N-1,1)$ to state $(N-k,0)$, there have to occur $k-1$ transmissions and k removals, so $2k-1$ events in total (stated differently: there are many routes from $(N-1,1)$ to $(N-k,0)$, but they all involve $2k-1$ steps). So $Q_{(N-k,0)}(l)$ is non-zero if and only if $l = 2k-1$.

xi) Multiply both the numerator and the denominator by N/α and use $R_0 = \beta N/\alpha$.

Exercise 1.35 We do not attempt to draw a three-dimensional analog of Figures 16.7 and 16.8, but instead write in self-explanatory notation

$$(n,m,k) \rightarrow \begin{cases} (n-1,m+1,k) \text{ with rate } \frac{\gamma mn}{n+m+k}; \\ (n,m-1,k) \quad \text{ with rate } (1-f)\alpha m; \\ (n,m-1,k+1) \text{ with rate } f\alpha m. \end{cases}$$

These rates add up to $\frac{\gamma mn}{n+m+k} + \alpha m$. We can now compute the probabilities with which each of the three possible outcomes occurs, as the relative contributions of the rates to the total rate (exactly as we did in Exercise 1.34-vi).

For $m > 1$ there are three ways to arrive at (n,m,k): 1) from $(n+1,m-1,k)$ by a transmission; 2) from $(n,m+1,k)$ by a removal that is fatal; 3) from $(n,m+1,k-1)$ by a removal that is non-fatal. These observations, and the fact that $R_0 = \gamma/\alpha$, show that the stated recurrence relation does the right bookkeeping. The absorbing states are now $(n,0,k)$ with $n+k \leq N$. The maximum number of events is still $2N-1$. And

$$\sum_{l=1}^{2N-1} Q_{(N-r,0,k)}(l) = \Pr\{r \text{ individuals in total are infected,}$$

$$k \text{ of which survived the disease caused by the infection}\}.$$

But, as in Exercise 1.34-x, it follows that $Q_{(N-r,0,k)}(l)$ is non-zero if and only if $l = 2r-1$.

Exercise 1.36 This is indeed the case when (1.11) applies (see Exercise 1.20-i). In the case of (1.22), a second parameter f is involved and we have to take possible differences in f into account when determining $s(\infty)$.

Exercise 1.37 Under the vaccination regime, a fraction v of all contacts is 'wasted' on protected individuals. So the expected number of secondary cases per primary case equals $(1-v)R_0$ and in order that this quantity lies below one we should have $v > 1 - 1/R_0$.

Exercise 1.38 If $z_\infty > 1/2$, it is quite likely that, even though $R_0 > 1$, the introduction of one infective into a virgin population leads to a minor outbreak only. Therefore, if one observes a few minor outbreaks, the conclusion that evidently $R_0 < 1$ is NOT warranted.

Exercise 1.39 By assumption, the chains of infection develop independently as long as there are only a few cases. So the probability that all chains are of the 'minor outbreak' type equals $(z_\infty)^k$.

Exercise 1.40 The probability of a major outbreak when k infected animals enter the herd is given by Exercise 1.39: $1 - (z_\infty)^k$. In the new situation n animals enter. The binomial probability that among these n animals there are exactly k infected animals is given by

$$\binom{n}{k} p^k (1-p)^{n-k}.$$

The probability of a major outbreak starting from k infecteds therefore is the product of this probability and $1 - (z_\infty)^k$. The expectation of the probability of a major outbreak, i.e., summing over all possibilities $k = 0, 1, \ldots, n$, is

$$\sum_{k=0}^{n} \binom{n}{k} p^k (1-p)^{n-k} (1 - (z_\infty)^k).$$

If we expand this and use the fact that the binomial probabilities sum to 1, we obtain

$$\sum_{k=0}^{n} \binom{n}{k} p^k (1-p)^{n-k} - \sum_{k=0}^{n} \binom{n}{k} (pz_\infty)^k (1-p)^{n-k}$$

$$= 1 - (pz_\infty + (1-p))^n = 1 - (1 - p(1 - z_\infty))^n.$$

Note that taking $p = 1$, i.e., all imported animals infected, reduces this to the result of Exercise 1.39.

Exercise 1.41 The first infected individual is expected to generate R_0 new cases; therefore, after one generation we expect $1 + R_0$ infected individuals. The R_0 first-generation cases are also expected to make R_0 new cases each in the second generation, etc. We therefore expect a total of

$$1 + R_0 + R_0^2 + R_0^3 + \cdots = \frac{1}{1 - R_0}$$

cases for $R_0 < 1$.

Exercise 1.42 i)

$$I_{k+1} + S_{k+1} - \frac{N}{R_0} \ln S_{k+1}$$

$$= S_k \left(1 - \exp \left(-R_0 \frac{I_k}{N} \right) \right) + S_k \exp \left(-R_0 \frac{I_k}{N} \right) - \frac{N}{R_0} \left(\ln S_k - R_0 \frac{I_k}{N} \right)$$

$$= I_k + S_k - \frac{N}{R_0} \ln S_k.$$

ii) In the relation

$$I_k + S_k - \frac{N}{R_0} \ln S_k = I_l + S_l - \frac{N}{R_0} \ln S_l$$

we take the limit $k \to \infty$ and $l \to -\infty$ and use that both I_k and I_l tend to zero, while S_l tends to N:

$$S_\infty - \frac{N}{R_0} \ln S_\infty = N - \frac{N}{R_0} \ln N.$$

Multiplying this relation by R_0/N we find, after a little rearrangement, the familiar final-size equation (1.11).

iii) We have nothing to add to what we said in the main text.

iv) In discrete-time formulations one should multiply (i.e., work with fractions) rather than subtract. It is quite possible that $\beta I_k S_k$ exceeds (for large β and suitable I_k, S_k) the quantity S_k, in which case we get negative numbers for quantities that ought to be positive.

Note that *for small I_k* we have

$$(1 - e^{-R_0 I_k/N}) = R_0 \frac{I_k}{N} + o(I_k),$$

so the quadratic term shows up by Taylor expansion for small numbers of infectives.

Exercise 1.43 Here μ is the probability per unit of time of dying and $a(t)$ is the expected infectivity at infection age t, conditional on survival.

If we integrate the differential equation for I, we find $I(t) = I_0 \exp(-\int_0^t \mu(\sigma)\, d\sigma)$, and so in the limit for $t \to \infty$: $I(\infty) = f I_0$. Integrating the differential equation for N we obtain $N(\infty) - N_0 = -\int_0^\infty \mu(\sigma) I(\sigma)\, d\sigma = \int_0^\infty \frac{dI}{dt}(\sigma)\, d\sigma = I(\infty) - I_0 = (f-1)I_0$. Considering dS/dN and using separation of variables, we arrive, exactly as in Exercise 1.28-iii, at $S(t)/S_0 = (N(t)/N_0)^{R_0/(1-f)}$.

Now we have to translate this information into the recurrence relation that corresponds to the 'next-generation' procedure. The relation

$$I_{k+1} = S_k - S_{k+1}$$

holds by definition. The relation $N(\infty) - N_0 = (f-1)I_0$ translates into

$$N_{k+1} = N_k - (1-f)I_k.$$

Combining this with $S_{k+1} = S_k(N_{k+1}/N_k)^{R_0/(1-f)}$ (which is the straightforward limit $t \to +\infty$ of an equation derived in the elaboration of Exercise 1.28), we deduce that

$$S_{k+1} = S_k \left(1 - (1-f)\frac{I_k}{N_k}\right)^{R_0/(1-f)}.$$

(So we can compute I_{k+1} even though S_{k+1} occurs on the right-hand side of the equation for I_{k+1}.)

It is actually helpful to also rewrite the relation between $S(t)/S_0$ and $N(t)/N_0$, derived above, in the form

$$\frac{S_{k+1}}{S_k} = \left(\frac{N_{k+1}}{N_k}\right)^{R_0/(1-f)},$$

because, by induction, this implies that

$$\frac{S_\infty}{S_0} = \left(\frac{N_\infty}{N_0}\right)^{R_0/(1-f)},$$

which, with $I_0 + S_0 = N_0$ and in the limit $I_0 \downarrow 0$, is (1.21) in minor disguise. Now define $R_0 = 0$ and $R_{k+1} = R_k + f I_k$; then we shall prove, by induction, that $N_k = S_k + I_k + R_k$. For $k = 0$ this is true by assumption. Suppose it holds for $k = l$; then $S_{l+1} + I_{l+1} + R_{l+1} = S_l + R_l + f I_l$, while $N_{l+1} = N_l - I_l + f I_l =$

$S_l + I_l + R_l - I_l + fI_l = S_l + R_l + fI_l$, so the relation holds for $k = l+1$. Likewise, we prove by induction that $R_{k+1} = f(N_0 - S_k)$. Again this is true for $k = 0$ by the assumption $R_0 = 0, N_0 = S_0 + I_0$. Suppose it is true for $k = l$. Then $R_{l+1} = R_l + fI_l = f(N_0 - S_{l-1}) + fI_l = f(N_0 - S_{l-1} + I_l) = f(N_0 - S_l)$, which shows that the relation is true for $k = l+1$. In the limit for $k \to \infty$ we find $N_\infty = S_\infty + R_\infty$ and $R_\infty = f(N_0 - S_\infty)$, which, in combination, and after division by N_0, yields (1.20).

Exercise 1.44 Recall from calculus (or take our word for it) that

$$\lim_{y \to \infty} \left(1 + \frac{x}{y}\right)^y = e^x.$$

In the present context take $y = -1/(1-f)$ in the recurrence relation for S and use the outcome in the recurrence relation for I.

16.2   ELABORATIONS FOR CHAPTER 2

Exercise 2.1 The $-\alpha I$ term in the equation for dI/dt represents that the length of the infectious period is exponentially distributed with parameter α. In other words, if $P_I(\tau)$ denotes the probability to be still infectious at time τ after infection then

$$\frac{dP_I}{d\tau} = -\alpha P_I, \quad P_I(0) = 1,$$

and consequently $P_I(\tau) = e^{-\alpha \tau}$.

Assume that death does not occur and that removed individuals are immune. We should read the term βSI as $\beta N \frac{S}{N} I$. If we consider just one infected individual, we delete the factor I. The factor S/N is the fraction of susceptibles in the total population, and is therefore declared to be the probability that a contact is with a susceptible (rather than with an infected or immune individual). The factor βN is in fact the product of two quantities: the number of contacts per unit of time and the probability of transmission, given a contact between an infectious and a susceptible individual. The number of contacts per unit of time we called c. So the probability of transmission, given a contact between an infectious individual and a susceptible individual, equals $\beta N/c$.

Let $A(\tau)$ be the probability that a susceptible is infected when it comes into contact with another individual that was itself infected τ units of time ago. Then

$$A(\tau) = \frac{\beta N}{c} P_I(\tau) = \frac{\beta N}{c} e^{-\alpha \tau}.$$

When death does occur, but we are still in the initial phase where the influence of the agent on population size can be neglected, the factor $\beta N/c$ is motivated exactly as above, but $P_I(\tau)$ should be interpreted as the probability to be alive and infectious at time τ after infection.

Exercise 2.2 i) Let $P_E(\tau)$ denote the probability to be still in the latency period at time τ after infection and $P_I(\tau)$ the probability to be in the infectious period at that time. Then

$$\frac{dP_E}{d\tau} = -\theta P_E, \qquad P_E(0) = 1,$$

$$\frac{dP_I}{d\tau} = \theta P_E - \alpha P_I, \qquad P_I(0) = 0,$$

from which it follows that $P_E(\tau) = e^{-\theta\tau}$ and

$$P_I(\tau) = \int_0^\tau e^{-\alpha(\tau-\sigma)}\theta P_E(\sigma)\, d\sigma$$

(this last formula is an instance of the variation-of-constants formula from the theory of ordinary differential equations (see e.g., Hale 1969), and has exactly the interpretation outlined in the hint). After some calculus, this expression becomes

$$P_I(\tau) = \frac{\theta}{\alpha-\theta}(e^{-\theta\tau} - e^{-\alpha\tau}).$$

In the preceding exercise it was already shown that

$$A(\tau) = \frac{\beta N}{c}P_I(\tau).$$

Combination of these two identities yields the desired result.

ii) We find that

$$R_0 = \beta N \frac{\theta}{\alpha-\theta}\left(\frac{1}{\theta} - \frac{1}{\alpha}\right) = \frac{\beta N}{\alpha},$$

and see that time delay does not influence R_0, which is a 'generation' quantity. Time delays do, however, influence r, which is a 'real-time' quantity.

iii) The formulation of this exercise is a bit vague. The idea is the following. Let some $A = A(\tau)$ be given. Extend the domain of definition of the function A by the convention that $A(\tau) = 0$ for $\tau < 0$.

Now consider a one-parameter family of 'shifted' versions of A that at τ assume the value $A(\tau - T_L)$, where the parameter $T_L \geq 0$ corresponds to the latency period (see Figure 16.9).

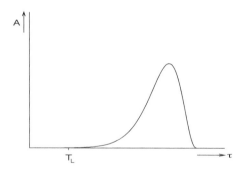

Figure 16.9: Typical shape of an infectivity function A, as a function of infection age τ for an immunizing infection, showing a latency period of length T_L.

Then, on the one hand,

$$\int_0^\infty A(\tau - T_L)\, d\tau = \int_{T_L}^\infty A(\tau - T_L)\, d\tau = \int_0^\infty A(\sigma)\, d\sigma$$

and so the value of R_0 is independent of the parameter T_L. On the other hand, we have

$$\int_0^\infty e^{-r\tau}A(\tau - T_L)\, d\tau = e^{-rT_L}\int_0^\infty e^{-r\sigma}A(\sigma)\, d\sigma$$

(so for fixed r this is a decreasing function of T_L, which implies that the root of the equation

$$1 = ce^{-rT_L} \int_0^\infty e^{-r\sigma} A(\sigma)\, d\sigma$$

shifts to the left, and so becomes smaller, if we increase T_L).

Exercise 2.3 The change is incorporated as a factor c_{reduced}/c, so, for example,

$$A(\tau) = \begin{cases} pc_{\text{reduced}}/c & \text{if } T_1 \leq \tau \leq T_2; \\ 0 & \text{otherwise.} \end{cases}$$

Then A is no longer a conditional probability, but also incorporates aspects of the contact process.

Exercise 2.4 i) When we use the word 'expected' in relation to $A(\tau)$, we refer to everything that might happen after some individual is infected, and, more precisely, we should take into account all that can occur in the interval of length τ after the moment of infection.

Only host individuals that are alive take part in the contact process and can transmit the agent to another host. So we have to incorporate the survival probability $\exp(-\mu\tau)$ as a factor in $A(\tau)$.

ii) The idea is to redo Exercise 2.2-iii, but now with

$$A(\tau) = e^{-\mu\tau}\widetilde{A}(\tau).$$

When 'shifting' to incorporate the effect of a latency period, we shift $\widetilde{A}$ while keeping the survival probability unchanged. That is, we consider the one-parameter family of functions that at time τ take the value $e^{-\mu\tau}\widetilde{A}(\tau - T_L)$, with $\widetilde{A}$ being defined to be zero for negative values of the argument. Then

$$R_0 = R_0(T_L) = c \int_0^\infty e^{-\mu\tau}\widetilde{A}(\tau - T_L)\, d\tau = e^{-\mu T_L} c \int_0^\infty e^{-\mu\sigma}\widetilde{A}(\sigma)\, d\sigma.$$

iii) Starting from

$$\frac{dP_I}{d\tau} = -(\alpha + \mu)P_I \quad \Rightarrow \quad P_I(\tau) = e^{-(\alpha+\mu)\tau},$$

the arguments presented in the elaboration of Exercise 2.1 yield

$$R_0 = \beta N \int_0^\infty e^{-(\alpha+\mu)\tau} d\tau = \frac{\beta N}{\alpha + \mu}.$$

iv) The probability to survive the latency period and to enter the infectious period equals $\theta/(\theta + \mu)$ (see Exercise 1.34-vi). If that happens, we are in the situation just considered in 2.4-iii.

v) The time scale of the infectious period is $1/\alpha$ and the time scale of life is $1/\mu$. When $1/\mu \gg 1/\alpha$, we can ignore death (here $\gg$ means is much larger than). We can also express this condition as $\mu \ll \alpha$. (When there is a latency period we also require that $1/\mu \gg 1/\theta$, i.e., $\mu \ll \theta$, where $1/\theta$ is the expected length of the latency period.)

Exercise 2.5 The picture in Figure 16.10 gives a useful symbolic representation of the assumptions.

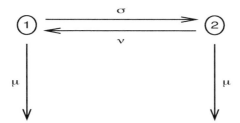

Figure 16.10: Schematic representation of a system with two states as specified in Exercise 2.5.

i) When $\mu = 0$ but $\nu > 0$, infected individuals will either always be infectious (if $\sigma = 0$) or will 'visit' the state 1 infinitely often. Each visit has an exponentially distributed duration with parameter σ. So the expected total time spent in 1 equals infinity.

ii) If you don't follow the hint, you need to do some calculations, which are presented here in self-explanatory notation.

$$\frac{dP_1}{d\tau} = -\mu P_1 - \sigma P_1 + \nu P_2, \qquad P_1(0) = 1,$$

$$\frac{dP_2}{d\tau} = -\mu P_2 + \sigma P_1 - \nu P_2, \qquad P_2(0) = 0,$$

$$\Rightarrow \frac{d}{d\tau}(P_1 + P_2) = -\mu(P_1 + P_2) \Rightarrow P_1(\tau) + P_2(\tau) = e^{-\mu\tau}.$$

Hence

$$\frac{dP_1}{d\tau} = -(\mu + \sigma + \nu)P_1 + \nu e^{-\mu\tau}, \qquad P_1(0) = 1.$$

Instead of solving this equation, we integrate it from 0 to $+\infty$ (note that one can deduce, with a bit of effort, from the equation that $P_1(\tau) \to 0$ for $\tau \to \infty$, as should be the case according to the interpretation) to obtain

$$-1 = -(\mu + \sigma + \nu) \int_0^\infty P_1(\tau)\,d\tau + \frac{\nu}{\mu}.$$

It follows readily that

$$\int_0^\infty P_1(\tau)\,d\tau = \frac{\mu + \nu}{\mu(\mu + \sigma + \nu)},$$

and the definition of h_1 is meant to imply the stated identity from this.

Exercise 2.6 Consider a continuous-time Markov chain with transition rate matrix Σ. Then $e^{\tau\Sigma}$ is the fundamental matrix solution; see e.g., Hale (1969). When applied to the unit vector e_j, this gives the vector of probabilities to be in the various states at time τ, given that we started in state j. Now integrate with respect to τ from 0 to ∞, and use that $\int_0^\infty e^{\tau\Sigma}\,d\tau = -\Sigma^{-1}$ whenever Σ is defective, in the sense that $\exp(\tau\Sigma)$ decreases exponentially at ∞. So if we take the ith component of $-\Sigma^{-1}e_j$, we obtain the expected total time spent in state i, given that we started in state j. But this is exactly $(-\Sigma^{-1})_{ij}$.

Exercise 2.7 i) The scheme of Figure 16.10 is represented by the matrix

$$\Sigma = \begin{pmatrix} -(\mu + \sigma) & \nu \\ \sigma & -(\mu + \nu) \end{pmatrix}.$$

Hence

$$-\Sigma^{-1} = \frac{1}{(\mu + \sigma)(\mu + \nu) - \nu\sigma} \begin{pmatrix} \mu + \nu & \nu \\ \sigma & \mu + \sigma \end{pmatrix}.$$

The starting vector is $\begin{pmatrix} 1 \\ 0 \end{pmatrix}$ and the infectivity vector is $(h_1 \ 0)^\top$, so

$$-h \cdot \Sigma^{-1} \Theta = \frac{h_1(\mu + \nu)}{(\mu + \sigma)(\mu + \nu) - \nu\sigma} = \frac{h_1(\mu + \nu)}{\mu(\mu + \nu + \sigma)}.$$

Another way of arriving at the same conclusion is to observe that the (1,1)-element of $-\Sigma^{-1}$ is the expected time spent in state 1, given that one starts in state 1.

ii) As a function of ν, the factor $\int_0^\infty A(\tau) \, d\tau$ of R_0 increases from the value $h_1/(\mu + \sigma)$ at $\nu = 0$ to the limit h_1/μ for $\nu \to \infty$.

Exercise 2.8 Look at Figure 2.2. The matrix Σ is now given by

$$\begin{pmatrix} -\sigma_1 & 0 & 0 \\ \sigma_1 & -\sigma_2 & 0 \\ 0 & \sigma_2 & -\alpha \end{pmatrix},$$

and one can use linear algebra to compute the inverse. The interpretation allows us, however, to find a more efficient procedure to calculate the relevant quantities. We start with certainty in state 1, but are also sure to arrive in state 2 (when exactly does not matter for an R_0 calculation). The expected duration of the period spent in state 2 equals $1/\sigma_2$ and we are certain to go to state 3 from there. The expected duration of the period spent in state 3 equals $1/\alpha$. So, from the values given for h_i, we find that

$$\int_0^\infty A(\tau) \, d\tau = \frac{1}{\sigma_2} + \frac{2}{\alpha}.$$

Exercise 2.9 The force of mortality that an individual experiences is proportional to the force of infection that it exerts. Now that we are working with a state concept at the i-level, this should hold for every state separately. Taken together, this translates as $\mu = qh$.

Exercise 2.10 In the deterministic case

$$\int_\Omega f(\xi) m(d\xi) = f(\bar{\xi}) = f \left(\int_\Omega m(d\xi) \right)$$

for every f, since m is the unit measure concentrated at the point $\bar{\xi}$. So, in particular, this holds for

$$f(\xi) = \exp \left(c \int_0^\infty a(\sigma; \xi) \, d\sigma \, (z - 1) \right).$$

Thus we find that

$$g(z) = e^{c \int_0^\infty a(\sigma; \bar{\xi}) \, d\sigma \, (z-1)},$$

which we rewrite as $g(z) = e^{R_0(z-1)}$, since, according to (2.5), $A(\tau) = a(\tau; \overline{\xi})$ and, according to (2.2), $R_0 = \int_0^\infty A(\tau)d\tau$.

Exercise 2.11 i) Jensen's inequality (see e.g., Rudin 1974) tells us that for a convex function φ we have

$$\varphi\left(\int_\Omega f(\xi)m(d\xi)\right) \le \int_\Omega (\varphi \circ f)(\xi)m(d\xi), \qquad (16.1)$$

when m is a positive measure and f is integrable with respect to m. We apply this here by taking $\varphi(x) = \exp(x(z-1))$ and $f(\xi) = c\int_0^\infty a(\sigma; \xi)\, d\sigma$. By Fubini's theorem (about conditions for interchanging the order of integration, see e.g., Kolmogorov and Fomin 1975) and (2.5), we have

$$\int_\Omega f(\xi)m(d\xi) = c\int_0^\infty \int_\Omega a(\sigma; \xi)m(d\xi)\, d\sigma = c\int_0^\infty A(\sigma)\, d\sigma = R_0,$$

so the left-hand side of (16.1) equals $e^{R_0(z-1)}$ while the right-hand side is, according to (2.6), equal to $g(z)$.

ii) The easiest deduction is the graphical argument in Figure 16.11, where we have drawn typical graphs of $y = g(z)$ and the graph of $y = e^{R_0(z-1)}$ and their intersection with the line $y = z$.

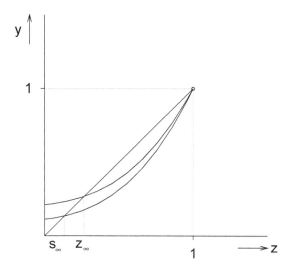

Figure 16.11: Typical graph of a relevant function $y = g(z)$ and of $y = e^{R_0(z-1)}$, and their intersections with the line $y = z$.

iii) Yes. When we repeat the 'experiment' of introducing the agent many times, the probability that we observe a major outbreak is less for the situation where there is variability in infectivity. More or less intuitively obvious: the very first infective may have low infectivity. (Once we have many cases the mean is decisive.) In the initial phase, the low-infectivity part of the distribution weighs more heavily than the high-infectivity part in determining whether or not a major outbreak occurs.

Exercise 2.12 We have

$$\frac{\partial R_0}{\partial c_1} = \frac{2c_1 N_1(c_1 N_1 + c_2 N_2) - N_1(c_1^2 N_1 + c_2^2 N_2)}{(c_1 N_1 + c_2 N_2)^2}$$

$$= \frac{(c_1^2 N_1 + c_2(2c_1 - c_2)N_2)N_1}{(c_1 N_1 + c_2 N_2)^2}.$$

Hence the condition for $\partial R_0/\partial c_1 < 0$ is

$$\frac{N_1}{N_2} < \left(\frac{c_2}{c_1} - 2\right)\frac{c_2}{c_1},$$

or, equivalently,

$$\frac{c_2}{c_1} > 1 + \sqrt{1 + \frac{N_1}{N_2}}.$$

Exercise 2.13 New cases occur in the ratio

$$\frac{c_1 N_1}{c_1 N_1 + c_2 N_2} : \frac{c_2 N_2}{c_1 N_1 + c_2 N_2} = 1 : 10.$$

The ratio among those responsible for transmission can be read from the two terms in the formula for R_0, and so equals

$$\frac{c_1^2 N_1}{c_1 N_1 + c_2 N_2} : \frac{c_2^2 N_1}{c_1 N_1 + c_2 N_2} = 1 : 1.$$

Exercise 2.14 Suppose the contact pattern (i.e., the distribution of contacts over the various types) is τ-independent, but that infectivity is a τ-dependent scalar multiplication factor. Then the distribution of new cases with respect to type will depend on the contact pattern only (while the number may depend on τ), and so will be the same for the generation and real-time perspective.

Exercise 2.15 Write

$$(1 - e^{-\gamma T})^i = \sum_{j=0}^{i} \binom{i}{j} (-1)^j e^{-j\gamma T}$$

and use that

$$\gamma \int_0^\infty e^{-\gamma(1+j+N-i)T} dT = \frac{1}{1+j+N-i}$$

to find that we need to prove the identity

$$\sum_{j=0}^{i} \binom{N}{i} \binom{i}{j} (-1)^j \frac{1}{1+j+N-i} = \frac{1}{N+1}.$$

A little rewriting reveals that this identity is equivalent to the one given in the hint.

16.3 ELABORATIONS FOR CHAPTER 3

Exercise 3.1 When $t = 1$ and $c = 5$ we have the Poisson distribution with mean 5. It follows that the probability that an individual has four contacts equals $5^4 e^{-5}/4! \approx 0.1755$.

Exercise 3.2 The contact rate with a specific other individual is c/N. Since there are $\binom{N+1}{2}$ pairs and each pair has contact rate c/N the overall contact rate equals $\binom{N+1}{2}c/N = (N+1)c/2$. Note that this indeed corresponds to $N + 1$ individuals having contact at rate c, since all contacts are counted twice.

Exercise 3.3 i) A has successful contact with B at rate cp/N. The number of (successful) contacts during the infectious period (of length $\Delta T = t$) is Poisson distributed with mean cpt/N. It follows that there will be no such contact with probability $e^{-cpt/N}$.

ii) $e^{-2cpt/N}$.

iii) $E(e^{-cp\Delta T/N}) = \int_0^\infty e^{-cpt/N}\alpha e^{-\alpha t}dt = \alpha \int_0^\infty e^{-(cp/N+\alpha)t} = \alpha/(\alpha + cp/N)$. Similarly $E(e^{-2cp\Delta T/N}) = \alpha/(\alpha + 2cp/N)$.

iv) The product of the probabilities of the two single events equals $(\alpha/(\alpha + cp/N))^2$. This is clearly not the same as the probability of the multiple event $\alpha/(\alpha + 2cp/N)$. The two events are hence dependent. The well-known fact that $(1+2\epsilon) < (1+\epsilon)^2$ for any $\epsilon > 0$ implies that the multiple event has larger probability. This means that the two events are positively correlated: knowing that one of the events occurs increases the chance that the other also occurs.

Exercise 3.4 Initially one individual is infected. In what follows we specify the specific chain of infection, i.e., the number of infected individuals in each 'generation.' For example, $1 \to 2 \to 1$ means that the first individual infected two individuals, these two individuals infected one in the next generation who did not infect anyone (the 0 is left out in the chain notation). We then have $P(Y = 2) = P(1 \to 2) + P(1 \to 1 \to 1) = \binom{N}{2}(1 - \pi)^2 \pi^{N-2}(\pi^2)^{N-2} + \binom{N}{1}(1 - \pi)^1 \pi^{N-1} \binom{N-1}{1}(1 - \pi)^1 \pi^{N-2}\pi^{N-2}$. For $P(Y = 3)$ we only give the chains and leave it to the reader to compute the probability of each chain: $P(Y = 3) = P(1 \to 3) + P(1 \to 2 \to 1) + P(1 \to 1 \to 2) + P(1 \to 1 \to 1 \to 1)$.

Exercise 3.5 i) Since $X \equiv x$ we have $p_k = (\gamma x)^k e^{-\gamma x}/k!$, i.e., the offspring distribution is Poisson distributed with mean γx.

ii) When X is exponentially distributed with rate α

$$p_k = \int_0^\infty \frac{(\gamma x)^k e^{-\gamma x}}{k!}\alpha e^{-\alpha x}dx \tag{16.2}$$

$$= \frac{\gamma^k \alpha}{(\gamma + \alpha)^{k+1}} \int_0^\infty (\gamma + \alpha)^{k+1} x^k e^{-(\gamma+\alpha)x}dx = \frac{\gamma^k \alpha}{(\gamma + \alpha)^{k+1}}, \tag{16.3}$$

where the last equality can be obtained by partial integration or by recognizing the expression inside the integral as the density of a gamma-distribution (which hence integrates to 1). We hence have that $p_k = \frac{\alpha}{\gamma+\alpha}\left(\frac{\gamma}{\gamma+\alpha}\right)^k$, $k = 0, 1, \ldots$. This means that the offspring distribution is geometric with parameter $p = \alpha/(\alpha+\gamma)$ and mean γ/α.

Exercise 3.6 $\sum_{k=0}^\infty \theta^k m^k e^{-m}/k! = e^{-m} \sum_{k=0}^\infty (\theta m)^k/k! = e^{-m}e^{\theta m} = e^{-m(1-\theta)}$.

Exercise 3.7 i) When $X \equiv x$ then $\phi_2(\gamma(1-\theta)) = E(e^{-\gamma(1-\theta)X}) = e^{-\gamma(1-\theta)x}$. As a consequence, the extinction probability z_∞ is the smallest solution to $\theta = e^{-\gamma x(1-\theta)}$. When $\gamma = 2$ and $x = 1$ the numerical solution equals $z_\infty \approx 0.2059$.

ii) When X is exponential with rate α, $\phi_2(s) = E(e^{-sX}) = \int_0^\infty e^{-sx} \alpha e^{-\alpha x} dx = \alpha/(\alpha+s)$, so $\phi_2(\gamma(1-\theta)) = \alpha/(\alpha+\gamma(1-\theta))$. The extinction probability z_∞ is hence the smallest solution to $\theta = \alpha/(\alpha + \gamma(1-\theta))$. It follows that $z_\infty = \min(1, \ \alpha/\gamma)$. So when $\gamma = 2$ and $\alpha = 1$ the extinction probability is $z_\infty = 0.5$.

Exercise 3.8 i) Let X and Y be independent Poisson variables with means λ and μ, respectively. Then $P(X + Y = k) = \sum_{i=0}^k P(X = i)P(Y = k - i) = \sum_{i=0}^k \frac{\lambda^i e^{-\lambda}}{i!} \frac{\mu^{k-i} e^{-\mu}}{(k-i)!} = \frac{e^{-(\lambda+\mu)}}{k!} \sum_{i=0}^k \binom{k}{i} \lambda^i \mu^{k-i} = \frac{(\lambda+\mu)^k e^{-(\lambda+\mu)}}{k!}$, so $X + Y$ has Poisson distribution with mean $\lambda + \mu$. From this it follows by induction that the jth convolution of a Poisson variable with mean λ is Poisson distributed with mean $j\lambda$.

ii) Consider k independent geometric variables $X_1, \ldots, X_k$ with parameter p, so $P(X = k) = pq^k$, $k = 0, 1, \ldots$ $(q = 1 - p)$. Think of X as the number of failures before a success, when an experiment, which results in success with probability p, is repeated until the first success occurs. This means that $X_1 + \ldots + X_k$ is the number of failures before the kth success. For $X_1 + \ldots + X_k$ to equal j, we must have done $j + k$ experiments out of which j were failures and k were successes, and the location (ordering) of the successes is arbitrary except that the $(j + k)$th experiment must have been a success — otherwise we would have stopped earlier. This motivates the following expression for the distribution of the kth convolution: $P(X_1 + \ldots + X_k = j) = \binom{k+j-1}{k-1} p^k q^j$. This is known as the negative binomial distribution with parameters k and p.

Exercise 3.9 i) When the adult life-length is constant and equal to x, it follows that $P(Y_\infty = j) = \frac{((j+1)\gamma x)^j e^{-(j+1)\gamma x}}{j!} \frac{1}{j+1} = \frac{((j+1)\gamma x)^j e^{-(j+1)\gamma x}}{(j+1)!}$. This distribution is known as the Borel-Tanner distribution. When $R_0 = \gamma x > 1$ these probabilities, summed over all j, add up to the extinction probability z_∞. The remaining probability mass $1 - z_\infty$ is at $\{Y_\infty = \infty\}$, i.e., the branching process grows beyond all limits.[2]

ii) When the adult life-length is exponential with parameter α (and mean $1/\alpha$) the distribution of the total progeny equals $P(Y_\infty = j) = \binom{2j}{j} \pi^{j+1}(1 - \pi)^j/(j+1)$, $j = 0, 1, \ldots$, where $\pi := \alpha/(\alpha + \gamma)$. Also in this case, when $R_0 = \gamma/\alpha > 1$ the distribution sums to the extinction probability z_∞. The remaining probability mass $1 - z_\infty$ is at $\{Y_\infty = \infty\}$, i.e., the branching process grows beyond all limits.

Exercise 3.10 The probability that only susceptible individuals are contacted among the first k_N contacts equals

$$1(1 - \frac{1}{N})(1 - \frac{2}{N}) \ldots (1 - \frac{k_N - 1}{N}) \ = \ 1 - \sum_{i=1}^{k_N - 1} \frac{j}{N} + k_N^2 O(\frac{1}{N^2}) \qquad (16.4)$$

$$= \ 1 - \frac{k_N(k_N - 1)}{2N} + k_N^2 O(\frac{1}{N^2}), (16.5)$$

and the right-hand side clearly converges to 1 as $N \to \infty$ if $k_N = o(\sqrt{N})$.

Exercise 3.11 i) In the Reed-Frost model the infectious period is constant and equal to ΔT and the rate of producing new infections during the early stages is

---

[2]B. Von Bahr and A. Martin-Löf: Threshold limit theorems for some epidemic processes. *Adv. Appl. Prob.*, **12** (1980), 319–349.

equal to cp. This implies that the final size Y_N may be approximated by the total progeny with fixed life-length equal to ΔT and birth rate cp. From Exercise 3.9-i we can then deduce that the probability distribution of the final number of infected individuals, excluding the index case, is well approximated by

$$P(Y_N = j) \approx \frac{((j+1)cp\Delta T)^j e^{-(j+1)cp\Delta T}}{(j+1)!}, \qquad j = 0, 1, \ldots.$$

ii) When the infectious period is exponentially distributed with parameter α and the successful contact rate equals cp, then the distribution of final size in a minor outbreak in a large community is well approximated by the distribution of the total progeny of a branching process with exponential-α adult period and constant birth rate cp. So, as a consequence of Exercise 3.9-ii, we have that the final-size distribution is well approximated by

$$P(Y_N = j) \approx \binom{2j}{j} \frac{p^{j+1} q^j}{j+1}, \qquad j = 0, 1, \ldots,$$

where $p = \alpha/(\alpha + cp) = 1 - q$.

Exercise 3.12 In Figure 16.12 we plot the final size probabilities for 0 infected up to 10 infected, both for the Reed-Frost model and for the 'general' stochastic epidemic and the given parameter values ($R_0 = 2$). The final-size probabilities are computed using the branching process approximation, so they only apply to a large population.

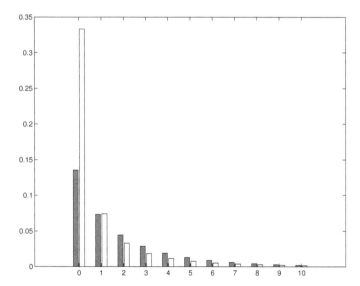

Figure 16.12: Plot of the probability functions for the final size up to 10 infected individuals (minor outbreak) for the 'general' stochastic epidemic (grey right bars) and the Reed-Frost model (black left bars). In both cases parameter values are chosen such that $R_0 = 2$.

For the Reed-Frost we have $P(Y = 0) = 0.1353$, $P(Y = 1) = 0.0733$, $P(Y = 2) = 0.0446$, $P(Y = 3) = 0.0286$, and for the general stochastic epidemic $P(Y =$

$0) = 0.3333$, $P(Y = 1) = 0.0741$, $P(Y = 2) = 0.0329$, $P(Y = 3) = 0.0183$. In Figure 16.13 the accumulated probabilities, i.e., the distribution functions, are plotted.

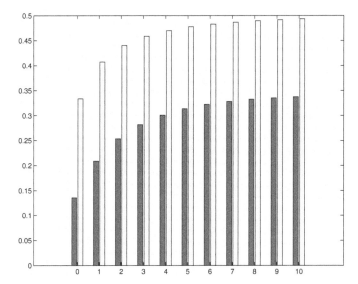

Figure 16.13: Plot of the accumulated probabilities, i.e., the distribution functions, of the final size, up to 10 infected individuals (minor outbreak). The Reed-Frost model are the black left (lower) bars and the general stochastic epidemic are grey right (higher) bars. In both cases parameter values are chosen such that $R_0 = 2$ and the community size N is assumed to be large enough for the branching process approximation to be valid.

From this we see that the 'general' stochastic epidemic systematically has higher accumulated probabilities than the Reed-Frost model. This agrees well with the previous observation that the 'general' epidemic had higher probability for a minor outbreak. From these two observations it seems plausible that the Reed-Frost model gives stochastically *larger* outbreaks than the 'general' stochastic epidemic having equal R_0.

Exercise 3.13 The numerical proportions escaping infection are $\sigma(R_0 = 1.1) = 0.8238$, $\sigma(R_0 = 1.5) = 0.4172$, $\sigma(R_0 = 2.0) = 0.2032$, $\sigma(R_0 = 3) = 0.0595$. The fraction that gets infected is 1 minus this value. It was noted that the final fraction not getting infected σ is equal to the probability of a minor outbreak in the Reed-Frost model z_∞. A plot of the proportion escaping infection in case of a major outbreak, as a function of R_0, is hence given by the lower curve in Figure 3.3 on page 57. In Figure 16.14 below we plot the proportion that does get infected, so it is simply 1 minus the proportion just mentioned.

It is worth noting that, contrary to the *probability* of a minor/major outbreak, which depends on the distribution of the infectious period (see e.g., Figure 16.13), the final size in case of a major outbreak only depends on R_0, and not on the form of the infectious period distribution or on the latency period. In particular, for

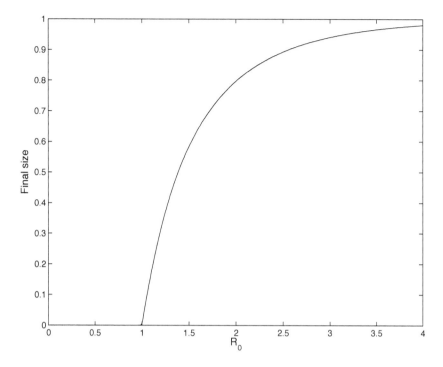

Figure 16.14: Plot of the proportion getting infected (final size), in case a major outbreak occurs, as a function of R_0.

equal R_0 the Reed-Frost model and the 'general' stochastic epidemic give the same limiting final size in case of a major outbreak.

Exercise 3.14 For chicken-pox, the fraction getting infected would be around 99.99%, and for polio 99.75%. Note that this assumes that everyone in the population is susceptible (and that the community mixes homogeneously).

Exercise 3.15 $P(Q > x+y|Q > x) = P(Q > x+y)/P(Q > x)$. If Q is exponential with parameter λ we have $P(Q > z) = \int_z^\infty \lambda e^{-\lambda x} dx = e^{-\lambda z}$. It hence follows that $P(Q > x + y|Q > x) = e^{-\lambda(x+y)}/e^{-\lambda x} = e^{-\lambda y} = P(Q > y)$, which is what we wanted.

Exercise 3.16 Suppose $Y = i$. According to equation (3.17) this implies that $\mathfrak{I}(j) > Q_{(j)}$, $j < i$ and $\mathfrak{I}(i) \leq Q_{(j)}$. Put another way, $\mathfrak{I}(j)$ (the infectious force from the initial infective and the first $j - 1$ infected) is enough to infect at least j individuals among the susceptibles, $j < i$, but $\mathfrak{I}(i)$ does not infect any further individual. This is equivalent to $\mathfrak{Q}(\mathfrak{I}(j)) > j - 1$, $j < i$, and $\mathfrak{Q}(\mathfrak{I}(i)) \leq i$.

Exercise 3.17 The expected number of cases equals $Nx^* = 6420$ and the standard deviation, taking the square root of the variance in equation (3.19), equals 75.2. The total number infected is hence expected to lie in the interval 6420 ± 150, with 95% probability (i.e., the mean $\pm 1.96\times$ the standard deviation, as the number of cases is normally distributed.)

16.4   ELABORATIONS FOR CHAPTER 4

Exercise 4.1 Of course we can linearize the full system (4.3) around the 'virgin' state and compute the eigenvalues. But this kind of biological invasion problem always allows a short cut. The point is that the right-hand side of the differential equation for I has a factor I and a factor $\beta S - \mu - \alpha$. Linearization amounts to replacing S by N in the second factor, and leads to the decoupled linear equation

$$\frac{dI}{dt} = (\beta N - \mu - \alpha)I.$$

Hence we have stability if $\beta N - \mu - \alpha < 0$ (equivalently: $R_0 < 1$) and instability if $\beta N - \mu - \alpha > 0$ (equivalently: $R_0 > 1$).

Exercise 4.2 For $I \neq 0, dI/dt = 0$ requires $\beta \overline{S} - \mu - \alpha = 0 \Rightarrow \overline{S} = (\mu + \alpha)/\beta$. We can rewrite this as $\overline{S}/N = (\mu + \alpha)/\beta N = 1/R_0$. The same observation also shows that $S = (\mu + \alpha)/\beta = N/R_0$ is an isocline, so along orbits the variable I takes its maxima and minima on this line. In particular, a steady state has to lie on this line. This is not at all surprising, since in steady state a case has to produce, on average, one secondary case and the expected number of secondary cases is R_0 multiplied by the reduction factor $\overline{S}/N$, which takes into account that only a fraction $\overline{S}/N$ of contacts is with susceptibles. See also Figure 16.15, where we have drawn part of a trajectory of the system (4.3).

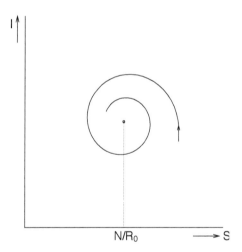

Figure 16.15: Typical trajectory of the system given in (4.3) converging to the endemic steady state.

Note that if we can estimate $\overline{S}/N$ (from blood samples taken at random), we can estimate $R_0 = N/\overline{S}$.

Exercise 4.3 Setting $dS/dt = 0$ in (4.3) we find

$$\overline{I} = \frac{B - \mu \overline{S}}{\beta \overline{S}} = \frac{\mu}{\beta}\left(\frac{B}{\mu \overline{S}} - 1\right) = \frac{\mu}{\beta}\left(\frac{N}{\overline{S}} - 1\right) = \frac{\mu}{\beta}(R_0 - 1).$$

Exercise 4.4 We calculate the Jacobi matrix and evaluate the elements for $S = \overline{S}$ and $I = \overline{I}$:

$$\frac{\partial}{\partial S}(B - \beta SI - \mu S) = -\beta I - \mu = -\beta\overline{I} - \mu,$$

$$\frac{\partial}{\partial I}(B - \beta SI - \mu S) = -\beta S = -\beta\overline{S},$$

$$\frac{\partial}{\partial S}((\beta S - \mu - \alpha)I) = \beta I = \beta\overline{I},$$

$$\frac{\partial}{\partial I}((\beta S - \mu - \alpha)I) = \beta S - \mu - \alpha = 0.$$

The corresponding matrix

$$\begin{pmatrix} -\beta\overline{I} - \mu & -\beta\overline{S} \\ \beta\overline{I} & 0 \end{pmatrix}$$

has trace $T = -\beta\overline{I} - \mu < 0$ and determinant $D = \beta^2\overline{S}\overline{I} > 0$, and so both roots of the characteristic equation $\lambda^2 - T\lambda + D = 0$ have negative real part (see e.g., Hale 1969). So both eigenvalues of the matrix have negative real part and, according to the principle of linearized stability, the endemic steady state is (locally asymptotically) stable. Global stability is established in Exercise 4.10.

Exercise 4.5 As we saw in the preceding exercise, the characteristic equation is $\lambda^2 + (\beta\overline{I} + \mu)\lambda + \beta^2\overline{S}\overline{I} = 0$. Dividing by μ^2 and using (4.4)–(4.6), we rewrite this as

$$\left(\frac{\lambda}{\mu}\right)^2 + (R_0 - 1 + 1)\frac{\lambda}{\mu} + (R_0 - 1)\frac{\beta}{\mu}\frac{N}{R_0}$$

$$= \left(\frac{\lambda}{\mu}\right)^2 + R_0\frac{\lambda}{\mu} + \frac{\alpha + \mu}{\mu}(R_0 - 1) = 0.$$

When $1/\mu \gg 1/\alpha$, we have $\alpha/\mu \gg 1$, and consequently we approximate the last term by $\frac{\alpha}{\mu}(R_0 - 1)$. The equation

$$y^2 + R_0 y + \frac{\alpha}{\mu}(R_0 - 1) = 0$$

has roots

$$y = \frac{-R_0 \pm \sqrt{R_0^2 - 4\frac{\alpha}{\mu}(R_0 - 1)}}{2}.$$

Using once more that $\alpha/\mu \gg 1$, we see that the expression under the square root is negative (we are considering the endemic steady state, so implicitly we have assumed that $R_0 > 1$) and that the roots are, in the first approximation,

$$\frac{\lambda}{\mu} = y = -\frac{R_0}{2} \pm i\sqrt{\frac{\alpha}{\mu}(R_0 - 1)}.$$

So the relaxation time $\frac{1}{|\text{Re}\lambda|}$ equals $\frac{2}{\mu R_0}$, while the frequency equals $\sqrt{\alpha\mu(R_0 - 1)}$, both in first approximation with respect to the small parameter μ/α. For $\mu \ll \alpha$, the relaxation time is of the order of $1/\mu$ but the period is of the order of $1/\sqrt{\mu}$, so the ratio between the two goes to infinity for $\mu \downarrow 0$. This means that we expect to see many oscillations while the deviation from steady state is damping out.

Exercise 4.6 In the notation from Chapter 2, we have

$$\frac{dP_S}{da} = -(\Lambda + \mu)P_S, \qquad P_S(0) = 1.$$

From this we have $P_S(a) = e^{-(\mu+\Lambda)a}$, and consequently the probability per unit time that the individual is infected at age a equals $\Lambda P_S(a) = \Lambda e^{-(\mu+\Lambda)a}$. Hence

$$\bar{a} = \frac{\int_0^\infty a\Lambda P_S(a)\,da}{\int_0^\infty \Lambda P_S(a)\,da} = \frac{1}{\mu + \Lambda}.$$

Because $\Lambda = \beta\bar{I}$, we have, by (4.6), that $\Lambda = \mu(R_0 - 1)$ and so $\bar{a} = \frac{1}{\mu R_0}$.

Exercise 4.7 For $\bar{a}$ ranging from 4 to 5 the period ranges, when computed from the formula, from 2.28 to 2.55 years. This is the right order of magnitude and, in fact, surprisingly close to the observed period of two years. (One should realize that seasonal effects, such as weather and the school system, force the actual period to be an integer, while in our model no such restriction or effect has been built in.)

Exercise 4.8 i) This is indeed already done: Exercise 2.4-iv.

ii) Setting $dI/dt = 0$, we find $\bar{E} = \frac{\alpha+\mu}{\theta}\bar{I}$, while $dE/dt = 0$ leads to $\beta\bar{S}\bar{I} = (\mu + \theta)\bar{E}$. Combining these two relations, we obtain, after dividing out a factor $\bar{I}$ (remember we are determining the endemic steady state, so we are not interested in $\bar{I} = 0$) that $\beta\bar{S} = (\mu + \theta)(\alpha + \mu)/\theta$ or, equivalently, $\bar{S} = N/R_0$. From $dS/dt = 0$, we find that $\bar{I} = \frac{B-\mu\bar{S}}{\beta\bar{S}} = \frac{\mu}{\beta}(R_0 - 1)$.

iii) The Jacobi matrix is derived analogously to the derivation in Exercise 4.4. To deduce the characteristic equation from the matrix is straightforward but tedious.

iv) This is an invitation, not an exercise!

Exercise 4.9 ii) In steady state, $\bar{\Lambda} = \frac{c}{N}\int_0^\infty A(\tau)\,d\tau\,\overline{\Lambda S} = \frac{R_0}{N}\overline{\Lambda S}$, so if $\bar{\Lambda} \neq 0$ then necessarily $\bar{S} = N/R_0$.

iii) The variation-of-constants formula says that $dx(t)/dt = m(t)x(t) + f(t)$ has as a solution that satisfies the initial condition $x(t_0) = x_0$ the expression

$$x(t) = e^{\int_{t_0}^t m(\sigma)\,d\sigma}x_0 + \int_{t_0}^t e^{\int_\tau^t m(\sigma)\,d\sigma}f(\tau)\,d\tau.$$

By letting $t_0 \to -\infty$, while assuming that $x(t) \to 0$ as $t \to -\infty$, we obtain the limiting form

$$x(t) = \int_{-\infty}^t e^{\int_\tau^t m(\sigma)\,d\sigma}f(\tau)\,d\tau,$$

which is what we applied here, with f the constant function assuming the value B, with $m(t) = -\mu - \Lambda(t)$ and $x(t) = S(t)$.

iv) In an endemic steady state we must have $B - \mu\bar{S} - \overline{\Lambda S} = 0$, which gives $\bar{\Lambda} = (B - \mu\bar{S})/\bar{S} = \mu(R_0 - 1)$. If we identify $\bar{\Lambda}$ with $\beta\bar{I}$ in (4.6), this is exactly the same identity.

v) To derive the equation for $y(t)$, we only have to expand

$$\Lambda(t - \tau)S(t - \tau) = \overline{\Lambda S} + \bar{\Lambda}x(t - \tau) + \bar{S}y(t - \tau) + \text{ h.o.t.}$$

(h.o.t. means higher-order terms). To derive the equation for $x(t)$, write

$$e^{-\int_{t-\sigma}^{t} \Lambda(s)\, ds} = e^{-\overline{\Lambda}\sigma} e^{-\int_{t-\sigma}^{t} y(s)\, ds} = e^{-\overline{\Lambda}\sigma} \left(1 - \int_{t-\sigma}^{t} y(s)\, ds + \text{h.o.t.} \right)$$

and apply partial integration to arrive at the form as stated.

If we substitute as an ansatz the given expressions for $x(t)$ and $y(t)$, we find, after dividing out a factor $\exp(\lambda t)$, the system of equations

$$x_0 = -\frac{N}{R_0} \int_0^\infty e^{-(\mu R_0 + \lambda)\sigma}\, d\sigma\, y_0,$$

$$y_0 = \frac{c}{R_0} \int_0^\infty A(\tau) e^{-\lambda \tau}\, d\tau\, y_0 + \frac{c}{N} \mu(R_0 - 1) \int_0^\infty A(\tau) e^{-\lambda \tau}\, d\tau\, x_0,$$

which is, for given λ, a homogeneous linear system of two equations in two unknowns. So, in order for there to be a non-trivial solution, the corresponding matrix should be singular. The characteristic equation is precisely the condition for the matrix to be singular.

vi) On the one hand, for $R_0 = 1$ the characteristic equation reads

$$1 = c\overline{A}(\lambda).$$

On the other hand, we can rewrite $R_0 = 1$ as $c\overline{A}(0) = 1$. So $\lambda = 0$ is a root. When λ is restricted to be real, $\overline{A}$ is a monotonically decreasing function of λ, so in particular $\overline{A}(\lambda) < 1$ for $\lambda > 0$. Now suppose $c\overline{A}(\lambda) = 1$. Then necessarily $\text{Im}\overline{A}(\lambda) = 0$ and

$$c\overline{A}(\lambda) = c \int_0^\infty e^{-\text{Re}\lambda\tau} \cos(\text{Im}\lambda\tau)\, A(\tau)\, d\tau.$$

But then in fact, under very mild conditions on A, we have that $c|\overline{A}(\lambda)| < c\overline{A}(\text{Re}\, \lambda)$ if $\text{Im}\, \lambda \neq 0$, which implies that no root can lie in the right half-plane or on the imaginary axis (other than $\lambda = 0$).

vii) The monotonicity of $\lambda \mapsto \overline{A}(\lambda)$ for real λ shows that the root $\lambda = 0$ must shift to the left when we increase R_0 a bit. So, for R_0 slightly bigger than 1, all roots lie in the left half-plane.

viii) In fact $\overline{A}(\lambda) \to 0$ if $|\lambda| \to \infty$ while λ is restricted to the right half-plane. This is immediately clear if $\text{Re}\, \lambda \to +\infty$ and follows from the Riemann-Lebesgue lemma if $\text{Re}\, \lambda$ stays bounded (see e.g., Rudin 1974; the key idea behind this lemma is that oscillations become so rapid that positive and negative contributions cancel each other).

ix) We argue by contradiction. That is, we assume that $\lambda = i\omega$ is a root. Then necessarily

$$1 = c\frac{|\overline{A}(i\omega)|}{R_0} \left| \frac{i\omega + \mu}{i\omega + \mu R_0} \right|.$$

But in fact the two inequalities

$$c\left|\overline{A}(i\omega)\right| = c\left| \int_0^\infty A(\tau) e^{-i\omega\tau}\, d\tau \right| \leq c \int_0^\infty A(\tau) \left| e^{-i\omega\tau} \right|\, d\tau \leq R_0$$

and

$$\left| \frac{i\omega + \mu}{i\omega + \mu R_0} \right| = \sqrt{\frac{\omega^2 + \mu^2}{\omega^2 + \mu^2 R_0^2}} < 1$$

show that the right-hand side is less than one — a contradiction.

x) This is a conclusion that follows by combining vii), viii) and ix) with the principle of linearized stability and the exponential decay of solutions of linear equations when all roots of the characteristic equation lie strictly to the left of the imaginary axis.[3]

Exercise 4.10 If we calculate the derivative of V with respect to t we obtain

$$
\begin{aligned}
\frac{dV}{dt} &= \frac{\partial V}{\partial S}\frac{dS}{dt} + \frac{\partial V}{\partial I}\frac{dI}{dt} \\
&= \left(1 - \frac{\overline{S}}{S}\right)(B - \beta SI - \mu S) + \left(1 - \frac{\overline{I}}{I}\right)I(\beta S - \alpha - \mu) \\
&= (S - \overline{S})\left(\frac{B}{S} - \beta I - \mu + \beta I - \beta \overline{I}\right) \\
&= (S - \overline{S})\left(\frac{B}{S} - \frac{B}{\overline{S}}\right) \\
&= -\frac{B}{S\overline{S}}(S - \overline{S})^2 = -\frac{\mu R_0}{S}\left(S - \frac{N}{R_0}\right)^2,
\end{aligned}
$$

where we have used $\overline{S} = (\alpha + \mu)/\beta$ and $B - \beta\overline{S}\overline{I} - \mu\overline{S} = 0$. We conclude that $dV/dt < 0$ except on the line $S = N/R_0$, where $dV/dt = 0$. At this line (which is an I-nullcline) we have $dI/dt = 0$ and $dS/dt = B - (\mu + \beta I)N/R_0$. So clearly $dS/dt > 0$ for $I < \overline{I}$ and $dS/dt < 0$ for $I < \overline{I}$ (with $\overline{I}$ the steady state value for I given explicitly by (4.6)). We conclude that orbits cannot 'stay' on the line, unless we consider the steady state. The LaSalle invariance principle (see Hale 1969), which is a very useful generalization of Lyapunov's stability theorem, then allows us to conclude that all orbits that stay bounded do converge to the steady state. The boundedness of orbits, on the other hand, is a direct consequence of our modeling assumptions (a constant population birth rate, while the per capita death rate is constant). Mathematically this is reflected in the invariance of the region $\{(S, I) : S \geq 0, I \geq 0$ and $S + I \leq N\}$ (note that $d/dt(S + I) = B - \mu(S + I) - \alpha I \leq B - \mu(S + I)$).

Exercise 4.11 We first explain $\pi_j = \tau_j/L$. The point is that the fractions π_j sum to 1 and that the constant of proportionality does not depend on j (this is somewhat implicit in the formulation of the microcosm principle), so the mean sojourn times should sum to L. The strength of the principle is that it yields relationships of a general nature that are independent of the details of the model specification. That we apply it here to the simple models is meant as a training in the use of the principle. Once you understand it, you may use it in far more complex situations where explicit calculations are often cumbersome (to put it mildly). Before starting with the elaboration, we make one more remark: sometimes it is helpful to number the i-states, and so j refers to a number, at other times it is helpful to indicate i-states by letters that carry information, and then j may take 'values' such as S, I, E. This should not lead to confusion.

For the system (4.3) we have $\pi_S = \overline{S}/N = 1/R_0$ while $\tau_S = \frac{1}{\mu + \beta\overline{I}} = \frac{1}{\mu + \mu(R_0 - 1)} = \frac{1}{\mu R_0}$, so indeed $\pi_S = \frac{\tau_S}{L}$. Next consider $j = I$ and observe that $\pi_I = \frac{\overline{I}}{N} = \frac{\mu(R_0 - 1)}{\beta N} = $

---

[3]The elaboration of this exercise can be much simplified, as shown in Theorem 3.1 of D. Breda, O. Diekmann, W. de Graaf, A. Pugliese and R. Vermiglio: On the formulation of epidemic models (an appraisal of Kermack and McKendrick). *J. Biol. Dynamics*, (2012).

$\frac{\mu(R_0-1)}{(\alpha+\mu)R_0}$. Now comes a subtlety. We have to distinguish between the mean time spent as an infective and the mean time spent as an infective, *given* that the individual does indeed become infected. The latter equals $1/(\mu+\alpha)$, while the former equals $\frac{\beta\bar{I}}{\mu+\beta\bar{I}}\frac{1}{\mu+\alpha}$, where the first factor is the probability that a newborn susceptible individual is infected during its life (see Exercise 1.34-vi for explanation, if needed). In the microcosm principle as formulated in the exercise the quantity τ_I refers to a 'mean' computed over newborn individuals. So we should check that

$$\frac{1}{L}\frac{\beta\bar{I}}{\mu+\beta\bar{I}}\frac{1}{\mu+\alpha} = \pi_I,$$

which is straightforward (use (4.6)).

Next consider the system (4.11). Again $\pi_S = \bar{S}/N = 1/R_0$ while $\tau_S = 1/(\mu+\beta\bar{I}) = 1/(\mu R_0)$, so $\pi_S = \tau_S/L$. Concerning the i-state E we have

$$\pi_E = \frac{\bar{E}}{N} = \frac{\alpha+\mu}{\theta}\frac{\bar{I}}{N} = \frac{\alpha+\mu}{\theta}\frac{\mu(R_0-1)}{\beta N},$$

which is indeed equal to $\frac{\beta\bar{I}}{\mu+\beta\bar{I}}\frac{1}{\theta+\mu}\mu$, since $\beta\bar{I} = \mu(R_0-1)$. Finally, $\pi_I = \frac{\bar{I}}{N} = \frac{\mu(R_0-1)}{\beta N}$. To arrive in state I, two conditions should be met, so we have two factors

$$\tau_I = \frac{\beta\bar{I}}{\mu+\beta\bar{I}}\frac{\theta}{\theta+\mu}\frac{1}{\mu+\alpha}.$$

Again the identity $\beta\bar{I} = \mu(R_0-1)$ is now all that is needed to verify that $\pi_I = \tau_I/L$.

Exercise 4.12 The underlying idea is that, quite generally under steady-state conditions, *rate of inflow × expected sojourn time = amount*. Three remarks are in order: i) it does not matter whether we measure 'amount' in numbers or in density, as long as we compute the rate in the corresponding units; ii) when individuals can return to the state under consideration, such as to the S-state in an SIS model, we should take the sojourn time of a *single* visit; iii) steady state is too strong a condition, in periodic conditions one can formulate an analog in terms of averages over the period (but in growing populations the rate of population growth has to be taken into account).

In this exercise we have to check that

$$\frac{\beta\overline{SI}}{\bar{I}} = \alpha+\mu,$$

which is indeed the case (combine (4.5) and (4.4)).

Exercise 4.13 If the incidence of a lethal disease is substantial, the demography will, in general, be influenced, and one should not model the inflow of newborn susceptibles as a constant.

The question is less rhetorical than one may now think: consider, for example, HIV and its spread in the subpopulation of homosexuals in the 1980s and 1990s. In that example one may, however, want to incorporate behavioral reactions to estimates of the prevalence.

Exercise 4.14 We number the years with the integers and denote by $n-$ the moment just before the reproduction event and by $n+$ the moment just after the reproduction event. We first describe the demography in as simple a manner as we can (constant number of newborns, fixed probability to survive one year):

$$
\begin{aligned}
S(n+) &= S(n-) + B, \\
S((n+1)-) &= S(n+)e^{-\mu},
\end{aligned}
$$

which combine into

$$
S((n+1)+) = S(n+)e^{-\mu} + B,
$$

and show convergence towards the globally stable steady state

$$
\overline{S}_+ = \frac{B}{1 - e^{-\mu}}.
$$

Sketch a cobweb picture (see e.g., Edelstein-Keshet 1988), plotting in the (x, y) plane the lines $y = x$ and $y = e^{-\mu}x + B$. Next we put the infectious agent onto the stage. The reproduction event is now described by

$$
\begin{aligned}
S(n+) &= S(n-) + B, \\
I(n+) &= I(n-),
\end{aligned}
$$

which means that all newborns are susceptible (no vertical transmission) and that the disease has no influence on fertility. To model the infection process in between reproduction events, we introduce the continuous time variable t, which runs from 0 to 1 (and so is reset to zero at every reproduction event; in other words, t measures time within a year just as a calendar does), and suppose that

$$
\begin{aligned}
\frac{dS}{dt} &= -\mu S - \beta S I, \\
\frac{dI}{dt} &= -\mu I + \beta S I - \alpha I.
\end{aligned}
$$

The notion of R_0 does not make sense in this setting, since even when we have a 'stroboscopic' steady state $\overline{S}_+$, the quantity S depends on time t and is not constant. Yet we can easily derive an invasion criterion (by which we mean a criterion that decides whether or not the agent will spread when initially rare). To this end, we consider

$$
\frac{dI}{dt} = (\beta \overline{S}(t) - \mu - \alpha)I,
$$

with

$$
\overline{S}(t) = \frac{B}{1 - e^{-\mu}} e^{-\mu t},
$$

and note that I will increase from year to year (when we take stock at the reproduction time) if and only if

$$
\beta \int_0^1 \overline{S}(t)\, dt - \mu - \alpha > 0
$$

or, equivalently,

$$
\frac{\beta B}{\mu(\mu + \alpha)} > 1.
$$

We refer to Roberts and Kao (1998)[4] for related work. Moreover, we mention that one can implement hybrid models in the numerical continuation and bifurcation package CONTENT[5] that can be found via a link on the home page of Yuri A. Kuznetsov http://www.staff.science.uu.nl/~kouzn101/. We fear that analytical methods are not powerful enough and that one needs such computational tools to proceed, and we therefore advocate their use. Likewise, we advocate hybrid models: we think they are unjustly neglected in the literature.[6]

Exercise 4.15 Consider

$$\frac{dS}{dt} = -\beta SI + \gamma I,$$
$$\frac{dI}{dt} = \beta SI - \gamma I.$$

Then $d/dt(S+I) = 0$; so $S+I = \text{constant} = N$, and we can eliminate one variable, for instance S, and deduce

$$\frac{dI}{dt} = (\beta N - \gamma - \beta I)I,$$

which is the famous logistic equation. An endemic steady state exists provided that

$$R_0 = \frac{\beta N}{\gamma} > 1,$$

and is globally stable.

When there is temporary immunity we may set

$$\frac{dS}{dt} = -\beta SI \qquad + \gamma R,$$
$$\frac{dI}{dt} = \beta SI - \alpha I,$$
$$\frac{dR}{dt} = \qquad + \alpha I - \gamma R,$$

for which $S + I + R = \text{constant} = N$. The system can therefore be reduced to the two-dimensional system

$$\frac{dS}{dt} = -\beta SI \qquad + \gamma(N - S - I),$$
$$\frac{dI}{dt} = \beta SI - \alpha I.$$

We find a positive endemic state

$$(\overline{S}, \overline{I}) = \left(\frac{\alpha}{\beta}, \frac{\gamma(N - \frac{\alpha}{\beta})}{\gamma + \alpha} \right),$$

---

[4]M.G. Roberts and R.R. Kao: The dynamics of an infectious disease in a population with birth pulses. *Math. Biosci.*, **149** (1998), 23–36.

[5]Y.A. Kuznetsov and V.V. Levitin (1997): CONTENT. A multiplatform environment for analysing dynamical systems. Dynamical Systems Laboratory, Centre for Mathematics and Computer Science, Amsterdam.

[6]But see H.T.M. Eskola and S.A.H. Geritz: On the mechanistic derivation of various discrete time population models. *Bull. Math. Biol.*, **69** (2007), 329–346; E. Pachepsky, R.M. Nisbet and W.W. Murdoch: Between discrete and continuous: consumer-resource dynamics with synchronized reproduction. *Ecology*, **89** (2008), 280–288.

provided that $R_0 = \beta N/\alpha > 1$. The corresponding Jacobi matrix

$$\begin{pmatrix} -\beta \overline{I} - \gamma & -\beta \overline{S} - \gamma \\ \beta \overline{I} & 0 \end{pmatrix}$$

has negative trace and positive determinant, so the endemic state is (locally asymptotically) stable.

Exercise 4.16 We consider the following system:

$$\begin{aligned}
\frac{dS}{dt} &= B\frac{S + (1 - \theta)I + R}{N} - \beta SI - \mu S, \\
\frac{dI}{dt} &= \theta B\frac{I}{N} + \beta SI - \mu I - \alpha I, \\
\frac{dR}{dt} &= -\mu R + \alpha I,
\end{aligned}$$

the idea being that a fraction θ of the offspring of an infectious individual is infected at birth. Here we use both B and N in the formulation. They are not independent parameters, however, but rather related to each other by $B = \mu N$. There are now two different transmission routes, and accordingly the interpretation yields the expression

$$R_0 = \frac{\beta N}{\mu + \alpha} + \frac{\theta B}{(\mu + \alpha)N} = \frac{\beta N}{\mu + \alpha} + \frac{\theta \mu}{\mu + \alpha},$$

which is the sum of two contributions. Likewise, we find, by looking at the equation for dI/dt, that

$$r = \frac{\theta B}{N} + \beta N - \mu - \alpha.$$

It follows from $d/dt(S+I+R) = B - \mu(S+I+R)$ that the set $\{(S, I, R) : S+I+R = N\}$ is attracting, and so we restrict our further analysis to that set. This means that we study the system

$$\begin{aligned}
\frac{dS}{dt} &= B(1 - \theta\frac{I}{N}) - \beta SI - \mu S, \\
\frac{dI}{dt} &= \theta B\frac{I}{N} + \beta SI - \mu I - \alpha I,
\end{aligned}$$

that is

$$\begin{aligned}
\frac{dS}{dt} &= B - \theta\mu I - \beta SI - \mu S, \\
\frac{dI}{dt} &= \theta\mu I + \beta SI - \mu I - \alpha I.
\end{aligned}$$

The endemic state has $\overline{S} = \frac{\mu + \alpha - \theta\mu}{\beta}$ (set $dI/dt = 0$), and $\overline{I} = \frac{B - \mu\overline{S}}{\theta\mu + \beta\overline{S}}$ (set $dS/dt = 0$) with corresponding Jacobi matrix

$$\begin{pmatrix} -\beta \overline{I} - \mu & -\theta\mu - \beta \overline{S} \\ \beta \overline{I} & 0 \end{pmatrix},$$

which has negative trace and positive determinant (provided $\overline{S} > 0$ and $\overline{I} > 0$, that is, provided $R_0 > 1$). So the endemic state is (locally asymptotically) stable. We refer to the book by Busenberg and Cooke (1993) for a wealth of information on vertical transmission and its modeling.

Exercise 4.17 This exercise touches the surface of an important and not yet fully explored problem area. See Section 4.5 and the references given there.

We first note that (from (4.5)) $\overline{S}/N = 1/R_0^1$ and next that, consequently, the expected number of secondary cases per primary case of infection by agent 2 will be $\overline{S}R_0^2/N = R_0^2/R_0^1$, which will exceed one if and only if $R_0^2 > R_0^1$.

The next question is: if $R_0 = \beta N/(\alpha + \mu$ (see (4.4)), with β a free parameter, but α related to β as depicted in Figure 4.1, for what value of β will R_0 be maximal? A straightforward computation shows that $dR_0/d\beta = 0$ is equivalent to $\alpha + \mu = \beta d\alpha/d\beta$. Now note that the tangent line to the graph of $y = \alpha(\beta) + \mu$ at the point $(\overline{\beta}, \alpha(\overline{\beta}) + \mu)$ (see Figure 16.16, where we have drawn the graph of $y = \alpha(\beta) + \mu$) is given by $y = \alpha(\overline{\beta}) + \mu + \frac{d\alpha}{d\beta}(\overline{\beta})(\beta - \overline{\beta})$ and that this tangent line runs through the origin if and only if $\alpha(\overline{\beta}) + \mu = \frac{d\alpha}{d\beta}(\overline{\beta})\overline{\beta}$.

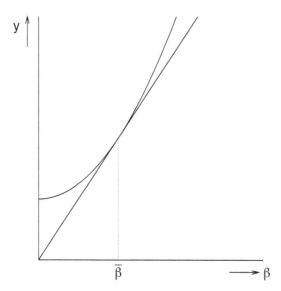

Figure 16.16: Schematic representation of a function $y = \alpha(\beta) + \mu$, where α is related to β in a manner given in Figure 4.1. The straight line is the tangent in the point $(\overline{\beta}, \alpha(\overline{\beta}) + \mu)$.

So the graphical criterion for extremality of R_0 is that the tangent line runs through the origin. Graphically, we also see that the extremum is a maximum: on the straight tangent line we have that β/y is constant, say c, and the fact that the graph lies above the line for $\beta \neq \overline{\beta}$ implies the inequality $\beta/(\alpha(\beta) + \mu) < c$ for $\beta \neq \overline{\beta}$, with equality for $\beta = \overline{\beta}$.

Within this setting the statement is correct. But in fact we have forced the statement to be true by assuming the trade-off between α and β as depicted in Figure 16.16. So the real question is: does such a trade-off exist? What is its mechanistic underpinning?

Exercise 4.18 i) If we assume that, for a certain fixed density, the number of 'contacts' per individual is constant, we should use $\gamma SI/N$ when the variables refer to numbers. When, in contrast, we assume that the curiosity of mice is such that they explore every corner of the cage, no matter how large it is, we should take βSI. We think that the last assumption is not warranted and that the first gives a more reasonable description until we go down to a really small cage (i.e., a really small number).

ii) From $dI/dt = 0$ we find that $\overline{S} = (\mu + \alpha)/\beta$, and from $dR/dt = 0$ that $\overline{R} = f\alpha\overline{I}/\mu$. Now set $dS/dt = 0$, insert the expression for $\overline{S}$ and solve for $\overline{I}$. This yields $\overline{I} = B/(\mu+\alpha) - \mu/\beta$. All that remains to be done is to rearrange the resulting expression for $\overline{S} + \overline{I} + \overline{R}$ a little.

iii) This computation is slightly more complicated than the preceding one. Setting $dI/dt = 0$, we find $\overline{S} = \frac{\mu+\alpha}{\gamma}\overline{N}$, while $dR/dt = 0$ yields $\overline{R} = f\alpha\overline{I}/\mu$, and $dI/dt = 0$ yields $\overline{I} = \frac{B}{\mu+\alpha} - \frac{\mu}{\mu+\alpha}\overline{S}$. This last equation, together with $\overline{S} = \frac{\mu+\alpha}{\gamma}\overline{N} = \frac{\mu+\alpha}{\gamma}(\overline{S} + \overline{I} + \frac{f\alpha}{\mu}\overline{I})$, can be considered as a system of two linear equations with two unknowns, $\overline{S}$ and $\overline{I}$. Standard manipulation to eliminate $\overline{I}$ then gives one equation for $\overline{S}$, with solution

$$\overline{S} = \left(1 - (1-f)\frac{\alpha}{\gamma}\right)^{-1}\frac{B}{\gamma}\left(1 + \frac{f\alpha}{\mu}\right).$$

Rather than computing $\overline{I}$ and $\overline{R}$ from this, we now recall that $\overline{N} = \gamma\overline{S}/(\mu+\alpha)$ and find the expression for $\overline{N}$ that is stated in the text.

Exercise 4.19 i) Instead of B births per unit of time into the susceptible class, there will be $(1-v)B$ births when a fraction v of the newborns is vaccinated. There is now a choice in determining where the vB vaccinated newborns end up. There are basically two options: they either enter the class of recovered and permanently immune individuals R, or they collectively form a new class of their own, say class V for vaccinated individuals, which then has differential equation $dV/dt = vB - \mu V$. We will see in Exercise 4.22 that it will depend on the situation which choice is appropriate. To determine the steady state value of susceptibles in the absence of the wild type, this does not matter. Instead of B/μ one now gets $(1-v)B/\mu$. Therefore

$$R_v = \frac{\beta}{\alpha+\mu}(1-v)\frac{B}{\mu} = (1-v)R_0,$$

leading to $1 - 1/R_0$ as the strict lower bound for the fraction to vaccinate in order to achieve $R_v < 1$.

ii) Note that, in contrast to scheme i), we now have a rate of vaccinating individuals, rather than a fraction to be vaccinated. The equation for dS/dt becomes $dS/dt = B - \beta SI - \mu S - vS$ and the same two options hold for where the vS individuals that get vaccinated per unit of time end up. The difference is that here individuals enter the susceptible class first and vaccination is one of three ways in which individuals can be removed from this class (the other two being death and getting infected), whereas in the scheme in 4.19-i) the vaccinated individuals never enter the susceptible class. The limits $v \to 1$ in scheme i) and $v \to \infty$ in scheme ii) coincide in the sense that in this case everyone is vaccinated at birth. The steady state susceptible population is given by $B/(\mu+v)$ and R_v is given by

$$R_v = \frac{\beta B}{(\alpha+\mu)(\mu+v)} = R_0\frac{\mu}{\mu+v},$$

which leads to the inequality $v > (R_0 - 1)\mu$ in order to eliminate the wild type.

Exercise 4.20 i) We consider the system

$$\frac{dS}{dt} = (1 - v)B - \mu S - \beta SI,$$

$$\frac{dI}{dt} = -\mu I + \beta SI - \alpha I,$$

but we still define R_0 as the expected number of secondary cases per primary case in the unvaccinated population, that is, we still take

$$R_0 = \frac{\beta N}{\alpha + \mu} = \frac{\beta B}{\mu(\alpha + \mu)}.$$

Setting $dI/dt = 0$, we find that in endemic steady state

$$\overline{S} = \frac{\mu + \alpha}{\beta} = \frac{N}{R_0},$$

so $\overline{S}/N$ does not change at all (which is not surprising, since the argument given in Exercise 4.2 still applies).

ii) Setting $dS/dt = 0$, we find, with a bit of manipulation,

$$\overline{I} = \frac{\mu}{\beta}((1 - v)R_0 - 1).$$

So the force of infection $\Lambda = \beta\overline{I} = \mu((1 - v)R_0 - 1)$ and the mean age at infection $\overline{a}$ (see Exercise 4.6) of the unvaccinated is given by

$$\overline{a} = \frac{1}{\mu + \Lambda} = \frac{1}{(1 - q)\mu R_0}.$$

We see that the force of infection decreases and the mean age at infection increases (by a factor $1/(1 - v)$) with v. For rubella this is a serious complicating factor that has to be taken into consideration when comparing various vaccination strategies; we will return to this in Section 9.6.

iii) The linearization around the steady state has the characteristic equation of Exercise 4.5 with R_0 replaced by $(1 - v)R_0$ everywhere. Our analysis of Exercise 4.5 remains valid as long as $(1 - v)R_0 > 1$. Both the relaxation time and the period of the oscillations increase with increasing q.

Exercise 4.21 We assume that in between vaccination events, the susceptible density develops according to

$$\frac{dS}{dt} = B - \mu S$$

when the population is free from infection. Hence

$$S(t) = N + (S_0 - N)e^{-\mu t},$$

with N the total population size, i.e., $N = B/\mu$. If we vaccinate a fraction p of the susceptible population at time T, we have that, after vaccination,

$$S(T) = (1 - p)\left(N + (S_0 - N)e^{-\mu T}\right).$$

So the infection-free dynamics are described by the recurrence relation

$$S(nT) = (1-p)\left(N + (S((n-1)T) - N)e^{-\mu T}\right),$$

which has the attracting fixed point

$$\overline{S} = \frac{(1-p)N(1-e^{-\mu T})}{1-(1-p)e^{-\mu T}}$$

(found by solving the fixed-point equation that is obtained by setting both $S(nT)$ and $S((n-1)T)$ equal to $\overline{S}$). In between two jumps, we accordingly then have

$$S(t) = N + (\overline{S} - N)e^{-\mu t}, \qquad (n-1)T \le t < nT,$$

with average value

$$\frac{1}{T}\int_0^T S(t)\,dt = N + (\overline{S} - N)\frac{1-e^{-\mu T}}{\mu T}.$$

We now first compute the effort E. Per event, there are

$$p\left(N + (\overline{S} - N)e^{-\mu T}\right)$$

vaccinations, and to obtain E we have to divide this amount by T:

$$E = \frac{p}{T}\left(N + (\overline{S} - N)e^{-\mu T}\right).$$

Now rewrite the fixed-point equation as

$$(\overline{S} - N)(1-e^{-\mu T}) = -p\left(N + (\overline{S} - N)e^{-\mu T}\right),$$

and find

$$E = (N - \overline{S})\frac{1-e^{-\mu T}}{T}.$$

Hence

$$\frac{1}{T}\int_0^T S(t)\,dt = N - \frac{E}{\mu},$$

which shows that the magnitude of the critical quantity $\frac{1}{T}\int_0^T S(t)\,dt$ depends on the vaccination effort E, but *not* on the vaccination period T. So, in particular, the vaccination effort required for elimination is independent of T, and is given by

$$E_{crit} = B\left(1 - \frac{1}{R_0}\right).$$

When we vaccinate a fraction v of all newborns, the vaccination effort is given by vB, and the condition for elimination $v > 1 - 1/R_0$ of Exercise 1.37 can be reformulated as

$$E = vB > B\left(1 - \frac{1}{R_0}\right).$$

We conclude that, for a constant transmission-rate constant β, pulse vaccination yields no advantage whatsoever. Of course, this may be different when, for instance, β is periodic and when the susceptible density fluctuates by other causes (and not only due to the vaccination campaigns). Indeed, the point is to level off the peaks in susceptible density as much as possible. We refer to Shulgin et al. (1998).[7]

Exercise 4.22 In contrast to the situation with a perfect vaccine, it now matters whether individuals that are successfully protected enter the class R of recovered and immune individuals, or enter a class of their own. The point is that it is possible that vaccine-induced immunity is less durable than naturally acquired immunity. In that case individuals in the vaccinated class would revert to the susceptible class (with rate w), but individuals in the removed class would not (or with a slower rate). For this exposition we choose to install a new class V. The extended system (4.3) becomes

$$\frac{dS}{dt} = (1 - \pi v)B - \beta SI - \mu S + wV,$$

$$\frac{dI}{dt} = \beta SI - \mu I - \alpha I + (1 - \psi)\beta VI,$$

$$\frac{dV}{dt} = \pi vB - \mu V - wV - (1 - \psi)\beta VI,$$

$$\frac{dR}{dt} = \alpha I - \mu R.$$

The steady state populations of susceptibles and vaccinated individuals in the absence of the wild-type agent are given by

$$\overline{V} = \frac{\pi vB}{\mu + w},$$

$$\overline{S} = \frac{(1 - \pi v)B + w\overline{V}}{\mu}.$$

Infected individuals can cause infections among susceptibles and, with reduced transmission probability upon contact but with the same contact rate, among vaccinated individuals. Therefore, the reproduction number R_v is, for the system above, given by

$$R_v = \frac{\beta \overline{S}}{\mu + \alpha} + \frac{(1 - \psi)\beta \overline{V}}{\mu + \alpha}.$$

If we now substitute the expressions for the infection-free steady state into the formula for R_v and remember that $R_0 = \frac{\beta B}{\mu(\mu + \alpha)}$, we obtain

$$R_v = R_0(1 - v\frac{\psi \pi \mu}{\mu + w}) = R_0(1 - v\vartheta)$$

where

$$\vartheta = \frac{\psi \pi \mu}{\mu + w}$$

summarizes the effects of the three ways of being imperfect, i.e., the impact of imperfection. The fraction to be vaccinated with such a vaccine in order to assure $R_v < 1$ is

$$v > \frac{1}{\vartheta}(1 - \frac{1}{R_0}).$$

---

[7]B. Shulgin, L. Stone and Z. Agur: Pulse vaccination strategy in the SIR epidemic model. *Bull. Math. Biol.*, **60** (1988), 1123–1148.

Note that $\vartheta \in [0,1]$, that $\vartheta = 1$ in case of a perfect vaccine, and that vaccination can never lead to elimination if $\vartheta < 1 - 1/R_0$. One can now study graphically what the effect of various levels of imperfection is on the minimal fraction of the population to be vaccinated.

We mention that a similar analysis is of course possible for the comparable system where transmission is modeled as $\gamma SI/N$ instead of βSI. This leads to the same correction factor ϑ.

Exercise 4.23 By adding the equations, we arrive at

$$
\begin{aligned}
N' &= bS + bR - \mu N - (1-f)\alpha I \\
&\Rightarrow \ N' = bN - bI - \mu N (1-f)\alpha I \\
&\Rightarrow \ N' = (b - \mu - (b + \alpha(1-f)))y)\, N.
\end{aligned}
$$

Next we find, using $S/N = 1 - y - z$, that

$$
\begin{aligned}
y' &= \frac{I'}{N} - \frac{I}{N}\frac{N'}{N} = \gamma(1 - y - z)y - \mu y - \alpha y \\
&\quad - y\left(b(1 - y - z) + bz - \mu - (1-f)\alpha y\right) \\
&\Rightarrow \ y' = y\left(\gamma(1 - y - z) - \alpha + \alpha(1-f)y + b(y-1)\right)
\end{aligned}
$$

and

$$
\begin{aligned}
z' &= \frac{R'}{N} - \frac{R}{N}\frac{N'}{N} = -\mu z + f\alpha y - z\left(b(1 - y - z) + bz - \mu - (1-f)\alpha y\right) \\
&\Rightarrow \ z' = y\left(f\alpha + \alpha(1-f)z\right) - bz(1-y).
\end{aligned}
$$

Exercise 4.24 Consider Figure 16.17, where we have drawn, for the system (4.14), the isoclines at which $y' = 0$ and the isoclines at which $z' = 0$, as well as part of a trajectory.

The isoclines at which $y' = 0$ are $y = 0$ and $\gamma z = \gamma - \alpha - b + \{b - \gamma + \alpha(1-f)\}y$, which is a straight line intersecting the line $y = 0$ at $z = z_1 = 1 - (\alpha + b)/\gamma$ and the line $y = 1$ at $z = z_2 = -\alpha f/\gamma$. Note that $z_1 < 1$ and $z_1 > 0$ if and only if $R_{0,\text{relative}} = \gamma/(\alpha + b) > 1$, while always $z_2 < 0$.

The triangle $\{(y,z) : 1 - y - z \geq 0, y \geq 0, z \geq 0\}$ is invariant (as it should be because of the interpretation!) since the z-axis is invariant; at the y-axis we have $dz/dt = f\alpha y > 0$, and on the line $z + y = 1$ we have (after writing out and simplifying)

$$
\frac{d}{dt}(y + z) = b(y - 1) < 0 \qquad \text{for } y < 1.
$$

The isocline at which $\dot{z} = 0$ is given by

$$
z = \frac{f\alpha}{b\frac{1-y}{y} - (1-f)\alpha},
$$

which is an increasing function of y (since $(1-y)/y$ is a decreasing function of y). So there is a unique intersection with the sloping $y' = 0$ isocline in the interior of the triangle whenever $R_{0,\text{relative}} > 1$. The $z' = 0$ isocline does not depend on γ. For fixed y the sloping $y' = 0$ isocline has a z value that increases with γ (differentiating

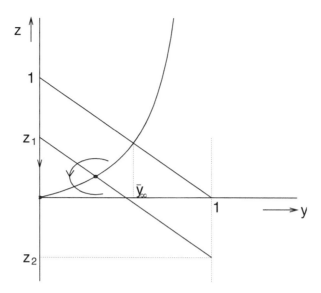

Figure 16.17: The isoclines at which $y' = 0$ and $z' = 0$ for the system given in (4.14), see the elaboration of Exercise 4.24 for details.

the equation for the line for fixed y with respect to γ, we find $z + \gamma dz/d\gamma = 1 - y$, which implies $dz/d\gamma > 0$). Hence both the y and z coordinates of the steady state are increasing functions of γ. For $\gamma \to \infty$, the sloping $y' = 0$ isocline approaches the line $z = 1 - y$, and the limiting value of the y coordinate of the steady state is therefore found by solving the equation

$$1 - y = \frac{f\alpha}{b\frac{1-y}{y} - (1-f)\alpha},$$

which is a quadratic equation in y that has $\bar{y}_\infty$ (see main text) as the root in the interval (0,1).

We first show that, for $R_{0,\text{relative}} = \gamma/(\alpha + b) < 1$, the origin $(y, z) = (0, 0)$ is globally stable. Indeed, we then have

$$
\begin{aligned}
y' &= y\left(\gamma(1 - y - z) - \alpha + \alpha(1 - f)y + b(y - 1)\right) \\
&\leq y\left((\alpha + b)(1 - y - z) - \alpha + \alpha y - \alpha f y + b(y - 1)\right) \\
&= y\left(-(\alpha + b)z - \alpha f y\right) \\
&\leq -\alpha f y^2,
\end{aligned}
$$

which shows that y decreases monotonically towards its limit zero. Since y tends to zero, the limiting equation for z is $dz/dt = -bz$, and consequently z also tends to zero for $t \to \infty$. In the rest of the elaboration we concentrate on the case $R_{0,\text{relative}} > 1$.

The Jacobi matrix of the non-trivial steady state is

$$
\begin{pmatrix}
y(-\gamma + \alpha(1 - f) + b) & -\gamma y \\
f\alpha + (1 - f)\alpha z + bz & (1 - f)\alpha y - b(1 - y)
\end{pmatrix},
$$

and we claim that its sign pattern is

$$\begin{pmatrix} - & - \\ + & - \end{pmatrix},$$

which then implies that the trace $T < 0$ and the determinant $D > 0$, and this, in turn, implies the (local asymptotic) stability of the steady state. The signs of the anti-diagonal elements are immediately clear, so we concentrate on the diagonal elements, starting with position $(1,1)$. The assumption $R_{0,\text{relative}} > 1$ implies $\gamma > b + \alpha$, or $-\gamma + b + \alpha < 0$, and hence certainly $-\gamma + b + \alpha - \alpha f < 0$. The element at position $(2,2)$ is negative for small y, but changes sign at $y = b/((1-f)\alpha + b)$. An easy calculation reveals that $\bar{y}_\infty$ is smaller than this value. Hence the $(2,2)$-element is negative for the relevant range of y values.

In fact, Yan Ping (personal communication), showed that one can easily prove that the steady state is globally stable: the function $B(y, z) = y^{-1}z^{-1}$ is a Dulac function in the interior of the triangle, i.e.,

$$\frac{\partial(BF_1)}{\partial y} + \frac{\partial(BF_2)}{\partial z} < 0$$

where F_1 is the right-hand side of the first equation in (4.14) and similarly F_2 for the second equation. The Bendixson-Dulac Theorem (see Hirsch and Smale 1974) then rules out that a non-trivial periodic orbit exists and hence, by the Poincaré-Bendixson Theorem, every orbit converges to the steady state.

It remains to calculate the critical value γ_c at which the growth rate changes sign. To do this, we append to the two isocline equations the condition that $N' = 0$ for $N \neq 0$. In other words, we consider the three equations

$$\begin{aligned} \gamma(1 - y - z) - \alpha + \alpha(1 - f)y + b(y - 1) &= 0, \\ y(f\alpha + \alpha(1 - f)z) - bz(1 - y) &= 0, \\ b - \mu - (b + \alpha(1 - f))y &= 0, \end{aligned}$$

and try to solve for y, z and γ while considering α, f, b and μ as known. The third equation gives

$$y = \frac{b - \mu}{b + \alpha(1 - f)}.$$

This we substitute in the second equation and then solve for z to find

$$z = \frac{\alpha f(b - \mu)}{\mu(b + \alpha(1 - f))}.$$

The first equation gives

$$\gamma_c = \frac{\alpha - \alpha(1 - f)y + b(1 - y)}{1 - y - z},$$

and upon substitution of the expressions for y and z we arrive at the expression for γ_c as given.

Exercise 4.25 As long as N is exponentially increasing, we may neglect the term K in the denominator and the system is very similar to system (4.12).[8] Now we

[8]For a detailed justification and elaboration see O. Diekmann and M. Kretzschmar: Patterns in the effects of infectious diseases on population growth. *J. Math. Biol.*, **29** (1991), 539–570.

look for steady states. Setting $I' = 0$, we find $-\mu + \gamma S/(K + N) - \alpha = 0$, which we rewrite as $(\gamma - \alpha - \mu)S - (\alpha + \mu)I = (\alpha + \mu)K$. Setting $dS/dt = 0$ (and using the relation we already have) we obtain $(b_1 - \mu)S + (b_2 - \alpha - \mu)I = 0$. These are two linear equations in the two unknowns S and I. The solution is given by

$$\overline{S} = \frac{(\alpha + \mu)K}{\gamma - \alpha - \mu - \frac{(\alpha+\mu)(\mu-b_1)}{b_2-\alpha-\mu}},$$

$$\overline{I} = \frac{\mu - b_1}{b_2 - \alpha - \mu}\overline{S},$$

but in order to be biologically meaningful, both of these expressions should be positive.

In order to have a growing population in the absence of the infective agent we assume $b_1 > \mu$. If an individual is infected right at birth, it is expected to produce $b_2/(\alpha+\mu)$ offspring. The condition that this is less than one is exactly the condition that the multiplication factor that relates $\overline{S}$ to $\overline{I}$ is positive. It remains to investigate the sign of the denominator in the expression for $\overline{S}$. The condition for this denominator to be positive turns out to be exactly the condition for the negativity of the joint growth rate of I and S in a model in which we omit K in the denominator. Again, we refer to Diekmann and Kretzschmar (1991) for details; the joint growth rate is explicitly given by $\frac{\gamma b_2}{\gamma - b_1 - \alpha + b_2} - \mu - \alpha$, which becomes zero for $\gamma = \gamma_c = \frac{(\mu+\alpha)(b_1+\alpha-b_2)}{\mu+\alpha-b_2}$; to find the growth rate, one has to use a transformation that decouples the equations for relative quantities from the equation for population size, just as we did when introducing (4.13).

Exercise 4.26 The function $C(N)$ should increase with N. In the foregoing exercise we had $C(N) = N/(K+N)$, which is bounded; in the next exercise we shall consider $C(N) = N$, which is unbounded. As we shall see, the difference matters. In the bounded case, everything proceeds essentially as in the foregoing exercise, except that it is more difficult (or, rather, impossible) to do explicit calculations when we do not have an explicit expression for the inverse C^{-1}.

Exercise 4.27 i) If we set $dI/dt = 0$, we obtain $\overline{S} = (\mu + \alpha)/\beta$, and substitution of this into the equation obtained by setting $dS/dt = 0$ leads to $(\mu + \alpha - b_2)\overline{I} = (b_1 - \mu)\overline{S}$. Because we assume that $b_1 > \mu$ (to have exponential growth when the agent is absent), positivity of $\overline{I}$ requires positivity of $\mu + \alpha - b_2$ or, equivalently $b_2/(\mu+\alpha) < 1$ (which has the interpretation that the expected number of offspring produced by an individual that is infected right at birth is less than one). The Jacobi matrix

$$\begin{pmatrix} b_1 - \mu - \beta\overline{I} & b_2 - \beta\overline{S} \\ \beta\overline{I} & 0 \end{pmatrix} \text{ has sign structure } \begin{pmatrix} - & - \\ + & 0 \end{pmatrix},$$

and hence has negative trace and positive determinant, and therefore the steady state is stable (to see the negativity of $b_1 - \mu - \beta\overline{I}$ rewrite it as $-b_2\overline{I}/\overline{S}$).

ii) We rewrite the equation for dS/dt as

$$\frac{dS}{dt} = \left(b_1 + b_2\frac{I}{S} - \mu - \beta I\right)S.$$

As long as $b_2/S - \beta > 0$, we can infer from this that $S(t)$ grows faster than $e^{(b_1-\mu)t}$. So, for large t, necessarily the opposite inequality has to hold and therefore

$S(t) \geq b_2/\beta$. If that is the case then

$$\frac{dI}{dt} = (\beta S - \mu - \alpha)I \geq (b_2 - \mu - \alpha)I,$$

and when $b_2 > \mu + \alpha$ this implies exponential growth of I. Returning to the equation for dS/dt in the form above, we then see that $dS/dt < 0$ if $S > b_2/\beta$. Hence $S(t)$ converges to b_2/β from above. Replacing S in the equation for dI/dt by this quantity, we infer that I grows exponentially at the rate $b_2 - \mu - \alpha$. In their pioneering work on host regulation, May and Anderson (1979)[9] investigated a variant of the system in this exercise.

Exercise 4.28 Note that $\frac{\bar{p}_2}{\bar{p}_1} = \frac{1}{2}(\alpha + \beta)d$ and $\frac{\bar{p}_1}{\bar{p}_0} = 2\alpha d$. Hence $\frac{\bar{p}_2 \bar{p}_0}{\bar{p}_1^2} = \frac{\alpha + \beta}{4\alpha}$, or, equivalently

$$\frac{\alpha}{\alpha + \beta} = \frac{\bar{p}_1^2}{4\bar{p}_0 \bar{p}_2}.$$

Exercise 4.29 The equation reads

$$s = \frac{1}{1 + \beta d} \cdot 1 + \frac{\beta d}{1 + \beta d}(1 + s)$$

and by solving it for s one obtains $s = 1 + \beta d$. Noting that the expected number of visits to state 2 is given by $s - 1$ we find that

$$T = s\frac{d}{1 + \beta d} + (s - 1)\frac{d}{2}$$

from which (4.20) follows at once.

Exercise 4.30 For an ICU of a given size it seems reasonable to assume that the force of infection on a susceptible patient equals $k\beta$, for some $\beta > 0$, if k patients are colonized (and hence infectious). Here we focus on transmission and have not yet incorporated the 'endogenous' route to colonization. The appropriate β may in fact depend on the size of the ICU and for an ICU of N beds it seems appropriate to take β as a decreasing function of N, for instance $\beta = \gamma/N$ for some 'universal' constant γ (recall the end of Section 1.3.2). But a more detailed underpinning would involve considerations about how staffing depends on size and how staff cohorts are organized (i.e., assignment of particular patients to particular health-care workers). Here we leave such issues aside and consider β as given.

For the rate of transition from 1 to 2 we have a contribution 2β because two susceptible patients can become colonized; for the rate of transition from 2 to 3 the contribution is also 2β, but in this case because two colonized patients contribute to the force of infection on the single susceptible patient.

The matrix A is now given by

$$\begin{pmatrix} -3\alpha & \frac{1}{d} & 0 & 0 \\ 3\alpha & -(2\alpha + 2\beta + \frac{1}{d}) & \frac{2}{d} & 0 \\ 0 & 2\alpha + 2\beta & -(\alpha + 2\beta + \frac{2}{d}) & \frac{3}{d} \\ 0 & 0 & \alpha + 2\beta & -\frac{3}{d} \end{pmatrix}.$$

[9]R.M. Anderson and R.M. May: Population biology of infectious diseases, Part 1. *Nature*, **280** (1979), 361–367; Part 2, **280**, 455–461.

Because the columns sum to zero, we have

$$\frac{d}{dt}(p_0 + p_1 + p_2 + p_3) = 0$$

and accordingly total probability is conserved and equals 1.

Exercise 4.31 The endogenous route contributes $(N - i)\alpha$ and the exogenous transmission route contributes $i(N - i)\beta$ because each of the i colonized patients contributes β to the force of infection on each of the $N - i$ susceptible patients. So the answer is $(N - i)\alpha + i(N - i)\beta$.

Exercise 4.32 Every day transitions from any state to any other state may occur. The difficulty is to express the various probabilities in terms of just a few interpretable parameters corresponding to recognized mechanisms. Why is this more difficult in the discrete-time setting? The reason is that several events may happen during the day (and in various orders), so there is an added combinatorial complexity. In fact there is no unique answer for the present exercise, as we will have to specify the assumptions in more detail and are bound to make choices while doing so.

In fact, the first choice we are going to make is that, by (debatable) assumption, discharge and admission take place at a fixed moment during the day, say at noon. We adopt the convention that our bookkeeping scheme concerns the state of the ICU at the moment immediately following a possible discharge, so at 12.05 h, say.

Suppose both patients are susceptible at 12.05 h. For ease of formulation, let us call the endogenous route to detectable colonization the α-mechanism, and the exogenous transmission route the β-mechanism. In principle it may happen that one patient becomes colonized by the α-mechanism and that, subsequently, the other patient becomes colonized by the β-mechanism. Clearly the time interval during which the β-mechanism may act depends on when exactly the first patient became infectious. By taking all possibilities into account one can derive an expression for the probability that this compound event happens. However, we shall sacrifice completeness in order to gain clarity: we simply assume that the event has probability zero.

Again suppose that both patients are susceptible at 12.05 h. At 11.55 h of the next day the number of colonized patients will be two with probability $(1 - e^{-\alpha})^2$; one with probability $2e^{-\alpha}(1 - e^{-\alpha})$; and zero with probability $(e^{-\alpha})^2$. At 12.05 h on that same (next) day it will be two with probability $(1 - \frac{1}{\delta})^2(1 - e^{-\alpha})^2$; one with probability

$$2\frac{1}{\delta}(1 - \frac{1}{\delta})(1 - e^{-\alpha})^2 + 2(1 - \frac{1}{\delta})e^{-\alpha}(1 - e^{-\alpha});$$

and zero with probability

$$(e^{-\alpha})^2 + \frac{2}{\delta}e^{-\alpha}(1 - e^{-\alpha}) + (\frac{1}{\delta})^2(1 - e^{-\alpha})^2.$$

Next, let us consider the case that we are in state 1 at 12.05 h, i.e., one patient is susceptible but the other is colonized. At 11.55 h on the next day, the state will be 1 with probability $\exp(-(\alpha + \beta))$, and it will be 2 with the complementary probability $1 - \exp(-(\alpha + \beta))$. Hence at 12.05 h on that same (next) day it will be 2 with probability $(1 - \frac{1}{\delta})^2(1 - e^{-(\alpha+\beta)})$; 1 with probability

$$2\frac{1}{\delta}(1 - \frac{1}{\delta})(1 - e^{-(\alpha+\beta)}) + (1 - \frac{1}{\delta})e^{-(\alpha+\beta)};$$

and 0 with probability

$$\frac{1}{\delta}e^{-(\alpha+\beta)} + (\frac{1}{\delta})^2(1 - e^{-(\alpha+\beta)}).$$

Finally, if we start out with two colonized patients (so in state 2), the situation can only change by discharge and admission. So at the same time the next day the state will be 2 with probability $(1 - \frac{1}{\delta})^2$; 1 with probability $2\frac{1}{\delta}(1 - \frac{1}{\delta})$; and 0 with probability $(\frac{1}{\delta})^2$.

This completes the calculation of the transition probabilities.

Exercise 4.33 We obtain $\hat{s} = 10\%$ and $\hat{i} \approx 0.024\%$, so only a very small fraction is infectious.

Exercise 4.34 In Figure 16.18 we have plotted the requested figures. It is seen that doubling or reducing R_0 by one half has a big effect on the time to extinction but doing the same thing with the length of the infectious period (keeping R_0 fixed) has an even greater impact. This is also seen from equation (4.61). If the infectious period is shortened by one half, four times as many individuals are needed to keep $E(T_Q)$ unchanged. The corresponding number when reducing R_0 is not as clear from the equation, but it can be shown that reducing R_0 by one half means about twice as many individuals are needed to keep $E(T_Q)$ unchanged.

Exercise 4.35 In Figure 16.19 we have plotted the approximate mean extinction time $E(T_Q)$ (specified by the right-hand side of equation (4.62)) as a function of v, the fraction of the population being vaccinated at birth. This is done for the case $R_0 = 10$, $\varepsilon = 1/3750$ and $1/\mu = 75$ for three different community sizes: $N = 100,000$ (left), $N = 1,000,000$ (center) and $N = 10,000,000$ (right). The critical vaccination coverage equals $v_c = 1 - 1/R_0 = 0.9$, so in all three situations the expected time to extinction is only relevant when $v < 0.9$. It is seen that when the population size is small, or even moderate, the time to extinction is quite short, even with little vaccination. However, when N equals ten million a substantial fraction has to be vaccinated before the extinction time does become small.

Exercise 4.36 Using equation (4.61) for the parameters defined for England we obtain $E(T_Q) = 945$ million years, while for Iceland we find $E(T_Q) = 3.80$ years. The numbers should be taken with a pinch of salt. Still, they clearly suggest that measles will go extinct in Iceland and stay endemic in England, just as was observed before the introduction of vaccination.

Exercise 4.37 When using (4.60) instead of (4.61), the corresponding numbers become $E(T_Q) = 5.71 \cdot 10^{12}$ for England and $E(T_Q) = 3.37$ for Iceland. The two approximations hence are the same for all practical purposes.

Exercise 4.38 The three characterizations of the critical community size yield $N_c^{(1)} = 15.1$ million, $N_c^{(2)} = 9.04$ million and $N_c^{(3)} = 3.23$ million. The three numbers are quite different, but still in the same ball park.

Exercise 4.39 If 90% of all new-born are vaccinated, the critical community size increases to $N_c^{(1)} = 420$ million, $N_c^{(2)} = 2531$ million and $N_c^{(3)} = 179$ million respectively. All three numbers have increased markedly, and they show that only in a very large community is it possible for measles to persist when 90% of all infants are vaccinated successfully. Again, the three numbers are in the same ball park, but their order relation has changed compared to the situation without vaccination.

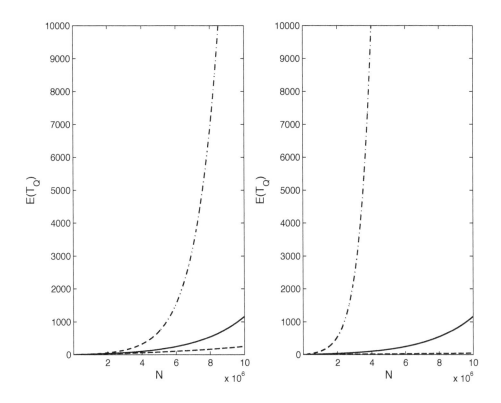

Figure 16.18: The approximation of the mean extinction time $E(T_Q)$ as a function of N. The solid line in both panels corresponds to $R_0 = 10$, $\varepsilon = 1/3750$ and $1/\mu = 75$. In the left panel the corresponding approximation is also given for $R_0 = 5$ (- -) and $R_0 = 20$ (- ·). In the right panel we have instead varied ε: $\varepsilon = 1/(2 \cdot 3750)$ (- -) and $\varepsilon = 2/3750$ (- ·).

Exercise 4.40 If we write (4.79) in the form

$$\frac{dt}{dx} = (1 - x)^{-1}$$

we find by integration that

$$t - \tilde{t} = \ln \frac{1 - x(\tilde{t})}{1 - x(t)}.$$

Next, note that

$$\ln \frac{1 - \bar{x}}{1 - x} + \ln \frac{1 - \phi(x)}{1 - \bar{x}} = \ln \frac{1 - \phi(x)}{1 - x}.$$

Exercise 4.41 The number of cycles follows a geometric distribution with mean $\frac{1 - F(1)}{F(1)}$. The duration of a cycle is i.i.d. with mean $\bar{T}$. Hence the expected time to extinction equals

$$\frac{1 - F(1)}{F(1)} \bar{T}.$$

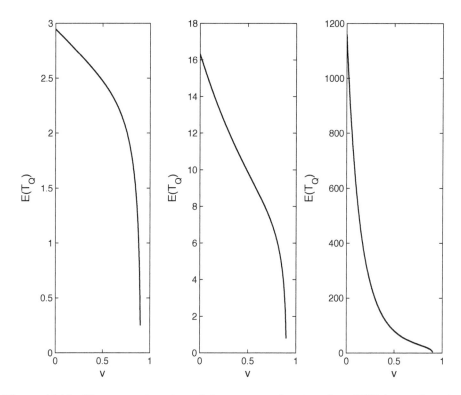

Figure 16.19: The approximation of the mean extinction time $E(T_Q)$ as a function of v for the case $R_0 = 10$, $\varepsilon = 1/3750$ and $1/\mu = 75$. The left panel is for the case $N = 100{,}000$, the center for $N = 1{,}000{,}000$ and the right panel for $N = 10{,}000{,}000$.

With relatively little extra effort one can compute the variance of the time to extinction (but note that, by adopting (4.77), we neglected the variance in the final size), see the paper by de Koeijer et al. (2008), quoted in the main text.

16.5 ELABORATIONS FOR CHAPTER 5

Exercise 5.1 The probability of the data, given λ, is

$$\prod_{i=1}^{n} \lambda (1-\lambda)^{T_i-1} = \lambda^n (1-\lambda)^{\sum\limits_{i=1}^{n} T_i - n}$$

and so the log-likelihood equals

$$n \log \lambda + \left(\sum_{i=1}^{n} T_i - n \right) \log(1-\lambda),$$

which has as derivative

$$\frac{n}{\lambda} - \frac{\sum\limits_{i=1}^{n} T_i - n}{1-\lambda},$$

which is zero for $\lambda = \lambda_{\text{mle}}$ with

$$\frac{1}{\lambda_{\text{mle}}} = \frac{1}{n}\sum_{i=1}^{n}T_i.$$

So just as in (5.7) we find that for the discrete-time variant, the MLE is also determined by the requirement that the expected LOS equals the average of the actually observed LOS.

The function $\lambda \mapsto \log(\lambda(1-\lambda)^{t-1}) = \log\lambda + (t-1)\log(1-\lambda)$ has derivative $1/\lambda - (t-1)/(1-\lambda)$ and second derivative $-1/\lambda^2 - (t-1)/(1-\lambda)^2$. Hence $I(\lambda) = 1/\lambda^2 + (\frac{1}{\lambda}-1)/(1-\lambda)^2$ (note that now the expectation in the definition of I does play a role). Hence

$$\frac{1}{nI(\lambda_{\text{mle}})} = \frac{n(\sum_{i=1}^{n}T_i - n)}{(\sum_{i=1}^{n}T_i)^3} = \frac{1}{n}\frac{(\frac{1}{n}\sum_{i=1}^{n}T_i - 1)}{(\frac{1}{n}\sum_{i=1}^{n}T_i)^3} = \frac{1}{n}\frac{\bar{T}-1}{\bar{T}^3}.$$

Exercise 5.2 Recall from Sections 3.2.2 and 5.1 the lack-of-memory of the exponential distribution. So if, in a thought experiment, we would manage to immediately un-infect the patient concerned, his expected remaining LOS would be $1/\lambda$. Hence we hold the infection responsible for the difference $1/\mu - 1/\lambda$. And if this difference is negative, we might be puzzled: does the infection have a *positive* effect? On second thought, we are reminded that 'discharge' is not at all positive if it is the result of death. So we are stimulated to look more carefully at the data, to see whether there is a correlation between the labels I and 'deceased' (see Chapter 14).

Exercise 5.3 Let $F(\tau)$ denote the probability that the patient is neither discharged nor infected before time τ, then our assumptions imply that $F(0) = 1$ and

$$\frac{dF}{d\tau} = -(\lambda + \sigma)F.$$

Hence

$$F(\tau) = e^{-(\lambda+\sigma)\tau}.$$

In the differential equation for F, the term λF corresponds to discharge and the term σF to infection. Hence the probability that the patient is discharged without being infected equals

$$\int_0^\infty \lambda F(\tau)d\tau = \frac{\lambda}{\lambda + \sigma},$$

while the complementary probability that the patient is infected before being discharged equals

$$\int_0^\infty \sigma F(\tau)d\tau = \frac{\sigma}{\lambda + \sigma}.$$

Exercise 5.4 Recall from the elaboration of the preceding exercise that the term λF corresponds to discharge, so is, modulo a normalization constant, the density that we want to determine. The normalization is, of course, achieved by dividing by the probability $\lambda/(\lambda+\sigma)$ that the patient is discharged with label 'not-infected.'

Exercise 5.5 This density is exactly the same as the one computed in the preceding exercise (so the distribution is exponential with parameter $\lambda + \sigma$). The reason is that the term σF, which corresponds to infection, is *proportional* to λF. So when performing the normalization the difference between the two is eliminated.

Exercise 5.6 The length of the period of stay *after* infection is, according to our assumptions, exponentially distributed with parameter μ. Because the lengths of the periods before and after infection are *independent* random variables, the density of the total LOS of a patient that is discharged with label 'infected' has density

$$\int_0^\tau (\lambda + \sigma)e^{-(\lambda+\sigma)t}\mu e^{-\mu(\tau-t)}dt = \frac{\mu(\lambda + \sigma)}{\lambda + \sigma - \mu}\left(e^{-\mu\tau} - e^{-(\lambda+\sigma)\tau}\right).$$

By independence the mean is simply the sum of the two means.

Exercise 5.7 The 'infected-not-infected' distinction follows a binomial distribution with parameter $\sigma/(\sigma + \lambda)$ (recall Exercise 5.3) and this explains the first three factors in the expression for L. The next N factors correspond to the conditional distribution considered in Exercise 5.4 and the final M factors to the conditional distribution considered in Exercise 5.6.

Exercise 5.8 We have

$$\log L = \log\binom{N+M}{M} + M\log\sigma + N\log\lambda + M\log\mu - (\lambda+\sigma)\sum_{j=1}^{N}T_j$$
$$+ \sum_{j=N+1}^{N+M}\log\frac{e^{-\mu T_j} - e^{-(\lambda+\sigma)T_j}}{\lambda + \sigma - \mu}.$$

(Note that the quotient

$$\frac{e^{-\mu T_j} - e^{-(\lambda+\sigma)T_j}}{\lambda + \sigma - \mu}$$

is always positive, while the denominator and the numerator may have any sign; this is the reason that we do not decompose the argument of the log function in the last summation.) Hence

$$\frac{\partial}{\partial\lambda}\log L = \sum_{j=N+1}^{N+M}\frac{\lambda+\sigma-\mu}{e^{-\mu T_j}-e^{-(\lambda+\sigma)T_j}}\left(\frac{T_j e^{-(\lambda+\sigma)T_j}}{\lambda+\sigma-\mu} - \frac{e^{-\mu T_j}-e^{-(\lambda+\sigma)T_j}}{(\lambda+\sigma-\mu)^2}\right)$$
$$+ \frac{N}{\lambda} - \sum_{j=1}^{N}T_j$$
$$= \frac{N}{\lambda} - \sum_{j=1}^{N}T_j + \sum_{j=N+1}^{N+M}\frac{T_j}{e^{(\lambda+\sigma-\mu)T_j}-1} - \frac{M}{\lambda+\sigma-\mu}$$
$$= \frac{N}{\lambda} - \frac{M}{\lambda+\sigma-\mu} - \sum_{j=1}^{N+M}T_j + \sum_{j=1}^{N+M}\frac{T_j e^{(\lambda+\sigma-\mu)T_j}}{e^{(\lambda+\sigma-\mu)T_j}-1}$$
$$= \frac{N}{\lambda} - \frac{M}{\lambda+\sigma-\mu} - \sum_{j=1}^{N+M}T_j + \sum_{j=1}^{N+M}\frac{T_j}{1-e^{-(\lambda+\sigma-\mu)T_j}}$$

and if we equate this to 0, we obtain the first equation of (5.18). The manipulations that convert $\partial\log L/\partial\sigma = 0$ into the second equation, and $\partial\log L/\partial\mu = 0$ into the third, are essentially the same, so we leave these to the reader.

Exercise 5.9 In the first scenario we have $N = 30$ and $\bar{Y}_{30} = 10/30 = 0.333$. From this we see that $\hat{R}_0 = -\log(0.667)/0.333 = 1.214$. For the standard error we use equation (5.24). If we believe that there is hardly any variation in infectious period (i.e., $r \approx 0$), the standard error equals $s.e.(\hat{R}_0) = 1/\sqrt{N\bar{Y}_{30}(1 - \bar{Y}_{30})} = 0.387$, whereas if the variation in the infectious period is large, say $r \approx 1$, then $s.e.(\hat{R}_0) = \sqrt{1 + 1 \cdot (1 - \bar{Y}_{30})\hat{R}_0}/\sqrt{N\bar{Y}_{30}(1 - \bar{Y}_{30})} = 0.546$. If the community size $N = 300$, but the same fraction gets infected, we still have $\hat{R}_0 = 1.214$, but the standard errors are reduced by a factor $1/\sqrt{10}$, because only N is changed (from 30 to 300). The resulting standard errors hence become 0.122 and 0.173, respectively. If we use the expressions for $\hat{v}_c$ and $s.e.(\hat{v}_c)$ for the case that $N = 300$ and $r = 0$, we obtain the estimates $\hat{v}_c = 0.822$, with $s.e.(\hat{v}_c) = 0.083$.

Exercise 5.10 i) There initially are $N^\diamond = Ns$ susceptible individuals. One such individual has overall contact rate with other susceptibles at rate $c^\diamond = cs$, and the probability of a contact being close enough is still p. The latent and infectious periods are unchanged. The epidemic within the susceptible community is hence just like the prototype stochastic epidemic model, but with $N^\diamond$ and $c^\diamond$ replacing N and c. An estimate of $R_0^\diamond = c^\diamond pE(T_I) = sR_0$ is hence given by $\hat{R}_0^\diamond = -\log(1 - \bar{Y}_N^\diamond)/\bar{Y}_N^\diamond$ ($R_0^\diamond$ may be called the *actual* reproduction number, taking into account immunity). Because the basic reproduction number R_0 satisfies $R_0 = R_0^\diamond/s$ we immediately have $\hat{R}_0 = -\log(1 - \bar{Y}_N^\diamond)/s\bar{Y}_N^\diamond$. Since $\hat{R}_0 = \hat{R}_0^\diamond/s$ it follows that the standard error for $\hat{R}_0$ satisfies $s.e.(\hat{R}_0) = s.e.(\hat{R}_0^\diamond)/s$, and $s.e.(\hat{R}_0^\diamond)$ is obtained from equation (5.24) but with $N^\diamond$, $\bar{Y}_N^\diamond$ and $\hat{R}_0^\diamond$.

ii) If there were 30 susceptibles out of which 10 became infected, but the community also had 30 immune individuals we have $N = 60$, $s = 0.5$, $\bar{Y}_N = 10/60 = 0.167$ and $\bar{Y}_N^\diamond = 10/30 = 0.333$. From this we deduce that $\hat{R}_0 = \hat{R}_0^\diamond/s = 1.2135/0.5 = 2.427$. The standard error equals $s.e.(\hat{R}_0) = s.e.(\hat{R}_0^\diamond)/0.5 = 0.387$ if there is no variation in the infectious period, and $0.546/0.5 = 1.092$ if the coefficient of variation $r \approx 1$ (as in the 'general' stochastic epidemic, i.e., with an exponentially distributed infectious period). The reason why R_0 is now estimated to be larger (cf. Exercise 5.9) is that in this situation half the contacts are 'wasted' from the 'point of view' of the infectious agent.

Exercise 5.11 There were $N = 639$ initially susceptible horses out of which $Y_N = 579$ were infected during the outbreak. Equation (5.23) on page 135 then gives the estimate $\hat{R}_0 = 2.61$. From equation (5.24) we find the standard error. For the case that $r = 0$ we get $s.e.(\hat{R}_0) = 0.136$, and for $r = 0.5$ this changes to $s.e.(\hat{R}_0) = 0.146$.

Exercise 5.12 $N = 150$ and $Y_N = 45$ so $\hat{R}_0 = 1.19$. Using equation (5.24) with $r = 1$ we find a standard error $s.e.(\hat{R}_0) = 0.114$. Since the estimator is approximately normally distributed a 95% confidence interval for R_0 is given by $1.19 \pm 1.96 \cdot 0.114 = (0.965, 1.67)$. It hence seems likely, but not absolutely certain, that R_0 is above threshold.

Exercise 5.13 We have that

$$\hat{\lambda} = \frac{Y}{\int_0^\tau I(s)\bar{S}(s)ds} = -\lambda \frac{Y}{M_1(\tau) - Y} \approx \lambda\left(1 + M_1(\tau)/Y\right),$$

where the approximation holds since Y will be of order N (more precisely, close to $N(1-\sigma)$ where σ was defined in Section 3.5.1) and $M_1(\tau)$ will be of order $\sqrt{N}$ since it is a zero mean martingale with variance $E(Y) \approx N(1 - \sigma)$. As a consequence, the variance of $\hat{\lambda}$ approximately equals $\frac{\lambda^2}{E(Y)^2} E(Y) = \frac{\lambda^2}{E(Y)}$, which implies that a standard error estimate is given by $\hat{\lambda}/\sqrt{Y}$.

Exercise 5.14 We have $N = 30$, $Y = 10$ and $\int_0^\infty I(s)ds = \sum_{i=0}^Y \Delta T_i = 4 + 3 + 5 + \ldots + 2 = 39$ (the sum of the infectious periods of all individuals). Next, we compute $\int_0^\infty I(s)\bar{S}(s)ds$. This is a sum where we add the terms $I(s)\bar{S}(s)$ multiplied by the duration of the corresponding interval. From the data we get: $30^{-1}(1 \cdot 30 \cdot 1 + 2 \cdot 29 \cdot 1 + 3 \cdot 28 \cdot 1 + \ldots + 1 \cdot 20 \cdot 1) = 925/30 = 30.833$. We now have all quantities in order to compute the requested estimates and their standard errors: $\hat{\lambda} = 10/30.8333 = 0.324$, $s.e.(\hat{\lambda}) = 0.324/\sqrt{10} = 0.102$, $1/\hat{\alpha} = 39/11 = 3.54$, $s.e.(1/\hat{\alpha}) = 3.54/\sqrt{11} = 1.068$, and finally $\hat{R}_0 = 0.324 \cdot 3.54 = 1.15$, $s.e.(\hat{R}_0) = (1.15/\sqrt{30})(30/10 + 30/11) = 0.502$.

From this we see that we are able to estimate $\lambda = cp$ and the mean of the infectious period $1/\alpha$ separately. With regard to $\hat{R}_0$, we point out that our estimate 1.15 is slightly different from the estimate obtained using only final size data (i.e., $Y = 10$) as we did in Exercise 5.9 (note that we used the same data $N = 30$ and $Y = 10$ there). In Exercise 5.9 we obtained the estimate 1.21 and the standard error, assuming the 'general' stochastic epidemic model applies, just as we do here, was slightly larger (0.546) than the present estimate 0.502 based on the more detailed data. In general, we expect to have higher precision (i.e., smaller standard errors) in our estimates the more detailed our data is. This is true, although the gain from observing an epidemic outbreak continuously in time, as opposed to only knowing its final size, is surprisingly small.

Exercise 5.15 We use equations (5.38)–(5.42) to estimate the parameters and their standard errors. From Table 5.1 we have $N = 119$ and $Y = 29$. We then compute $\int I(s)ds = 217$ and $\int I(s)\bar{S}(s)ds = 179.81$. Using this we get the following estimates with standard errors in parentheses: $\hat{\gamma} = 0.161$ (0.030), $1/\hat{\alpha} = 7.23$ (1.32) and $\hat{R}_0 = 1.17$ (0.30).

Chapter Seventeen

Elaborations for Part II

17.1 ELABORATIONS FOR CHAPTER 7

Exercise 7.2 i) We have $Kx = c_1 K\psi^{(1)} + c_2 K\psi^{(2)} = c_1\lambda_1\psi^{(1)} + c_2\lambda_2\psi^{(2)}$ and, by induction, $K^m x = c_1\lambda_1^m\psi^{(1)} + c_2\lambda_2^m\psi^{(2)}$.

ii) We can rewrite $K^m x$ as

$$K^m x = c_1\lambda_1^m \left[\psi^{(1)} + \frac{c_2}{c_1}\left(\frac{\lambda_2}{\lambda_1}\right)^n \psi^{(2)} \right]$$

(provided $c_1 \neq 0$) and since $|\lambda_2/\lambda_1| < 1$, the second term within the square brackets approaches zero for $m \to \infty$. The influence of the initial condition (the zeroth generation) is restricted to the value of c_1. When $\lambda_1 > 1$ we observe exponential growth, when $0 < \lambda_1 < 1$ exponential decline.

Exercise 7.5 i) The kth element of the vector $K^m e_j$, where e_j denotes the jth unit vector (which has a one at position j and zeros everywhere else), is precisely $(K^m)_{kj}$. So if K is irreducible, there exists an m for which this quantity is positive, meaning that, via a chain of infections of length m, the infectious individual with h-state j 'at birth,' produces cases with h-state k at birth, with positive probability. 'Eventually' thus means 'in some future generation.' When K is primitive, not only can we choose the length m of the chain independent of j and k, but in addition $(K^m)_{kj}$ is positive for all integer powers larger than this m. This means that for all generations after the mth, the expected number of cases with h-state k at birth is positive, no matter what the h-state at birth was of the ancestor that started the chain of infections. In that case we can take 'eventually' to mean 'after a specified number of generations.'

ii) Yes, you should agree. The key point is the precise specification of 'eventually'; otherwise the difference is just the context dependent interpretation of the positivity of $(K^m)_{kj}$.

iii) First let us concentrate on the sign structure and write $K = \begin{pmatrix} 0 & + \\ + & 0 \end{pmatrix}$.
It then follows that $K^2 = \begin{pmatrix} + & 0 \\ 0 & + \end{pmatrix}$ and $K^3 = \begin{pmatrix} 0 & + \\ + & 0 \end{pmatrix}$, and, by induction,
$K^{2m} = \begin{pmatrix} + & 0 \\ 0 & + \end{pmatrix}$ and $K^{2m+1} = \begin{pmatrix} 0 & + \\ + & 0 \end{pmatrix}$. The sign structure is periodic with period 2, and K is irreducible, but not primitive.

Now we will compute $\| K^m \|^{1/m}$. First we note that $K^2 = 10^3 I$, and consequently we have that $K^3 = 10^3 K$, $K^4 = 10^6 I$, $K^5 = 10^6 K, \ldots$. So we alternate between K and I, modulo a multiplication factor. From this observation, the following follows readily: for $m = 1, 2, \ldots, 10$ we find respectively $\| K^m \|^{1/m} = 10^2, 10^{3/2}, 10^{5/3}, 10^{3/2}, 10^{8/5}, 10^{3/2}, 10^{11/7}, 10^{3/2}, 10^{14/9}, 10^{3/2}$.

iv) The subpopulation of homosexual males is isolated from the rest, and so is the subpopulation of homosexual females. By 'isolated' we mean that if the agent is introduced only in that subpopulation, it stays in that subpopulation and does not spread to other subpopulations. The reason is the contact structure: in the absence of bisexuality, homosexual males only have contacts within their own group. Likewise the subpopulation of heterosexual individuals, comprising both the males and females, is isolated.

So 'reducible' means that we can subdivide the population into isolated groups or, in more mathematical terms, that we can decompose the next-generation matrix into uncoupled lower-dimensional matrices (or matrices that are coupled in one direction only, as in $\begin{pmatrix} + & 0 \\ + & + \end{pmatrix}$, where spread from 1 to 2 is possible, but spread from 2 to 1 is not possible).

Exercise 7.7 The matrix

$$K = \begin{pmatrix} \rho_1 c_1 & (1 - \rho_2)c_2 \\ (1 - \rho_1)c_1 & \rho_2 c_2 \end{pmatrix}$$

has as its characteristic equation $\lambda^2 - (\rho_1 c_1 + \rho_2 c_2)\lambda + (\rho_1 + \rho_2 - 1)c_1 c_2 = 0$ and consequently

$$R_0 = \frac{1}{2}\left(\rho_1 c_1 + \rho_2 c_2 + \sqrt{(\rho_1 c_1 + \rho_2 c_2)^2 - 4(\rho_1 + \rho_2 - 1)c_1 c_2}\right).$$

When $\rho_1 = \rho_2 = 1$, this, of course, yields $R_0 = \max\{c_1, c_2\}$. The other boundary extreme occurs when one of the ρ's is equal to zero. Suppose that $c_1 N_1 < c_2 N_2$; then this happens for $\rho_1 = 0$, $\rho_2 = 1 - \frac{c_1 N_1}{c_2 N_2}$. For the numbers as given in Section 2.2 we have $\rho_2 = 0.9$ and $R_0 \approx 15.5$ (while $\max\{c_1, c_2\} = c_1 = 100$). We did not formally check that R_0 is maximal for $\rho_1 = \rho_2 = 1$ and that it achieves its minimum for $\rho_1 = 0$, $\rho_2 = 1 - \frac{c_1 N_1}{c_2 N_2}$ when $c_1 N_1 < c_2 N_2$, but we are quite confident that this is nevertheless true.

Exercise 7.8 i) and ii) amount to translation of the verbal description into formulas. iii) First compute the inverse of Σ (see also the interlude below) and then use the definition

$$K_L = -T\Sigma^{-1}$$

$$= \begin{pmatrix} 0 & 0 & p\beta \\ 0 & 0 & (1-p)\beta \\ 0 & 0 & 0 \end{pmatrix} \begin{pmatrix} \frac{1}{\nu_1+\mu} & 0 & 0 \\ 0 & \frac{1}{\nu_2+\mu} & 0 \\ \frac{\nu_1}{(\nu_1+\mu)(\alpha+\mu)} & \frac{\nu_2}{(\nu_2+\mu)(\alpha+\mu)} & \frac{1}{\alpha+\mu} \end{pmatrix}$$

$$= \begin{pmatrix} \frac{p\beta\nu_1}{(\nu_1+\mu)(\alpha+\mu)} & \frac{p\beta\nu_2}{(\nu_2+\mu)(\alpha+\mu)} & \frac{p\beta}{\alpha+\mu} \\ \frac{(1-p)\beta\nu_1}{(\nu_1+\mu)(\alpha+\mu)} & \frac{(1-p)\beta\nu_2}{(\nu_2+\mu)(\alpha+\mu)} & \frac{(1-p)\beta}{\alpha+\mu} \\ 0 & 0 & 0 \end{pmatrix}.$$

Interlude: From a computational point of view, it seems easier to use mathematical software to compute K_L from T and Σ. We remark, however, that the only cumbersome step, i.e., computing the inverse of Σ, can be easily performed using the biological interpretation of $-\Sigma^{-1}$. As we noted in Exercise 2.6, and when motivating the definition of K_L in the present Section 7.2, the element $(-\Sigma^{-1})_{ij}$ is the expected time that an individual that presently has state j will spend in state

i during its entire future 'life' (in the epidemiological sense). In the above example this works out as follows. Individuals that are presently in state E_i will spend, on average, an amount of time $1/(\nu_i + \mu)$ in that state. The same individuals will spend on average an amount of time $\frac{\nu_i}{\nu_i+\mu}\frac{1}{\alpha+\mu}$ in state I, where the first factor is the probability that an individual actually changes its state from E_i to I, instead of leaving state E_i by dying, and the second factor is the average amount of time an individual who enters state I spends in state I. For completeness, we add that this is a general rule (already observed in Exercise 1.34-vi): for example when an individual can leave a state A in several ways, the probability of going to a particular state B is the product of the per-capita rate of changing from state A to state B and the average time spent in state A (sojourn time). The individuals in state E_i will spend no time at all in state E_j, with $j \neq i$, leading to zeroes for the appropriate elements. Finally, individuals that are presently in state I will spend no time at all in states E_1 and E_2, and will, on average, spend an amount of time $1/(\alpha + \mu)$ in state I. This leads to a full specification of $-\Sigma^{-1}$. See also the elaboration of Exercise 7.11.

Of the three infected states, only two are a state-at-infection: all individuals are 'epidemiologically born' in either state E_1 or E_2. If we therefore restrict K_L to its first two columns and rows we obtain K.

iv) Direct calculation gives the characteristic equation

$$-\lambda^3 + \lambda^2 \left(\frac{p\beta\nu_1}{(\nu_1 + \mu)(\alpha + \mu)} + \frac{(1-p)\beta\nu_2}{(\nu_2 + \mu)(\alpha + \mu)} \right) = 0$$

for K_L and

$$\lambda^2 - \lambda \left(\frac{p\beta\nu_1}{(\nu_1 + \mu)(\alpha + \mu)} + \frac{(1-p)\beta\nu_2}{(\nu_2 + \mu)(\alpha + \mu)} \right) = 0$$

for K, leading to the same dominant eigenvalue as given in the exercise.

Exercise 7.9 The appropriate matrices C and R for this system are given in the main text. Using the definition of K_S we then find

$$K_S = -R\Sigma^{-1}C = -\begin{pmatrix} 0 & 0 & \beta \end{pmatrix} \Sigma^{-1} \begin{pmatrix} p \\ 1-p \\ 0 \end{pmatrix}$$

$$= \left(\frac{p\beta\nu_1}{(\nu_1 + \mu)(\alpha + \mu)} + \frac{(1-p)\beta\nu_2}{(\nu_2 + \mu)(\alpha + \mu)} \right).$$

Because this is a one-dimensional 'matrix,' it is equal to its 'dominant eigenvalue,' leading to the expression for R_0 given in the main text.

Exercise 7.10 We first show that the NGM and the NGM with large domain have the same non-zero eigenvalues. Let v be an eigenvector of K with corresponding eigenvalue λ. Then $Kv = -\mathcal{E}^\top T\Sigma^{-1}\mathcal{E}v = \lambda v$. Multiply this identity by $\mathcal{E}$ to get $-\mathcal{E}\mathcal{E}^\top T\Sigma^{-1}\mathcal{E}v = \lambda\mathcal{E}v$. But $\mathcal{E}\mathcal{E}^\top T = T$, so $\mathcal{E}v$ is an eigenvector of K_L with corresponding eigenvalue λ, and the non-zero eigenvalues of K and K_L are the same. (Note that it is impossible that $\mathcal{E}v = 0$ because this would imply that $\lambda v = Kv = 0$, hence $v = 0$ as $\lambda \neq 0$.)

Next we show that the NGM with small domain and the NGM with large domain have the same non-zero eigenvalues. Let v be an eigenvector of K_S with corresponding eigenvalue λ. Then $K_S v = -R\Sigma^{-1}Cv = \lambda v$. Multiply this identity by C to get $-CR\Sigma^{-1}Cv = \lambda Cv$. But $CR = T$, so Cv is an eigenvector of K_L with corresponding eigenvalue λ. (One excludes the possibility of Cv being 0, as above.)

We have established that the matrices K, K_L and K_S have the same non-zero spectrum and hence the same dominant eigenvalue (as well as the same rank).

Exercise 7.11 We illustrate the general procedure by deriving an expression for element k_{11}. For the element k_{11}, we start with one individual with state-at-infection 1 (i.e., an individual that has just entered state E_1), and determine, by following that individual for the remainder of its infectious life, how many new cases of state-at-infection 1 it is expected to produce. Before the individual can infect, it has to survive the E_1 state and move to the I state. This happens with probability $\nu_1/(\nu_1 + \mu)$.

While in the I state, the individual is expected to produce new cases at a rate β, for an expected time $1/(\alpha + \mu)$. A fraction p of these will be new cases with state-at-infection 1. Multiplying these factors we obtain

$$k_{11} = \frac{\nu_1}{\nu_1 + \mu} \beta \frac{1}{\alpha + \mu} p.$$

Analogous reasoning gives the expressions for k_{12}, k_{21} and k_{22}.

Exercise 7.12 Before embarking on the exercise proper, we explain what 'the grower chooses these rates ... ' in the text of the exercise means. The dynamic equations for the population sizes in field and nursery in the absence of infection are

$$\frac{dN_1}{dt} = -\mu_1 N_1 + \zeta N_2,$$
$$\frac{dN_2}{dt} = \gamma N_1 - (\mu_2 + \zeta)N_2.$$

Hence, to achieve a stable situation we require that the matrix

$$\begin{pmatrix} -\mu_1 & \zeta \\ \gamma & -(\mu_2 + \zeta) \end{pmatrix}$$

has eigenvalue zero, which amounts to choosing the rates to satisfy the condition $\gamma\zeta = \mu_1/(\mu_2 + \zeta)$. Under this condition one can take for $\begin{pmatrix} N_1 \\ N_2 \end{pmatrix}$ any eigenvector corresponding to eigenvalue zero, so any multiple of $\begin{pmatrix} \zeta \\ \mu_1 \end{pmatrix}$.

If we now introduce infection into this population, there are two states-at-infection: a plant can either become infected while standing in the field, or when being planted or standing in the nursery. Because disease-induced mortality rates and transmission rate constants differ for these types, we have to take both into account. The states-at-infection are the only infected states in the way the system is defined, therefore K_L and K are equal and R_0 is the dominant eigenvalue of this 2×2 matrix. Number the states-at-infection 1 and 2, for plants that are in the field or in the nursery at the moment of infection, respectively. The transition matrix Σ is given by

$$\begin{pmatrix} -(\mu_1 + \rho_1) & \zeta \\ 0 & -(\mu_2 + \rho_2 + \zeta) \end{pmatrix}$$

and

$$-\Sigma^{-1} = \begin{pmatrix} \frac{1}{\mu_1+\rho_1} & \frac{\zeta}{(\mu_1+\rho_1)(\mu_2+\rho_2+\zeta)} \\ 0 & \frac{1}{\mu_2+\rho_2+\zeta} \end{pmatrix}.$$

For the transmission matrix the assumptions lead to

$$T = \begin{pmatrix} \beta_1 N_1 & 0 \\ p\gamma & \beta_2 N_2 \end{pmatrix},$$

since field plants are expected to produce $\beta_1 N_1$ new field infections per unit of time, and are expected to produce $p\gamma$ infected cuttings (i.e., individuals that are born with state-at-infection 2, by vertical transmission); infected nursery plants do not have contact with field plants, but are expected to produce $\beta_2 N_2$ new nursery infections per unit of time.

Multiplying the two ingredients in the right order, we find

$$K = K_L = \begin{pmatrix} \frac{\beta_1 N_1}{\mu_1+\rho_1} & \frac{\zeta\beta_1 N_1}{(\mu_1+\rho_1)(\mu_2+\rho_2+\zeta)} \\ \frac{p\gamma}{\mu_1+\rho_1} & \frac{\beta_2 N_2}{\mu_2+\rho_2+\zeta} + \frac{p\gamma\zeta}{(\mu_1+\rho_1)(\mu_2+\rho_2+\zeta)} \end{pmatrix}.$$

Finally,

$$R_0 = \frac{1}{2}(k_{11} + k_{22}) + \frac{1}{2}\sqrt{k_{11}^2 - 2k_{11}k_{22} + k_{22}^2 + 4k_{12}k_{21}}.$$

Exercise 7.13 i) There are four infected states in this system: latently infected animals (E), (horizontally) infectious animals (I), persistently infected (and infectious) animals (P), and finally, recovered and immune animals in class Z that can produce new infections by giving birth to an infected calf (vertical transmission). In all previous examples the recovered states did not occur in the infected subsystem because the individuals themselves are considered to be free from the infectious agent and (possibly temporarily) immune. These recovered states therefore did not play a role in the construction of the NGM and the calculation of R_0. In the BVDV-system this is different, because although the individual itself has indeed lost the infection, it is still carrying an infected fetus and can therefore still produce a new case.

Among the four infected states there are only two that are also states-at-infection: animals that have just been horizontally infected (entering class E; to be denoted type 1) and animals that have just been born persistently infected (entering class P; type 2). Horizontally infected animals can arise in two ways: by contact either with another horizontally infected animal or with a persistently infected animal. Persistently infected animals only arise in one way: through birth from a horizontally infected mother in immune class Z. Because persistently infected animals have an increased death rate, reduced fertility and a much higher horizontal (mass-action) transmission rate constant (for causing transient infections), compared with horizontally infected animals, it makes good sense to explicitly distinguish the two states.

ii) For the next-generation matrix K, we have to model four elements. The element k_{12} is straightforward, because it describes the expected number of horizontal infections (type 1), caused by a persistently infected individual (type 2) during its entire life. Life expectancy of a persistently infected animal is $1/(\mu + b)$, and the mass-action transmission rate is β_2. So, $k_{12} = \beta_2/(\mu + b)$. The element k_{22} describes the expected number of persistent infections caused by a persistently infected animal during its entire life. Since persistent infections directly caused by a given individual can only arise vertically through giving birth, we have that $k_{22} = (\mu - a)/(\mu + b)$, since all calves will be infected (by assumption). The element k_{11} describes the expected number of horizontal infections caused by one individual that has just become horizontally infected. The freshly infected individual survives with probability $\nu/(\mu + \nu)$ the E class to become infectious. While in the I class it is expected to produce β_1 new horizontal cases per unit of time, for an average of $1/(\gamma + \mu)$ time units. Therefore,

$$k_{11} = \frac{\beta_1 \nu}{(\mu + \nu)(\gamma + \mu)}.$$

It remains to characterize the element k_{21}. This describes the expected number of persistently infected animals arising from a horizontally infected animal during its entire infectious life. We note first that from the modeling assumptions in the text, it follows immediately that this number will be at most one (assuming one calf per pregnancy). The reason is that persistently infected calves can only be born if the mother enters immune state Z after which she becomes permanently immune and so cannot produce an infected calf again. Consider an animal that has just become horizontally infected. With probability $\nu/(\mu+\nu)$ the animal will survive the latency period, and with probability $\gamma/(\mu + \gamma)$ it will survive the subsequent infectious period and recover. With probability p_1 it will recover into class Z. Finally, the animal will survive in the recovered class long enough to give birth with probability $\alpha/(\mu + \alpha)$, and with probability p_2 the calf will be alive and persistently infected. Combining all terms gives

$$k_{21} = \frac{\alpha p_2}{\mu + \alpha} \frac{\gamma p_1}{\mu + \gamma} \frac{\nu}{\mu + \nu}.$$

iii) The transmission and transition matrices are, respectively,

$$T = \begin{pmatrix} 0 & \beta_1 & 0 & \beta_2 \\ 0 & 0 & 0 & 0 \\ 0 & 0 & 0 & 0 \\ 0 & 0 & p_2\alpha & \mu - a \end{pmatrix}$$

and

$$\Sigma = \begin{pmatrix} -(\nu + \mu) & 0 & 0 & 0 \\ \nu & -(\gamma + \mu) & 0 & 0 \\ 0 & p_1\gamma & -(\alpha + \mu) & 0 \\ 0 & 0 & 0 & -(\mu + b) \end{pmatrix}.$$

Therefore,

$$-\Sigma^{-1} = \begin{pmatrix} \frac{1}{\nu+\mu} & 0 & 0 & 0 \\ \frac{\nu}{\nu+\mu}\frac{1}{\gamma+\mu} & \frac{1}{\gamma+\mu} & 0 & 0 \\ \frac{\nu}{\nu+\mu}\frac{p_1\gamma}{\gamma+\mu}\frac{1}{\alpha+\mu} & \frac{p_1\gamma}{\gamma+\mu}\frac{1}{\alpha+\mu} & \frac{1}{\alpha+\mu} & 0 \\ 0 & 0 & 0 & \frac{1}{\mu+b} \end{pmatrix}$$

and finally

$$
K_L = -T\Sigma^{-1} = \begin{pmatrix}
\frac{\nu\beta_1}{(\nu+\mu)(\gamma+\mu)} & \frac{\beta_1}{\gamma+\mu} & 0 & \frac{\beta_2}{\mu+b} \\
0 & 0 & 0 & 0 \\
0 & 0 & 0 & 0 \\
\frac{\nu p_1 \gamma p_2 \alpha}{(\nu+\mu)(\gamma+\mu)(\alpha+\mu)} & \frac{p_1\gamma p_2\alpha}{(\gamma+\mu)(\alpha+\mu)} & \frac{p_2\alpha}{\alpha+\mu} & \frac{\mu-a}{\mu+b}
\end{pmatrix}.
$$

Our main interest is K. Note that T has non-zero elements in two rows only, hence K is two-dimensional. The two states-at-infection are the horizontally infected E state and the vertically (and persistently) infected P state. Define

$$
\mathcal{E} = \begin{pmatrix}
1 & 0 \\
0 & 0 \\
0 & 0 \\
0 & 1
\end{pmatrix}.
$$

Then the NGM is given by

$$
K = -\mathcal{E}^\top T\Sigma^{-1}\mathcal{E} = \begin{pmatrix}
\frac{\nu\beta_1}{(\nu+\mu)(\gamma+\mu)} & \frac{\beta_2}{\mu+b} \\
\frac{p_2\alpha\nu p_1\gamma}{(\nu+\mu)(\gamma+\mu)(\alpha+\mu)} & \frac{\mu-a}{\mu+b}
\end{pmatrix}.
$$

Exercise 7.14 It seems most natural to designate only the production of free virus as reproduction, and to consider the event of free virus entering a target cell as a transition in the state of the virus. That view leads to the choice

$$
T = \begin{pmatrix} 0 & 0 \\ p & 0 \end{pmatrix}, \quad \Sigma = \begin{pmatrix} -(\mu+d) & k\hat{X} \\ 0 & -(k\hat{X}+c) \end{pmatrix},
$$

and hence to

$$
-\Sigma^{-1} = \frac{1}{(\mu+d)(k\hat{X}+c)} \begin{pmatrix} k\hat{X}+c & k\hat{X} \\ 0 & \mu+d \end{pmatrix}
$$

and

$$
-T\Sigma^{-1} = \begin{pmatrix} 0 & 0 \\ \frac{p}{\mu+d} & \frac{k\hat{X}}{k\hat{X}+c}\frac{p}{\mu+d} \end{pmatrix}
$$

so that

$$
R_0 = \frac{k\hat{X}}{k\hat{X}+c}\frac{p}{\mu+d}.
$$

The alternative is to also call the event of free virus entering a target cell a reproduction event. In that case there are no state transitions, but there is only birth and death. We then have

$$
T = \begin{pmatrix} 0 & k\hat{X} \\ p & 0 \end{pmatrix}, \quad \Sigma = \begin{pmatrix} -\mu-d & 0 \\ 0 & -k\hat{X}-c \end{pmatrix}
$$

and we obtain

$$
-T\Sigma^{-1} = \begin{pmatrix} 0 & \frac{k\hat{X}}{k\hat{X}+c} \\ \frac{p}{\mu+d} & 0 \end{pmatrix}
$$

and finally

$$
R_0 = \sqrt{\frac{k\hat{X}}{k\hat{X}+c}\frac{p}{\mu+d}}.
$$

So, exactly as in the case of an infectious agent transmitted by a vector, it matters what we choose to call 'reproduction.' The difference is relatively harmless though, if one only wishes to characterize threshold behavior: if in one view $R_0 = 1$, then so it is in any other view!

Exercise 7.15 The next-generation matrix K has been specified in Exercise 7.13. When $\beta_1 = 0$ we can then write the formula for the dominant eigenvalue of a 2×2-matrix as:

$$R_0 = \frac{1}{2}B + \frac{1}{2}\sqrt{B^2 + 4A},$$

with $B := k_{22}$ and $A := k_{21}k_{12}$. Cherry et al. (1998; quoted in Exercise 7.13) derive that the infection-free steady state of their eight-dimensional system is unstable if the quantity $A + B$ is larger than 1. This is a valid statement, but the quantity does not have the individual-level biological interpretation of R_0. It is equivalent only in indicating the stability of the infection-free steady state. We have the following series of simple implications (where we use that $A > 0$ and $2 - B > 0$):

$$
\begin{aligned}
R_0 &= \frac{1}{2}B + \frac{1}{2}\sqrt{B^2 + 4A} > 1 \\
&\Leftrightarrow \sqrt{B^2 + 4A} > 2 - B \\
&\Leftrightarrow B^2 + 4A > 4 - 4B + B^2 \\
&\Leftrightarrow A + B > 1.
\end{aligned}
$$

Exercise 7.16 The perturbed matrix

$$\begin{pmatrix} 10^3 & \varepsilon_{12} \\ \varepsilon_{21} & 1 \end{pmatrix}$$

has as its characteristic equation $\lambda^2 - 1001\lambda + (1000 - \varepsilon_{12}\varepsilon_{21}) = 0$, with roots

$$\lambda_\pm = \frac{1001}{2} \pm \frac{999}{2}\sqrt{1 + \frac{4\varepsilon_{12}\varepsilon_{21}}{998001}},$$

and therefore $R_0 = \lambda_d = \lambda_+ \approx 1000 + \varepsilon_{12}\varepsilon_{21}10^{-3}$. The eigenvector $x = (x_1, x_2)^\top$ has to satisfy

$$10^3 x_1 + \varepsilon_{12} x_2 = \lambda_+ x_1 \approx 10^3 x_1 + \varepsilon_{12}\varepsilon_{21}10^{-3}x_1,$$

from which it follows that

$$x_2 \approx \varepsilon_{21}10^{-3}x_1.$$

Therefore if $x_1 + x_2 = 1$ then $x_1 = O(1)$ while $x_2 = O(\varepsilon_{21})$, and we conclude that new cases predominantly have h-state 1 at birth and only a very small fraction has h-state 2. The reason is that most effective contacts are within the 1-group and the coupling is a very loose one. The point of the remark is that we see this reflected in an order-of-magnitude difference between the components of the normalized eigenvector corresponding to R_0.

Exercise 7.17 When $k_{ij} = a_i b_j$, we can write $(K\phi)_i = a_i \sum_{j=1}^n b_j\phi_j$, and therefore $K\phi = (\sum_{j=1}^n b_j\phi_j)a$ or, in words, for arbitrary ϕ the vector $K\phi$ is a multiple of the vector a (using a bit more jargon, we can also express this by saying that the range of K is spanned by a). Hence eigenvectors have to be a multiple of a. Since $Ka = (\sum_{j=1}^{n} b_j a_j)a$, the corresponding (one and only non-zero) eigenvalue equals $\sum_{j=1}^{n} b_j a_j$. By definition, this is also R_0.

If $k(\xi, \eta) = a(\xi)b(\eta)$, the range of K is spanned by the function a, and, following exactly the same line of argument, one arrives at the expression $R_0 = \int_\Omega b(\eta)a(\eta)\,d\eta$.

If $\Lambda(\eta)(\omega) = \alpha(\omega)b(\eta)$, the range of K is spanned by the measure α, and hence $R_0 = \int_\Omega b(\eta)\alpha(d\eta)$.

Exercise 7.18 We have

$$R_0 = \sum_{j=1}^{n} b_j a_j = \frac{\sum_{j=1}^{n} c_j^2 N_j}{\sum_{k=1}^{n} c_k N_k}.$$

Normalize the N_i such that $\sum_{k=1}^{n} N_k = 1$ (this keeps the expression for R_0 unchanged, since the N_i occur linearly in both numerator and denominator). Then $E(c) = \sum_{k=1}^{n} c_k N_k$ is the mean, while the variance is $V(c) = \sum_{k=1}^{n}(c_k - E(c))^2 N_k = \sum_{k=1}^{n} c_k^2 N_k - E(c)^2$. We find

$$R_0 = \frac{E(c)^2 + V(c)}{E(c)} = E(c) + \frac{V(c)}{E(c)}.$$

Exercise 7.19 Under the assumption that $k_{ij} = a_i b_j + c_j \delta_{ij}$, the eigenvalue problem $K\psi = \lambda\psi$, written out in components, reads

$$a_i \sum_{j=1}^{n} b_j \psi_j = (\lambda - c_i)\psi_i.$$

Now multiply this identity by $b_i(\lambda - c_i)^{-1}$ and sum over i to obtain

$$\sum_{i=1}^{n} \frac{b_i a_i}{\lambda - c_i} \sum_{j=1}^{n} b_j \psi_j = \sum_{i=1}^{n} b_i \psi_i,$$

which shows that necessarily $\sum_{j=1}^{n} b_j \psi_j = 0$ or that

$$\sum_{i=1}^{n} \frac{b_i a_i}{\lambda - c_i} = 1.$$

For the dominant eigenvalue, the first possibility is ruled out, since all contributions to the sum are positive (see Theorem 7.6). As a function of a real variable λ, the sum $\sum_{i=1}^{n} b_i a_i/(\lambda - c_i)$ is strictly decreasing on the interval $(\max c_i, \infty)$. Let Q_0 denote the value of this sum for $\lambda = 1$. Then the sum assumes the value 1 to the right of $\lambda = 1$ if and only if $Q_0 > 1$. As the value of λ for which the sum equals 1 must be R_0, we have verified the threshold property.

A derivation of the third case goes as follows. In terms of $\Lambda(\eta)(\omega) = a(\omega)b(\eta) + c(\eta)\delta_\eta(\omega)$, the eigenvalue problem $Km = \lambda m$ reads

$$\alpha(\omega) \int_\Omega b(\eta)m(d\eta) + \int_\omega c(\eta)m(d\eta) = \lambda m(\omega),$$

where $\omega \subset \Omega$. Now integrate the function $\xi \mapsto b(\xi)/(\lambda - c(\xi))$ over Ω with respect to both the left- and right-hand sides to obtain

$$\int_\Omega \frac{b(\xi)}{\lambda - c(\xi)} \alpha(d\xi) \int_\Omega b(\eta)m(d\eta) + \int_\Omega \frac{b(\xi)c(\xi)}{\lambda - c(\xi)} m(d\xi) = \lambda \int_\Omega \frac{b(\xi)}{\lambda - c(\xi)} m(d\xi),$$

or, after a little rearrangement,

$$\int_\Omega \frac{b(\xi)}{\lambda - c(\xi)} \alpha(d\xi) \int_\Omega b(\eta)m(d\eta) = \int_\Omega b(\xi)m(d\xi).$$

The rest of the argument is identical to the one presented above.

Exercise 7.20 We have

$$(Km)(\omega) = \int_\Omega \Lambda(\eta)(\omega)m(d\eta) = \sum_{i=1}^n \alpha_i(\omega) \int_\Omega b_i(\eta)m(d\eta),$$

which shows that the range of K is spanned by $\{\alpha_i\}_{i=1}^n$. If $m = \sum_j c_j \alpha_j$ then $Km = \sum_i d_i \alpha_i$, with

$$d_i = \sum_{j=1}^n \int_\Omega b_i(\eta)\alpha_j(d\eta)c_j,$$

so the coefficients satisfy $d = Lc$, where matrix L has elements $l_{ij} = \int_\Omega b_i(\eta)\alpha_j(d\eta)$. Necessarily then, non-zero eigenvalues of K are in one-to-one correspondence with non-zero eigenvalues of L.

Exercise 7.21 We have

$$\begin{aligned}
(Km)(i,\widetilde{\omega}) &= \sum_{j=1}^n \int_{\widetilde{\Omega}} \Lambda_i(j,\zeta)(\widetilde{\omega})m(j,d\zeta) \\
&= \alpha_i(\widetilde{\omega}) \sum_{j=1}^n \int_{\widetilde{\Omega}} b_{ij}(\zeta)m(j,d\zeta),
\end{aligned}$$

which shows that the range of K is spanned by the measures α_i on $\widetilde{\Omega}$. If $m(i,\widetilde{\omega}) = c_i \alpha_i(\widetilde{\omega})$ then $(Km)(i,\widetilde{\omega}) = d_i \alpha_i(\widetilde{\omega})$, with $d_i = \sum_{j=1}^n \int_{\widetilde{\Omega}} b_{ij}(\zeta)\alpha_j(d\zeta)c_j$, which shows that the coefficients transform according to a matrix L with $l_{ij} = \int_{\widetilde{\Omega}} b_{ij}(\zeta)\alpha_j(d\zeta)$. Necessarily, R_0 is the dominant eigenvalue of this matrix.

Exercise 7.22 Let $\Omega = \{1, 2, \ldots, n\}$. We have

$$k_{ij} = \int_0^\infty h \sum_{l=1}^n c_{il} P(\tau, l, j) \, d\tau,$$

with

$$P(\tau, l, j) = (e^{\tau\Sigma})_{lj},$$

and hence

$$\int_0^\infty P(\tau, l, j) \, d\tau = -(\Sigma^{-1})_{lj}.$$

Exercise 7.23

$$R_0^D = \frac{\beta}{\mu + \sigma} = \frac{pc}{\mu + \sigma}.$$

Exercise 7.24 The rates at which an individual has contacts with '0' and '+' individuals are respectively cN_0/N and cN_+/N. If an individual is D-infected but not d-infected and it has contact with a D-susceptible that is not d-infected, the probability of transmission is p. If the D-infected individual is also d-infected, then this success ratio is increased by a factor w, and if the D-susceptible individual is also d-infected then there is an increase with a factor v. Hence

$$T = \frac{pc}{N} \begin{pmatrix} N_0 & N_0 w \\ N_+ v & N_+ vw \end{pmatrix}.$$

Exercise 7.25 Although not explicitly stated in the text, the idea is to describe demographic turnover as in the beginning of Section 4.2. This means that we have a population birth rate μN and that all newborns are susceptible to d. Hence the differential equations read

$$\begin{aligned}
\frac{dN_0}{dt} &= \mu N - \mu N_0 - \zeta N_0 + \gamma N_+, \\
\frac{dN_+}{dt} &= -\mu N_+ + \zeta N_0 - \gamma N_+,
\end{aligned}$$

and $N_0 + N_+ = N$. Using this last equation and setting the right-hand side of the differential equation for N_0 equal to zero, we arrive at $\mu(N-N_0)-\zeta N_0+\gamma(N-N_0) = 0$, or $N_0 = \frac{\gamma+\mu}{\zeta+\gamma+\mu}N$. Hence $N_+ = N - N_0 = \frac{\zeta}{\zeta+\gamma+\mu}N$.

Exercise 7.26 To obtain Σ from the description in the previous exercise, we have to do two things: i) leave out the birth term, since we now concentrate on one particular individual, and ii) increase the death rate to $\mu+\sigma$, since we now consider a D-infected individual. Hence

$$\Sigma = \begin{pmatrix} -\zeta - \mu - \sigma & \gamma \\ \zeta & -\gamma - \mu - \sigma \end{pmatrix}$$

and $\det \Sigma = (\zeta + \mu + \sigma)(\gamma + \mu + \sigma) - \gamma\zeta = (\mu + \sigma)(\zeta + \mu + \sigma + \gamma)$, and so

$$-\Sigma^{-1} = \frac{1}{(\mu + \sigma)(\mu + \sigma + \gamma + \zeta)} \begin{pmatrix} \mu + \sigma + \gamma & \gamma \\ \zeta & \mu + \sigma + \zeta \end{pmatrix}.$$

In this system the states-at-infection coincide with the infected states. Therefore $K = K_L$ and we conclude that $K = -T\Sigma^{-1}$ with T given in Exercise 7.24 and Σ as above.

Exercise 7.27 The determinant of T equals zero, so the columns are linearly dependent (indeed, the second is w times the first). So the range is spanned by the first column, i.e., by $\begin{pmatrix} N_0 \\ N_+ v \end{pmatrix}$. Stated differently: in the initial phase new D-cases arise in the ratio $N_0 : N_+ v$, with respect to being d-free or d-infected. The key assumption that underlies this result is that the success ratio is enlarged by the product vw when both individuals are d-infected. This product structure reflects the assumption that the states of the susceptible individual and the infectious individual have independent influence on the transmission probability.

Because K is the product of T and another matrix, the range of K is spanned by the same vector that spans the range of T. The range of K is therefore also one-dimensional and hence $\det K = 0$. The NGM with small domain is the restriction of K to that range and K_S is therefore a one-dimensional matrix.

Exercise 7.28 In Exercise 7.27 we showed that the column space of T is spanned by the vector $\binom{N_0}{N_+v} = C$. The row space is spanned by the vector $\begin{pmatrix} 1 & w \end{pmatrix}$ (indeed the first row is pcN_0/N times that vector, and the second row is pcN_+v/N times that vector). We have $T = CR$, with $R = \frac{pc}{N}\begin{pmatrix} 1 & w \end{pmatrix}$. By definition of the NGM with small domain we then have

$$K_S = -R\Sigma^{-1}C = -\frac{pc}{N}\begin{pmatrix} 1 \\ w \end{pmatrix} \cdot \Sigma^{-1}\begin{pmatrix} N_0 \\ N_+v \end{pmatrix},$$

which is equal to its dominant eigenvalue because K_S is one-dimensional (i.e., a scalar). Alternatively, one can reason in terms of the two-dimensional NGM K. Because it has a one-dimensional range it has one eigenvector and the prime candidate for this eigenvector is of course $\phi = C = \binom{N_0}{N_+v}$. For this choice of ϕ we then have $K\phi = \lambda\phi$, with the only (and hence dominant) non-zero eigenvalue

$$\lambda = -\frac{pc}{N}\begin{pmatrix} 1 \\ w \end{pmatrix} \cdot \Sigma^{-1}\phi = -\frac{pc}{N}\begin{pmatrix} 1 \\ w \end{pmatrix} \cdot \Sigma^{-1}\begin{pmatrix} N_0 \\ N_+v \end{pmatrix}.$$

Exercise 7.29 We identify R_0 with the only non-zero eigenvalue of K found in the preceding exercise. Before elaborating on the explicit expression, we note that for $v = w = 1$ we should retrieve the expression $R_0 = pc/(\mu + \sigma)$ that we found in Exercise 7.23, since in that case it is totally irrelevant for D-transmission whether or not an individual is d-infected.

Since

$$-\Sigma^{-1}\begin{pmatrix} N_0 \\ N_+v \end{pmatrix} = \frac{1}{(\mu+\sigma)(\mu+\sigma+\gamma+\zeta)}\begin{pmatrix} (\mu+\sigma+\gamma)N_0 + \gamma N_+v \\ \zeta N_0 + (\mu+\sigma+\zeta)N_+v \end{pmatrix},$$

we find that

$$
\begin{aligned}
R_0 &= -\frac{pc}{N}\begin{pmatrix} 1 \\ w \end{pmatrix} \cdot \Sigma^{-1}\begin{pmatrix} N_0 \\ N_+v \end{pmatrix} \\
&= \frac{pc}{(\mu+\sigma)(\mu+\sigma+\gamma+\zeta)}\left\{ (\mu+\sigma+\gamma)\frac{N_0}{N} + \gamma v\frac{N_+}{N} \right. \\
&\qquad \left. + w\zeta\frac{N_0}{N} + (\mu+\sigma+\zeta)vw\frac{N_+}{N} \right\}.
\end{aligned}
$$

If we now substitute the expressions for N_0 and N_+ derived in Exercise 7.25, take out the common factor $(\mu+\gamma+\zeta)^{-1}$ from the expression in $\{\cdots\}$ and rearrange terms, we obtain

$$R_0 = pc\frac{(\mu+\gamma)(\mu+\sigma+\gamma+\zeta w) + \zeta vw(\frac{\gamma}{w}+\mu+\sigma+\zeta)}{(\mu+\gamma+\zeta)(\mu+\sigma)(\mu+\sigma+\gamma+\zeta)}.$$

For $v = w = 1$ this does indeed reduce to $pc/(\mu+\sigma)$, as required.

Exercise 7.30 i) We rewrite F as

$$F = \frac{\gamma + \mu}{\gamma + \mu + \zeta} \frac{\mu + \sigma + \gamma + \zeta w}{\mu + \sigma + \gamma + \zeta} + \frac{\zeta}{\gamma + \mu + \zeta} \frac{v\gamma + vw\mu + vw\sigma + vw}{\mu + \sigma + \gamma + \zeta}$$

and note that F is a convex combination of two numbers that are both at least 1, because all parameters are positive and $v, w \geq 1$. This implies that $F \geq 1$.

ii) Take the limit $\gamma \to \infty$ or $\zeta \to 0$ in the expression for F. We indeed expect a disease d of which infected individuals are cured very rapidly (i.e., have a very short infectious period $1/\gamma$), and with a very low infectious pressure on the population (i.e., a very small force of infection ζ), not to make much impact on the transmission of D.

iii) Note first that $F = 1$, if $v = w = 1$. Consider $F = F(\gamma, \zeta)$, for the case $v, w > 1$. To keep track of all parameters, we define (with apologies for the terrible notation, but try and come up with something better for quick occasional analysis like this) $f(w) := \mu + \sigma + \gamma + \zeta w$, and $\phi := v(w\mu + w\sigma + \gamma + \zeta w)$ and note that $f(1) < f(w) < \phi < vwf(1)$ and $wf(1) > f(w)$ and that $df(w)/d\gamma = 1$, $df(w)/d\zeta = w$, $d\phi/d\gamma = v$, $d\phi/d\zeta = vw$. Then

$$F = \frac{(\gamma + \mu)f(w) + \zeta\phi}{(\gamma + \mu + \zeta)f(1)}.$$

Therefore

$$\frac{\partial F}{\partial \gamma} = \frac{(\gamma + \mu + \zeta)f(1)[f(w) + \gamma + \mu + \zeta v] - [(\gamma + \mu)f(w) + \zeta\phi][f(1) + \gamma + \mu + \zeta]}{(\gamma + \mu + \zeta)^2 f^2(1)},$$

and so $\partial F/\partial \gamma < 0$, using the inequalities that hold for f and ϕ.

Similarly,

$$\frac{\partial F}{\partial \zeta} = \frac{(\gamma + \mu + \zeta)f(1)[w\gamma + \mu + \phi + \zeta vw] - [(\gamma + \mu)f(w) + \zeta\phi][f(1) + \gamma + \mu + \zeta]}{(\gamma + \mu + \zeta)^2 f^2(1)},$$

and so $\partial F/\partial \zeta > 0$, using the same inequalities.

iv) We can rewrite F as

$$F = \frac{(\mu + \gamma)(\frac{\mu}{\zeta} + \frac{\sigma}{\zeta} + \frac{\gamma}{\zeta} + w)}{\zeta(\frac{\mu}{\zeta} + \frac{\gamma}{\zeta} + 1)(\frac{\mu}{\zeta} + \frac{\sigma}{\zeta} + \frac{\gamma}{\zeta} + 1)} + \frac{vw(\frac{\gamma}{w\zeta} + \frac{\mu}{\zeta} + \frac{\sigma}{\zeta} + 1)}{(\frac{\mu}{\zeta} + \frac{\gamma}{\zeta} + 1)(\frac{\mu}{\zeta} + \frac{\sigma}{\zeta} + \frac{\gamma}{\zeta} + 1)},$$

which leads to $F \to vw$ for $\zeta \to \infty$.

Exercise 7.31 i) We can rewrite F as

$$F = \frac{(\mu + \gamma)}{(\mu + \gamma + \zeta)} \frac{(\mu + \sigma + \gamma + \zeta v)}{(\mu + \sigma + \gamma + \zeta)} + \frac{\zeta v^2}{(\mu + \gamma + \zeta)} \frac{(\frac{\gamma}{v} + \mu + \sigma + \zeta)}{(\mu + \sigma + \gamma + \zeta)},$$

which gives the desired result by substituting the expression for N_+/N from Exercise 7.25.

ii) What we have in mind here is the following. The natural death rate μ is usually much smaller than 1 ($1/\mu$ is the life expectancy), but the cure rate γ is likely to be greater than one ($1/\gamma$ is the average time before the patient is cured of d) and therefore $\gamma \gg \mu$. For disease D we had HIV in mind, where the time period from initial infection to death by the disease can be long and σ is therefore substantially smaller than one, but $\sigma > \mu$.

iii) Start from the expression obtained in i) and divide both numerators and denominators by $\gamma + \mu$:

$$F = \left(1 - \frac{N_+}{N}\right) \frac{(1 + \frac{\sigma}{\gamma+\mu} + \frac{\zeta v}{\gamma+\mu})}{(1 + \frac{\sigma}{\gamma+\mu}\frac{\zeta}{\gamma+\mu})} + \frac{N_+}{N} v^2 \frac{(\frac{\gamma}{v(\gamma+\mu)} + \frac{\mu}{\gamma+\mu} + \frac{\sigma}{\gamma+\mu} + \frac{\zeta}{\gamma+\mu})}{(1 + \frac{\sigma}{\gamma+\mu}\frac{\zeta}{\gamma+\mu})}.$$

Now use the approximations in ii) and take out the common factor $(1 + \frac{\zeta}{\gamma+\mu})^{-1} = 1 - \frac{N_+}{N}$ to find

$$F \approx \left(1 - \frac{N_+}{N}\right) \left\{ \left(1 - \frac{N_+}{N}\right)\left(1 + v\frac{\zeta}{\gamma+\mu}\right) + \frac{N_+}{N}v + \frac{N_+}{N}v^2\frac{\zeta}{\gamma+\mu} \right\}.$$

Now note that $(1 - \frac{N_+}{N})\frac{\zeta}{\gamma+\mu} = \frac{N_+}{N}$ and expand the parentheses:

$$F \approx (1 - \frac{N_+}{N})\left\{1 - \frac{N_+}{N} + 2\frac{N_+}{N}v\right\} + \left(\frac{N_+}{N}\right)^2 v^2$$

and

$$F \approx 1 + \left(\frac{N_+}{N}\right)^2 (1 - 2v + v^2) + 2v\frac{N_+}{N} - 2\frac{N_+}{N},$$

which finally leads to the desired approximation.

Exercise 7.32 Since we consider invasion of D, the steady-state values N_0 and N_+ pertaining to d do not change. Nor does the matrix T change (this matrix basically describes the transmission probabilities of D as a function of the status with respect to d of the two individuals, one D-infectious and the other D-susceptible, involved in a contact).

However, in the matrix Σ we should replace γ by γz, since this matrix pertains to a D-infected individual and describes its transitions with respect to d. So we get

$$
\begin{aligned}
R_0 &= -\frac{pc}{N}\begin{pmatrix}1\\w\end{pmatrix} \cdot \Sigma^{-1}\begin{pmatrix}N_0\\N_+v\end{pmatrix}\\
&= \frac{pc}{(\mu+\sigma)(\mu+\sigma+\gamma z+\zeta)}\left\{(\mu+\sigma+\gamma z)\frac{N_0}{N} + \gamma z v\frac{N_+}{N} + w\zeta\frac{N_0}{N}\right.\\
&\quad \left. +(\mu+\sigma+\zeta)vw\frac{N_+}{N}\right\}\\
&= \frac{pc}{(\mu+\sigma)(\mu+\sigma+\gamma z+\zeta)(\zeta+\gamma+\mu)}\left\{(\mu+\sigma+\gamma z)(\gamma+\mu) + \gamma z v\zeta\right.\\
&\quad \left. +w\zeta(\gamma+\mu) + (\mu+\sigma+\zeta)vw\zeta\right\}
\end{aligned}
$$

and finally

$$R_0 = pc\frac{(\mu+\gamma)(\mu+\sigma+\gamma z+\zeta w) + \zeta vw(\frac{\gamma z}{w}+\mu+\sigma+\zeta)}{(\mu+\gamma+\zeta)(\mu+\sigma)(\mu+\sigma+\gamma z+\zeta)}.$$

Exercise 7.33 To understand the matrix, one has to keep in mind that the components of a four-vector correspond, in this order, to the states $(f, 0)$, $(f, +)$, $(m, 0)$ and $(m, +)$, and that of the two indices of a matrix element the second corresponds to a D-infectious individual and the first to a D-susceptible individual. With these bookkeeping conventions, the elements of T are a direct translation of the assumptions in the text. Concerning the range of T, simply observe that the two listed vectors are, respectively, the third and first column of the matrix T, that the fourth column is w_m times the third, while the second column is w_f times the first (and that the first and third column are clearly linearly independent). A similar observation can be made concerning the rows.

Exercise 7.34 The matrix $K = K_L$ has, in self-explanatory notation, the structure

$$K = K_L = \begin{pmatrix} 0 & T_{fm} \\ T_{mf} & 0 \end{pmatrix} \begin{pmatrix} -\Sigma_f^{-1} & 0 \\ 0 & -\Sigma_m^{-1} \end{pmatrix}.$$

From Exercise 7.33 we see that the auxiliary matrices R and C, in terms of two-vectors (with '0' indicating a vector of zeros in R), are given by

$$R = \begin{pmatrix} 0 & \binom{1}{w_m} \\ \binom{1}{w_f} & 0 \end{pmatrix}, \qquad C = \begin{pmatrix} 0 & \psi_1 \\ \psi_2 & 0 \end{pmatrix}.$$

This leads to

$$K_S = R \begin{pmatrix} -\Sigma_f^{-1} & 0 \\ 0 & -\Sigma_m^{-1} \end{pmatrix} C,$$

which leads to the matrix given in the exercise. Consequently, $R_0 = \sqrt{k_{fm}k_{mf}}$, and all that remains to be done is to give explicit expressions for Σ_f^{-1} and Σ_m^{-1} to find a completely explicit expression for R_0.

Exercise 7.35 We have

$$N_{(i,0)} = \frac{\gamma + \mu}{\gamma + \mu + i\zeta} N_i, \qquad N_{(i,+)} = \frac{i\zeta}{\gamma + \mu + i\zeta} N_i,$$

$$\Sigma_i = \begin{pmatrix} -i\zeta - \mu - \sigma & \gamma \\ i\zeta & -\gamma - \mu - \sigma \end{pmatrix}.$$

Exercise 7.36 The number of contacts per unit of time between $(i, \cdot)$-individuals and $(j, \cdot)$-individuals is given by

$$\frac{cijN_iN_j}{\sum_k kN_k}.$$

When we leave out the factor N_j, we get the per j-capita number of contacts per unit of time with $(i, \cdot)$-individuals.

Exercise 7.37 The operator K acts on a countable vector of two-vectors to produce again such a countable vector of two-vectors. The element K_{ij} of the matrix representation tells us how the jth two-vector of the original contributes to the ith two-vector of the image. We arrive at K_{ij} from the expression in Exercise 7.36 by realizing that a fraction $N_{(i,0)}/N_i$ of contacts with $(i, \cdot)$-individuals is with $(i,0)$-individuals etc., and by taking the probability of transmission and its various enhancement factors into account. From the expression for K_{ij}, one sees at once that the range of K has the ith component spanned by $\binom{N_{(i,0)}}{vN_{(i,+)}}$, with the corresponding coefficient obtained from the inner product of $\binom{1}{w}$ with $\frac{pp}{\sum_k kN_k} \sum_{j=0}^{\infty} j\Sigma_j^{-1}\phi_j$.

Exercise 7.38 The expression for R_0 now follows at once from the characterization of the range and the expression for the coefficient as given at the end of the preceding exercise.

Exercise 7.40 i) $D = (\rho + \mu)(\sigma + 2\mu) - \rho(\sigma + \mu) = \mu(\sigma + 2\mu) + \rho\mu = \mu(\rho + \sigma + 2\mu)$.
The formula for $-\Sigma^{-1}$ is just the general formula $\begin{pmatrix} a & b \\ c & d \end{pmatrix}^{-1} = \frac{1}{D}\begin{pmatrix} d & -b \\ -c & a \end{pmatrix}$
applied to this particular case.

   ii) The element (i,j) of $-\Sigma^{-1}$ is the expected time that an individual that currently has state j will spend in state i in the future. By summing over i, we get the expected future lifespan. Both columns of the matrix $\begin{pmatrix} \sigma + 2\mu & \sigma + \mu \\ \rho & \rho + \mu \end{pmatrix}$ sum to $\sigma + 2\mu + \rho$, so, taking the factor $1/D$ into account, we deduce that the expected future lifespan equals $1/\mu$, irrespective of the current state. Since being single or paired has, in this model, no influence on the survival probability, this outcome was predictable and serves more as a check that we did not make mistakes.

   The expected time to be single is found in the first row and equals $\frac{1}{\mu}\frac{\sigma+2\mu}{\rho+\sigma+2\mu}$ if the current state is single, and $\frac{1}{\mu}\frac{\sigma+\mu}{\rho+\sigma+2\mu}$ otherwise.

   iii) We want to calculate the expected number Q of future partners of an individual that is single.

   a) The expected duration of one partnership equals $1/(\sigma+2\mu)$ and the expected total time of being paired for an individual that is currently single equals ρ/D. Since the durations of subsequent partnerships are independent and identically distributed, the expected number Q of partners equals the quotient $\rho(\sigma + 2\mu)/D$ of these two quantities.

   b) A second derivation is based on the embedded discrete time process of state transitions (recall Exercise 1.34-vi). When an individual is single, its next state will be 'paired' with probability $\rho/(\rho + \mu)$, and in that case we have to add 1 to the number of partners. Return to the 'single' state is necessary for acquiring a new partner (in this model; admittedly, reality is less restrictive in its possibilities). With probability $(\sigma + \mu)/(\sigma + 2\mu)$, the next state will be 'single.' But then we are back to where we started, and the expected number of future partners is again Q. Hence

$$Q - \frac{\rho}{\rho + \mu}\left(1 + \frac{\sigma + \mu}{\sigma + 2\mu}Q\right),$$

and solving this equation once more leads to $Q = \rho(\sigma + 2\mu)/D$.

Exercise 7.41 i) If we are in state 2, the probability that the next state transition brings us to state 3 equals $h/(h + \sigma + 2\mu)$.

   ii) If we are in state 3, the probability that the next state transition brings us to state 1 equals $(\sigma + \mu)(\sigma + 2\mu)$.

   iii) If we concentrate on the distinction between being paired or single and suppress all information related to the infection, we are exactly in the situation considered in Exercise 7.40. (A key point here is that infected individuals have exactly the same ρ, σ and μ as non-infected individuals. In particular, an additional death rate due to the infection would necessitate that we redo the earlier calculation.) To make this explicit we have to lump the current states 2 and 3 into one state, which is possible since the outgoing arrows of 2 and 3, including the corresponding rates, are identical.

iv) It follows from the results derived above that

$$R_0 = \frac{\sigma + \mu}{\sigma + 2\mu} \frac{\rho(\sigma + 2\mu)}{D} \frac{h}{h + \sigma + 2\mu} = \frac{\rho h(\sigma + \mu)}{\mu(\rho + \sigma + 2\mu)(h + \sigma + 2\mu)}.$$

Exercise 7.42 i) Given that the current state is 1, the probability that the next transition will bring the system to state 2 equals $\rho/(\rho + \mu)$. From state 2, we will go with probability $h/(h + \sigma + 2\mu)$ to state 3 (in which case we are sure to arrive in 3) and with probability $(\sigma + \mu)/(h + \sigma + 2\mu)$ back to state 1 (in which case the probability to eventually arrive in 3 is again P_{13}). Hence

$$P_{13} = \frac{\rho}{\rho + \mu} \left(\frac{h}{h + \sigma + 2\mu} + \frac{\sigma + \mu}{h + \sigma + 2\mu} P_{13} \right),$$

and by solving this equation we find

$$P_{13} = \frac{h\rho}{(h + \mu)\rho + (h + \sigma + 2\mu)\mu}.$$

ii) In order to return to 3 from 3, one has to go through 1. The probability to go from 3 to 1 equals $(\sigma + \mu)/(\sigma + 2\mu)$. Hence

$$P_{33} = \frac{\sigma + \mu}{\sigma + 2\mu} P_{13}.$$

iii) A newly infected individual is in state 3 and is, by definition, expected to produce R_0 secondary cases. When it returns to state 3, it has produced one secondary case and is, by the Markov property, expected to produce R_0 more. Hence $R_0 = P_{33}(1 + R_0)$. So

$$
\begin{aligned}
R_0 &= \frac{P_{33}}{1 - P_{33}} = \frac{(\sigma + \mu)P_{13}}{\sigma + 2\mu - (\sigma + \mu)P_{13}} \\
&= \frac{(\sigma + \mu)h\rho}{(\sigma + 2\mu)[(h + \mu)\rho + (h + \sigma + 2\mu)\mu] - (\sigma + \mu)h\rho} \\
&= \frac{(\sigma + \mu)h\rho}{\mu[(h + \mu)\rho + (h + \sigma + 2\mu)\mu] + (\sigma + \mu)\mu(\rho + h + \sigma + 2\mu)} \\
&= \frac{(\sigma + \mu)h\rho}{\mu[(h + \mu)\rho + (h + \sigma + 2\mu)(\sigma + 2\mu) + \rho(\sigma + \mu)]} \\
&= \frac{(\sigma + \mu)h\rho}{\mu(h + \sigma + 2\mu)(\sigma + 2\mu + \rho)}.
\end{aligned}
$$

iv) It follows from $R_0 = P_{33}/(1 - P_{33})$ that $R_0 > 1 \Leftrightarrow P_{33} > 1/2$. When returning to state 3, an infectious individual has 'cloned' itself once or, in other words, the number of infected individuals in state 3 has doubled from 1 to 2. So if this event has probability $P_{33} > 1/2$, we have a net increase, and exponential growth of the subpopulation of infecteds is bound to happen within a deterministic context.

Exercise 7.43 The individual that just entered state 1 will return to state 3 with probability P_{13}. If this happens, there are in fact two individuals in state 3. Each of these enters state 1 with probability $(\sigma + \mu)/(\sigma + 2\mu)$. Hence

$$\widetilde{R}_0 = 2 \frac{\sigma + \mu}{\sigma + 2\mu} P_{13} = 2 P_{33}.$$

Exercise 7.44 i) By dividing both the numerator and the denominator by σ^2, we arrive at

$$R_0 = \frac{(1 + \frac{\mu}{\sigma})\frac{h}{\sigma}\rho}{\mu(\frac{h}{\sigma} + 1 + \frac{2\mu}{\sigma})(1 + \frac{2\mu}{\sigma} + \frac{\rho}{\sigma})},$$

which for $\sigma, h \to \infty$ with $h/\sigma \to p$, has the limit

$$R_0 = \frac{\rho p}{\mu(p+1)} = \frac{\rho q}{\mu},$$

with $q = p/(p+1)$.

ii) When $h = p\sigma$ for fixed p, we have

$$R_0 = \frac{(\sigma + \mu)p\rho}{\mu(p + 1 + \frac{2\mu}{\sigma})(\sigma + 2\mu + \rho)}.$$

Now note that both $\frac{\sigma+\mu}{\sigma+2\mu+\rho}$ and $\frac{1}{p+1+\frac{2\mu}{\sigma}}$ are increasing functions of σ.

iii) When $\sigma \gg \mu$, the expected duration of a partnership is approximately $1/\sigma$. Consequently, the probability that transmission occurs during a partnership of an infectious and a susceptible individual is approximately h/σ. In ii) we have taken h/σ to be constant while σ varied. In that case the effect of increasing σ is primarily that an individual is expected to have more partners and hence, since the success ratio per partner is approximately constant, is expected to produce more secondary cases. If, however, h is taken to be fixed (i.e., independent of σ), an increase of σ has the effect that transmission is less likely during the shorter period that a partnership lasts. So then it is not clear how R_0 depends on σ.

In the present class of models we think of transmission opportunities on two levels: on the one hand formation and dissociation of pairs and on the other hand (sexual) contacts within partnerships. When analyzing the dependence on parameters, one has to realize that there may be opposing effects on the different levels. This can create results that at first sight are counterintuitive (such as a decrease of R_0 when σ is increased). When making comparative statements, therefore, the choice and implications of the gauging should be clearly stated and analyzed. This is a non-trivial issue that all too often does not get the attention it deserves.

Exercise 7.45 i) When the disease causes an additive death rate κ, the scheme of transitions is as depicted in Figure 17.1.

To compute R_0, we follow the method of Exercise 7.42. This yields

$$P_{13} = \frac{\rho}{\rho + \mu + \kappa}\left(\frac{h}{h + \sigma + 2\mu + \kappa} + \frac{\sigma + \mu}{h + \sigma + 2\mu + \kappa}P_{13}\right),$$

which gives

$$P_{13} = \frac{\rho h}{(\rho + \mu + \kappa)(h + \sigma + 2\mu + \kappa) - \rho(\sigma + \mu)}$$

$$= \frac{\rho h}{(\mu + \kappa)(h + \sigma + 2\mu + \kappa) + \rho(h + \mu + \kappa)}$$

and

$$P_{33} = \frac{\sigma + \mu + \kappa}{\sigma + 2\mu + 2\kappa}P_{13}.$$

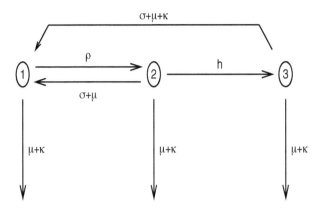

Figure 17.1: Schematic representation of the transitions in the Markov model with three states, as in Figure 7.2, but with a disease-related additional death rate κ.

Now recall that
$$R_0 = P_{33}(1 + R_0) \Rightarrow R_0 = \frac{P_{33}}{1 - P_{33}},$$
which leads to
$$\begin{aligned}
R_0 &= \frac{(\sigma + \mu + \kappa)P_{13}}{\sigma + 2\mu + 2\kappa - (\sigma + \mu + \kappa)P_{13}} \\
&= \frac{(\sigma + \mu + \kappa)\rho h}{(\sigma + 2\mu + 2\kappa)[(\mu + \kappa)(h + \sigma + 2\mu + \kappa) + \rho(h + \mu + \kappa)] - (\mu + \sigma + \kappa)\rho h} \\
&= \frac{(\sigma + \mu + \kappa)\rho h}{(\mu + \kappa)[(\sigma + 2\mu + 2\kappa)(h + \sigma + 2\mu + \kappa) + \rho(h + \sigma + 2\mu + 2\kappa)]}.
\end{aligned}$$

ii) In the scheme in Figure 17.2, the states 1–3 have the same meaning as before and 4 is the state of mourning.

Here we have assumed that this state results whenever the partner dies, irrespective of the cause of death (a different model would be to assume that state 4 is only visited after the death of the partner from the disease). To compute R_0, one can use, as before,
$$R_0 = P_{33}(1 + R_0),$$
but now with
$$P_{33} = \left(\frac{\sigma}{\sigma + 2\mu + 2\kappa} + \frac{\mu + \kappa}{\sigma + 2\mu + 2\kappa} \frac{\alpha}{\alpha + \mu + \kappa} \right) P_{13}$$
where P_{13} is calculated from
$$\begin{aligned}
P_{13} &= \frac{\rho}{\rho + \mu + \kappa} \left\{ \frac{h}{h + \sigma + 2\mu + \kappa} + \left(\frac{\sigma}{h + \sigma + 2\mu + \kappa} \right. \right. \\
&\quad \left. \left. + \frac{\mu}{h + \sigma + 2\mu + \kappa} \frac{\alpha}{\alpha + \mu + \kappa} \right) P_{13} \right\}.
\end{aligned}$$

(Note that for $\alpha \to \infty$ we recover the earlier set of identities.) We refrain from the derivation of the explicit expression for R_0.

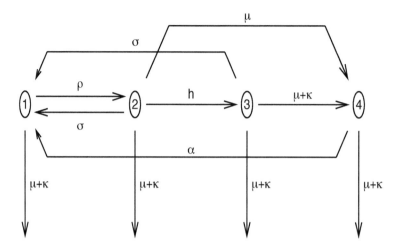

Figure 17.2: Schematic representation of a Markov system with four states, similar to that of Figure 17.1, but with a fourth state representing a state of mourning after the death of the partner.

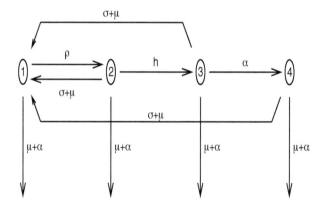

Figure 17.3: Schematic representation of a Markov system with four states, similar to that of Figure 17.1, but with a fourth state representing a state in which the partner is (temporarily) immune.

Exercise 7.46 The scheme is now as in Figure 17.3.

Here 4 means 'paired to an immune individual.' Since the arrows going out of 3 and 4 are exactly the same, we may actually lump states 3 and 4 without any loss of information. Applying the computational scheme as before, we find the explicit expression

$$R_0 = \frac{h\rho(\mu + \sigma)}{(\mu + \alpha)(\alpha + h + 2\mu + \sigma)(\alpha + 2\mu + \rho + \sigma)}.$$

We should like to draw your attention to a subtle difference between the setting of this exercise and that of Exercise 7.38-i. In the present situation an increase of α leads to a decrease of R_0, as one can deduce immediately from the explicit expression. In the situation of Exercise 7.38-i, the additional death rate has two effects: it shortens the period of infectiousness of the individual we consider (which is just the same as loss of infectiousness), but it may also bring about an early death of an infected partner, thus creating the potential for an increased number of partners. This is reflected in the fact that κ appears in the numerator of the expression for R_0 in Exercise 7.45-i. A more refined analysis establishes that R_0 is still a decreasing function of κ (in other words, the first effect is the more dominant one).

Exercise 7.47 The demographic part of the model is described by the scheme in Figure 17.4.

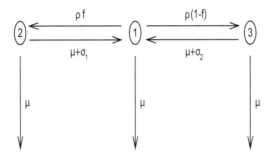

Figure 17.4: Schematic representation of a Markov system with three states, representing single individuals (state 1), casual pairs (state 3) and long-term relationships (state 2).

When we superimpose the transmission of an infectious agent, we may split both 2 and 3 into two states; see Figure 17.5.

In Figure 17.5, the individuals in states marked with a have a partner that is susceptible and those in a state marked with b have a partner that is infectious. So there are now two different states, 2b and 3b, in which an individual can begin its life as an infected and we have to set up a 2×2 next-generation matrix and compute R_0 as its dominant eigenvalue. We will see that the matrix has one-dimensional range, the underlying reason being that a newly infected individual necessarily has to pass through state 1 in order to infect other individuals.

An individual in state 2b will with probability $\pi_1 = (\mu + \sigma_1)/(2\mu + \sigma_1)$ go to state 1 before dying. For an individual in state 3b, the corresponding probability is $\pi_2 = (\mu + \sigma_2)/(2\mu + \sigma_2)$. Consider an infected individual in state 1. Let R_s be the expected number of serious partners that it will infect and R_c the expected number casual partners that it will infect. Then the 2×2 next-generation matrix is

$$M = \begin{pmatrix} \pi_1 R_s & \pi_2 R_s \\ \pi_1 R_c & \pi_2 R_c \end{pmatrix}.$$

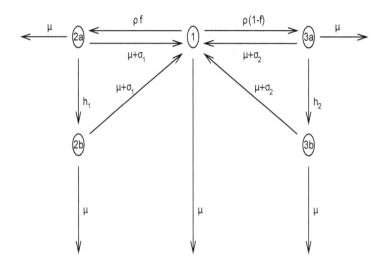

Figure 17.5: Schematic representation of the system from Figure 17.4, now including the infection-related states, where a label a denotes a susceptible partner and label b denotes an infectious partner.

Clearly $Mx = (\pi_1 x_1 + \pi_2 x_2)\binom{R_s}{R_c}$, that is, the range of M is spanned by $\binom{R_s}{R_c}$. Hence we have

$$R_0 = \pi_1 R_s + \pi_2 R_c.$$

It remains to calculate R_s and R_c.

To an individual in state 1, three things can happen. It can die, in which case it will make no further 'victims.' With probability $\rho f/(\rho + \mu)$ it forms a serious partnership and with probability $\rho(1-f)/(\rho + \mu)$ a casual partnership. In the first case it will transmit the infectious agent to its partner with probability $h_1/(h_1 + 2\mu + \sigma_1)$, and it will return to state 1 with probability $(\mu + \sigma_1)/(2\mu + \sigma_1)$ (note that the distinction 2a/2b is irrelevant for the probabilities of dying or returning to 1, so for computing the probability of return to state 1 we can lump 2a and 2b). In the second case we have the same expressions, but with index 1 replaced by index 2.

We now combine this information into the identities

$$R_s = \frac{\rho f}{\rho + \mu}\left(\frac{h_1}{h_1 + 2\mu + \sigma_1} + \frac{\mu + \sigma_1}{2\mu + \sigma_1}R_s\right) + \frac{\rho(1-f)}{\rho + \mu}\frac{\mu + \sigma_2}{2\mu + \sigma_2}R_s,$$

$$R_c = \frac{\rho f}{\rho + \mu}\frac{\mu + \sigma_1}{2\mu + \sigma_1}R_c + \frac{\rho(1-f)}{\rho + \mu}\left(\frac{h_2}{h_2 + 2\mu + \sigma_2} + \frac{\mu + \sigma_2}{2\mu + \sigma_2}R_c\right).$$

After solving these equations and inserting the result into the expression for R_0, we find that R_0 is given by

$$\frac{\rho F}{(\rho + \mu)(2\mu + \sigma_1)(2\mu + \sigma_2) - \rho f(\mu + \sigma_1)(2\mu + \sigma_2) - \rho(1 - f)(2\mu + \sigma_1)(\mu + \sigma_2)}$$

with

$$F = f\frac{h_1(\mu + \sigma_1)(2\mu + \sigma_2)}{h_1 + 2\mu + \sigma_1} + (1 - f)\frac{h_2(2\mu + \sigma_1)(\mu + \sigma_2)}{h_2 + 2\mu + \sigma_2}.$$

The present example is ideal for illustrating the fact that one can sometimes meaningfully define another threshold quantity $\tilde{R}_0$, that has a similar interpretation but yet is different from R_0. Consider an infected individual in state 1. Let $\tilde{R}_0$ be the expected number of infected individuals (including the originally considered individual itself) entering state 1 after one potential pair formation event. Then

$$\tilde{R}_0 = \frac{\rho f}{\rho + \mu}\left(1 + \frac{h_1}{h_1 + 2\mu + \sigma_1}\right)\frac{\mu + \sigma_1}{2\mu + \sigma_1} + \frac{\rho(1 - f)}{\rho + \mu}\left(1 + \frac{h_2}{h_2 + 2\mu + \sigma_2}\right)\frac{\mu + \sigma_2}{2\mu + \sigma_2}$$

where the first term is the contribution from a serious pair being formed and the second term is that from a casual pair being formed. One easily checks that $R_0 = 1 \Leftrightarrow \tilde{R}_0 = 1$.

To check the calculations, we may take $\sigma_1 = \sigma_2$ and $h_1 = h_2$ and recover the expression for R_0 derived in Exercise 7.41. As a small research program, one might study the effect of neglecting variations in types of pairs by comparing R_0 as computed in the present exercise with the expression of Exercise 7.41 with $1/\sigma = f/\sigma_1 + (1 - f)/\sigma_2$ and h equal to h_1(expected time in pair of type 2)/(expected time in pair) $+ h_2$(expected time in pair of type 3)/(expected time in pair).

Exercise 7.48 The point is the phenomenological parameter ρ, which is the probability per unit of time with which a single acquires a partner. When we consider males and females, we have to impose consistency. If the sex ratio is 1:1, this is automatically guaranteed. If we have a different sex ratio, however, we should also have different ρ's for males and females.

Exercise 7.49 First linearize the system at the infection-free steady state $(N, 0)$. We focus on the equation for I only and obtain $x'(t) = (\gamma(t) - \alpha)x(t)$. The solution $x(t)$ of this equation is given by

$$x(0)\exp\left(\int_0^t (\gamma(s) - \alpha)ds\right) = x(0)\exp\left((\gamma_0 - \alpha)t + \frac{\gamma_1\gamma_0}{2\pi}\sin(2\pi t)\right) = \Phi(t)x(0).$$

Hence

$$E := \Phi(1) = \exp(\gamma_0 - \alpha) \tag{17.1}$$

and we see that the solutions decay to zero if and only if $\gamma_0 - \alpha < 0$, i.e., the infection-free steady state is stable when $\gamma_0/\alpha < 1$ and unstable when $\gamma_0/\alpha > 1$.

Exercise 7.50 When there is temporal variation that affects epidemiological ingredients, it will matter for the potential number of secondary cases produced by a given infected individual *when* exactly that individual became infected. This means that the 'epidemiological life' of the individual will depend on the moment of 'epidemiological birth.' In other words: individuals are not born (epidemiologically speaking) in the same way, and the concept of a 'generation' of infected individuals then becomes questionable. Because the definition of R_0 is directly dependent on the generation view, and in particular the biological interpretation is intimately linked to the generation concept, we see that, under temporal variation, a threshold quantity for invasion is unlikely to have the same biological meaning as R_0.

Exercise 7.51 Quite in general, if M is a matrix and c a scalar, then the eigenvalues of cM are of the form $c\lambda$ with λ an eigenvalue of M (i.e., $\sigma(cM) = c\sigma(M)$). We now present more details for the present special case. Let x be the dominant eigenvector of K, i.e., $Kx = R_0 x$. For K_v we then have: $K_v x = (1-v)Kx = (1-v)R_0 x$. The vector x is therefore also an eigenvector of K_v, with eigenvalue $(1-v)R_0$. This must necessarily be the dominant eigenvalue of K_v. One can see this easily when we suppose that $R_v > (1-v)R_0$ and let y be the dominant eigenvector of K_v related to R_v. Then $K_v y = R_v y \iff (1-v)Ky = R_v y \iff Ky = R_v y/(1-v)$, and we conclude that y is also an eigenvector of K, with eigenvalue $R_v/(1-v) > R_0$, a contradiction. Therefore $R_v = (1-v)R_0$. An alternative proof makes use of the definition of $R_0 = \lim_{n\to\infty} \|K^n\|^{1/n}$, and $R_v = \lim_{n\to\infty} \|(K_v)^n\|^{1/n} = \lim_{n\to\infty} \|(1-v)^n(K)^n\|^{1/n}$ from Section 7.1, and properties of the norm.

Exercise 7.52 The next-generation matrix K_v in the vaccinated population is given by

$$K_v = \begin{pmatrix} 0 & (1-v)k_{12} \\ k_{21} & 0 \end{pmatrix}.$$

Let R_v be its dominant eigenvalue: $R_v = \sqrt{(1-v)k_{21}k_{12}} = \sqrt{1-v}R_0$. Therefore, in order to prevent outbreaks ($R_v < 1$) we need to vaccinate at least a fraction $1 - 1/R_0^2$ of the host population.

Exercise 7.53 The next-generation matrix in the presence of vaccination is given by

$$K_v = \begin{pmatrix} 0 & (1-v)k_{12} \\ k_{21} & k_{22} \end{pmatrix}$$

and its leading eigenvalue is therefore $R_v = \frac{1}{2}[k_{22} + \sqrt{k_{22}^2 + 4(1-v)k_{12}k_{21}}]$. By rewriting $R_v = 1$, while assuming $k_{22} < 1$, we get the desired expression for v_c. By substituting $v = 0$ in the expression for K_v, we find that R_0 satisfies the equation $R_0(R_0 - k_{22}) = k_{12}k_{21}$. Finally substitute this for $k_{12}k_{21}$ in the expression for v_c.

Exercise 7.54 If $k_{22} > 1$ this means that the type 2 alone is able to sustain the infectious agent, or at least to allow it to invade successfully. In that case $R_v > 1$ no matter how much control effort is imposed on type 1; the infectious agent can cycle in the vector population even in the absence of any susceptible hosts. The borderline case $k_{22} = 1$ leads to $R_v \geq 1$, and accordingly we can still conclude in that case that it is impossible to bring R_v below 1 by vaccinating hosts.

Exercise 7.55 Starting from one infected individual of type 1 (generation zero), the first generation is given by Ke. Before tracing the infection into the second generation we project the first generation on types $2, \ldots, n$, giving $(I-P)Ke$ as the starting distribution for the second generation. The second generation is then given by $K(I-P)Ke$. Of these we get, after projection, the vector $(I-P)K(I-P)Ke = ((I-P)K)^2 e$ that forms the starting point for the third generation. The cumulative number of type 1 individuals after three generations is then $e^\top Ke + e^\top K(I-P)Ke + e^\top K((I-P)K)^2 e$. This process continues and leads to the stated result.

Exercise 7.56 If the spectral radius of $(I - P)K$ is larger than 1, this means that the types $2, \ldots, n$ are able to sustain the growth of the infectious agent, in the absence of type 1 individuals. Type 1 individuals are therefore not necessary for successful invasion by the infectious agent into this system and no amount of control effort targeted at type 1 alone will be able to prevent an outbreak in the population. One could also say that, collectively, the types $2, \ldots, n$ form, from the point of view of type-1 individuals, a reservoir for the infectious agent.

Exercise 7.57 One way to show property i) is to show that $1 - R_0$ and $1 - T$ always have the same sign. It therefore makes sense to try and establish an equality involving these expressions. Starting from the matrix M we argue as follows

$$
\begin{aligned}
M \quad &:= \quad PK \sum_{j=0}^{\infty} ((I - P)K)^j \\
&= \quad PK \left(I + \sum_{j=0}^{\infty} ((I - P)K)^j] (I - P)K \right) \\
\iff \quad & M - PK = M(I - P)K \\
\iff \quad & PK - MPK = M - MK.
\end{aligned}
$$

Then let w be the positive eigenvector of K corresponding to the dominant eigenvalue R_0. So

$$
R_0(I - M)Pw = (1 - R_0)Mw.
$$

Because Pw and Mw are non-negative it must now hold that $1 - \rho(M)$ and $1 - R_0$ have the same sign (see e.g., Minc 1988).

For property 2 we first define the type-reproduction number T_v in a population where the vaccination strategy v is implemented:

$$
T_v = e^\top P K_v (I - (I - P)K_v)^{-1} e
$$

where K_v is the next-generation matrix in the vaccinated population. In the present case, the first row of K_v has elements $(1 - v)k_{1j}$ for $j = 1, \ldots, n$, and K_v is identical to the basic uncontrolled matrix K for all other elements. Because the first row is annihilated by the action of $(I - P)$, we see that

$$
\begin{aligned}
T_v \quad &= \quad e^\top P K_v (I - (I - P)K)^{-1} e \\
&= \quad (1 - v)e^\top P K (I - (I - P)K)^{-1} e \\
&= \quad (1 - v)T.
\end{aligned}
$$

Property 1 ensures that outbreaks in the vaccinated population will be prevented if $T_v < 1$, which translates into $v > 1 - 1/T$.

Exercise 7.58 We only present the full matrix case, as the other calculations are special cases of this. Note that the interpretation of a host-vector system does not play any role. We have

$$
K = \begin{pmatrix} k_{11} & k_{12} \\ k_{21} & k_{22} \end{pmatrix}, \quad P = \begin{pmatrix} 1 & 0 \\ 0 & 0 \end{pmatrix}.
$$

Then

$$
\begin{aligned}
M &= PK(I-(I-P)K)^{-1} = PK\left(I - \begin{pmatrix} 0 & 0 \\ k_{21} & k_{22} \end{pmatrix}\right)^{-1} \\
&= PK\begin{pmatrix} 1 & 0 \\ \frac{k_{21}}{1-k_{22}} & \frac{1}{1-k_{22}} \end{pmatrix} \\
&= P\begin{pmatrix} k_{11} + \frac{k_{12}k_{21}}{1-k_{22}} & \frac{k_{12}}{1-k_{22}} \\ k_{21} + \frac{k_{22}k_{21}}{1-k_{22}} & \frac{k_{22}}{1-k_{22}} \end{pmatrix} = \begin{pmatrix} k_{11} + \frac{k_{12}k_{21}}{1-k_{22}} & \frac{k_{12}}{1-k_{22}} \\ 0 & 0 \end{pmatrix}.
\end{aligned}
$$

Therefore
$$
T = e^\top M e = \rho(M) = k_{11} + \frac{k_{12}k_{21}}{1 - k_{22}}.
$$

17.2   ELABORATIONS FOR CHAPTER 8

Exercise 8.1 i) When $\Delta T \equiv 1$ the unconditional offspring distribution is also Poisson: $X_{fm} \sim Po(cp_{fm}\pi_f)$ and $X_{mf} \sim Po(cp_{mf}\pi_m)$.

ii) When $\Delta T \sim Exp(1)$ we compute the offspring by first conditioning on $\Delta T = t$ and next using that we know that the conditional distribution is Poisson:

$$
\begin{aligned}
P(X_{fm} = k) &= \int_0^\infty P(X_{fm} = k|\Delta T = t)f_{\Delta T}(t)dt \\
&= \int_0^\infty \frac{e^{-cp_{fm}\pi_f t}(cp_{fm}\pi_f t)^k}{k!} e^{-t}dt \\
&= \frac{1}{cp_{fm}\pi_f + 1}\left(\frac{cp_{fm}\pi_f}{cp_{fm}\pi_f + 1}\right)^k C
\end{aligned}
$$

with
$$
C := \int_0^\infty \frac{(cp_{fm}\pi_f + 1)^{k+1}t^k e^{-(cp_{fm}\pi_f+1)t}}{k!}dt,
$$

You can easily verify that the integral C equals 1 by direct integration, or note that the expression under the integral is a gamma density, $\Gamma(k, cp_{fm}\pi_f + 1)$, that is integrated over the whole sample space. It follows that X_{fm} has a geometric distribution, with $p = 1/(cp_{fm}\pi_f + 1)$. Similarly one can show that X_{mf} has a geometric distribution with $p = 1/(cp_{mf}\pi_m + 1)$.

Exercise 8.2 In Exercise 8.1 it was shown that the offspring distribution is geometric. Our numerical values imply that $cp_{fm}\pi_f = 2$ and $cp_{mf}\pi_m = 1$, respectively. From this we get the equations

$$
q_m = \sum_{k=0}^\infty q_f^k \frac{1}{3}\left(\frac{2}{3}\right)^k = \frac{1}{1 + 2(1 - q_f)}
$$

$$
q_f = \sum_{k=0}^\infty q_m^k \frac{1}{2}\left(\frac{1}{2}\right)^k = \frac{1}{1 + (1 - q_m)},
$$

where, for each equation, the last equality follows from the relation $\sum_{k=0}^\infty x^k = 1/(1-x)$, for $|x| < 1$. Solving this pair of equations we find the positive solution $q_m = 2/3 \approx 0.667$ and $q_f = 3/4 = 0.75$. As for the case when $\Delta T \equiv 1$ we have

$q_m < q_f$. We can also conclude that changing from $\Delta T \equiv 1$ to an infectious period with positive variance actually increases the q's (and so decreases the probability of a major outbreak). The same phenomenon has already been observed for the prototype stochastic epidemic model (see Section 3.3.2, as well as Exercise 2.11).

Exercise 8.3 As the epidemic systematically affects more unvaccinated than vaccinated individuals (compared with the ratio in which they occur), the proportion of vaccinated individuals among the susceptibles will gradually increase. As we ignore this effect for the time being, we have to add the adjective 'early.'

Exercise 8.4 Because of the one-dimensional range for a two-dimensional mapping, the eigenvalues are 0 and R_v, and hence their sum equals R_v. Since the trace of a matrix (the sum of the diagonal terms) always equals the sum of the eigenvalues (counting multiplicity), we find the stated expression.

The alternative is to apply the matrix to the vector $(1 - v, vf)^\top$ that spans the range. This yields the vector

$$(1 - v + vf\phi)R_0 \begin{pmatrix} 1-v \\ vf \end{pmatrix},$$

and hence $R_v = (1 - v + vf\phi)R_0$.

Exercise 8.5 The incidence $i(t)$ is now a vector with two components, the first corresponding to unvaccinated individuals and the second to vaccinated individuals. Exactly as (1.9) was derived in Section 1.2.3, we deduce that

$$i(t) = \int_0^\infty cA(\tau) \begin{pmatrix} 1-v & (1-v)\phi \\ vf & vf\phi \end{pmatrix} i(t - \tau) \, d\tau.$$

Upon making the ansatz that $i(t) = e^{rt}i^s$ for the (as yet unknown) stable distribution vector i^s, it follows that the matrix

$$\int_0^\infty cA(\tau)e^{-r\tau} \, d\tau \begin{pmatrix} 1-v & (1-v)\phi \\ vf & vf\phi \end{pmatrix}$$

should have dominant eigenvalue one.

Referring back to Exercise 2.14, we see that the time dependence enters in the same way for any susceptible-infective combination (i.e., any position in the matrix) and that consequently i^s equals the stable distribution in a generation perspective. The one-dimensional range property allowed us to calculate the latter explicitly as $\begin{pmatrix} 1-v \\ vf \end{pmatrix}$, modulo a normalization factor. Inserting this expression for i^s (or using the trace argument of Exercise 8.4), we find that r_v should satisfy the equation

$$1 = \frac{R_v \int_0^\infty e^{-r_v\tau} A(\tau) \, d\tau}{\int_0^\infty A(\tau) \, d\tau}.$$

Exercise 8.6 To get started, do Exercise 1.7 (or read its elaboration). We introduce the symbol $\tilde{q}^1(k_1, k_2)$ to denote the probability that a type 1 infected individual infects in total k_1 type 1 individuals and k_2 type 2 individuals, and the symbol $\tilde{q}^2(k_1, k_2)$ to denote the corresponding probability for an infected individual of type 2. By so-called first-step analysis (as it is used more generally in the theory of

Markov processes; see the end of Section 4.6, Section 7.8, and for a systematic exposition e.g., Taylor and Karlin 1984), we arrive at the consistency conditions

$$\pi_i = \sum_{k_1,k_2=0}^{\infty} \tilde{q}^i(k_1, k_2)\pi_1^{k_1}\pi_2^{k_2}, \quad i = 1, 2.$$

To proceed, we have to determine the $\tilde{q}^i(k_1, k_2)$ from the assumptions about the contact process and the probability of transmission. To do so, we make a short cut based on the ideas presented in (the elaboration of) Exercises 1.10 and 1.11. That is, we note that the numbers of infected individuals of the two types are *independent* random variables with a Poisson distribution, with parameters that are easily expressed in terms of R_0, v, f and ϕ. In fact, we have

$$\tilde{q}^i(k_1, k_2) = q^{i,1}(k_1)q^{i,2}(k_2),$$

with

$$
\begin{aligned}
q^{1,1}(k_1) &= \frac{((1-v)R_0)^{k_1}e^{-(1-v)R_0}}{k_1!}, \\
q^{1,2}(k_2) &= \frac{(vfR_0)^{k_2}e^{-vfR_0}}{k_2!}, \\
q^{2,1}(k_1) &= \frac{((1-v)\phi R_0)^{k_1}e^{-(1-v)\phi R_0}}{k_1!}, \\
q^{2,2}(k_2) &= \frac{(vf\phi R_0)^{k_2}e^{-vf\phi R_0}}{k_2!}.
\end{aligned}
$$

Evaluation of the sums now leads to the equations (8.5).

Exercise 8.7 If we write $\pi_i = e^{A_i}$ then $A_2 = \phi A_1$, so we have $\pi_2 = \pi_1^{\phi}$, which reflects that the infectivity of type 2 individuals is reduced by a factor ϕ.

Exercise 8.8 We start with the last question: if we exposed the population to infectious material from outside and one individual were to become infected, this would with probability $(1-v)/(1-v+vf)$ be a type 1 individual and with probability $vf/(1-v+vf)$ a type 2 individual. In the first case, we would get a major outbreak with probability $1 - \pi_1$ and in the second case one with probability $1 - \pi_2$. So, modulo the normalizing factor $(1 - v + vf)^{-1}$, ξ is the probability of a major outbreak, given that one individual is infected as a result of uniform exposure to infectious material coming from outside the population.

The definition of ξ and the equations (8.5) imply that $\pi_1 = e^{-R_0\xi}$ and $\pi_2 = e^{-\phi R_0\xi}$, and upon substitution of these relations into the defining relation for ξ, equation (8.6) for ξ is obtained.

Exercise 8.9 The probability to escape is the zero term of a Poisson distribution with an intensity parameter involving the total force of infection exerted during the entire epidemic, which equals $R_0((1-v)(1-\sigma_1) + \phi v(1-\sigma_2))$ in the case of type 1 individuals. This quantity is reduced by a factor f in the case of type 2 individuals. Thus we arrive at the equations (8.7).

Exercise 8.10 We first note that $\sigma_2 = \sigma_1^f$, which reflects that the susceptibility of type 2 individuals is reduced by a factor f. In terms of θ, the equations (8.7) can be rewritten as $\sigma_1 = e^{-R_0\theta}$ and $\sigma_2 = e^{-fR_0\theta}$, and upon substitution of these relations into the defining relation for θ, equation (8.8) for θ is obtained. The quantity θ is (proportional to) the size of a major outbreak when we measure 'size' in terms of the output of infectious material (indeed, note the factor ϕ in the second term).

Exercise 8.11 The equations (8.6) and (8.8) for ξ and θ, respectively, are identical except for an interchange of f and ϕ. To express this more precisely, we first incorporate the dependence on f and ϕ into the notation by writing $\xi = \xi(f, \phi)$ and $\theta = \theta(f, \phi)$, and then note that $\xi(f, \phi) = \theta(\phi, f)$.

Let k with $0 < k < 1$ be given and define

$$J(f) := \xi(f, \frac{k}{f}) \, \theta(f, \frac{k}{f})$$

for $k \leq f \leq 1$. The goal is to minimize J.

Our first observation is that

$$J(1) = \xi(1, k) \, \theta(1, k) = \theta(k, 1) \, \xi(k, 1) = J(k),$$

or, in words, that the boundary values are equal. In fact, we have the more general symmetry relation

$$J(f) = \xi(f, \frac{k}{f}) \, \theta(f, \frac{k}{f}) = \theta(\frac{k}{f}, f) \, \xi(\frac{k}{f}, f) = J(\frac{k}{f}),$$

which implies that J has an extreme in the 'midpoint' $f = \sqrt{k}$. Is this a maximum or a minimum?

The first attempt is to find an intuitive argument that yields the answer. Alas, this is in vain. The second attempt is to calculate and investigate the derivative of J (as well as the second derivative). Alas, this is also in vain. The third attempt is to look at extreme situations: very large R_0 on the one hand, and $R_0(1 - v + kv)$ only slightly greater than 1 on the other hand.

For $R_0 \to \infty$, we have $\xi \to 1 - v + fv$ (since the exponentials in equation (8.6) go to zero) and $\theta \to 1 - v + \phi v$ (for an analogous reason). So, $\xi\theta \to (1 - v)^2 + kv^2 + (f + k/f)v(1 - v)$. As one easily verifies, the function $f \mapsto f + k/f$ has a *minimum* at $f = \sqrt{k}$. We conclude that for large values of R_0, the best vaccines are those that divide their reduction potential equally over susceptibility and infectivity.

Next, let us examine the critical situation characterized by $R_0(1 - v + kv) = 1$ (see Exercise 8.4). In that situation both θ and ξ are small and we may approximate the exponentials by the first few terms in their Taylor expansion. This leads to

$$\xi = (1 - p)\left(R_0\xi - \frac{1}{2}R_0^2\xi^2\right) + fp\left(\phi R_0\xi - \frac{1}{2}R_0^2\phi^2\xi^2\right) + \text{h.o.t.},$$

and, after dividing by ξ, neglecting the h.o.t and solving for ξ, to

$$\xi = \frac{(1 - v + kv)R_0 - 1}{1 - v + kv\phi} \frac{2}{R_0^2} + \cdots.$$

Likewise we have (by replacing ϕ by f)

$$\theta = \frac{(1 - v + kv)R_0 - 1}{1 - v + kvf} \frac{2}{R_0^2} + \cdots.$$

The product $\theta\xi$ therefore has, in leading order, all f-dependence in the factor

$$(1 - v + kv\phi)(1 - v + kvf) = (1 - v)^2 + k^3v^2 + k\left(f + \frac{k}{f}\right)v(1 - v)$$

in the denominator. Since $f \mapsto f + k/f$ has a minimum at $f = \sqrt{k}$, the product $\theta\xi$ has a maximum at $f = \sqrt{k}$. We conclude that, near criticality, vaccines that put all their reduction potential into either susceptibility or infectivity (irrelevant which of the two) are better than vaccines that have a 'mixed' strategy.

The final conclusion is that, within the setting considered in this section, it depends on quantitative details what type of vaccine performs best.

As an encore, we look briefly at the special case $v = 1$ (everybody vaccinated). Then

$$\xi = f(1 - e^{-\phi R_0\xi}) \Rightarrow \phi\xi = k(1 - e^{-\phi R_0\xi})$$

and

$$\theta = \phi(1 - e^{-f R_0\theta}) \Rightarrow f\theta = k(1 - e^{-f R_0\theta}),$$

which shows that both $\phi\xi$ and $f\theta$ depend only on k and R_0 (and not on f and ϕ separately) and are in fact equal. Hence $\xi\theta = (\phi\xi)(f\theta)/k$ is independent of f when $f\theta = k$. That is, if all individuals are vaccinated, it makes no difference whether the vaccine reduces infectivity or susceptibility.

Exercise 8.13 The operator

$$(K_r\phi)(\xi) = N(\xi)a(\xi)\int_\Omega\int_0^\infty b(\tau,\eta)e^{-\lambda\tau}\,d\tau\,\phi(\eta)\,d\eta$$

has a one-dimensional range spanned by $N(\cdot)a(\cdot)$, and hence the one and only non-zero eigenvalue equals

$$\int_\Omega\int_0^\infty b(\tau,\eta)e^{-\lambda\tau}\,d\tau\,N(\eta)a(\eta)\,d\eta,$$

and r is that value of λ for which this expression equals 1.

Exercise 8.21 Define

$$I(t,\xi) := \int_{-\infty}^t e^{-\alpha(\xi)(t-\tau)}i(\tau,\xi)\,d\tau,$$

which is the total size of the infective sub-population with h-state ξ at time t. Differentiation of this expression with respect to t gives

$$\frac{dI}{dt} = i(t,\xi) - \alpha(\xi)I(t,\xi), \tag{17.2}$$

where $i(t,\xi)$ is given by (8.10), in the nonlinear version. If we substitute the assumption on A into the nonlinear version of (8.10), we obtain

$$i(t,\xi) = S(t,\xi)\int_0^\infty\int_\Omega \beta(\xi,\eta)e^{-\alpha(\eta)\tau}i(t-\tau,\eta)\,d\eta\,d\tau.$$

Exchange of the order of integration using Fubini's Theorem (see e.g., Rudin 1974) and substitution of $\tau \leftrightarrow t - \tau$ leads to

$$i(t,\xi) = S(t,\xi)\int_\Omega \beta(\zeta,\eta)\int_{-\infty}^t e^{-\alpha(\eta)(t-\tau)}i(\tau,\eta)\,d\tau\,d\eta.$$

So, together with the definition of $I(t, \eta)$, we have

$$i(t, \xi) = S(t, \xi) \int_\Omega \beta(\xi, \eta) I(t, \eta) \, d\eta. \tag{17.3}$$

Finally, substitution of (17.2) into (17.3) gives the differential equation for $I(t, \xi)$ we were looking for:

$$\frac{dI}{dt}(t, \xi) = S(t, \xi) \int_\Omega \beta(\xi, \eta) I(t, \eta) \, d\eta - \alpha(\xi) I(t, \xi).$$

Exercise 8.22 Define

$$I(t, a) := \int_0^a i(t - \tau, a - \tau) e^{-\int_{a-\tau}^a \alpha(\sigma) \, d\sigma} \, d\tau,$$

and find by differentiation

$$\frac{\partial I}{\partial t} + \frac{\partial I}{\partial a} = i(t, a) - \alpha(a) I(t, a). \tag{17.4}$$

If we substitute the assumption on A into the nonlinear version of (8.10), we obtain

$$i(t, a) = S(t, a) \int_0^\infty \int_0^\infty \beta(a, \eta + \tau) e^{-\int_\eta^{\eta + \tau} \alpha(\sigma) \, d\sigma} i(t - \tau, \eta) \, d\eta \, d\tau.$$

Substitution of $a' = \eta + \tau$ gives

$$i(t, a) = S(t, a) \int_0^\infty \int_\tau^\infty \beta(a, a') e^{-\int_{a'-\tau}^{a'} \alpha(\sigma) d\sigma} i(t - \tau, a' - \tau) da' d\tau,$$

and then, using Fubini's Theorem,

$$i(t, a) = S(t, a) \int_0^\infty \beta(a, a') \int_0^{a'} e^{-\int_{a'-\tau}^{a'} \alpha(s) \, ds} i(t - \tau, a' - \tau) \, d\tau \, da'.$$

We can rewrite this as

$$i(t, a) = S(t, a) \int_0^\infty \beta(a, a') I(t, a') \, da'$$

and substitute the result into (17.4) to obtain the partial differential equation we were looking for.

Exercise 8.23 $\partial S(t, x)/\partial t$ is the number of new cases per unit of time per unit of space (if S is a spatial density) at time t, with h-state x at the moment of becoming infected. The right-hand side is the product of two factors, the first being the density $S(t, x)$ of susceptibles at time t with h-state x. So the second factor is, by definition, the force of infection, i.e., the probability for a susceptible with h-state x to become infected at time t.

The modeling assumption is that the force of infection is the sum of contributions of individuals that themselves became infected τ units of time earlier (so at time $t - \tau$), with state-at-infection η. The size of that contribution is measured by the modeling ingredient $A(\tau, x, \eta)$, which has to be specified in more detail (and eventually even quantitatively, perhaps) in order to make the model concrete and applicable to a specific infectious agent in a specific 'host' population. It is the expected force of infection exerted on individuals with h-state x by individuals with state-at-infection η, and current infection-age τ.

The equation may serve to describe, for example, a fungal disease in an agricultural crop, by simply interpreting x and η as referring to spatial position (and to let S indeed be a spatial density). A special feature in that case is that the h-state is static (plants do not move), and that consequently it may make sense to let $A(\tau, x, \eta)$ only depend on the spatial distance $|x - \eta|$ and not on x and η separately (see Chapter 10 for elaboration).

Inflow of new susceptibles is *not* included, since we equate the incidence to $\partial S(t, x)/\partial t$, requiring that there are no changes in S due to causes other than infection. So, for the same reason, return to the class of susceptibles by loss of immunity after having experienced infection is also *not* included.

Exercise 8.24 Divide both sides of (8.21) by $S(t, x)$ and integrate over time from $-\infty$ to t. The left-hand side then becomes

$$\int_{-\infty}^{t} \frac{1}{S(\sigma, x)} \frac{\partial S}{\partial t}(\sigma, x)\, d\sigma \;=\; \int_{-\infty}^{t} \left[\frac{d}{dt} \ln S(t, x)\right]_{t=\sigma} d\sigma$$

$$= \ln S(t, x) - \ln S(-\infty, x) \;=\; \ln \frac{S(t, x)}{S(-\infty, x)},$$

while for the right-hand side we have

$$\int_{-\infty}^{t} \int_{\Omega} \int_{0}^{\infty} A(\tau, x, \xi) \frac{\partial S}{\partial t}(\sigma - \tau, \xi)\, d\tau\, d\xi\, d\sigma$$

$$= \int_{-\infty}^{t} \int_{\Omega} \int_{0}^{\infty} A(\tau, x, \xi) \frac{d}{d\sigma} S(\sigma - \tau, \xi)\, d\tau\, d\xi\, d\sigma$$

$$= \int_{\Omega} \int_{0}^{\infty} A(\tau, x, \xi) \int_{-\infty}^{t} \frac{d}{d\sigma} S(\sigma - \tau, \xi)\, d\sigma\, d\tau\, d\xi$$

$$= \int_{\Omega} \int_{0}^{\infty} A(\tau, x, \xi)\{S(t - \tau, \xi) - S(-\infty, \xi)\}\, d\tau\, d\xi.$$

Together this yields equation (8.23), which, by taking the limit $t \to \infty$, implies (8.22).

Exercise 8.25 Take exponentials of both sides of (8.22) after having slightly rewritten the right-hand side, and you arrive at (8.24). The point of this exercise, however, is to show that one may derive (8.24) directly by employing the arguments introduced in Section 1.3.1.

The left-hand side of (8.24) is the fraction of individuals with h-state x that escape infection, which we equate to the probability to escape infection. The right-hand side is the zero term of a Poisson distribution with mean

$$\int_{\Omega} \int_{0}^{\infty} A(\tau, x, \eta)\, d\tau\, S(-\infty, \eta) \left\{1 - \frac{S(\infty, \eta)}{S(-\infty, \eta)}\right\} d\eta,$$

which is indeed the total force of infection exerted on individuals with h-state x, given that for any η a fraction $S(\infty, \eta)/S(-\infty, \eta)$ fell victim. Thus it is shown that (8.24) is indeed a consistency condition.

Exercise 8.26 R_0 was defined in Chapter 7; see in particular Section 7.3. The question should be reformulated as: how does the kernel $k(x, \eta)$ of the integral operator K relate to the present modeling ingredients $S(-\infty, x)$ and $A(\tau, x, \eta)$? The answer is

$$k(x, \eta) = S(-\infty, x) \int_0^\infty A(\tau, x, \eta) \, d\tau.$$

The linear integral operator with this kernel has R_0 as its dominant positive eigenvalue (assuming the spectral radius is indeed an eigenvalue).

If we introduce

$$u(x) := -\ln \frac{S(\infty, x)}{S(-\infty, x)},$$

we can rewrite (8.24) as

$$u(x) = \int_\Omega k(x, \eta) \left(1 - e^{-u(\eta)}\right) d\eta. \tag{17.5}$$

On the basis of the interpretation we expect that (8.24) has a solution $S(\infty, x) < S(-\infty, x)$ when $R_0 > 1$ (note that we, of course, consider $S(-\infty, x)$ as given and $S(\infty, x)$ as unknown). This translates into (17.5) having a positive solution for $R_0 > 1$. What kind of arguments enable us to check our expectations?

In the rest of this elaboration we shall sketch the essence of such arguments, without going into the technical details that are involved in complete and rigorous mathematical proofs.

Two key observations are as follows:

- $u \mapsto 1 - e^{-u}$ is an increasing (and bounded) function.

- $1 - e^{-u} \geq (1 - \varepsilon)u$ for ε positive and small and $0 \leq u \leq \overline{u}$, where $\overline{u} = \overline{u}(\varepsilon)$ is the positive root of $1 - e^{-u} = (1 - \varepsilon)u$. (Draw a picture for yourself; use the fact that the graph of $u \mapsto 1 - e^{-u}$ has tangent line $u \mapsto u$ in the origin.)

We can rewrite (17.5) as

$$u = Lu,$$

where L is the nonlinear integral operator defined by

$$(Lu)(x) = \int_\Omega k(x, \eta) \left(1 - e^{-u(\eta)}\right) d\eta.$$

For functions defined on Ω we shall use the partial order relation $\phi \geq \psi$ (defined by $\phi \geq \psi$ if and only if $\phi(x) \geq \psi(x)$ for all $x \in \Omega$, or, for almost all $x \in \Omega$ in the case that we work with integrable functions that are not necessarily defined for all $x \in \Omega$). From now on we restrict attention to positive (i.e., non-negative) functions.

The operator L is monotone, i.e., $\phi \geq \psi \Rightarrow L\phi \geq L\psi$ (this is a direct consequence of the positivity of k and the monotonicity of $u \mapsto 1 - e^{-u}$, see the first observation). Bounded monotone sequences converge. This can be extended to sequences of functions under an additional technical condition (compactness) related to the precise form of convergence that one wants to consider. On the one hand, in the present situation $L_1(\Omega)$-convergence is most natural, but, on the other hand, uniform convergence, i.e., $C(\Omega)$-convergence, is more easy to work with; we shall not dwell on this, but confine ourselves to the remark that one needs a minor technical condition on the kernel k that guarantees that the integral operator

L is compact in the function space considered. We want to generate a monotone sequence by applying the operator L repeatedly, i.e., by considering the iteration scheme

$$u_{n+1} = Lu_n.$$

Since L is (assumed to be) continuous, a convergent sequence u_n yields a fixed point of L (just pass to the limit in $u_{n+1} = Lu_n$), i.e., a solution of (17.5). To generate a monotone sequence, we only have to get started, i.e., to find u_0 such that

$$u_1 := Lu_0 \geq u_0,$$

since the monotonicity of L then guarantees, by induction, that $u_{k+1} \geq u_k$ for all k. We now concentrate on finding u_0. It is here that the assumption $R_0 > 1$ and the key second observation are crucial.

Choose ε positive but small enough to guarantee that $(1 - \varepsilon)R_0 > 1$. Let ϕ be the eigenfunction of the linear integral operator K corresponding to the eigenvalue R_0. Assume that ϕ is bounded (this is another minor assumption on the kernel k, as a rule subsumed under earlier ones). Now choose $u_0 = \delta\phi$ with δ such that $u_0(x) \leq \bar{u} = \bar{u}(\varepsilon)$. Then

$$(Lu_0)(x) = \int_\Omega k(x,\eta)\left(1 - e^{-u_0(\eta)}\right) d\eta \geq (1 - \varepsilon)\int_\Omega k(x,\eta)u_0(\eta)\, d\eta,$$

which leads to

$$(Lu_0)(x) \geq (1 - \varepsilon)\delta \int_\Omega k(x,\eta)\phi(\eta)\, d\eta = (1 - \varepsilon)R_0\delta\phi(x) = (1 - \varepsilon)R_0 u_0(x) \geq u_0(x).$$

For $u_0 = \delta\phi$, therefore, the iteration scheme $u_{n+1} = Lu_n$ yields a non-decreasing sequence. Since $1 - e^{-u} \leq 1$ for $u \geq 0$, boundedness is immediate:

$$u_n(x) \leq \int_\Omega k(x,\eta)d\eta.$$

This ends our sketch of the proof of the implication $R_0 > 1 \Rightarrow$ '(8.22) has a non-trivial solution.'

What about the uniqueness of the non-trivial solution of (8.22)? To study uniqueness, the notion of *sublinearity* as introduced by Krasnoselskiĭ is very helpful.[1] Let $\tilde{u}$ be any positive number. On the interval $(0, \tilde{u})$ the graph of $u \mapsto 1 - e^{-u}$ lies above the straight line through the origin and $(\tilde{u}, 1 - e^{-\tilde{u}})$. Analytically this is expressed by

$$1 - e^{-\theta\tilde{u}} > \theta(1 - e^{-\tilde{u}}) \quad \text{for } 0 < \theta < 1.$$

Now let u and w be two strictly positive solutions of (17.5). Let $\tilde{u}$ be a common upper bound for u and w. Define $\theta = \inf_{x\in\Omega} u(x)/w(x)$. Assume that the infimum is actually attained at some point $x = \bar{x}$ (alternatively, assume that Ω is compact and that u and w are continuous). Suppose that $\theta < 1$. Then we arrive at a contradiction by noting that

$$u(\bar{x}) = \int_\Omega k(\bar{x},\eta)\left(1 - e^{-u(\eta)}\right) d\eta \geq \int_\Omega k(\bar{x},\eta)\left(1 - e^{-\theta w(\eta)}\right) d\eta,$$

---

[1]M.A. Krasnoselskiĭ: *Positive Solutions of Operator Equations.* Noordhoff, Groningen, 1964, (Sections 6.1.3 and 6.1.7 in that book).

so

$$u(\overline{x}) > \theta \int_\Omega k(\overline{x}, \eta) \left(1 - e^{-w(\eta)}\right) d\eta = \theta w(\overline{x}).$$

Therefore $\theta = 1$ or $u(x) \geq w(x)$. But, as we can consider $w(x)/u(x)$ in exactly the same way, we also find $u(x) \leq w(x)$, and hence can conclude that $u(x) = w(x)$.

Exercise 8.27 i) $B(x)$ is the rate at which individuals with h-state x are born. We are supposing here that the h-state is static! Individuals with h-state x die from 'natural' causes at a per capita rate $\mu(x)$. The incidence of individuals with h-state x at time t is $i(t, x)$.

ii) The infection-free steady state is $\widetilde{S}(x) = \frac{B(x)}{\mu(x)}$.

iii) R_0 is the dominant eigenvalue of the linear integral operator with kernel

$$k(x, \eta) = \widetilde{S}(x) \int_0^\infty A(\tau, x, \eta) \, d\tau.$$

iv) An endemic steady state is characterized by the two equations

$$
\begin{aligned}
B(x) - \mu(x)\overline{S}(x) &= \overline{i}(x), \\
\overline{i}(x) &= \frac{\overline{S}(x)}{\widetilde{S}(x)} \int_\Omega k(x, \eta)\overline{i}(\eta) \, d\eta,
\end{aligned}
$$

which, by elimination of $\overline{S}(x)$ and use of the expression for $\widetilde{S}(x)$, combine into the equation

$$\overline{i}(x) = \left(1 - \frac{\overline{i}(x)}{B(x)}\right) \int_\Omega k(x, \eta)\overline{i}(\eta) \, d\eta$$

for the unknown $\overline{i}$ with $0 \leq \overline{i}(x) \leq B(x)$.

v) Writing the previous equation as $\overline{i} = L\overline{i}$, we see that the linearization of L, thus defined, at the trivial (infection-free) fixed point $\overline{i} = 0$ gives the linear integral operator K, which has R_0 as its dominant eigenvalue. So the situation is reminiscent of the situation considered in the preceding exercise. A major difference, however, is that L is *not* monotone in the present case. One has to resort to more complicated arguments.[2] medskip

Exercise 8.28 What is the probability that the 'line' originating with a 'child' of an individual of type x goes extinct? If that child has type ξ, this probability equals, by definition, $\pi(\xi)$. We do not know the type of the child, however; we only know the probability distribution $m(x, \cdot)$. Accordingly, the probability that the line goes extinct is

$$\int_\Omega \pi(\xi)m(x, d\xi).$$

If there are k children, the probability that all k lines will go extinct is simply the kth power of this quantity, by (assumed) independence. The identity (8.25) is now self-evident.

Exercise 8.29 The aim is to relate p_k and m to the modeling ingredients $S(-\infty, \cdot)$ and A. The quantity $f(y, x)$ is the expected number of 'children' with type y produced by an individual of type x (in the initial phase, where the reduction of

---

[2]R.D. Nussbaum: *The Fixed Point Index and Some Applications*. Séminaire de Mathématiques Supérieures. Les Presses de l'Université Montreal, 1985.

susceptibles by infection can still be ignored). If we integrate this quantity with respect to y, we get the total expected number of offspring. An assumption made throughout is that contacts (and hence transmissions) are made according to a Poisson process. As the Poisson process is completely specified by one parameter, the mean, we can determine $p_k(x)$ from this mean, and the result is (8.26).

The interpretation of $f(\cdot, x)$ and $m(x, \cdot)$ implies at once that f is the density of m, apart from the normalization that is incorporated in m, but not in f. Hence (8.27) should hold.

If we substitute (8.26) and (8.27) into (8.25), a factor $(\int_\Omega f(y, x)\, dy)^k$ in $p_k(x)$ cancels the same factor in the denominator due to the normalization of m, and (8.28) results.

Exercise 8.30 The assumption (8.29) says that the expected force of infection acting on individuals of type x exerted by individuals that had state η at the moment of becoming infected factorizes as a product, with one factor depending only on the characteristic x of the potential victim and one factor depending only on the characteristics η and τ, the time elapsed since infection, of the infective considered. Recalling the function f and the measure m as introduced in Exercise 8.29, we find that the distribution of the h-state of newly infected individuals has density $y \mapsto a(y)S(-\infty, y)/\int_\Omega a(\eta)S(-\infty, \eta)\, d\eta$, for all x (the function $y \mapsto f(y, x) = a(y)S(-\infty, y)/\int_0^\infty b(\tau, x)\, d\tau$ has an x-dependent factor, but this is eliminated when we normalize).

Exercise 8.31 The idea behind the ansatz is that the force of infection acting on individuals with h-state x factorizes into an x-specific factor $a(x)$ and a function of time. Somewhat loosely, we may think of this function of time as the amount of infectious material 'in the air' at time t. The function $w(t)$ then measures the cumulative amount of infectious material to which individuals have been exposed. By integration (after dividing both sides by $S(t, x)$), we find

$$S(t, x) = S(-\infty, x)e^{-a(x)w(t)}.$$

If we substitute this into (8.23) and use (8.29), we find (8.31). The key point is that the unknown $w(t)$ in (8.31) is a *scalar* quantity. In fact (8.31) is a so-called nonlinear renewal equation (which is a Volterra integral equation of convolution type), written in time-translation-invariant form. To obtain the more traditional form, split the interval of τ integration into $[0, t)$ and $[t, \infty)$ and pretend that the contribution of $[t, \infty)$ yields a 'known' function $f(t)$ (since, indeed, it gives the contribution to the cumulative quantity $w(t)$ of those individuals that were infected before time zero).

Exercise 8.32 The kernel k also factorizes:

$$k(x, \eta) = a(x)S(-\infty, x)\int_0^\infty b(\tau, \eta)\, d\eta.$$

Hence the integral operator K has one-dimensional range and we obtain for R_0 the explicit expression

$$R_0 = \int_\Omega \left(\int_0^\infty b(\tau, \eta)d\tau \right) a(\eta)S(-\infty, \eta)\, d\eta.$$

Consider the right-hand side of the final-size equation (8.33) as a function $g = g(w(\infty))$ of $w(\infty)$. For small values of w we have $g(w) \sim R_0 w$ (since $1 - e^{-aw} \sim aw$

for $w \to 0$). So, when $R_0 > 1$, the graph of g starts out above the 45° line. For $w \to \infty$ we find that g tends to a constant (since $e^{-aw} \to 0$ for $w \to \infty$) and therefore, for large enough values of w, the graph of g lies below the 45° line. It follows that the graph of g must cross the 45° line. This geometrical observation translates into the conclusion that (8.33) must have a positive solution $w(\infty)$.

Exercise 8.33 The appropriate ansatz is that

$$\frac{S(\infty, x)}{S(-\infty, x)} = e^{-a(x)w(\infty)}.$$

Once we substitute this into (8.24) and use, in addition, (8.29), we obtain (8.33).

Exercise 8.34 The first part is basic substitution. Then put $\partial S/\partial t = 0$ in (8.35), to find

$$\overline{S}(x) = \frac{B(x)}{\mu(x) + a(x)\overline{v}}.$$

In the steady-state version of (8.35) we can factor out $\overline{v}$. Combining these observations, we obtain (8.36). Call the right-hand side of (8.36) a function $g = g(\overline{v})$ of $\overline{v}$. Then $g(0) = R_0$, where R_0 is given by the expression derived in Exercise 8.32 with $S(-\infty, \eta)$ replaced by $\overline{S}(\eta)$. Moreover, g is a strictly decreasing function that tends to zero at infinity. Hence (8.36) has a unique positive solution when $R_0 > 1$.

Exercise 8.35 With the assumption (8.37), we can write the consistency condition (8.25) as

$$\pi(x) = \sum_{k=0}^{\infty} p_k(x) \left(\int_{\Omega} \pi(\eta) v(d\eta) \right)^k = \sum_{k=0}^{\infty} p_k(x) z^k,$$

which is (8.39). So z should equal the integral of the function of x on the right-hand side against the measure v, that is

$$z = \sum_{k=0}^{\infty} z^k \int_{\Omega} p_k(x) v(dx),$$

which is (8.38).

Exercise 8.36 Since constant (i.e., independent of x and η) factors can be incorporated in either a or b, we can still add a normalization condition on either a and b to (8.29). In the present case we chose (8.40). Next, recall that, under the assumption (8.29), we have for f as introduced in Exercise 8.29 that

$$f(y, x) = S(-\infty, y)a(y) \int_0^{\infty} b(\tau, x) \, d\tau$$

and hence we have

$$\int_{\Omega} f(y, x)dy = \int_0^{\infty} b(\tau, x) \, d\tau. \tag{17.6}$$

Using this information in (8.27), we find

$$m(x, \omega) = \int_{\omega} S(-\infty, y)a(y) \, dy,$$

which is indeed independent of x and so can be identified with $v(\omega)$ of the preceding exercise, giving (8.41). Finally, inserting (17.6) into the expression (8.26) for $p_k(x)$, we find (8.42).

Exercise 8.37 i) The expected number of s-types a given s-type has infectious contacts with during its infectious period equals $(\lambda_{ss}/N) \cdot 1 \cdot N_s = \alpha_s \alpha_s \pi_s$; the first factor is the rate, the '1' is the mean length of the infectious period and N_s is the number of s-individuals. The same type of reasoning applies to the other type of transmissions among standard and core individuals The resulting matrix of mean offspring is given by

$$\begin{pmatrix} \alpha_s^2 \pi_s & \alpha_s \alpha_c \pi_c \\ \alpha_c \alpha_s \pi_s & \alpha_c^2 \pi_c \end{pmatrix}.$$

The largest eigenvalue to this matrix is given by $\alpha_s^2 \pi_s + \alpha_c^2 \pi_c$ (cf. Section 7.4.1).

ii) In the first situation we have a homogeneous community with contact rate 1.5 and non-random infectious period of length 1. The probability of a minor outbreak q is then given by the smallest solution to

$$q = e^{-R_0(1-q)}$$

as shown in Sections 1.2.2 and 3.3 (in particular Exercise 3.7). When $R_0 = 1.5$ the solution equals $q = 0.4172$, so the probability of a major outbreak equals $1 - q = 0.5828$.

For the intermediate case we have to do a bit more work. In a similar fashion as was done for the simplistic STD model for a heterosexual community (Section 8.1.1) we have to solve a system of two equations similar to (8.2) and (8.3), but more complicated since here every type has offspring of every type. In the present situation the two equations become

$$q_s = E\left(q_s^{X_{ss}} q_c^{X_{sc}}\right),$$
$$q_c = E\left(q_s^{X_{cs}} q_c^{X_{cc}}\right),$$

where X_{ij} denotes the random number of type j individuals that an i-individual has infectious contacts with during the infectious period. Since the infectious periods are non-random these random variables are independent and Poisson distributed. Their means are $E(X_{ss}) = \alpha_s^2 \pi_s = 0.875$, $E(X_{sc}) = \alpha_s \alpha_c \pi_c \approx 0.247$, $E(X_{cs}) = \alpha_c \alpha_s \pi_s \approx 2.22$ and $E(X_{cc}) = \alpha_c^2 \pi_c = 0.625$. For a Poisson variable X with mean λ it holds that $E(\pi^X)) = e^{-\lambda(1-\pi)}$. From this it follows that the equations become:

$$q_s = e^{-\alpha_s^2 \pi_s(1-q_s)} e^{-\alpha_s \alpha_c \pi_c(1-q_c)} = e^{-0.875(1-q_s)} e^{-0.247(1-q_c)},$$
$$q_c = e^{-\alpha_c \alpha_s \pi_s(1-q_s)} e^{-\alpha_c^2 \pi_c(1-q_c)} = e^{-2.22(1-q_s)} e^{-0.625(1-q_c)}.$$

Solving this numerically we find the outbreak probabilities $1 - q_s = 0.433$ and $1 - q_c = 0.762$ respectively. Because it was assumed that the outbreak was initiated by one infected core individual, the outbreak probability equals 0.762.

Finally, the other extreme situation is where all infections are among core individuals. We can then neglect standard individuals completely and the model says that a (core) individual infects on average $(\alpha_C^2/N) \cdot 1 \cdot N_c = \sqrt{15}^2 \cdot 0.1 = 1.5$ other (core) individuals. Since the infectious period is non-random we get the equation $q = e^{-R_0(1-q)}$ for a minor outbreak, just like in the homogeneous case. The probability of a major outbreak hence equals 0.583.

This means that the probability of a major outbreak is highest somewhere in-between the two extreme cases, i.e., where the core group has higher transmission but when also standard individuals take part in transmission.

iii) The system of equations for the final fractions infected is identical to that for the probability of a major outbreak. (This is only true when the infectious period is non-random — otherwise the two systems differ.) This means that we have already computed the solutions. For the first situation when there is no core group both types get infected equally much: $x_s^* = x_c^* = x^* = 0.5828$. For the intermediate case we get $x_s^* = 0.433$, $x_c^* = 0.762$ and the overall fraction infected hence equals $0.9 \cdot 0.433 + 0.1 \cdot 0.762 = 0.466$. Finally, in the extreme case where only core group individuals are affected by the outbreak we get $x_s^* = 0$ and $x_c^* = 0.5828$, which implies an overall fraction infected of $0.1 \cdot x_c^* = 0.0583$. To summarize, the total outbreak size becomes smaller if the community becomes more heterogeneous.

17.3   ELABORATIONS FOR CHAPTER 9

Exercise 9.1 i) The distribution function of the length of life is $1 - \mathcal{F}_d(a)$ and the corresponding probability density function is therefore $-d\mathcal{F}_d(a)/da = +\mu(a)\mathcal{F}_d(a)$. Hence the expected length of life equals $\int_0^\infty a\mu(a)\mathcal{F}_d(a)\, da$. The expected length of life is also the mean age at death.

ii) The stable age distribution is (cf. (9.1)) $Ce^{-ra}\mathcal{F}_d(a)$; hence the mean age of those dying at any particular moment in time equals

$$\frac{C \int_0^\infty a\mu(a)e^{-ra}\mathcal{F}_d(a)\, da}{C \int_0^\infty \mu(a)e^{-ra}\mathcal{F}_d(a)\, da},$$

and the factors C cancel.

iii) The growth of a population positively affects the relative frequency of the younger age groups and this perseveres in samples of individuals at the point of death.

iv) On the one hand, $\int_0^\infty a\mu e^{-\mu a}\, da = -ae^{-\mu a}|_0^\infty + \int_0^\infty e^{-\mu a}\, da = 1/\mu$, the life expectancy. On the other hand, we have $\int_0^\infty \mu e^{-(r+\mu)a}\, da = \mu/(r+\mu)$ and $\int_0^\infty a\mu e^{-(r+\mu)a}\, da = -a\frac{\mu}{r+\mu}e^{-(r+\mu)a}|_0^\infty + \frac{\mu}{r+\mu}\int_0^\infty e^{-(r+\mu)a}\, da = \frac{\mu}{(r+\mu)^2}$, so the mean age of those dying at any particular moment in time equals $1/(r+\mu)$.

Exercise 9.2 When r is large, $N(a)$ given by (9.1) is decreasing rapidly with age. When a population census is represented graphically, one often uses two adjacent histograms, one for males and one for females. Hence the description 'pyramid.'

Exercise 9.3 R_0 is the number of children that a newborn individual is expected to beget in its entire life:

$$R_0 = \int_0^\infty \beta(a)\mathcal{F}_d(a)\, da.$$

Exercise 9.4 i) For sexual contacts, kissing, shaking hands, etc., yes. But, for example, for parents (and day-care center employees) cleaning children's noses and changing diapers, no; for blood transfusion certainly no.

ii) A contact may be 'defined' as any event generating a transmission opportu-nity, so this has, by definition, a non-zero transmission probability. Some types of contact, like sexual intercourse or shaking hands, may be defined in other terms, unrelated to any agent, and then it makes sense to introduce the 'probability of transmission, given a contact' as a separate factor. But 'hugging' occurs with variable intensity and duration, and one would expect the probability of transmission of influenza to be a monotone function of both.

In any case, an asymmetric probability of transmission (such as for STDs, where male → female transmission is sometimes easier than female → male transmission) can spoil the strict symmetry requirement that stems from bookkeeping considerations.

Exercise 9.5 i) An individual infected at age α will be alive τ units of time later with probability $\mathcal{F}_d(\alpha + \tau)/\mathcal{F}_d(\alpha)$ and then has infectivity $h(\tau, \alpha)$. At that time it is aged $\alpha + \tau$ and therefore makes $c(a, \alpha + \tau)N(a)$ contacts per unit of time with individuals of age a. Since 'infectivity' is used here in the sense of 'probability of transmission, given a contact with a susceptible,' we only have to integrate the product with respect to τ to obtain the expected number of secondary cases with age a at the moment of becoming infected.

ii) If $c(a, \alpha) = f(a)g(\alpha)$ then $k(a, \alpha) = f(a)N(a)\psi(\alpha)$, with (compare (9.3))

$$\psi(\alpha) = \int_0^\infty h(\tau, \alpha)g(\alpha + \tau)\frac{\mathcal{F}_d(\alpha + \tau)}{\mathcal{F}_d(\alpha)} \, d\tau.$$

We see that K has a one-dimensional range spanned by fN, and consequently R_0 is given by

$$R_0 = \int_0^\infty \psi(\alpha)f(\alpha)N(\alpha) \, d\alpha,$$

which is (9.2).

When $f = g$, we can interpret this quantity as age-specific social activity, and then the assumption on c means that the contact intensity of an individual is proportional to its social activity and to the social activity of the entire population and that contacts are made according to supply, without any selection of age. When $f \neq g$, one could try a similar line of reasoning, while distinguishing between active and passive contact (or, in any case, a role difference between the two individuals involved in a contact). Transmission via blood transfusion could be a concrete example.

iii) Now let $c(a, \alpha) = \sum_{k=1}^n f_k(a)g_k(\alpha)$. Define

$$\psi_k(\alpha) = \int_0^\infty h(\tau, \alpha)g_k(\alpha + \tau)\frac{\mathcal{F}_d(\alpha + \tau)}{\mathcal{F}_d(\alpha)} \, d\tau;$$

then

$$(K\phi)(a) = \sum_{k=1}^n f_k(a)N(a) \int_0^\infty \psi_k(\alpha)\phi(\alpha) \, d\alpha,$$

from which we see that the range of K is spanned by $\{f_k N\}_{k=1}^n$. So let $\phi(a) = \sum_{j=1}^n c_j f_j(a)N(a)$; then

$$(K\phi)(a) = \sum_{k=1}^n d_k f_k(a)N(a),$$

with $d_k = \sum_{j=1}^{n} c_j \int_0^\infty \psi_k(\alpha) f_j(\alpha) N(\alpha) \, d\alpha$. So, in other words, the vector c is transformed into the vector $d = Mc$ where M is the matrix with elements

$$m_{ij} = \int_0^\infty \psi_i(\alpha) f_j(\alpha) N(\alpha) \, d\alpha.$$

Necessarily then, R_0 is the dominant eigenvalue of M.

Exercise 9.6 The 'short-disease approximation' amounts to $\psi(\alpha) = H(\alpha) g(\alpha)$ and $\psi_k(\alpha) = H(\alpha) g_k(\alpha)$. Otherwise, the expressions for R_0 in Exercise 9.5-ii and for the matrix element m_{ij} in Exercise 9.5-iii remain identical.

Exercise 9.7 i)
$$\mathcal{F}_d(a) = \begin{cases} 1, & a \le M; \\ 0, & a > M. \end{cases}$$

ii) The uniform mixing implies that $c(a, \alpha) = c_m$, a constant.

iii) This third assumption allows us to compute $h(\tau, \alpha)$. In fact, h is independent of α and given explicitly by
$$h(\tau, \alpha) = \beta e^{-\gamma \tau},$$

where β is the constant infectivity during the infectious period. Applying the formula (9.2) we find

$$R_0 = \int_0^\infty \int_0^\infty \beta e^{-\gamma \tau} c_m \frac{\mathcal{F}_d(\alpha + \tau)}{\mathcal{F}_d(\alpha)} \, d\tau \, N(\alpha) \, d\alpha,$$

which, upon substitution of the formula for $\mathcal{F}_d$ and of $N(a) = v$ for $0 \le a \le M$, while $N(a) = 0$ for $a > M$, simplifies to

$$\begin{aligned} R_0 &= \int_0^M \int_0^{M-\alpha} \beta v e^{-\gamma \tau} c_m \, d\tau \, d\alpha = \beta c_m v \int_0^M \frac{1 - e^{-\gamma(M-\alpha)}}{\gamma} \, d\alpha \\ &= \frac{\beta c_m v}{\gamma} \left(M - \frac{1 - e^{-\gamma M}}{\gamma} \right). \end{aligned}$$

Exercise 9.8 The difficulty is: what does 'same age' mean? Born in the same year, month, week, day, hour, minute, second, ... ? In the continuous-time formulation, we idealize and pretend that we can even go beyond milliseconds. But clearly such a distinction is meaningless when we aim to describe social behavior. So, biologically speaking, this does not make any sense. This answer provides a good motivation for the next section.

Exercise 9.9 i) We have assumed that the feed is randomly distributed over the herd, and so i) follows directly from the demographic steady state (imposed by the farmer) by taking the age-dependent susceptibility $\beta(a)$ into account. Note that this factor describes not only the actual susceptibility, but also the age-dependent exposure to infection (i.e., the share of potentially infected feed in an animal's diet).

ii) The important parameters are age-dependent, which would call for an infinite number of states-at-infection. The fact that the individuals infected by feed have an age distribution (at 'epidemiological birth') that is independent of the age of the infector, allows us to consider them as having one fixed (but stochastic) state-at-infection (cf. Sections 7.4.1 and 7.4.3). So it is possible to work with just two states-at-infection: animals that become infected by feed (type 1; distributed over all ages as described by the density (9.4)), and maternally infected newborns (type 2).

iii) We first look at element k_{21}, i.e., we take an animal that was just infected through feed and want to estimate the number of infected calves it will produce during the remainder of its life. We know that its age at 'birth' is distributed according to (9.4). The probability that the animal will survive (natural mortality) from the moment of infection to infection age τ is $\mathcal{F}_d(a+\tau)/\mathcal{F}_d(a)$. Define the probability that it will still be infectious at infection age τ as $\mathcal{F}_c(\tau)$, and the infectivity at that infection age is $\gamma(\tau)$. Finally, the animal has probability per unit of time $b(a+\tau)$ of producing offspring at infection age τ, and with probability m newborns will be infected. If we integrate over all infection ages, and integrate over all ages at infection, according to (9.4), we obtain

$$k_{21} = \frac{\int_0^\infty \beta(a) \int_0^\infty mb(a+\tau)\mathcal{F}_d(a+\tau)\mathcal{F}_c(\tau)\gamma(\tau)\,d\tau\,da}{\int_0^\infty \beta(a)\mathcal{F}_d(a)\,da}.$$

For element k_{12}, we have to count the expected number of feed infections that are caused by one maternally infected animal. For these animals, true age is equal to infection age. The probability to survive and remain infectious until age τ is $\mathcal{F}_d(\tau)\mathcal{F}_c(\tau)$ and the rate at which animals are culled is $\mu(\tau)+\nu(\tau)$. The animal, once rendered into feed, will come into 'contact' with animals whose age distribution is given by $\overline{S}(a)N$, with susceptibility $\beta(a)$. So, we obtain

$$k_{12} = \int_0^\infty \beta(a)\overline{S}(a)N\,da \int_0^\infty (\mu(\tau)+\nu(\tau))\mathcal{F}_d(\tau)\mathcal{F}_c(\tau)\frac{\gamma(\tau)}{N}\,d\tau.$$

The other two elements of K are derived with completely analogous reasoning. We find

$$k_{22} = \int_0^\infty mb(\tau)\mathcal{F}_d(\tau)\mathcal{F}_c(\tau)\gamma(\tau)\,d\tau$$

for the expected number of maternally infected animals produced by a maternally infected animal. Finally, for k_{11} we start with an individual that has just become infected by feed, its age at that moment is distributed as (9.4). The animal that survives to infection age τ, should be infectious at that infection age, with infectivity $\gamma(\tau)$. The death rate is $\mu(a+\tau)+\nu(\tau)$ at infection age τ. We are then in the situation as for element k_{12} to establish the animals with which it comes into 'contact' once it has been rendered and turned into feed. These contact animals have age distribution $\overline{S}(\alpha)N$ with susceptibility $\beta(\alpha)$. We then get

$$k_{11} = \frac{\int_0^\infty \beta(\alpha)\overline{S}(\alpha)N\,d\alpha \int_0^\infty \beta(a) \int_0^\infty (\mu(a+\tau)+\nu(\tau))\mathcal{F}_d(a+\tau)\mathcal{F}_c(\tau)\frac{\gamma(\tau)}{N}\,d\tau\,da}{\int_0^\infty \beta(a)\mathcal{F}_d(a)da}.$$

Of course, many of the convenient assumptions in this exercise need to be abandoned if one wants to use R_0 to study, for example, the effects of changes in the rendering process on the spread of the agent. We encourage the reader to think about how one could include differences in infectivity between the types and differences in success of the rendering process.

The above expressions can be simplified a little bit by making use of the relation $\overline{S}(a) = \mathcal{F}_d(a)/\int_0^\infty \mathcal{F}_d(\alpha)d\alpha$.

iv) For convenience, we assume that $\beta(a)$ now only describes susceptibility and not diet. In the expressions for k_{11} we add

$$\int_0^\infty w\beta(a)\overline{S}(a)N \int_0^\infty \frac{\mathcal{F}_d(a+\tau)}{\mathcal{F}_d(a)}\mathcal{F}_c(\tau)\frac{\gamma(\tau)}{N}\, d\tau\, da,$$

which can be summarized by substituting $\mu(a)+\nu(a)+w$ in the appropriate place in the expression for k_{11} in iii). This substitution also takes care of the necessary change in k_{12}. The other elements of K remain unchanged.

Exercise 9.10 i) By looking at k_{11}, we find

$$A(\alpha,\tau) = (\mu(\alpha+\tau)+\nu(\tau))\frac{\mathcal{F}_d(\alpha+\tau)}{\mathcal{F}_d(\alpha)}\mathcal{F}_c(\tau)\frac{\gamma(\tau)}{N}.$$

ii) By substituting the ansatz into the integral equation, and canceling the factor $e^{\lambda t}$ on both sides, we find

$$f(a) = \overline{S}(a)\beta(a)\int_0^\infty \int_0^\infty A(\alpha,\tau)f(\alpha)e^{-\lambda\tau}\, d\tau\, d\alpha.$$

Now define an operator M, acting on a function f, by the right-hand side of this equation, and we see that $f(\cdot)$ will be an eigenfunction of M corresponding to eigenvalue 1. Now note that the operator satisfies the criterion of separable mixing, and therefore has a one-dimensional range. From Section 7.4.1, we then know that the eigenfunction corresponding to the non-zero eigenvalue is given by $\beta(\cdot)\overline{S}(\cdot)$. If we substitute this into the relation $f = Mf$ above, the term $\beta(a)\overline{S}(a)$ appears on both sides and hence cancels, and we find the condition (9.5). In this condition only λ is an unknown quantity, and the growth rate r can be characterized as the real solution of this equation.

iii) If we define a (continuous) function $g(\lambda)$ by the left-hand side of (9.5), we note that $g(0) = R_0$ (for feed transmission only, i.e., $R_0 = k_{11}$), that $g' < 0$ and that $\lim_{\lambda\to\infty}g(\lambda) = 0$. Therefore there exists a unique value r such that $g(r) = 1$, and we see that $r > 0 \Leftrightarrow R_0 > 1$ (recall Figure 1.2). Since the value is unique, we can use, for example, Newton iteration to find the unique zero of the function $g(\lambda) - 1$ to calculate r numerically.

Exercise 9.11 Let the number of new cases (i.e., with infection age 0) arising at time t with age a be given by

$$I(a,t,0) = \begin{pmatrix} I_1(a,t,0) \\ I_2(t,0) \end{pmatrix} \tag{17.7}$$

where $I_1(a,t,0)$ describes the new cases arising through feed contamination, and where $I_2(t,0)$ describes new cases arising through vertical transmission (necessarily starting with age 0, which explains why age is suppressed in notation). The real-time evolution, linearized around the infection-free demographic steady state, is generated by the equation

$$I(a,t,0) = \begin{pmatrix} \overline{S}(a)\beta(a) & 0 \\ 0 & 1 \end{pmatrix}\int_0^\infty \int_0^\infty \Xi(\alpha,\tau)I(\alpha,t-\tau,0)d\tau d\alpha$$

where

$$\Xi(\alpha,\tau) = \begin{pmatrix} \theta_{11}(\alpha,\tau) & \theta_{12}(0,\tau) \\ \theta_{21}(\alpha,\tau) & \theta_{22}(0,\tau) \end{pmatrix}$$

and where $\beta(a)\theta_{12}(0,\tau)$ is the expected contribution to infectivity acting through feed on susceptibles of age a of an infected individual that was itself infected vertically (i.e., at age 0), τ time units ago; $\beta(a)\theta_{11}(\alpha,\tau)$ has the same interpretation for an individual that was itself infected through feed at age α; $\theta_{22}(0,\tau)$ and $\theta_{21}(\alpha,\tau)$ are the expected number of infected offspring produced through vertical transmission by vertically infected and feed infected animals, respectively.

Consider an exponential solution

$$I(a,t,0) = \begin{pmatrix} X_1(a) \\ X_2 \end{pmatrix} e^{\lambda t} = X(a)e^{\lambda t} \tag{17.8}$$

to this integral equation. If we substitute this into the integral equation for I we obtain

$$X(a) = \begin{pmatrix} \overline{S}(a)\beta(a) & 0 \\ 0 & 1 \end{pmatrix} \int_0^\infty \int_0^\infty \Xi(\alpha,\tau)X(\alpha)e^{-\lambda\tau}d\tau d\alpha.$$

We can now define a positive operator K_λ acting on a two-vector $\phi(a) = (\phi_1(a), \phi_2)^\top$ as

$$(K_\lambda\phi)(a) = \begin{pmatrix} \overline{S}(a)\beta(a) & 0 \\ 0 & 1 \end{pmatrix} \int_0^\infty \int_0^\infty \Xi(\alpha,\tau)e^{-\lambda\tau}d\tau\phi(\alpha)d\alpha.$$

The exponential growth rate r is then defined as the value λ such that the operator K_λ has dominant eigenvalue 1, to be determined numerically. In the BSE example we have assumed separable mixing in age and effectively

$$(K_\lambda\phi)(a) = \begin{pmatrix} \overline{S}(a)\beta(a) & 0 \\ 0 & 1 \end{pmatrix} c(\phi,\lambda)$$

where $c(\phi,\lambda)$ is a constant vector depending on ϕ and λ.

Exercise 9.12 In the short-disease approximation we have for this special case that

$$\psi_k(\alpha) = H(\alpha)\sum_{l=1}^n c_{kl}\chi_{I_l}(\alpha),$$

and hence

$$\begin{aligned} m_{ij} &= \int_0^\infty H(\alpha)\sum_{l=1}^n c_{il}\chi_{I_l}(\alpha)\chi_{I_j}(\alpha)N(\alpha)\,d\alpha \\ &= c_{ij}\int_{I_j} H(\alpha)N(\alpha)\,d\alpha. \end{aligned}$$

Exercise 9.13 For small Λ we have

$$\Lambda(a)e^{-\int_0^a \Lambda(\alpha)\,d\alpha} \approx \Lambda(a),$$

and hence the right-hand side of (9.7) can be approximated by its linearization (as an operator transforming $\Lambda = \Lambda(a)$ into another function of a)

$$(\tilde{K}\Lambda)(a) = \int_0^\infty \int_0^\infty h(\tau,\alpha)c(a,\alpha+\tau)\Lambda(\alpha)N(\alpha)\frac{\mathcal{F}_d(\alpha+\tau)}{\mathcal{F}_d(\alpha)}d\tau d\alpha.$$

Putting $\phi(a) = \Lambda(a)N(a)$ and $(K\phi)(a) = N(a)(\tilde{K}\Lambda)(a)$, we see that K thus defined is indeed the next-generation operator K with the kernel from Exercise 9.5-i. (Note that the relation between Λ and ϕ makes sense, since Λ is a force of infection and ϕ a number density.) Equality of left- and right-hand sides for the linearization would imply $R_0 = 1$. We refer to (the elaboration of) Exercises 8.27-v and 8.26 for a sketch of the proof that $R_0 > 1$ guarantees that a non-trivial solution $\Lambda > 0$ exists.

Exercise 9.14 In the formula (9.2) we replace $N(a)$ by $N(a)\mathcal{F}_v(a)$, to incorporate that only non-vaccinated individuals, and those for whom vaccination failed, are susceptible.

Exercise 9.15 If $\mathcal{F}_v(a) = 1 - v$ for $a > 0$ then $R_v = (1 - v)R_0$ and so $R_v < 1$ corresponds to $v > 1 - 1/R_0$.

Exercise 9.16 Next let $\mathcal{F}_v(a) = 1$ for $0 < a < a_v$; and $\mathcal{F}_v(a) = 1 - v$ for $a > a_v$, then

$$R_v = R_0 - v \int_{a_v}^{\infty} \psi(\alpha)f(\alpha)N(\alpha) \, d\alpha,$$

where $\psi(\alpha)$ is defined by (9.3), and so $R_v < 1$ corresponds to

$$v > \frac{R_0 - 1}{\int_{a_v}^{\infty} \psi(\alpha)f(\alpha)N(\alpha) \, d\alpha}.$$

Exercise 9.17 i) This is Exercise 13.7-i, the special case $n = 1$, but now with $N(a)$ replaced by $N(a)\mathcal{F}_v(a)$.

ii) The short-disease approximation amounts to replacing $\psi(\alpha)$, given by (9.3), by $H(\alpha)g(\alpha)$.

17.4   ELABORATIONS FOR CHAPTER 10

Exercise 10.1 Define $u(t, x) = w(t, x)e^{\kappa t}$; then $u_t = w_t e^{\kappa t} + \kappa u$, and consequently (10.1) can be rewritten in the form

$$w_t = D\triangle w$$

(the terms κu cancel, and then $e^{\kappa t}$ can be factored out). As you can find in any text book on partial differential equations (PDE; see e.g., Courant and Hilbert 1962), the so-called fundamental solution of this equation is

$$w(t, x) = \frac{1}{4\pi Dt} \exp\left(-\frac{|x|^2}{4Dt}\right).$$

For completeness, however, we verify that w thus defined satisfies the PDE. To do so, we first compute $\partial w(t, x)/\partial x_1$:

$$\frac{\partial}{\partial x_1} \frac{1}{4\pi Dt} \exp\left(-\frac{x_1^2 + x_2^2}{4Dt}\right) = -\frac{2x_1}{4Dt}w(t, x).$$

Hence

$$\frac{\partial^2}{\partial x_1^2} \frac{1}{4\pi Dt} \exp\left(-\frac{x_1^2 + x_2^2}{4Dt}\right) = -\frac{2}{4Dt}w(t, x) + \frac{4x_1^2}{(4Dt)^2}w(t, x).$$

By symmetry, we have

$$\frac{\partial^2}{\partial x_2^2} \frac{1}{4\pi Dt} \exp\left(-\frac{x_1^2 + x_2^2}{4Dt}\right) = -\frac{2}{4Dt} w(t, x) + \frac{4x_2^2}{(4Dt)^2} w(t, x),$$

and consequently

$$D\triangle w = -\frac{1}{t} w + \frac{|x|^2}{4Dt^2} w.$$

Taking the derivative with respect to t we obtain

$$\frac{\partial}{\partial t} \frac{1}{4\pi Dt} \exp\left(-\frac{|x|^2}{4Dt}\right) = -\frac{1}{4\pi Dt^2} \exp\left(-\frac{|x|^2}{4Dt}\right) + \frac{|x|^2}{4Dt^2} \frac{1}{4\pi Dt} \exp\left(-\frac{|x|^2}{4Dt}\right)$$

$$= -\frac{1}{t} w + \frac{|x|^2}{4Dt^2} w,$$

which upon combination with the expression for $D\triangle w$ establishes that w satisfies the PDE.

Exercise 10.2 This is most easily achieved by applying a transformation to polar coordinates:

$$\int_{-\infty}^{\infty} \int_{-\infty}^{\infty} e^{-a(x_1^2 + x_2^2)} dx_1 dx_2 = \int_0^{\infty} \int_0^{2\pi} e^{-ar^2} r \, d\varphi \, dr = 2\pi \frac{-1}{2a} e^{-ar^2} |_0^{\infty} = \frac{\pi}{a}.$$

So when $a = \frac{1}{4Dt}$ this yields $4\pi Dt$. But w has an additional factor $\frac{1}{4\pi Dt}$ and we conclude that $\int_{\mathbb{R}^2} w(t, x) \, dx = 1$. Hence $\int_{\mathbb{R}^2} u(t, x) \, dx = e^{\kappa t}$.

Exercise 10.3 For $t \downarrow 0$ we have $\frac{1}{4\pi Dt} \uparrow \infty$. However, when $|x| \geq \varepsilon$, the factor $\exp(-\frac{|x|^2}{4Dt})$ goes to zero much faster than $\frac{1}{4\pi Dt}$ goes to infinity, and hence $\lim_{t \downarrow 0} u(t, x) = 0$ for $|x| \geq \varepsilon$.

Exercise 10.4 Define $\xi = \xi(t, x) = x \cdot \nu - ct = x_1 \nu_1 + x_2 \nu_2 - ct$; then $\partial \xi / \partial t = -c$ and $\partial \xi / \partial x_1 = \nu_1$, $\partial \xi / \partial x_2 = \nu_2$. Hence, by the chain rule,

$$\frac{\partial}{\partial t} w(\xi) = -cw'(\xi), \quad \frac{\partial}{\partial x_1} w(\xi) = \nu_1 w'(\xi), \quad \frac{\partial^2}{\partial x_1^2} w(\xi) = \nu_1^2 w''(\xi),$$

and by symmetry

$$\frac{\partial^2}{\partial x_2^2} w(\xi) = \nu_2^2 w''(\xi).$$

Inserting (10.3) into (10.1) and using these identities, we find $-cw' = Dw'' + kw$ (since, ν being a unit vector, $\nu_1^2 + \nu_2^2 = 1$), which is (10.4). Equation (10.4) is a second-order, linear, homogeneous differential equation. Standard theory (see e.g., Hale 1969) tells us that we should look for two (linearly independent) solutions. Substitution of $w(\xi) = e^{\lambda \xi}$ into (10.4) yields (10.5), as the factor $e^{\lambda \xi}$ cancels. Hence (10.6) holds.

Exercise 10.5 For uniform (i.e., x-independent) solutions, (10.1) predicts exponential growth at rate k. If we consider a traveling wave with minimal velocity, however, and measure the growth rate at an arbitrary position, we find that it is twice as large. The mechanism is spillover: population density is substantially larger to the left than it is to the right, and so diffusion contributes to net growth rate at any position.

Exercise 10.6 For w we find the equation

$$Dw'' + (c - \theta)w' + kw = 0,$$

and so we should have $c - \theta \geq c_0$, which amounts to $c \geq c_0 + \theta$. If we replace σ by $-\sigma$, that is, if we look for solutions of the form

$$u(t, x) = w(-\sigma \cdot x - ct),$$

we find for w the equation

$$Dw'' + (c + \theta)w' + kw = 0,$$

and so we should have that $c + \theta \geq c_0$, which amounts to $c \geq c_0 - \theta$. If $\theta > c_0$, the minimal wave speed in the direction $-\sigma$ is *negative*! This has to be interpreted as lack of spread in this direction. In other words, the growth of the species is confined to a region that is blown off towards infinity in the σ-direction.

Exercise 10.7 We first perform the transformation $\eta = x - \xi$ of the integration variable in (10.8):

$$u(t, x) = \int_0^\infty \int_{\mathbb{R}^2} B(\tau, |\eta|) \, u(t - \tau, x - \eta) \, d\eta \, d\tau$$

(here we have used that $(-1)^2 = 1$). Now we substitute (10.9) and obtain

$$w(x \cdot \nu - ct) = \int_0^\infty \int_{\mathbb{R}^2} B(\tau, |\eta|) \, w(x \cdot \nu - \eta \cdot \nu - ct + c\tau) \, d\eta \, d\tau.$$

Next, call $x \cdot \nu - ct = \theta$ and write

$$w(\theta) = \int_0^\infty \int_{\mathbb{R}^2} B(\tau, |\eta|) \, w(\theta - \eta \cdot \nu + c\tau) d\eta d\tau.$$

The final step consists of introducing a tailor-made coordinate system. We supplement the unit vector ν by the orthogonal unit vector $\nu^\perp$, which is chosen such that the determinant of the matrix $(\nu \nu^\perp)$ equals one. Let α and σ be coordinates relative to the basis defined by ν and $\nu^\perp$, i.e., let

$$\eta = \alpha\nu + \sigma\nu^\perp;$$

then

$$w(\theta) = \int_0^\infty \int_{-\infty}^\infty \int_{-\infty}^\infty B(\tau, \sqrt{\alpha^2 + \sigma^2}) \, w(\theta - \alpha + c\tau) \, d\alpha \, d\sigma \, d\tau.$$

With the change of variable $\zeta = \alpha - c\tau$, we obtain from this

$$w(\theta) = \int_0^\infty \int_{-\infty}^\infty \int_{-\infty}^\infty B(\tau, \sqrt{(\zeta + c\tau)^2 + \sigma^2}) \, w(\theta - \zeta) \, d\zeta \, d\sigma \, d\tau,$$

which leads to

$$w(\theta) = \int_{-\infty}^\infty V_c(\zeta) \, w(\theta - \zeta) \, d\zeta$$

after interchanging the order of integration.

Exercise 10.8 All you have to do is to cancel the common factor $e^{\lambda\theta}$ on both sides of the equality.

Exercise 10.9 Substituting (10.11) into (10.13), we find

$$L_c(\lambda) = \int_{-\infty}^{\infty} e^{-\lambda\zeta} \int_0^{\infty} \int_{-\infty}^{\infty} B(\tau, \sqrt{(\zeta + c\tau)^2 + \sigma^2}) \, d\sigma \, d\tau \, d\zeta,$$

which, upon reversing the substitution $\alpha = \zeta + c\tau$ made earlier, can be written as

$$L_c(\lambda) = \int_0^{\infty} \int_{-\infty}^{\infty} \int_{-\infty}^{\infty} e^{-\lambda\alpha} e^{\lambda c\tau} B(\tau, \sqrt{\alpha^2 + \sigma^2}) \, d\sigma \, d\alpha \, d\tau,$$

which, essentially, is (10.14). By putting $\lambda = 0$ and reversing the coordinate transformation $\eta = \alpha\nu + \sigma\nu^{\perp}$, we arrive at the first identity of (10.15), while the second is a direct consequence of the interpretation of B and R_0. If we differentiate with respect to λ, we get two terms, one involving an additional factor $-\alpha$ and the other involving an additional factor $c\tau$. The first of these terms vanishes by symmetry considerations (whatever the function f, $\int_{-\infty}^{\infty} \alpha f(\alpha^2) \, d\alpha = 0$, since $\int_{-\infty}^0 \alpha f(\alpha^2) \, d\alpha = \int_{\infty}^0 -\sigma f(\sigma^2) \, d(-\sigma) = -\int_0^{\infty} \sigma f(\sigma^2) \, d\sigma$). These observations, together with those leading to (10.15), give the equality in (10.16), and the inequality (10.17) is then a direct consequence of the non-negativity of B and the (implicitly!) assumed non-negativity of c.

If we differentiate twice with respect to λ, we obtain three terms involving additional factors α^2, $-2\alpha c\tau$ and $c^2\tau^2$, respectively. For reasons of symmetry, the middle of these factors vanishes, and we arrive at the conclusion that this second derivative is positive.

The assertion (10.18) is a straightforward consequence of the fact that all the c dependence is concentrated in the factor $\exp(\lambda c\tau)$.

For $c = 0$, the convex function $\lambda \mapsto L_c(\lambda)$ achieves its minimum at $\lambda = 0$ and, when $R_0 > 1$, therefore $L_c(\lambda) > 1$ for all λ. By the property (10.19) the set $\{c :$ 'there exists $\lambda < 0$ such that $L_c(\lambda) < 1$'$\}$ is non-empty and, moreover, contains $[\bar{c}, \infty)$ whenever it contains $\bar{c}$. Hence this set is either the whole line or a half-line. Since zero does not belong to the set, it must be a half-line. In line with earlier notation, we call the boundary point c_0.

Exercise 10.10 L_c is a convex function of λ that, for $c = c_0$, 'touches' the level one, but does not dip below this level. Equation (10.12) reads $L_c(\lambda) = 1$ and equation (10.18) says we should have a minimum when varying λ. Together therefore, the two equations state that the minimum should be one. To make more precise assertions, we have to distinguish the case where the minimum is actually attained for a finite value of λ from the case where L_c has, as a function of λ, an infimum at minus infinity. This involves the behavior of $L_c(\lambda)$ for $\lambda \to -\infty$. This behavior is, in turn, determined by the competition between the factor $\exp(-\lambda\alpha)$, tending to $+\infty$ for $\alpha > 0$, and the factor $\exp(\lambda c\tau)$, tending to zero. Hence it is determined by the support of the function B (i.e., the interval(s) where $B > 0$). We think that it is helpful to be aware of the possibility that c_0 is characterized by $\lim_{\lambda \to -\infty} L_{c_0}(\lambda) = 1$, but that this is somewhat exceptional and that it makes good pragmatic sense to try and determine c_0 and λ_0 from $L_c(\lambda) = 1$ and (10.18).

Exercise 10.11 Ignoring the reduction of the density of susceptibles, the incidence is the product of the force of infection A and the host population density S_0. So, with the right interpretation of the word 'offspring' in the definition of B, we arrive at (10.21). In order for B thus defined to only depend on $|x - \xi|$, it is necessary that, among other things, S_0 is constant as a function of x.

17.5   ELABORATIONS FOR CHAPTER 11

Exercise 11.1 An age representation is not applicable to macroparasitic infections since infected individuals can receive additional doses of the infectious agent after the start-up dose. In the microparasite case the fast reproduction after the start-up dose will cause further incoming doses of the same agent to go 'unnoticed' in the large amount already built up inside the host. Consequently, we can neglect this influence. In the macroparasite case the additional doses make the crucial difference between various infected individuals, and both the number and size of the doses received and the timing of these re-infections determines the impact of the parasite on the host and the contribution of the host to the spread of infection in the population.

Exercise 11.2 i) Invasion in the 'virgin' situation is still characterized as a 'best case' (from the point of view of the parasite). In other words, we are in the situation where the transmission process can proceed optimally. For microparasites the key point is that no infectious material enters hosts that are already infected. For macroparasites one should interpret 'adverse conditions' as anything that hinders development of parasites. This means, for example, that parasites do not compete with each other for resources within the host, or interact with the immune system of the host, but can each optimally produce eggs to be shed into the environment. In other words, there are no density-dependent feedback effects acting on the parasite in any of the stages of its life cycle, i.e., there is no nonlinear interaction between parasites, or between parasites and the environment they inhabit (including hosts).

ii) In the invasion phase there are no density-dependent constraints, and parasite densities in all stages of the life cycle will be very low. Also there will be few infected hosts, and the chances of infecting the same host more than once will be rather small in the invasion phase. This can be a problem if the parasite species has obligatory sexual reproduction in the host, since in that case at least one male and one female parasite need to be present within the same individual at the same time. In the very beginning this is obviously not the case, which implies that the epidemic can never take off. In practice, parasites are both distributed (as eggs) and picked up (as larvae) in clumps, thus increasing the probability that in new infections both sexes are present.

iii) The steady states are given by the intersections of the isocline $I = \mu N m / A$ with the isocline $I = C(m)N/(\delta + C(m))$, i.e., by the intersections of a straight line with a sigmoidal curve (by assumption) in the (I, m) plane that both pass through the origin. For small A, the straight line increases too steeply to intersect the sigmoid isocline for positive values of m. When we increase A, the straight line turns clockwise. It will first touch the sigmoid isocline (for some critical value of A) and then intersect it in exactly two positions, if A is increased further. So, taking the zero steady state into account, the number of steady states is one for A small and three for A above the critical level.

R_0 equals zero (for all values of A), since $C'(0) = 0$, and so in the linearization around the zero steady state no new infected snails are produced.

The critical value of A is the value for which the isoclines touch. It corresponds to a saddle-node bifurcation (see e.g., the book by Kuznetsov 1998) of endemic steady states. The lower (in terms of I) steady state is a saddle point, and its stable manifold separates the domains of attraction of the infection-free steady state and the stable endemic steady state.

Exercise 11.3 $\triangle t$ should be small. We assume that the probability is, to a good approximation, equal to a probability per unit time (prescribed by the modeling assumptions) multiplied by the length of the time interval. Does the hint make sense? The assumption is really that these 'rates' are waiting-time-independent.

Exercise 11.4 The key events are death and 'birth' of a parasite within the host (where 'birth' is interpreted as 'entering the host'). Death of a parasite has probability $\mu \triangle t + o(\triangle t)$. Hence the probability that one of the parasites dies is $i\mu\triangle t + o(\triangle t)$. The parasite load increases by one with probability $\beta\triangle t + o(\triangle t)$.

Exercise 11.5 The quantity $p_i(t)$ decreases with rate $\beta p_i(t)$ due to a transition $i \to i+1$ corresponding to a larva entering the host and turning into an adult parasite. Likewise, $p_i(t)$ increases with rate $\beta p_{i-1}(t)$ due to the same event happening to a host carrying up to that moment $i-1$ parasites.

The quantity $p_i(t)$ decreases with rate $i\mu p_i(t)$ because of one of the i inhabitant parasites dying. Likewise, $p_i(t)$ increases with rate $(i+1)\mu p_{i+1}(t)$ due to the same event happening within a host carrying up to that moment $i+1$ parasites.

Combining the above statements, one arrives at (11.1), if $i \geq 1$. In the case of $p_0(t)$, i.e., the hosts carrying no parasites, no parasite can die and there are no hosts with one less parasite. Thus (11.2) is derived from (11.1) by setting $i = 0$ and adopting the convention that $p_{-1}(t) = 0$ for all t.

Exercise 11.6 Each adult parasite, by assumption, produces larvae with rate λ. The total number of parasites in the population is given by $\sum_{i=1}^{\infty} i p_i(t) =: P(t)$, and therefore the larval population increases with rate $\lambda P(t)$. Larvae die in the environment with per capita rate ν, and larvae have contact with hosts according to the law of mass action, i.e., with a rate proportional to $\hat{N}L(t)$. Since no other things are assumed to happen to larvae, we obtain equation (11.3) to describe the changes in the larval population. The term $-\theta\hat{N}L(t)$ should correspond to $\sum_{i=0}^{\infty} \beta p_i(t) = \beta\hat{N}$ so we should have $\beta = \theta\hat{N}$.

Exercise 11.7 The number $p_i(t)$ decreases with rate $(d + i\kappa)p_i(t)$ due to death of hosts. The number $p_i(t)$ increases with rate $b\sum_{i=0}^{\infty} p_i(t)$ from reproduction of hosts. Otherwise the equations remain unchanged.

Exercise 11.8 If you throw a die, the probability that the outcome is 1 is $\frac{1}{6}$. The law-of-large numbers now states that throwing a die many times in succession will indeed give the outcome 1 in about one-sixth of the attempts, with 'about' becoming more and more 'precise' as the number of throws increases. Imagine now a large population from which we draw individuals at random and check if their parasite load happens to be i. If there are sufficiently many individuals, the probability to find an individual with i parasites will be 'about' the same as the fraction of the population with i parasites.

Exercise 11.9 If we differentiate the defining relation of $g(t, z)$ with respect to t, we find

$$
\begin{aligned}
\frac{\partial g}{\partial t}(t, z) &= \sum_{i=0}^{\infty} \frac{dp_i}{dt}(t)z^i \\
&= -\beta g(t, z) - \mu \sum_{i=0}^{\infty} ip_i(t)z^i + \mu \sum_{i=0}^{\infty}(i+1)p_{i+1}(t)z^i + \beta \sum_{i=0}^{\infty} p_{i-1}(t)z^i \\
&= \beta(z-1)g(t, z) - \mu(z-1)\frac{\partial g}{\partial z}(t, z).
\end{aligned}
$$

Exercise 11.10 If we differentiate (11.7) with respect to z, we obtain

$$
\frac{\partial^2 g}{\partial t \partial z} = \beta(z-1)\frac{\partial g}{\partial z} + \beta g - \mu(z-1)\frac{\partial^2 g}{\partial z^2} - \mu\frac{\partial g}{\partial z}.
$$

Now note that $\partial g(t, z)/\partial z|_{z=1} = P(t)$ and $g(t, 1) = N(t)$. If we evaluate the above PDE in $z = 1$ and substitute these relations, we find the desired result.

Exercise 11.11 If we combine (11.4) and (11.5), we obtain

$$
\begin{aligned}
\frac{dN}{dt} &= \sum_{i=0}^{\infty} \frac{dp_i}{dt} \\
&= b\sum_{i=0}^{\infty} p_i(t) - (\beta + d)\sum_{i=0}^{\infty} p_i(t) - (\mu + \kappa)\sum_{i=1}^{\infty} ip_i(t) \\
&\quad + \mu\sum_{i=0}^{\infty}(i+1)p_{i+1}(t) + \beta\sum_{i=1}^{\infty} p_{i-1}(t) \\
&= (b-d)N - \beta\sum_{i=0}^{\infty} p_i(t) - (\mu + \kappa)\sum_{i=1}^{\infty} ip_i(t) \\
&\quad + \mu\sum_{j=1}^{\infty} jp_j(t) + \beta\sum_{j=0}^{\infty} p_j(t) \\
&= (b-d)N - \kappa\sum_{i=1}^{\infty} ip_i(t) = (b-d)N - \kappa P.
\end{aligned}
$$

In the same way, we can derive a differential equation for P:

$$
\begin{aligned}
\frac{dP}{dt} &= \sum_{i=0}^{\infty} i \frac{dp_i}{dt} = \sum_{i=1}^{\infty} i \frac{dp_i}{dt} \\
&= -(\beta + d) \sum_{i=1}^{\infty} ip_i - (\mu + \kappa) \sum_{i=1}^{\infty} i^2 p_i + \mu \sum_{i=1}^{\infty} i(i+1)p_{i+1} + \beta \sum_{i=1}^{\infty} ip_{i-1} \\
&= -(\beta + d) \sum_{i=1}^{\infty} ip_i - (\mu + \kappa) \sum_{i=1}^{\infty} i^2 p_i + \mu \sum_{j=2}^{\infty} (j-1)jp_j + \beta \sum_{k=0}^{\infty} (k+1)p_k \\
&= -dP - \beta P + \beta P + \beta N - \mu p_1 - \mu \sum_{i=2}^{\infty} i^2 p_i + \mu \sum_{j=2}^{\infty} j^2 p_j \\
&\quad -\mu \sum_{j=2}^{\infty} jp_j - \kappa \sum_{i=1}^{\infty} i^2 p_i \\
&= \beta N - dP - \mu P - \kappa \sum_{i=1}^{\infty} i^2 p_i.
\end{aligned}
$$

Exercise 11.12 For details on the method of characteristics for solving quasilinear first-order differential equations see a textbook on PDEs (e.g., Courant and Hilbert 1962).

In the present case we have the equation (11.7) and find that the characteristic equations are

$$
\begin{aligned}
\frac{dt}{ds} &= 1, \\
\frac{dz}{ds} &= \mu(z-1), \\
\frac{dg}{ds} &= \beta(z-1)g.
\end{aligned}
$$

Assume $t(0) = 0$ and $z(0) = z_0$; then the first equation gives $t = s$, and the second equation can be solved by the variation-of-constants formula. This formula gives the solution of the ODE $x'(t) = ax(t) + b(t)$ as $x(t) = e^{at}x_0 + \int_0^t e^{a(t-s)}b(s)\,ds$. We therefore find

$$
z(t) = e^{\mu t}(z_0 - 1) + 1.
$$

Define an operator

$$
T(t, x) := e^{\mu t}(x - 1) + 1;
$$

then we can interpret the solution $z(t)$ for a given initial value z_0 as the result of the operator $T(t, \cdot)$ acting on z_0. By the uniqueness of solutions, $T(\cdot, x)$ has the semigroup property (see Section 6.1): $T(t_2, T(t_1, x)) = T(t_2 + t_1, x)$. Calculating back in time from $z(t)$, we then find $T(-t, z) = T(-t, T(t, z_0)) = T(0, z_0) = z_0$. In this way, we can express z_0 in terms of z (rather than the other way around).

Finally, we can now 'solve' the third equation in terms of z:

$$
\begin{aligned}
g(t, z) &= g(0, z_0)e^{\beta \int_0^t (z(a)-1)\, da} \\
&= g_0(z_0) \exp\left(\beta \int_0^t T(a, z_0)\, da - \beta t \right) \\
&= g_0(T(-t, z)) \exp\left(\beta \int_0^t T(a, T(-t, z))\, da - \beta t \right) \\
&= g_0(T(-t, z)) \exp\left(\beta \int_0^t T(a - t, z)\, da - \beta t \right).
\end{aligned}
$$

Therefore we find that the solution to (11.7) is given by

$$
g(t, z) = g_0(T(-t, z)) \exp\left(\frac{\beta}{\mu}(z - 1)(1 - e^{-\mu t}) \right),
$$

which gives the desired result if we re-substitute the expression for $T(-t, z)$.

For $t \to \infty$ we have convergence of g towards

$$
g(\infty, \tau) = g_0(1) \exp\left(\frac{\beta}{\mu}(z - 1) \right),
$$

which is, modulo normalization, the generating function of the Poisson distribution with parameter β/μ (see Exercise 3.6). This is not at all surprising: we consider hosts with an eternal life and immigrating parasites at rate β that die at per capita rate μ.

Exercise 11.13 i) We have

$$
g'(t, 1) := \frac{\partial}{\partial z} g(t, z)|_{z=1} = \frac{\partial}{\partial z} \sum_{i=0}^{\infty} r_i(t) z^i |_{z=1} \tag{17.9}
$$

$$
= \sum_{i=0}^{\infty} i r_i(t) z^{i-1}|_{z=1} = \sum_{i=0}^{\infty} i r_i(t) \tag{17.10}
$$

$$
= \frac{1}{N(t)} \sum_{i=0}^{\infty} i p_i(t) = \frac{P(t)}{N(t)}. \tag{17.11}
$$

ii) On the one hand, we have $\sum_{i=0}^{\infty} i^2 p_i(t) = N(t) \sum_{i=0}^{\infty} i^2 r_i(t)$. On the other hand, we have $N(t)(g''(t, 1) + g'(t, 1)) = N(t)(\sum_{i=0}^{\infty} i(i - 1)r_i(t) + \sum_{i=0}^{\infty} i r_i(t)) = N(t) \sum_{i=0}^{\infty} i^2 r_i(t)$. The variance is calculated as $E(i^2) - E(i)^2$, where E denotes 'expectation.' Therefore, the variance is given by $\sum_{i=0}^{\infty} i^2 r_i(t) - (\sum_{i=0}^{\infty} i r_i(t))^2 = g''(t, 1) + g'(t, 1) - (g'(t, 1))^2$. Finally, we find that $\sum_{i=0}^{\infty} i^2 r_i(t) = g''(t, 1) + g'(t, 1) - (g'(t, 1))^2 + (g'(t, 1))^2 = $ variance $+$ mean$^2 = $ mean(mean $+$ variance/mean).

iii) From ii), we know that $\sum_{i=0}^{\infty} i^2 p_i = N(g''(1) + g'(1))$ (where we have dropped t since now g does not depend on time), so it suffices to calculate $g''(1)$ and $g'(1)$. First calculate the derivatives

$$
\begin{aligned}
g'(z) &= -k\left(1 + \frac{m}{k}(1 - z)\right)^{-(k+1)}\left(-\frac{m}{k}\right) = m\left(1 + \frac{m}{k}(1 - z)\right)^{-(k+1)} \\
g''(z) &= -(k+1)m\left(1 + \frac{m}{k}(1 - z)\right)^{-(k+2)}\left(-\frac{m}{k}\right) \\
&= \frac{k+1}{k}m^2\left(1 + \frac{m}{k}(1 - z)\right)^{-(k+2)},
\end{aligned}
$$

and then substitute $z = 1$ to find $g'(1) = m$ and $g''(1) = \frac{k+1}{k}m^2$. From $m = P/N$ and the relation derived in ii), we now immediately find

$$\sum_{i=0}^{\infty} i^2 p_i = N\left(\left(\frac{P}{N}\right)^2 \frac{k+1}{k} + \frac{P}{N}\right).$$

Exercise 11.14 We can arrive at the expression in two ways. A heuristic argument would go as follows: in a steady state the number of larvae entering hosts per unit time, βN, should equal the number of larvae produced per unit time, λP, multiplied by the fraction $\theta N/(\nu + \theta N)$ of these that eventually make their way into a host. A more formal argument calculates the quasi-steady state by putting $dL/dt = 0$ in (11.3) and using the definition $\beta N = \theta NL$.

Exercise 11.15 i) One adult parasite produces λ larvae per unit time and will, on average and in a density-independent situation, survive for $1/(\mu+d+\kappa)$ time units. Each larva that is produced has a probability $\theta N/(\nu + \theta N)$ of becoming an adult. This gives the expression for R_0. When the density of hosts N becomes large (for example when the host population grows exponentially, i.e., when $b > d$), larvae will increasingly more likely be removed from the environment by ingestion by a host (θN) than by death (ν). Therefore for large N most larvae are expected to make it to adulthood.

ii) Yes, since for R_0 it does not matter what the duration of the larval stage is. The only important thing is the fraction of larvae that eventually reach the adult stage.

Exercise 11.16 We first restate the system of ODEs

$$\frac{dN}{dt} = (b-d)N - \kappa P, \tag{17.12}$$

$$\frac{dP}{dt} = P\left(\frac{\lambda N}{c+N} - (\mu + d + \kappa) - \kappa\frac{k+1}{k}\frac{P}{N}\right),$$

with $b - d > 0$, by assumption, and where we have put $c := \nu/\theta$.

i) From Exercise 11.15-i, we know that $R_0 = \lambda/(\mu + d + \kappa)$ in this situation and that therefore the parasite cannot be sustained in the host population if $\lambda < \mu + d + \kappa$. And indeed, in that case, dP/dt equals P times a negative quantity, so the parasite population will go extinct, $P \to 0$, and (11.8) shows that N will eventually increase exponentially with rate $b - d$.

ii) First note that, if we assume as an ansatz that N eventually grows exponentially, we can put asymptotically $N/(c + N) = 1$. We make this assumption and check afterwards that N indeed grows exponentially for large t. From i), we see that if $\lambda > \mu + d + \kappa$, the invasion will be successful and P will start to grow away from 'zero.' Initially, N will grow exponentially, but this is not guaranteed if P increases fast enough. To see whether that happens, we derive a differential equation for the mean $m(t) := P(t)/N(t)$. We have

$$\frac{dm}{dt} = \frac{\lambda PN - (\mu + d + \kappa)PN - \kappa\frac{k+1}{k}P^2 - (b-d)PN + \kappa P^2}{N^2}$$

$$= (\lambda - (b + \mu + \kappa))m - \frac{\kappa}{k}m^2.$$

This differential equation has two steady states:

$$\overline{m} = 0,$$
$$\overline{m} = \frac{k}{\kappa}(\lambda - (b + \mu + \kappa)).$$

One sees immediately that for $\lambda < b + \mu + \kappa$ there really is only one biologically feasible steady state, $\overline{m} = 0$, and that this steady state is stable, since it is clear from the differential equation that small perturbations from $\overline{m} = 0$ decay to zero. Therefore for $\mu + d + \kappa < \lambda < b + \mu + \kappa$ the parasite population will grow, but not fast enough to keep up with the growing host population N, and the parasites will be diluted as expressed by $P/N \to 0$. The host will asymptotically grow exponentially with rate $b - d$, and therefore the assumption $N/(c + N) = 1$ was justified.

iii) For $\lambda > b + \mu + \kappa$ the differential equation for m has two steady states, and the picture of dm/dt is qualitatively like that of Figure 10.3. For $\lambda > b + \mu + \kappa$, therefore, the steady state $\overline{m} = k(\lambda - (b + \mu + \kappa))/\kappa$ is stable and we conclude that now P does grow fast enough to keep up with a growing N. Since P will behave asymptotically as $\overline{m}N$ we have that, for large t,

$$\frac{dN}{dt} \sim (b - d - \kappa\overline{m})N.$$

We see that again the assumption that $N/(c + N) = 1$, asymptotically, is justified as long as

$$b - d - \kappa\overline{m} > 0.$$

If we substitute the expression for $\overline{m}$ into this inequality, we can rewrite it as an inequality for λ, and we find that N will grow exponentially as long as

$$-\lambda k + bk + \mu k + \kappa k + b - d > 0,$$
$$-\lambda + b + \mu + \kappa + \frac{b - d}{k} > 0,$$
$$-\lambda + \mu + d + \kappa + b - d + \frac{b - d}{k} > 0,$$
$$\mu + d + \kappa + (b - d)\frac{k + 1}{k} > \lambda.$$

For the growth rate of P we have asymptotically

$$\frac{dP}{dt} \sim (\lambda - (\mu + d + \kappa) - \kappa\frac{k + 1}{k}\overline{m})P$$
$$= (\lambda - (\mu + d + \kappa) - \kappa\overline{m} - \frac{\kappa}{k}\overline{m})P$$
$$= (b - d - \kappa\overline{m})P,$$

where we have substituted the expression for $\overline{m}$ in the final term. We conclude that P and N grow with the same exponential rate.

iv) Putting $dN/dt = 0$ and $dP/dt = 0$, we find

$$(b - d)\overline{N} = \kappa\overline{P},$$
$$\frac{\lambda\overline{N}}{c + \overline{N}} = (\mu + d + \kappa) - \kappa\frac{k + 1}{k}\frac{\overline{P}}{\overline{N}}$$

for the steady-state values $\overline{N}$ and $\overline{P}$, with $\overline{P} > 0$. After substitution of $\kappa\overline{P}$ from the first equation into the second one and rearranging the terms, we find

$$\overline{N} = \frac{c\lambda_2}{\lambda - \lambda_2}, \quad \overline{P} = \frac{b - d}{\kappa}\overline{N},$$

where

$$\lambda_2 := \mu + d + \kappa - (b - d)\frac{k + 1}{k},$$

and we conclude that the endemic steady state is biologically feasible only if $\lambda > \lambda_2$, i.e., in the region where both P and N cease to grow exponentially for large t. We will skip the stability analysis of this endemic steady state.

Exercise 11.17 The preferred choice is to multiply the contributions. The reason is that the probability per unit of time (of producing offspring) is reduced and that the assumption of independent influence of each parasite would lead to inconsistency when we add contributions. Indeed, if we would add the reductions, there is a possibility that the birth rate could become negative, which would make the model badly posed from a biological point of view.

Exercise 11.18 i) Stage 2 refers to the adult stage where reproduction takes place, stage 1 individuals are then the larvae (see the definition of the life cycle preceding the exercise). The rate d_1 is then the loss rate of larvae: $d_1 = \theta N + \nu$ (loss of larvae through ingestion by hosts and loss by death); the rate of loss of adults is $d_2 = \mu + d + \kappa$ (loss of adult parasites by death and loss due to death of the host, both natural and parasite-induced). The rate m_1 is the maturation rate of larvae, i.e., the rate at which larvae become adults. In the present set-up this is equal to the rate of becoming ingested by a host: $m_1 = \theta N$; the rate m_2 is the rate of production of larvae: $m_2 = \lambda$. We see from Exercise 11.15 that

$$R_0 = \frac{m_1 m_2}{d_1 d_2}.$$

ii) First look at i) again. We see that the quotient m_1/d_1 is to be interpreted as the fraction of larvae that reach the adult stage (the rate of ingestion m_1 multiplied by the average life span $1/d_1$). The quotient m_2/d_2 is to be interpreted as the number of larvae produced per adult (the number of larvae produced per adult per unit time multiplied by the average life span of an adult). If there are n stages in the life cycle, we obtain n of these quotients, each giving either the fraction of the population in the current stage that reaches the next stage or the number of individuals of the next stage produced per individual of the present stage. Summarizing both in one statement, one can say that one parasite of stage i is expected to produce m_i/d_i individuals of stage $i + 1$. The product of these quotients then gives the expected number of individuals of any stage i produced per individual of stage i, i.e., the product equals R_0.

Exercise 11.19 The text of the exercise is so detailed that an elaboration is superfluous (just in case: recall Section 7.2).

Exercise 11.20 We are as a first step interested in characterizing $z(t, k)$, the incidence at time t of infected hosts born with k parasites, as a fraction of the population. The probability that an individual becomes infected with k parasites by an infected individual who currently has i parasites is given by cq_{ki}. This infecting individual was itself born with some number of parasites $j \geq i$, τ units of time ago. Therefore

$$c \sum_{i=1}^{j} x(i,j,\tau) q_{ki} =: A(\tau, k, j)$$

is the probability that an individual who was born with j parasites infects an individual with k parasites, τ units of time later. Taking into account all possible birth states j and infection ages τ, and realizing that the fraction of the population that was born with j parasites and that currently has infection age τ is given by $z(t - \tau, j)$, we arrive at

$$z(t, k) = \int_0^\infty \sum_{j \geq 1} (c \sum_{i=1}^{j} x(i,j,\tau) q_{ki}) z(t - \tau, j) d\tau.$$

The fraction of the population that at time t carries i parasites is found by summing over the fraction of the population that was born with $j \geq i$ parasites at any time τ before t and that has lost $j - i$ parasites in the time interval τ. Again, the fraction of the population that was born with j parasites a time τ ago is given by $z(t - \tau, j)$. In symbols, we find

$$r_i(t) = \int_0^\infty \sum_{j \geq i} x(i,j,\tau) z(t - \tau, j) \, d\tau.$$

If parasites die independently of each other, we have that each parasite has probability $e^{-\mu\tau}$ to still be alive after a time τ, and a probability $1 - e^{-\mu\tau}$ to have died in the meantime. The probability to have i parasites left out of j initial parasites after a time τ is then given by a binomial distribution

$$x(i,j,\tau) = \binom{j}{i} e^{-i\mu\tau} (1 - e^{-\mu\tau})^{j-i},$$

i.e., of the j parasites, i must have survived and $j - i$ must have died before τ, and these i remaining parasites can be chosen from the original j parasites in $\binom{j}{i}$ ways.

Now rewrite $r_i(t)$ as

$$r_i(t) \quad = \quad \int_{-\infty}^{t} \sum_{j \geq i} x(i,j,t-\sigma) z(\sigma, j) \, d\sigma$$

$$= \quad \sum_{j \geq i} \int_{-\infty}^{t} x(i,j,t-\sigma) z(\sigma, j) \, d\sigma,$$

and therefore

$$\frac{dr_i}{dt}(t) \quad = \quad \sum_{j \geq i} x(i,j,0) z(t,j) - \sum_{j \geq i} x(i,j,\infty) z(-\infty, j)$$

$$+ \sum_{j \geq i} \int_{-\infty}^{t} \frac{d}{dt} x(i,j,t-\sigma) z(\sigma, j) \, d\sigma$$

$$= \quad z(t, i) + \sum_{j \geq i} \int_{-\infty}^{t} \frac{d}{dt} x(i,j,t-\sigma) z(\sigma, j) \, d\sigma.$$

We consider the second term first, and obtain (using the chain rule)

$$
\sum_{j \geq i} \int_{-\infty}^{t} \left(-i\mu \binom{j}{i} e^{-i\mu(t-\sigma)} (1 - e^{-\mu(t-\sigma)})^{j-i} \right.
$$

$$
\left. + (j-i)\mu e^{-\mu(t-\sigma)} \binom{j}{i} e^{-i\mu(t-\sigma)} (1 - e^{-\mu(t-\sigma)})^{j-i-1} \right) d\sigma
$$

$$
= -i\mu r_i(t) + (i+1)\mu \int_{-\infty}^{t} \frac{j-i}{i+1} \binom{j}{i} e^{-(i+1)\mu(t-\sigma)} (1 - e^{-\mu(t-\sigma)})^{j-(i+1)} d\sigma
$$

$$
= -i\mu r_i(t) + (i+1)\mu \int_{-\infty}^{t} \binom{j}{i+1} e^{-(i+1)\mu(t-\sigma)} (1 - e^{-\mu(t-\sigma)})^{j-(i+1)} d\sigma
$$

$$
= -i\mu r_i(t) + (i+1)\mu r_{i+1}(t).
$$

From the integral equation for z, for the first term we obtain

$$
z(t,i) = c \int_{0}^{\infty} \sum_{j \geq 1} \left(\sum_{k=1}^{j} x(k,j,\tau) q_{ik} \right) z(t-\tau, j) \, d\tau
$$

$$
= c \int_{0}^{\infty} \sum_{k \geq 1} \sum_{j=k}^{\infty} x(k,j,\tau) z(t-\tau, j) \, d\tau \, q_{ik}
$$

$$
= c \sum_{k \geq 1} r_k(t) q_{ik}.
$$

Putting everything together, we find that

$$
\frac{dr_i}{dt}(t) = -i\mu r_i(t) + (i+1)\mu r_{i+1}(t) + c \sum_{j \geq 1} r_j(t) q_{ij}.
$$

Exercise 11.21 i) Each parasite is assumed to have c (indirect) contacts with susceptibles per unit time and at each contact on average η parasites become established in the contacted host. So, each parasite is expected to produce $c\eta$ new parasites per unit time. The life expectancy of a parasite is $1/\mu$, so that we find

$$
R_0 = \frac{c\eta}{\mu}.
$$

ii) We have from (11.13) that the elements of U are given by

$$
U_{ij} = cq_{ij} - i\mu\delta_{ij} + (i+1)\mu\delta_{i+1,j},
$$

where δ_{ij} is the Kronecker delta. The elements of the transpose $U^{\top}$ are then given by exchanging the indices: $U_{ij}^{\top} = U_{ji}$. We now look at the vector $\phi(i) = i$ and calculate

$$
(U^{\top}\phi)(i) = \sum_{j=0}^{\infty} U_{ij}^{\top} j = \sum_{j=0}^{\infty} j(cq_{ji} - j\mu\delta_{ji} + (j+1)\mu\delta_{j+1,i})
$$

$$
= c \sum_{j=0}^{\infty} jq_{ji} - i^2\mu + (i-1)i\mu = ic\eta - i\mu = i(c\eta - \mu),
$$

and we see that the vector $\phi(i) = i$ is a formal eigenvector of $U^\top$ and therefore $c\eta - \mu$ also belongs to the spectrum of U. We cannot conclude that it is also an *eigenvalue* of U, since we have not specified the space on which U acts (in particular, the behavior at infinity is a delicate point).

iii) Using the $x(i, j, \tau)$ given in Exercise 11.23, we obtain

$$A_{kj} = \int_0^\infty A(\tau, k, j)\, d\tau = c \int_0^\infty \sum_{i=1}^j x(i,j,\tau) q_{ki}\, d\tau$$

$$= c \sum_{i=1}^j q_{ki} \int_0^\infty \binom{j}{i} e^{-i\mu\tau}(1 - e^{-\mu\tau})^{j-i}\, d\tau.$$

Now the integral is the expected amount of time that an infected individual carries i parasites, when parasites die independently of each other with rate $1/\mu$. The expectation is therefore $1/(i\mu)$ (as can also be checked by working out the integral, carrying out repeated partial integrations). Therefore

$$A_{kj} = c \sum_{i=1}^j q_{ki} \frac{1}{i\mu} = \frac{c}{\mu} \sum_{i=1}^j \frac{q_{ki}}{i}$$

and

$$(K\phi)(k) = \sum_{j\geq 1} A_{kj}\phi(j) = \frac{c}{\mu} \sum_{j\geq 1} \sum_{i=1}^j \frac{q_{ki}}{i}\phi(j).$$

iv) The elements of $K^\top$ are given by $(K^\top)_{kj} = A_{jk}$. We check that the vector $\phi(j) = j$ is indeed an eigenvector of $K^\top$:

$$(K^\top\phi)(k) = \sum_{j\geq 1} j A_{jk}$$

$$= \frac{c}{\mu} \sum_{j\geq 1} j \sum_{i=1}^k \frac{q_{ji}}{i}$$

$$= \frac{c}{\mu} \sum_{j\geq 1} j \left(q_{j1} + \frac{1}{2}q_{j2} + \cdots + \frac{1}{k}q_{jk} \right)$$

$$= \frac{c}{\mu} \left(\sum_{j\geq 1} j q_{j1} + \frac{1}{2}\sum_{j\geq 1} j q_{j2} + \cdots + \frac{1}{k}\sum_{j\geq 1} j q_{jk} \right)$$

$$= \frac{c}{\mu} \left(\eta + \frac{1}{2}2\eta + \cdots + \frac{1}{k}k\eta \right) = k\frac{c\eta}{\mu}.$$

We conclude that $c\eta/\mu$ is indeed the dominant eigenvalue, since it corresponds to a strictly positive eigenvector. This shows again that

$$R_0 = \frac{c\eta}{\mu}.$$

17.6   ELABORATIONS FOR CHAPTER 12

Exercise 12.1 For mass action, the contacts per unit time per individual rise linearly with the population density. For low densities this might be a reasonable description of the contact process. For higher densities, however, one can hardly expect that a population twice as dense will lead to twice as many contacts for a given infective, except possibly when transmission is via aerosols. One reason is surely that contacts take time, whereas mass action assumes them to be instantaneous, and therefore only a limited number of contacts can occur per unit of time. We expect the contacts to saturate at high population density. Another effect that certainly causes saturation is satiation, which plays a role in sexual contacts, but also in blood meals taken by mosquitoes.

Exercise 12.2 i) This fraction is $C(N)S$, and the probability that a contact is with an infective equals I/N.

ii) The general idea is that both the fraction of the individuals and the fraction of time for one individual are equal to the probability for an individual to be engaged in a contact. Let $F(N)$ be the average fraction of the time that an individual spends having contact in a population of size N. We show that $F(N) = C(N)$.

Consider the time interval $[0, T]$. Consider Figure 17.6, where the contact patterns for a few individuals are drawn. Here a thin line indicates that the individual is single in that time interval and a thick line indicates that the individual is engaged in a contact during that period.

The idea now is to count contacts in two ways: by integrating 'horizontally' over time and 'vertically' over individuals. Let $\chi_i(t, N)$ be the indicator function of individual i, taking the value 1 for those times t when the individual is involved in a contact, and the value 0 for those times t when the individual is single. First we calculate the fraction of time that this individual i is engaged in contacts:

$$\frac{1}{T} \int_0^T \chi_i(t, N) \, dt.$$

To calculate $F(N)$, we now have to average this over all individuals

$$F(N) = \frac{1}{N} \sum_{i=1}^{N} \frac{1}{T} \int_0^T \chi_i(t, N) \, dt.$$

For $C(N)$ we first count vertically in Figure 17.6 by fixing a time t and calculating first the fraction of the population that is engaged in a contact at that time,

$$\frac{1}{N} \sum_{i=1}^{N} \chi_i(t, N),$$

and proceed to average this fraction over all t:

$$C(N) = \frac{1}{T} \int_0^T \frac{1}{N} \sum_{i=1}^{N} \chi_i(t, N) \, dt.$$

By interchanging summation and integration, we see that $F(N) = C(N)$.

Exercise 12.3 Some reasonable properties are i) $C(N) > 0$, for $N > 0$; ii) $C(\cdot)$ non-decreasing; iii) $C(N)$ linear for small N; iv) $C(N)$ constant for large N. The situation that $C(N)$ is linear in N describes interaction governed by mass action. Examples where C would be approximately constant as a function of N probably include sexual contacts and blood meals taken by mosquitoes.

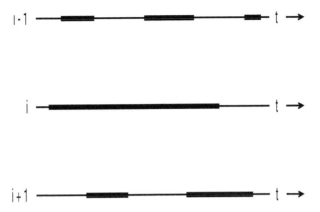

Figure 17.6: Example of time axis for three individuals $i-1, i$ and $i+1$, indicating when these individuals are involved in a contact (thick line segments) and when not (thin line segments).

Exercise 12.4 Denote the search time by T_s. Then $T_s = T - T_h Z$. Also, we can express Z as $Z = aNT_s$ by Holling's assumption, where a is the proportionality constant, interpreted as the effective search rate (also called 'search efficiency'). So, $T - T_h Z = Z/aN$, which we can rewrite as

$$Z = \frac{aNT}{1 + aNT_h}.$$

The functional response, i.e., the number of prey caught per predator per unit of time, is then given by

$$\frac{Z}{T} = \frac{aN}{1 + aNT_h}.$$

Exercise 12.5 The point is that we pretend that N is constant when deriving $\bar{p}_0$. The idea is that N varies slowly, that p_0 adapts quickly to the current value of N, and that therefore we obtain a good approximation by putting $p_0 = \bar{p}_0$, while subsequently studying the slow dynamics of N. The adjectives 'quasi' and 'pseudo' reflect that we cheated and that, actually, N is not constant and $\bar{p}_0$ not a true steady state.

We find from $dp_0/dt = 0 = -aN\bar{p}_0 + e\bar{p}_1$ and $\bar{p}_0 + \bar{p}_1 = 1$ (predators are assumed to be either searching for prey or busy handling prey) that

$$\bar{p}_0 = \frac{e}{e + aN} = \frac{1}{1 + aNT_h}.$$

Now note (you might recall the basic idea of Exercise 12.2) that the number of prey caught per predator per unit of time is given by $aN\bar{p}_0$. We therefore again find that the functional response is given by

$$aN\bar{p}_0 = \frac{aN}{1 + aNT_h}.$$

Exercise 12.6 In the predator-prey formulation, there are two types of individuals that are treated differently: prey and predator (and the predator can be in two states). The prey is considered to be abundant relative to the predator, and, on the short time scale, the predation process consequently does not significantly affect the prey abundance. In the infective-susceptible formulation, there is only one type of individual, which can be in two states: susceptible and infective. The susceptibles therefore do not play the role of the prey, since they are time-limited in the same way as the infectives and they are not much more abundant. The infectives therefore do not encounter susceptibles at their total density, since a fraction of the susceptibles will be engaged in contacts themselves, and only contacts with 'free' susceptibles can be initiated. The consequences are that the Holling argument in Exercises 12.4 and 12.5 does not apply and that the formula (12.2) will need a correction to take into account that the available susceptibles do not have density N.

Exercise 12.7 The following is the basis for the equations given: if one pair dissolves, with probability per unit of time σ, it gives rise to two singles (hence the factor 2); singles meet other singles to form pairs according to mass-action kinetics with reaction rate constant ρ. There are two singles needed to make one pair (hence the factor $1/2$).

Exercise 12.8 We have

$$\frac{dX}{dt} = \frac{dS_1}{dt} + \frac{dI_1}{dt} = -\rho(S_1 + I_1)(S_1 + I_1) + 2\sigma(S_2 + M + I_2)$$
$$= -\rho X^2 + 2\sigma P$$

and

$$\frac{dP}{dt} = \frac{dS_2}{dt} + \frac{dM}{dt} + \frac{dI_2}{dt} = \frac{1}{2}\rho S_1^2 - \sigma S_2 + \rho S_1 I_1 - \sigma M - \beta M$$
$$+ \frac{1}{2}\rho I_1^2 - \sigma I_2 + \beta M$$
$$= \frac{1}{2}\rho(S_1^2 + 2S_1 I_1 + I_1^2) - \sigma(S_2 + M + I_2)$$
$$= \frac{1}{2}\rho X^2 - \sigma P.$$

Exercise 12.9 Setting the right-hand side of the differential equations equal to zero, both times we find the same equation, viz., $\rho X^2 - 2\sigma P = 0$. This reflects the conservation of individuals, and so we have to supplement this equation by the normalization condition $X + 2P = N$. The resulting quadratic equation in X reads (with $\nu := \rho/\sigma$)

$$\nu X^2 + X - N = 0.$$

It has a unique positive solution

$$\overline{X} = \frac{-1 + \sqrt{1 + 4\nu N}}{2\nu}.$$

If we substitute the result into $\overline{P} = \frac{1}{2}(N - \overline{X})$ we find

$$\overline{P} = \frac{1 + 2\nu N - \sqrt{1 + 4\nu N}}{4\nu}.$$

Using $X + 2P = N$, we see that the system (12.4) is fully described by the one-dimensional ODE $dX/dt = -\rho X^2 + \sigma(N - X)$, and a graphical argument (draw a graph of the right-hand side) now shows that all relevant solutions (i.e., those starting with $0 \le X \le N$) converge to $\overline{X}$ for $t \to \infty$. So the steady state is a global attractor.

Exercise 12.10

$$\frac{dS}{dt} = \frac{dS_1}{dt} + \frac{dM}{dt} + 2\frac{dS_2}{dt} = -\rho S_1^2 + 2\sigma S_2 - \rho S_1 I_1 + \sigma M + \rho S_1^2 - 2\sigma S_2$$
$$+ \rho S_1 I_1 - \sigma M - \beta M = -\beta M,$$

$$\frac{dI}{dt} = \frac{dI_1}{dt} + \frac{dM}{dt} + 2\frac{dI_2}{dt} = -\rho S_1 I_1 - \rho I_1^2 + 2\sigma I_2 + \sigma M + \rho I_1^2 - 2\sigma I_2 2\beta M$$
$$+ \rho S_1 I_1 - \sigma M - \beta M = \beta M.$$

And, indeed, the 'transformation' of susceptibles into infectives occurs in mixed pairs at rate β (and only in mixed pairs), so the result was predictable.

Exercise 12.11 The verification is a matter of substitution of the expressions in (12.7) into the right-hand side of the system (12.3) and noting that this indeed yields zero if we ignore the βM terms. The logic is very simple. The pair formation process is completely independent of the (static, by assumption) S-I distinction, and so each category, according to this distinction, occupies the relevant fraction of singles or pairs. Indeed, $\overline{S}_1$ is just the susceptible fraction of singles and $\overline{I}_1$ is the infective fraction of singles. Pairs occur in three categories: {SS}, {SI}, and {II}. Consider, as an example, the {SI} category. We find $\overline{M} = 2\overline{P}\frac{S}{N}\frac{I}{N}$, because the first or the second member of the pair can be susceptible, provided that the second or the first respectively is infective.

Exercise 12.12 Since $\overline{M} = 2\overline{P}\frac{S}{N}\frac{I}{N}$, we have that

$$C(N) = \frac{2\overline{P}}{N} = \frac{1 + 2\nu N - \sqrt{1 + 4\nu N}}{2\nu N}.$$

Multiplying the right-hand side by

$$\frac{1 + 2\nu N + \sqrt{1 + 4\nu N}}{1 + 2\nu N + \sqrt{1 + 4\nu N}},$$

we can check that the two expressions for $C(N)$ in (12.8) are equal to each other. It is easier to check the properties listed in the elaboration of Exercise 12.3 when using the second expression for $C(N)$.

Property i) is now self-evident. For property ii) we calculate the derivative

$$\frac{dC}{dN} = \frac{2\nu(1 + 2\nu N + \sqrt{1 + 4\nu N}) - 2\nu N\left(2\nu + 2\nu(1 + 4\nu N)^{-1/2}\right)}{(1 + 2\nu N + \sqrt{1 + 4\nu N})^2},$$

which, after some reordering of terms, gives

$$\frac{dC}{dN} = 2\nu\left(1 + \sqrt{1 + 4\nu N} - \frac{1}{\sqrt{1 + 4\nu N}}\right) \ge 0.$$

For property iii), note that for N much smaller than 1 we have that $1 + 2\nu N + \sqrt{1 + 4\nu N} \approx 2$ and therefore that $C(N) \approx \nu N$. Finally, for property iv) we calculate the limit

$$\lim_{N \to \infty} C(N) = \lim_{N \to \infty} \frac{2\nu}{\frac{1}{N} + 2\nu + \frac{\sqrt{1 + 4\nu N}}{N}} = \frac{2\nu}{2\nu} = 1.$$

Exercise 12.13 The model formulation of the exercise leads to the following system of differential equations:

$$
\begin{aligned}
\frac{dX_0}{dt} &= \varepsilon_1 X_1 - \varepsilon_0 X_0, \\
\frac{dX_1}{dt} &= -\varepsilon_1 X_1 + \varepsilon_0 X_0 - \rho X_1^2 + 2\sigma P, \\
\frac{dP}{dt} &= \frac{1}{2}\rho X_1^2 - \sigma P.
\end{aligned}
$$

Carrying out the same analysis as in Exercise 12.9, we express the steady state $\overline{P}$ as $\overline{P} = \frac{1}{2}\nu\overline{X}_1^2$. The consistency condition is now given by $\overline{X}_0 + \overline{X}_1 + 2\overline{P} = N$. The steady states $\overline{X}_0$ and $\overline{X}_1$ are obtained by putting the right-hand sides of the corresponding differential equations equal to zero. This leads to the equations

$$
\overline{X}_0 = \frac{\varepsilon_1}{\varepsilon_0}\overline{X}_1
$$

and

$$
\nu\overline{X}_1^2 + \left(\frac{\varepsilon_1}{\varepsilon_0} + 1\right)\overline{X}_1 - N = 0.
$$

The positive solution to this quadratic equation is

$$
\overline{X}_1 = \frac{-\left(\frac{\varepsilon_1}{\varepsilon_0} + 1\right) + \sqrt{\left(\frac{\varepsilon_1}{\varepsilon_0} + 1\right)^2 + 4\nu N}}{2\nu},
$$

and therefore

$$
\begin{aligned}
\overline{P} &= \frac{1}{2}(N - \left(\frac{\varepsilon_1}{\varepsilon_0} + 1\right)\overline{X}_1) \\
&= \frac{\left(\frac{\varepsilon_1}{\varepsilon_0} + 1\right)^2 - \left(\frac{\varepsilon_1}{\varepsilon_0} + 1\right)\sqrt{\left(\frac{\varepsilon_1}{\varepsilon_0} + 1\right)^2 + 4\nu N} + 2\nu N}{4\nu},
\end{aligned}
$$

and finally, using $C(N) = 2P/N$,

$$
C(N) = \frac{\left(\frac{\varepsilon_1}{\varepsilon_0} + 1\right)^2 - \left(\frac{\varepsilon_1}{\varepsilon_0} + 1\right)\sqrt{\left(\frac{\varepsilon_1}{\varepsilon_0} + 1\right)^2 + 4\nu N} + 2\nu N}{2\nu N}.
$$

We see that by either taking $\varepsilon_1 = 0$ or the limit $\varepsilon_0 \to \infty$ (i.e. by letting the resting period be infinitesimally short), we recover the expression derived in Exercise 12.9.

Exercise 12.14 Let $\overline{X}$ be the singles available for contact. If we repeat the Holling argument from Exercise 12.4, with X instead of N (and ρ instead of a), we find

$$
Z = \rho\overline{X}T_s = \rho\overline{X}(T - ZT_h).
$$

With $Y := Z/T$, this leads to

$$
Y = \rho\overline{X}(1 - YT_h) \Rightarrow Y = \frac{\rho\overline{X}}{1 + \rho T_h\overline{X}}.
$$

Since T_h is the average duration of a contact, we can write $\rho T_h = \nu$ to connect to the notation of Exercise 12.9.

Now note that, in contrast to the predator-prey case, the singles are themselves time-limited and the available singles out of the population of size N are a fraction of N determined by the fraction of the time that individuals are not participating in a contact. The total time that an individual is expected to be in contact is ZT_h (i.e., the number of contacts per individual times their average duration). Therefore the time not spent in contacts is $T - ZT_h$ and the fraction of time not spent in contacts is $(T - ZT_h)/T$. Therefore

$$\overline{X} = N \left(\frac{T - ZT_h}{T} \right) = N(1 - YT_h).$$

So actually the relation for Y is not $Y = \rho \overline{X}(1 - YT_h)$ but

$$Y = \rho N(1 - YT_h)(1 - YT_h). \qquad (17.13)$$

Now note that

$$\overline{P} = \frac{1}{2} NYT_h, \qquad (17.14)$$

since $YT_h = ZT_h/T$ is the fraction of time spent in contact for a given individual, which is identical to the fraction of the population engaged in contact at any given time (Exercise 12.2), i.e., NYT_h gives the average number of individuals in the population that is engaged in contact at any particular time. The number of pairs is then half this number. If we substitute the expression (12.7) into (12.8), we obtain

$$\overline{P} = \frac{1}{2} N^2 (1 - YT_h)^2 \rho T_h = \frac{1}{2} \nu \overline{X}^2.$$

Together with the conservation condition $2\overline{P} + \overline{X} = N$, this gives the same equations from which we calculated $C(N)$ in Exercise 12.8.

Exercise 12.15 Start a clock at the moment that a pair consisting of a susceptible and an infective is created, and let t be the time according to this clock. To calculate the probability of transmission within this pair, we have to know the probability that transmission occurs before the pair reaches age t and take the weighted average of that over all t, where the weight is given by the probability that the pair lasts exactly t units of time. Given that the transmission rate is β during the period of existence of the pair, we see that $e^{-\beta t}$ is the probability that transmission has not occurred before time t, and therefore $1 - e^{-\beta t}$ is the probability that transmission has occurred by that time. Since pairs dissolve with exponential rate σ, the probability density function for pair duration is given by $\sigma e^{-\sigma t}$. So, the weighted average is

$$
\begin{aligned}
\int_0^\infty \left(1 - e^{-\beta t}\right) \sigma e^{-\sigma t}\, dt &= \sigma \int_0^\infty \left(e^{-\sigma t} - e^{-(\sigma + \beta)t}\right) dt \\
&= \sigma \left(\frac{1}{\sigma}\right) - \sigma \left(\frac{1}{\sigma + \beta}\right) = \frac{\beta}{\sigma + \beta} \\
&= \frac{b}{1 + b},
\end{aligned}
$$

where we have defined $b := \beta/\sigma$. We conclude that one should, when taking the limit $\sigma \to \infty$, let the infection rate β go to infinity as well, but in such a way as to keep their ratio b constant. In that way, the pair duration becomes infinitesimally short, while at the same time the infection probability per pair remains constant.

Exercise 12.16 We have seen in the previous exercise that the right way to take the limit towards infinitesimally short-lived pairs is to let both σ and β go to infinity, while keeping their ratio b constant. We therefore rewrite the transmission term with $C(N)$ given by (12.8) as

$$
\begin{aligned}
\beta C(N) \frac{SI}{N} &= \frac{2\beta \frac{\rho}{\sigma} N}{1 + 2\frac{\rho}{\sigma} N + \sqrt{1 + 4\frac{\rho}{\sigma} N}} \frac{SI}{N} \\
&= \frac{2\rho b N}{1 + 2\frac{\rho}{\sigma} N + \sqrt{1 + 4\frac{\rho}{\sigma} N}} \frac{SI}{N}.
\end{aligned}
$$

Now take the limit

$$
\lim_{\sigma,\beta \to \infty} \frac{2\rho b N}{1 + 2\frac{\rho}{\sigma} N + \sqrt{1 + 4\frac{\rho}{\sigma} N}} \frac{SI}{N} = 2\rho b N \frac{SI}{N} = \widehat{\beta} SI,
$$

where $\widehat{\beta} := 2\rho b$ is the new transmission rate constant.

Exercise 12.18 For the incidence dS_i/dt among i-individuals, we have the expression

$$
\frac{dS_i}{dt} = - \left(\sum_{j=1}^{n} \pi_{ij} \phi_{ij} \frac{I_j}{N_j} \right) \frac{S_i}{N_i},
$$

since, of the ϕ_{ij} contacts per unit of time, a fraction $\frac{I_j}{N_j} \frac{S_i}{N_i}$ is between susceptible individuals of type i and infective individuals of type j.

Exercise 12.19 Per low-tide period, the number of individuals that sunbathe at site k while being based at site j equals $\rho_{kj} N_j$. The probability that such an individual has contact with an individual based at site i is proportional to

$$
\frac{\rho_{ki} N_i}{\sum_l \rho_{kl} N_l}.
$$

Hence ϕ_{ij} is proportional to

$$
\sum_k \frac{\rho_{ki} N_i \rho_{kj} N_j}{\sum_l \rho_{kl} N_l},
$$

where the constant of proportionality measures the number of 'contacts' (e.g., lying within a distance of two meters from each other) per low tide period.

Exercise 12.20 The total number of contacts per unit of time between members of the aggregated groups l and r is found by adding contacts over all possible combinations between individuals of the subgroups out of which l and r are composed. This rate equals

$$
\sum_{i \in s_l} \sum_{j \in s_r} c_{ij} N_i N_j.
$$

Therefore the average contact rate is given by

$$
\ddot{c}_{lr} = \frac{\sum_{i \in s_l} \sum_{j \in s_r} c_{ij} N_i N_j}{\widehat{N}_l \widehat{N}_r}.
$$

If $c_{ij} = v_i v_j$, we obtain

$$
\begin{aligned}
\widehat{c}_{lr} &= \frac{\sum_{i \in s_l} \sum_{j \in s_r} v_i v_j N_i N_j}{\widehat{N}_l \widehat{N}_r} = \frac{\sum_{i \in s_l} v_i N_i \sum_{j \in s_r} v_j N_j}{\widehat{N}_l \widehat{N}_r} \\
&= \frac{\sum_{i \in s_l} v_i N_i}{\widehat{N}_l} \frac{\sum_{j \in s_r} v_j N_j}{\widehat{N}_r} =: \widehat{v}_l \widehat{v}_r.
\end{aligned}
$$

Exercise 12.21 i) Per unit time, the total number of contacts between type-i and type-j individuals equals $c_{ij} N_i N_j$. So, the total number of contacts of all type-i individuals equals, on the one hand, $\sum_j c_{ij} N_i N_j$, and, on the other hand, $c_i N_i$. Therefore $c_i = \sum_j c_{ij} N_j$. From the c_i contacts, only $c_{ii} N_i$ are within the own group, which amounts to a fraction

$$
q_i = \frac{c_{ii} N_i}{c_i}.
$$

ii) We have to show that $\widehat{R}_0 = \hat{c} \le R_0$, with

$$
\hat{c} = \frac{c_1 N_1 + c_2 N_2}{N_1 + N_2}
$$

and

$$
R_0 = \frac{c_1^2 N_1 + c_2^2 N_2}{c_1 N_1 + c_2 N_2}.
$$

The following chain of equivalent inequalities leads from the one that we wish to prove to one that we know is true:

$$
\begin{aligned}
\frac{c_1 N_1 + c_2 N_2}{N_1 + N_2} &\le \frac{c_1^2 N_1 + c_2^2 N_2}{c_1 N_1 + c_2 N_2}, \\
(c_1 N_1 + c_2 N_2)^2 &\le (N_1 + N_2)(c_1^2 N_1 + c_2^2 N_2), \\
c_1^2 N_1^2 + 2 c_1 c_2 N_1 N_2 + c_2^2 N_2^2 &\le c_1^2 N_1^2 + c_1^2 N_1 N_2 + c_2^2 N_1 N_2 + c_2^2 N_2^2, \\
c_1^2 + c_2^2 - 2 c_1 c_2 &\ge 0.
\end{aligned}
$$

iii) The matrix M has entries

$$
m_{ij} = N_i c_{ij} p T
$$

and SMS^{-1} therefore has entries

$$
(SMS^{-1})_{ij} = \frac{1}{\sqrt{N_i}} m_{ij} \sqrt{N_j} = p T c_{ij} \sqrt{N_i} \sqrt{N_j}.
$$

Because $c_{ij} = c_{ji}$, we conclude that $(SMS^{-1})_{ij} = (SMS^{-1})_{ji}$.

With

$$
v_i := \frac{\sqrt{N_i}}{(\sum_{j=1}^n N_j)^{1/2}}
$$

we find that

$$
\begin{aligned}
(v, SMS^{-1}v) &= \frac{\sum_{i=1}^n \sqrt{N_i} \sum_{j=1}^n p T c_{ij} \sqrt{N_i} \sqrt{N_j} \sqrt{N_j}}{\sum_{j=1}^n N_j} \\
&= p T \frac{\sum_{i=1}^n c_i N_i}{\sum_{j=1}^n N_j} = p T \bar{c}.
\end{aligned}
$$

It follows that the dominant eigenvalue of the symmetric matrix SMS^{-1} is larger than $pT\bar{c}$. Finally, the eigenvalues of SMS^{-1} coincide with those of M.

iv) Partition the indices $\{1, \ldots, n\}$ into k mutually disjoint subsets $s_1, \ldots, s_k$. Define

$$\widehat{N}_l \quad := \quad \sum_{i \in s_l} N_i,$$

$$\widehat{c}_l \quad := \quad \frac{\sum_{i \in s_l} c_i N_i}{\widehat{N}_l},$$

$$\widehat{c}_{lr} \quad := \quad \frac{\sum_{i \in s_l} \sum_{j \in s_r} c_{ij} N_i N_j}{\widehat{N}_l \widehat{N}_r}.$$

Then $\widehat{c}_{rl} = \widehat{c}_{lr}$. Let $\widehat{M}$ denote the $k * k$-matrix with entries

$$\widehat{m}_{lr} = pT\widehat{N}_l \widehat{c}_{lr}$$

and let $\widehat{R}_0$ denote the dominant eigenvalue of $\widehat{M}$. We want to prove that $R_0 \geq \widehat{R}_0$.

Let the rescaling $\widehat{S} : R^k \to R^k$ be defined by

$$(\widehat{S}x)_r = \frac{x_r}{\sqrt{\widehat{N}_l}}.$$

As in part iii) it follows that $\widehat{S}\widehat{M}\widehat{S}^{-1}$ is symmetric. Let $\widehat{\psi}$ denote the positive eigenvector of $\widehat{S}\widehat{M}\widehat{S}^{-1}$ corresponding to $\widehat{R}_0$ and normalised such that $\left\|\widehat{\psi}\right\|_{\ell_2} = 1$. Then $\widehat{R}_0 = (\widehat{\psi}, \widehat{S}\widehat{M}\widehat{S}^{-1}\widehat{\psi})$ or, written out in detail,

$$\begin{aligned}
\widehat{R}_0 &= \sum_{l=1}^{k} \widehat{\psi}_l \sum_{r=1}^{k} pT\widehat{c}_{lr} \sqrt{\widehat{N}_l} \sqrt{\widehat{N}_r} \widehat{\psi}_r \\
&= pT \sum_{l=1}^{k} \sum_{r=1}^{k} \widehat{c}_{lr} \sqrt{\widehat{N}_l} \sqrt{\widehat{N}_r} \widehat{\psi}_l \widehat{\psi}_r
\end{aligned}$$

For $i = 1, \ldots, n$ we define $r(i) \in \{1, \ldots, k\}$ by the requirement that $i \in s_{r(i)}$. Define a vector v of length n by

$$v_i = \frac{\sqrt{N_i}}{\sqrt{\widehat{N}_{r(i)}}} \widehat{\psi}_{r(i)}$$

We claim that $\|v\|_{\ell_2} = 1$. Indeed

$$\|v\|_{\ell_2}^2 = \sum_{i=1}^{n} v_i^2 = \sum_{i=1}^{n} \frac{N_i}{\widehat{N}_{r(i)}} \widehat{\psi}_{r(i)}^2 = \sum_{l=1}^{k} \frac{\sum_{i \in s_l} N_i}{\widehat{N}_l} \widehat{\psi}_l^2 = \sum_{l=1}^{k} \widehat{\psi}_l^2 = 1.$$

Finally we compute

$$
\begin{aligned}
(v, SMS^{-1}v) &= \sum_{i=1}^{n} \frac{\sqrt{N_i}}{\sqrt{\widehat{N}_{r(i)}}} \widehat{\psi}_{r(i)} \sum_{j=1}^{n} pTc_{ij}\sqrt{N_i}\sqrt{N_j} \frac{\sqrt{N_j}}{\sqrt{\widehat{N}_{r(j)}}} \widehat{\psi}_{r(j)} \\
&= pT \sum_{r=1}^{k}\sum_{l=1}^{k} \frac{\widehat{\psi}_r}{\sqrt{\widehat{N}_r}} \sum_{i\in s_r}\sum_{j\in s_l} c_{ij}N_i N_j \frac{\widehat{\psi}_l}{\sqrt{\widehat{N}_l}} \\
&= pT \sum_{r=1}^{k}\sum_{l=1}^{k} \widehat{c}_{lr}\sqrt{\widehat{N}_l}\sqrt{\widehat{N}_r}\widehat{\psi}_r\widehat{\psi}_l
\end{aligned}
$$

and, as we saw above, the right-hand side equals $\widehat{R}_0$. We conclude that the largest eigenvalue of SMS^{-1}, and hence the largest eigenvalue of M, exceeds $\widehat{R}_0$.

Exercise 12.22 The next-generation matrix for the two-group case becomes

$$
\begin{pmatrix} 10q & 30(1-q) \\ 10(1-q) & 30q \end{pmatrix},
$$

and the dominant eigenvalue is given by

$$
R_0 = 20q + 10\sqrt{4q^2 - 6q + 3},
$$

whereas $\widehat{R}_0 = 10 \times 2 = 20$. Analyzing the function

$$
f(q) := 20q + 10\sqrt{4q^2 - 6q + 3} - 20
$$

we find that f is zero at $p = 1/2$ and that $f < 0$ for all $p < 1/2$.

Exercise 12.24 We have $\phi(\theta) = e^{-\theta}$, $m = 1$, $n = 2$ and $c_{\mathrm{L}}p = 0.5$. Inserting this in equation (12.24) for $k = 0$ we get

$$
P_0^{(1,2)} = \binom{2}{0}\left(e^{-2c_{\mathrm{L}}p}\right)^1 = e^{-1} \approx 0.3678.
$$

Using this and equation (12.24) for $k = 1$ we get

$$
P_1^{(1,2)} = \binom{2}{1}\left(e^{-c_{\mathrm{L}}p}\right)^2 - \binom{2}{1}\left(e^{-c_{\mathrm{L}}p}\right)^1 P_0^{(1,2)} = 2e^{-1}\left(1 - e^{-0.5}\right) \approx 0.2895.
$$

Finally, using equation (12.24) for $k = 2$ ($= n$) we get

$$
P_2^{(1,2)} = 1 - P_0^{(1,2)} - P_1^{(1,2)} \approx 0.3426.
$$

The last equation should come as no surprise: equation (12.24) for $k = n$ always equals

$$
P_n^{(m,n)} = 1 - \sum_{i=0}^{n-1} P_i^{(m,n)}
$$

as it should.

Exercise 12.25 We have the same situation as above with the exception that $\phi(\theta) = 1/(1 + \theta)$. Using equation (12.24) recursively we then get

$$P_0^{(1,2)} = \binom{2}{0} \left(\frac{1}{1 + 2c_\mathrm{L} p} \right)^1 = \frac{1}{2} = 0.5,$$

$$P_1^{(1,2)} = \binom{2}{1} \left(\frac{1}{1 + c_\mathrm{L} p} \right)^2 - \binom{2}{1} \left(\frac{1}{1 + c_\mathrm{L} p} \right)^1 P_0^{(1,2)} = \frac{2}{9} \approx 0.2222,$$

$$P_2^{(1,2)} = 1 - P_0^{(1,2)} - P_1^{(1,2)} = \frac{5}{18} \approx 0.2778.$$

Exercise 12.26 Because all households are of size three we have $\pi_3 = 1$ (and all other $\pi_j = 0$) and $h = 3$. Using the household outcome probabilities $P_i^{(1,2)}$ from Exercise 12.24 we then get

$$\mu_3 = 1 + \sum_{i=0}^{2} i P_i^{(1,2)} \approx 1.975.$$

Because $c_\mathrm{G} p = 1$ and $E(\Delta T) = 1$ we get $R_* = 1 \cdot 1 \cdot \mu_3 \cdot 3/3 \approx 1.975$.

Exercise 12.27 Similar to the previous exercise, but with the household outcome probabilities $P_i^{(1,2)}$ from Exercise 12.25 we get $\mu_3 = 1 + \sum_{i=0}^{2} i P_i^{(1,2)} = 16/9 \approx 1.778$, and $R_* = 1 \cdot 1 \cdot \mu_3 \cdot 3/3 \approx 1.778$.

Exercise 12.28 From the examples it seems as if more randomness in ΔT decreases R_*. This is also the typical situation but need not be the case for each choice of distribution. For certain families of distributions, increasing the variance of ΔT while keeping $E(\Delta T)$ fixed results in decreasing R_*. The examples given above present a case in point. The exponential distribution $Exp(1)$ coincides with the Γ-distribution $\Gamma(1, 1)$ and the limit of the Γ-distributions $\Gamma(n, n)$ as n tends to infinity gives a point mass at 1 corresponding to $\Delta T \equiv 1$. If $\Delta T \sim \Gamma(n, n)$ we have $E(\Delta T) = 1$ and $V(\Delta T) = 1/n$, and for this family of distributions it has been shown that R_* increases monotonically with n, i.e., decreases with the variance.

Exercise 12.29 We have $\pi_3 = 1$ and $c_\mathrm{G} p E(\Delta T) = 1$ which inserted in (12.27) gives

$$p_{3,0} = c^{-3u}.$$

To compute $p_{3,1}$ we use $P_0^{(1,2)}$ from Exercise 12.24 and (12.28)

$$p_{3,1} = \binom{3}{1} \left(1 - e^{-u} \right)^1 e^{-2u} P_0^{(1,2)} = 1.1036 \left(1 - e^{-u} \right) e^{-2u}.$$

In order to compute $p_{3,2}$ we need $P_0^{(2,1)}$, which has not yet been derived. From (12.24) with $\phi(\theta) = e^{-\theta}$ we obtain $P_0^{(2,1)} = e^{-1}$. Inserting this in (12.28) gives us

$$p_{3,2} = \binom{3}{1} \left(1 - e^{-u} \right)^1 e^{-2u} P_1^{(1,2)} + \binom{3}{2} \left(1 - e^{-u} \right)^2 e^{-u} P_0^{(2,1)}$$

$$= 0.8685 \left(1 - e^{-u} \right) e^{-2u} + 1.1036 \left(1 - e^{-u} \right)^2 e^{-u}.$$

Finally, in a similar fashion we get

$$p_{3,3} = 1.0281 \left(1 - e^{-u} \right) e^{-2u} + 1.8964 \left(1 - e^{-u} \right)^2 e^{-u} + \left(1 - e^{-u} \right)^3.$$

These four probabilities all depend on the over-all proportion infected u. But this proportion must satisfy $u = 3^{-1}\sum_{i=0}^{3} ip_{3,i}$ which hence specifies u. Solving the equation numerically gives $u \approx 0.707$. Inserting this in the equations for $p_{3,i}$ results in the final size distribution $p_{3,0} = 0.121$, $p_{3,1} = 0.136$, $p_{3,2} = 0.247$, $p_{3,3} = 0.496$. The interpretation of this is the following. If an infectious disease is spread in a large community with households of size 3, where the disease is spread according to the model and given parameters, then, in case a major outbreak occurs, approximately the proportion $p_{3,0} \approx 0.121$ of the households will have no infected, the proportion $p_{3,1} \approx 0.136$ will have 1 infected and so on.

Exercise 12.30 We have the same situation as in the previous exercise with the only difference that we should use the household outcome probabilities from Exercise 12.25. Performing a similar analysis gives

$$p_{3,0} = e^{-3u},$$

$$p_{3,1} = 1.5\left(1 - e^{-u}\right)e^{-2u},$$

$$p_{3,2} = \frac{2}{3}\left(1 - e^{-u}\right)e^{-2u} + \frac{4}{3}\left(1 - e^{-u}\right)^2 e^{-u},$$

$$p_{3,3} = \frac{5}{6}\left(1 - e^{-u}\right)e^{-2u} + \frac{5}{3}\left(1 - e^{-u}\right)^2 e^{-u} + \left(1 - e^{-u}\right)^3.$$

As in the previous exercise $u = 3^{-1}\sum_{i=0}^{3} ip_{3,i}$ gives an equation for finding u numerically, which in our case gives $u = 0.641$. Inserting this in the equations for $p_{3,i}$ results in the final size distribution $p_{3,0} = 0.147$, $p_{3,1} = 0.197$, $p_{3,2} = 0.245$, $p_{3,3} = 0.411$.

Exercise 12.31 Consider a newly infected k-individual. Its expected number of 'offspring' is $(k - 1)\bar{q}$, since, in the initial phase, exactly one of its acquaintances, viz., the one by which it was infected itself, is not susceptible. Note that we assume independence between the type k and the amount of infectious material that is disseminated. Such an assumption deserves scrutiny when, for instance, k is the number of concurrent sexual partners and the agent is transmitted during sexual contact (the point being that there could be a correlation between sexual activity and skin injuries or ulcerations that could enhance transmission).

Each of the offspring has probability ν_m to be of type m. So the distribution of type at 'birth' is given by ν and is, notably, independent of the type of the 'mother.' To calculate R_0, we simply average the number of offspring with respect to ν.

Note that apparently we also assume independence between the type and the susceptibility along any particular connection. This assumption may be questionable in certain contexts for the reasons explained above. The overall susceptibility is, of course, proportional to k. This explains the occurrence of ν_k in the formula (12.31). The factor $k - 1$ then incorporates the dependence of infectivity on type, and our crucial network assumption guarantees that the types have independent influence on the likelihood of a connection. From Section 7.4.1 we know that an explicit formula for R_0 is then to be expected.

Exercise 12.32 We have

$$R_0 = \bar{q}\sum_{k=1}^{\infty}\frac{k^2 - k}{E(D)}\mu_k = \bar{q}\left(\frac{V(D) + E(D)^2}{E(D)} - 1\right)$$
$$= \bar{q}\left(E(D) - 1 + \frac{V(D)}{E(D)}\right).$$

In Exercise 7.18 we derived $R_0 = \text{mean} + \text{variance/mean}$. The slight difference is that in the present context, the first 'mean' is replaced by 'mean'-1 to reflect that, by becoming infected, the transmission opportunities diminish by one.

Exercise 12.33 A k-individual begets j offspring with probability

$$p_{jk} = \binom{k-1}{j} \bar{q}^j (1-\bar{q})^{k-1-j}, \qquad 0 \le j \le k-1.$$

By definition,

$$g_b(z) = \sum_{j=0}^{\infty} \left(\sum_{k=j+1}^{\infty} \nu_k p_{jk} \right) z^j.$$

Hence, interchanging the order of summation, we find

$$g_b(z) = \sum_{k=1}^{\infty} \nu_k \sum_{j=0}^{k-1} p_{jk} z^j = \sum_{k=1}^{\infty} \nu_k (1 - \bar{q} + \bar{q}z)^{k-1}.$$

Clearly

$$g_b'(1) = \sum_{k=1}^{\infty} (k-1)\bar{q}\nu_k = R_0.$$

Exercise 12.34 The probability of extinction can be related to π_b immediately after the first reproduction. Indeed, it equals π_b^j when the first individual begets j offspring. Hence it equals

$$\sum_{k=1}^{\infty} \mu_k \sum_{j=0}^{k} \bar{q}^j (1-\bar{q})^{k-j} \pi_b^j = \sum_{k=1}^{\infty} \mu_k (1 - \bar{q} + \bar{q}\pi_b)^k.$$

Exercise 12.35 On the one hand, we have

$$
\begin{aligned}
1 - \eta &= \pi_b = g_b(\pi_b) = \sum_{k=1}^{\infty} \nu_k (1 - \eta\bar{q})^{k-1} \\
&= \sum_{k=1}^{\infty} \nu_k \left(1 - (k-1)\eta\bar{q} + \frac{1}{2}(k-1)(k-2)\eta^2\bar{q}^2 + \cdots \right) \\
&= 1 - R_0\eta + \frac{1}{2}\eta^2\bar{q}^2 \sum_{k=3}^{\infty} (k-1)(k-2)\nu_k + \cdots .
\end{aligned}
$$

Rearranging this equality (i.e., eliminating the 1 on both sides and dividing by η) we find

$$R_0 - 1 = \frac{1}{2}\eta^2\bar{q}^2 \sum_{k=3}^{\infty} (k-1)(k-2)\nu_k + \cdots .$$

On the other hand, we have

$$
\begin{aligned}
s_\infty &= \sum_{k=1}^{\infty} \mu_k \left(1 - k\eta\bar{q} + \frac{1}{2}(k)(k-1)\eta^2\bar{q}^2 + \cdots \right) \\
&= 1 - E(D)\eta\bar{q} + \cdots ,
\end{aligned}
$$

which implies that

$$1 - s_\infty = E(D)\eta\bar{q} + \cdots .$$

Combining the two expressions, we can eliminate the first-order term in η and obtain (12.39), i.e.,

$$1 - s_\infty = \frac{2E(D)}{\bar{q}\sum_{k=3}^\infty (k-1)(k-2)\nu_k}(R_0 - 1) + \cdots .$$

Exercise 12.36 The probability that a (k, ξ)-individual begets j offspring equals

$$\binom{k-1}{j}(1 - q(\xi))^{k-1-j}(q(\xi))^j.$$

Hence,

$$g_f(z) = \sum_{j=0}^\infty z^j \sum_{k=1}^\infty \nu_k \binom{k-1}{j} \int_\Omega (1 - q(\xi))^{k-1-j}(q(\xi))^j m(d\xi),$$

which, after changing the order of summation and integration to, from left to right, summation over k, integration over ξ and summation over j yields, by the binomial expansion formula,

$$g_f(z) = \sum_{k=1}^\infty \nu_k \int_\Omega (1 - q(\xi) + zq(\xi))^{k-1} m(d\xi).$$

Because g_f is the generating function, the probability that the line of descent from any infected individual is finite equals π_f, the root in the interval $(0, 1)$ of the equation $\pi = g_f(\pi)$. But when we speak of the probability of a minor outbreak, we have in mind that we start with infecting an arbitrary individual from 'outside.' This has two effects: the probability distribution of type k is given by $\{\mu_k\}$, not by ν_k, and a k-type individual can infect all of its k acquaintances, not just $k - 1$. Hence the probability of a minor outbreak equals

$$\sum_{k=1}^\infty \mu_k \sum_{j=0}^k \pi_f^j \binom{k}{j} \int_\Omega (1-q(\xi))^{k-j}(q(\xi))^j m(d\xi) = \sum_{k=1}^\infty \mu_k \int_\Omega (1-q(\xi)+\pi_f q(\xi))^k m(d\xi).$$

Exercise 12.37 By differentiation, we find

$$g_f'(z) = \sum_{k=1}^\infty (k-1)\nu_k \int_\Omega (1 - q(\xi) + zq(\xi))^{k-2} q(\xi)m(d\xi),$$

and hence $g_f'(1) = \sum_{k=1}^\infty (k-1)\nu_k \bar{q} = R_0$.

Exercise 12.38 From

$$\frac{d\mathcal{F}}{d\tau} = -\beta\mathcal{F}, \quad 1 \le \tau \le \xi,$$

$$\mathcal{F}(\tau, \xi) = 1, \quad 0 \le \tau \le 1,$$

we deduce that
$$\mathcal{F}(\tau,\xi) = e^{-\beta \min(\tau-1,\xi-1)},$$
and hence
$$q(\xi) = 1 - \mathcal{F}(\infty,\xi) = 1 - e^{-\beta(\xi-1)}.$$
Therefore
$$\bar{q} = \int_2^3 \left(1 - e^{-\beta(\xi-1)}\right) d\xi = 1 - \frac{1}{\beta}\left(e^{-\beta} - e^{-2\beta}\right).$$

As the expected length of the infectious period is $3/2$, the reduction factor equals $\bar{q}2/(3\beta)$.

Exercise 12.39 Now $\mathcal{F}(\tau,\xi) = exp(-\beta \min(\tau,\xi))$. Hence

$$\begin{aligned}
\overline{\mathcal{F}}(\tau) &= \int_0^\infty e^{-\beta \min(\tau,\xi)} \alpha e^{-\alpha\xi}\, d\xi = \int_0^\tau e^{-\beta\xi}\alpha e^{-\alpha\xi}\, d\xi + \int_\tau^\infty e^{-\beta\xi}\alpha e^{-\alpha\xi}\, d\xi \\
&= \frac{\alpha}{\alpha+\beta}\left(1 - e^{-(\beta+\alpha)\tau}\right) + e^{-(\beta+\alpha)\tau}.
\end{aligned}$$

Exercise 12.40 The probability of transmission to any susceptible acquaintance, as a function of the time τ elapsed since the infected individual was itself infected, is given by
$$1 - \overline{\mathcal{F}}(\tau).$$

The infected individual has with probability ν_k exactly $k-1$ susceptible acquaintances. So the rate of producing 'offspring' is given by

$$\sum_{k=1}^\infty \nu_k(k-1)\frac{d}{d\tau}(1 - \overline{\mathcal{F}}(\tau)),$$

and if we integrate $e^{-r\tau}$ over this quantity, the result should be one if r is to be the growth rate. This is exactly what equation (12.45) states.

Exercise 12.41 The total number of male relationships equals $\sum_i iN_m\pi_i^{(m)} = N_m\mu_m$, where μ_m is the average number of partnerships of males. Similarly, $\sum_i iN_f\pi_i^{(f)} = N_f\mu_f$, and as a consequence, the relationship between partnership distributions of men and women must satisfy $N_m\mu_m = N_f\mu_f$. So if $N_f = N_m$ then $\mu_f = \mu_m$ must hold.

Exercise 12.42 i) The number of male partners such a female has is distributed according to the size-biased distribution, i.e., $P(\tilde{X}_f = i) = \tilde{\pi}_i^{(f)} = i\pi_i^{(f)}/\sum_r r\pi_r^{(f)}$.
  ii) We get

$$\begin{aligned}
E(\tilde{X}_f) &= \sum_i i\tilde{\pi}_i^{(f)} = \sum_i i\frac{i\pi_i^{(f)}}{E(X_f)} = \frac{E(X_f^2)}{E(X_f)} = \frac{V(X_f) + (E(X_f))^2}{E(X_f)} = \\
&= E(X_f) + \frac{V(X_f)}{E(X_f)}.
\end{aligned}$$

  iii) Similarly, $P(\tilde{X}_m = i) = \tilde{\pi}_i^{(m)}$, and $E(\tilde{X}_m) = E(X_m) + V(X_m)/E(X_m)$.

Exercise 12.43 i) Given that $\tilde{X}_f = i$ she will infect each of the $i-1$ uninfected partners independently with the same probability $p^{(mf)}$. As a consequence, the total number she infects, Y_f, is binomially distributed with parameters $i - 1$ and $p^{(mf)}$, denoted $Y_f | \tilde{X}_f = i \sim Bin(i - 1, p^{(mf)})$. This random variable has expected value $E(Y_f | \tilde{X}_f = i) = (i - 1)p^{(mf)}$.

ii) Similarly, given that $\tilde{X}_m = i$ he will infect $Y_m | \tilde{X}_m = i \sim Bin(i - 1, p^{(fm)})$ number of females. Its expected values equals $E(Y_m | \tilde{X}_m = i) = (i - 1)p^{(fm)}$.

Exercise 12.44 i) We have that $E(\tilde{X}_f) = E(X_f) + V(X_f)/E(X_f)$ and that $E(Y_f | \tilde{X}_f = i) = (i - 1)p^{(mf)}$. From this it follows that

$$E(Y_f) = p^{(mf)}(E(\tilde{X}_f) - 1) = p^{(mf)} \left(E(X_f) + \frac{V(X_f)}{E(X_f)} - 1 \right)$$

$$= p^{(mf)} \left(E(X_f) + \frac{V(X_f) - E(X_f)}{E(X_f)} \right).$$

ii) Using identical arguments we get

$$E(Y_m) = p^{(fm)} \left(E(X_m) + \frac{V(X_m) - E(X_m)}{E(X_m)} \right).$$

Exercise 12.45 The eigenvalues of K are defined by the solutions λ and to the determinant equation $|\Lambda - \lambda I| = 0$. Since we have a 2×2 matrix with 0's on the diagonal and off-diagonal elements $E(Y_m)$ and $E(Y_f)$, the determinant equation equals $\lambda^2 - E(Y_m)E(Y_f) = 0$. It follows that the solutions are given by $\lambda = \pm \sqrt{E(Y_m)E(Y_f)}$, so the largest eigenvalue equals $\sqrt{E(Y_m)E(Y_f)}$.

Inserting our expressions for $E(Y_m)$ and $E(Y_f)$ we end up with the following expression for the basic reproduction number:

$$R_0 = \sqrt{p^{(mf)}p^{(fm)} \left(E(X_m) + \frac{V(X_m) - E(X_m)}{E(X_m)} \right) \left(E(X_f) + \frac{V(X_f) - E(X_f)}{E(X_f)} \right)}.$$

$$(17.15)$$

Exercise 12.46 i) We have the same situation as in the initial branching process approximation, only now the matrix K is transformed to

$$K_v = \begin{pmatrix} 0 & p^{(fm)}(1 - v)E(Y_m - 1) \\ p^{(mf)}(1 - v)E(Y_f - 1) & 0 \end{pmatrix}.$$

As a consequence, the new reproduction number equals

$$R_v = (1 - v)\sqrt{p^{(mf)}p^{(fm)} \left(E(X_m) - \frac{V(X_m) - E(X_m)}{E(X_m)} \right)}$$

$$\times \sqrt{\left(E(X_f) - \frac{V(X_f) - E(X_f)}{E(X_f)} \right)}$$

$$= (1 - v)R_0.$$

ii) We have $E(X_m) = E(X_f) = 3$ and $p^{(mf)} = p^{(fm)} = 0.8$. Depending on whether $V(X_m) = V(X_f)$ equals 0, 3 or ∞ we get from (17.15) that $R_0 = 1.6$, $R_0 = 2.4$ and $R_0 = \infty$, respectively. This implies that $R_v = (1 - v)1.6$, $(1 - v)2.4$

and $(1 - v) \cdot \infty$. The last expression is formally not well defined but should be interpreted as equal to infinity when $v < 1$ and equal to 0 when $v = 1$. As a consequence, the critical vaccination coverage for the three scenarios are $v_c = 0.375$, $v_c = 0.583$ and $v_c = 1$. We see that the variance (i.e., the variation in number of sex-partners) plays a crucial role for the possibility to prevent outbreaks — it is not only the average number of partners that is important.

Exercise 12.47 If each S-individual takes part in k pairs, then there are in total $k[S]$ pairs that involve at least one S-individual. A fraction $[I]/N$ of these pairs is then with an I-individual if pairs have been made between the N individuals in the population at random.

Exercise 12.48 We argue in much the same way as in Exercise 12.47. To calculate the number of triples $[XYZ]$ we start with the number of pairs of type $[XY]$. If we regard the Y-individual in that pair then this individual is linked to $k - 1$ other individuals in addition to the specific X-individual it is paired with. In total the $[XY]$ pairs therefore have $(k - 1)[XY]$ links available to form additional ties to the Y-individual of these pairs. We are only interested in that fraction of these ties that connect to a Z-individual, because only these lead to an $[XYZ]$ triple. The total number of links to Y-individuals is given by $k[Y]$ and of these a fraction $[YZ]/k[Y]$ is with a Z-individual in a randomly formed network.

Exercise 12.49 In the main text we have described the four ways in which $[SI]$ can change. In an $[SSI]$ triple the central S-individual can become infected by the I-individual with rate $\theta[SSI]$, leading to the contribution $\theta \frac{k-1}{k} \frac{[SS][SI]}{[S]}$ in the pair-approximation case. A similar reasoning holds for $[ISI]$ triples, where the first I-individual can infect the central S-individual with rate $\theta[ISI]$, leading to the contribution $-\theta \frac{k-1}{k} \frac{[SI]^2}{[S]}$, where we use the symmetry that an IS-pair is equal to an SI-pair. The minus-sign expresses that this change results in the loss of an SI-pair. The remaining two options do not involve triples: simple transmission within the pair at rate $\theta[SI]$ and recovery of the I-individual in the pair contributing $-\alpha[II]$. Putting all these together and rewriting the result then leads to the right-hand side of $d[SI]/dt$. The equations for $[SS]$ and $[II]$ arise by similar reasoning.

Exercise 12.50 The pairing model takes at least some of the possible correlations between individuals into account. This means that, unlike the standard SIR model, it will lead — in a rudimentary way — to local saturation during the outbreak because of local depletion of susceptibles. This depletion is due to the static nature of the network: once an individual becomes infected, the infection must come from one of its acquaintances, and therefore this newly infected possibly can no longer realize the full potential of secondary cases. These effects accumulate quickly at the focus of spread. In general, the pairing model is therefore expected to show outbreaks with a slower initial growth rate and with a lower peak in incidence.

Chapter Eighteen

Elaborations for Part III

18.1 ELABORATIONS FOR CHAPTER 13

Exercise 13.1 By definition, R_0 is a generation property: it gives, after a sufficiently large number of generations, the per-generation growth factor of the infected population, when circumstances remain fixed as they were at the moment of introduction. One immediately sees three issues: generation data are not observed, it depends on the situation studied what 'sufficiently large' is (even when such a 'quantity' could be well-defined), and circumstances as a rule start to change immediately after an outbreak has been discovered, for example because of control actions. With regard to the generation data: what are observed are as a rule detections (i.e., individuals expressing symptoms). There is a possibly unknown, but definitely agent-host-specific, incubation period distribution that convolutes the infection process into the observed process (we will return to this). With regard to the changing circumstances, an additional problem is that typically, as we have seen in Chapter 12, individuals predominantly have contacts to a limited subset of all susceptibles (i.e., spatially or socially local contacts), and therefore saturation effects will start to operate early on in an outbreak. In addition, the definition calls for introduction into a fully susceptible population. This will seldom be the case, for example because of the intricacies of cross-immunity.

Exercise 13.2 There is, as a rule, substantial heterogeneity among individuals causing marked differences in the time course and severity of an infection, as well as in the subset of the population to which different individuals have contacts. This heterogeneity manifests in potentially important differences in the value of epidemiological parameters. If one regards a small sample, then this might introduce bias, if the sample does not consist of randomly chosen 'independent' individuals. How much this influences results of course depends on the host-agent-environment system studied; particular systems can have narrow distributions for several epidemiological ingredients, making small samples less problematic. However, even in that case the sample can be biased by the fact that one generally only observes symptomatic individuals and sub-clinically infected individuals will be missed (and also symptoms vary in clarity). In addition, early on in an outbreak stochastic effects are likely to play a dominant role in determining which individuals enter the sample (the order of infections is not necessarily the same as the order of detections). For specific parameters there are additional possibilities for bias. For example, when clarity of symptoms influences the probability to be in a sample, and infectiousness and symptoms are correlated, then this introduces bias in an estimate of infectiousness, leading to an overestimate because weakly infectious individuals are likely to be missed. Another example is the estimate of incubation time. One aspect that introduces bias here is when samples are drawn from a growing epidemic. Early on in the epidemic one is likely to observe primarily the individuals who progress to symptoms rapidly, leading to an underestimate of the incubation period. This phenomenon has had a notable impact on early estimates of the incubation period of HIV. Finally, in epidemiology textbooks one can find ways to correct for bias that is introduced by the fact that diagnostic tests are often not perfect enough in distinguishing infected from non-infected individuals. This could play a role, for example, when using final-size data to estimate R_0, where these data are then influenced by the specificity/sensitivity of the diagnostic test used.

Exercise 13.3 We translate the ingredients of (8.24) into the symbolic language of the present section, by looking at the interpretations. First of all, we have $s(-\infty, i) = \pi_i s_i N$, the susceptible population of type i before the outbreak, and $s(\infty, i) = (1 - p_i)s_i \pi_i N$, the susceptibles of type i left after the outbreak. We put $\int A_{ij}(\tau)d\tau = k_{ij}$. Equation (8.24) then translates into:

$$\frac{(1 - p_i)\pi_i s_i N}{\pi_i s_i N} = e^{-\sum_{j=1}^{n} k_{ij}\pi_j s_j N(1 - \frac{(1-p_j)\pi_j s_j N}{\pi_j s_j N})},$$

which can be simplified to (13.1).

Exercise 13.5 i) Let $\mathcal{F}(a)$ denote the probability to be alive at age a, then

$$\mathcal{F}(a) = \mathcal{F}_d(a) \left\{ \mathcal{F}_i(a) + \theta(1 - \mathcal{F}_i(a)) \right\}.$$

ii) Let $P(a)$ denote the probability to be alive at age a, given that one does not die of disease-unrelated causes. Then $P(a)$ equals the conditional probability

$$P(a) = \frac{\mathcal{F}(a)}{\mathcal{F}_d(a)} = \mathcal{F}_i(a) + \theta(1 - \mathcal{F}_i(a)).$$

iii) Conversely, if we consider the probability to be alive at age a, given that one does not die of the disease (or, equivalently, in a situation in which the infectious agent is eliminated), then this is just $\mathcal{F}_d(a) = \mathcal{F}(a)/P(a)$ (which is also obtained by putting $\theta = 1$, of course).

iv) For constant force of infection Q we have $\mathcal{F}_i(a) = e^{-Qa}$, and hence

$$\frac{1}{P(a)} = \frac{1}{e^{-Qa} + \theta(1 - e^{-Qa})}$$

so, $1/P(a) \to 1/\theta$ for $a \to \infty$. Now recall that the solution of the logistic differential equation $dx/dt = rx(1 - x/K)$ with initial condition $x(0) = x_0$ is given by

$$x(t) = \frac{1}{e^{-rt}\frac{1}{x_0} + \frac{1}{K}(1 - e^{-rt})},$$

and observe that $1/P(a)$ is of this form with $x_0 = 1$, $K = 1/\theta$ and $r = Q$.

Exercise 13.6 The probability that an individual is both alive and uninfected at age a is $\mathcal{F}_i(a)\mathcal{F}_d(a)$, and such individuals have probability per unit of time $\Lambda(a)$ of becoming infected. Hence we want to compute the expected value of the stochastic variable 'age at infection,' which has probability density function

$$\frac{\Lambda(a)\mathcal{F}_i(a)\mathcal{F}_d(a)}{\int_0^\infty \Lambda(\alpha)\mathcal{F}_i(\alpha)\mathcal{F}_d(\alpha)\, d\alpha},$$

where the denominator serves to achieve the right normalization. This is then exactly the quantity $\bar{a}$.

When $\mathcal{F}_d(a) = 1$ for $0 \leq a < M$, and $\mathcal{F}_d(a) = 0$ for $a \geq M$, we have

$$\int_0^M a\Lambda(a)\mathcal{F}_i(a)\,da = -a\mathcal{F}_i(a)\big|_0^M + \int_0^M \mathcal{F}_i(a)\,da = -M\mathcal{F}_i(M) + \int_0^M \mathcal{F}_i(a)\,da,$$

which leads to the second expression for $\bar{a}$.

Exercise 13.7 i) Under the assumption on c, the right-hand side of (7.5) is of the form

$$\sum_{k=1}^n f_k(a) \int_0^\infty \int_0^\infty h(\tau, \alpha) g_k(\alpha + \tau) \Lambda(\alpha) S(\alpha) \frac{\mathcal{F}_d(\alpha + \tau)}{\mathcal{F}_d(\alpha)}\,d\tau\,d\alpha,$$

which shows that any solution Λ must be of the form

$$\Lambda(a) = \sum_{j=1}^n d_j f_j(a).$$

Substituting this expression into the equation we find that the coefficients d_k should satisfy the nonlinear system of equations

$$d_k = \sum_{j=1}^n d_j \int_0^\infty \int_0^\infty h(\tau, \alpha) g_k(\alpha + \tau) f_j(\alpha) \frac{\mathcal{F}_d(\alpha + \tau)}{\mathcal{F}_d(\alpha)} N(\alpha) e^{-\sum_{i=1}^n d_i \int_0^\alpha f_i(\sigma)d\sigma}\,d\tau\,d\alpha.$$

In the special case $n = 1$ we can divide both sides by $d_1 = d$ (thus canceling the trivial solution $d = 0$) and arrive at the simple form

$$1 = \int_0^\infty \int_0^\infty h(\tau, \alpha) g(\alpha + \tau) f(\alpha) \frac{\mathcal{F}_d(\alpha + \tau)}{\mathcal{F}_d(\alpha)} N(\alpha) e^{-d \int_0^\alpha f(\sigma)\,d\sigma}\,d\tau\,d\alpha$$

which has, because of the monotonicity of the right-hand side as a function of d, a unique positive solution provided that $R_0 > 1$.

ii) In the short-disease approximation, the system simplifies to

$$d_k = \sum_{j=1}^n d_j \int_0^\infty H(\alpha) g_k(\alpha) f_j(\alpha) N(\alpha) e^{-\sum_{i=1}^n d_i \int_0^\alpha f_i(\sigma)\,d\sigma}\,d\alpha$$

and, for $n = 1$, this reduces to

$$1 = \int_0^\infty H(\alpha) g(\alpha) f(\alpha) N(\alpha) e^{-d \int_0^\alpha f(\sigma)\,d\sigma}\,d\alpha$$

while with interval decomposition we find

$$d_k = \sum_{j=1}^n c_{kj} d_j \int_{I_j} H(\alpha) N(\alpha) e^{-\sum_{i=1}^n d_i \int_0^\alpha \chi_{I_i}(\sigma)\,d\sigma}\,d\alpha.$$

Note that effectively the summation in the exponent is only up to j, which corresponds to $\mathcal{F}_i(a)$ depending only on Λ up to a.

iii) The question is: suppose one can measure the d_i, can one determine from these the c_{ij}? This would amount to determining n^2 unknowns from n data, which is impossible for $n > 1$. We can improve a bit if we require symmetry, i.e., $c_{ij} = c_{ji}$, since that assumption reduces the number of unknowns to $n(n+1)/2$, but this still exceeds n even for $n = 2$.

iv) According to Anderson and May (1991), the element (i, j) of this matrix is the probability per unit time that an infective in age class j infects a susceptible in age class i. Apart from an infectivity factor, this corresponds to the matrix element c_{ij}.

v) As discussed in iii), symmetry is the first assumption that suggests itself. An obvious possibility is to choose $c_{ij} = b_i b_j$ for some $b_1, \ldots, b_n$. We refer to the cited literature for other possible choices.

Exercise 13.8 i) Suppose one can, from serological data, determine the fraction $S(a)/N(a)$ of susceptibles for various values of a. According to the assumptions, we have $S(a) = N(a)e^{-\Lambda a}$, so $\frac{1}{a} \ln \frac{N(a)}{S(a)}$ should be approximately constant, and its mean value yields an estimate for Λ.

An alternative is to determine the mean age at infection $\bar{a}$ from data and to put $\Lambda = \bar{a}^{-1}$. If, however, $\mathcal{F}_d(a) = e^{-\mu a}$ then we already found in Exercise 4.6 (formula (4.10)), that $\bar{a} = 1/(\mu + \Lambda)$ and hence $\Lambda = 1/\bar{a} - \mu$.

ii) Assume that f is identically one, g is constant, h does not depend on α and $\mathcal{F}_d(\alpha + \tau)/\mathcal{F}_d(\alpha) = e^{-\mu\tau}$. The equation derived in Exercise 13.7-i then simplifies to (note that d is replaced by Q now)

$$1 = D \int_0^\infty e^{-(r+\mu+Q)\alpha} \, d\alpha = \frac{D}{r+\mu+Q},$$

where $D = C \int_0^\infty h(\tau)g e^{-\mu\tau} \, d\tau$, with C from (7.1). Similarly, the expression (9.2) for R_0 simplifies to

$$R_0 = D \int_0^\infty e^{-(r+\mu)\alpha} \, d\alpha = \frac{D}{r+\mu}.$$

Hence $R_0 = (r + \mu + Q)/(r + \mu) = 1 + Q/(r + \mu)$, a formula that allows one to estimate R_0 from the observable quantities r, μ and Q. Inserting $Q = 1/\bar{a} - \mu$ and putting $r = 0$, we recover the formula

$$R_0 = \frac{1}{\mu\bar{a}} = \frac{L}{\bar{a}} = \frac{\text{life expectancy}}{\text{mean age at infection}}$$

derived in Exercise 4.6. By now we are much better aware of all assumptions that are required to arrive at this identity.

iii) We have $1 = \int_0^\infty \psi(\alpha)N(\alpha)e^{-Q\alpha} \, d\alpha$ and $R_0 = \int_0^\infty \psi(\alpha)N(\alpha) \, d\alpha$. If ψ is (approximately) constant we can eliminate ψ from these equations and find the desired result.

iv) The second formula of Exercise 13.6 implies that, under the assumptions, $\bar{a} = 1/Q$. The result from iii) can be rewritten, using (9.1), as

$$R_0 = \frac{\int_0^\infty e^{-ra}\mathcal{F}_d(a) \, da}{\int_0^\infty e^{-(r+Q)a}\mathcal{F}_d(a) \, da},$$

which, given the choice of survival function, can be written as

$$R_0 = \frac{\int_0^L e^{-ra} \, da}{\int_0^L e^{-(r+Q)a} \, da} = \frac{\frac{1}{r} - \frac{e^{-rL}}{r}}{\frac{1}{r+Q} - \frac{1}{r+Q}e^{-(r+Q)L}} = \frac{\frac{1}{r} - \frac{1}{r}(1 - rL + o(r))}{\frac{1}{r+Q} - \frac{1}{r+Q}e^{-(r+Q)L}}.$$

For $r \approx 0$, this leads to

$$R_0 \approx \frac{QL}{1 - e^{-QL}},$$

which, for $QL \gg 1$, finally gives

$$R_0 \approx QL = \frac{L}{a}.$$

The condition $QL \gg 1$ is not too difficult to meet for human hosts, since life expectancy is usually of the order of 50 years or more.

Exercise 13.9 i) We start from the conservation equation (13.4) and substitute into that the short-disease approximation and the assumption on c. This immediately leads to the conclusion that $\lambda(a)$ is equal to a scalar times $f(a)$.

ii) We start from (13.9) and substitute (13.3) and the assumption on c. We get

$$\begin{aligned} R_0 l(\alpha) &= \int_0^\infty l(a) k(a, \alpha) da \\ &= \int_0^\infty l(a) f(a) N(a) \int_0^\infty h(\tau, \alpha) g(\alpha + \tau) e^{-\mu\tau} d\tau da \\ &= \hat{f} \psi(\alpha) \end{aligned}$$

where $\hat{f} = \int_0^\infty l(a) f(a) N(a) da$ is a scalar and where $\psi(\alpha)$ is given by (13.7). In the short-disease approximation we have that $\psi(\alpha)$ is proportional to $g(\alpha)$ and the result follows.

iii) This is now an easy substitution using i) and ii) and (13.10).

Exercise 13.10 We first take the natural logarithm on both sides of (13.12):

$$\begin{aligned} \ln L &= \ln \left(\prod_{i=0}^{n-1} \binom{n_i}{m_i} \right) + \ln \left(\prod_{i=0}^{n-1} G_i^{m_i} \right) + \ln \left(\prod_{i=0}^{n-1} (1 - G_i)^{n_i - m_i} \right) \\ &= \sum_{i=0}^{n-1} \ln \left(\binom{n_i}{m_i} \right) + \sum_{i=0}^{n-1} m_i \ln(G_i) + \sum_{i=0}^{n-1} (n_i - m_i) \ln(1 - G_i). \end{aligned}$$

Then take the partial derivative with respect to G_i to obtain

$$\frac{\partial \ln L}{\partial G_i} = \frac{m_i}{G_i} + \frac{n_i - m_i}{1 - G_i}.$$

If we equate this partial derivative to zero in order to find the extreme values, we obtain

$$(1 - \hat{G}_i) m_i = \hat{G}_i (n_i - m_i),$$

leading to the estimator $\hat{G}_i$ for G_i

$$\hat{G}_i = \frac{m_i}{n_i}.$$

This makes intuitive sense because it is the fraction seropositive observed in the sample for interval i. If we use the defining relation $G_i = 1 - \exp(-\lambda_i(a_{i+1} - a_i))$ (see main text) and equate the left-hand side to the estimator for G_i, we can rewrite the resulting expression to obtain the estimate for $\hat{\lambda}_i$ given in the exercise.

Exercise 13.11 i) In the early phase we have $dI/dt \approx (\beta - \gamma)I = (R_0 - 1)\gamma I$. Therefore we get that $I(t)$ grows approximately exponentially with rate $(R_0 - 1)\gamma$ in the early phase of an outbreak described by the simple SIR model. If we let t_d denote the doubling time of the incidence, we get that

$$2I(t_0) = I(t_0)e^{(R_0-1)\gamma t_d},$$

which we can rewrite as the desired result.

ii) If the incidence grows as e^{rt}, then the cumulative incidence g is given by

$$g(t) = g(t_0) + \frac{c}{r}(e^{rt} - e^{rt_0})$$

for some (unknown) constant c. Suppose we observe g at three points in time, say t_0, t_1 and t_2, with $t_2 > t_1 > t_0$. We can use this information to eliminate c and write

$$\frac{g(t_2) - g(t_1)}{g(t_1) - g(t_0)} = \frac{e^{rt_2} - e^{rt_1}}{e^{rt_1} - e^{rt_0}} = \frac{e^{r(t_2-t_1)} - 1}{1 - e^{-r(t_1-t_0)}}$$

where r is now the only unknown quantity. The right-hand side is a monotone increasing function of r with limit $(t_2 - t_1)/(t_1 - t_0)$ for $r \downarrow 0$, and going to infinity for $r \to \infty$. Therefore, this equation has a unique solution r if the left-hand side exceeds $(t_2 - t_1)/(t_1 - t_0)$. This solution can be obtained numerically.

If one has an estimate for γ available, one can then produce an estimate for R_0 by using $r = (R_0 - 1)\gamma$.

Note that the values of r and the doubling time depend on the choice of the time points of observation. The estimate is therefore possibly very rough and can be influenced by reporting delays, reporting accuracy and choice of time points in the early phase.

Exercise 13.12 The first part of the exercise starts with the result of Exercise 2.2-i, leading to

$$f(t) = \frac{\alpha\theta}{\alpha - \theta}\left(e^{-\theta t} - e^{-\alpha t}\right).$$

We then have, using (13.16), the relation

$$R_0^{\exp}\int_0^\infty e^{-rt}\frac{\alpha\theta}{\alpha - \theta}\left(e^{-\theta t} - e^{-\alpha t}\right)dt = 1.$$

By working out the integral and rewriting we find

$$R_0^{\exp} = 1 + r\left(\frac{1}{\theta} + \frac{1}{\alpha}\right) + \frac{r^2}{\theta\alpha}.$$

Then realize that, in the exponential case, the mean generation interval is simply $T_G = T_E + T_I$, and that $T_E = 1/\theta$, and $T_I = 1/\alpha$. If we substitute these in the expression above, we find the estimator given in the exercise.

18.2   ELABORATIONS FOR CHAPTER 15

Exercise 15.1 i) We have observed $X = 2$ from $Bin(n = 3, \pi)$, so the likelihood is given by: $L(\pi; X = 2) = \binom{3}{2}\pi^2(1 - \pi)^1$. Maximizing this with respect to π is equivalent to maximizing the log-likelihood

$$\ell(\pi) = \ln\left(\binom{3}{2}\right) + 2\ln(\pi) + \ln(1 - \pi).$$

We find the π that maximizes the (log-)likelihood by differentiating and equating the derivative to 0 (and checking that it is a maximum rather than a minimum). The derivative of the log-likelihood, also known as the score-function, equals $\ell'(\pi) = 2/\pi - 1/(1-\pi)$. Equating this to 0 gives us the maximum likelihood (ML) estimate $\hat\pi = 2/3 \approx 0,67$. For general n and $X = x$ the ML estimate is $\hat\pi = x/n$, i.e., the observed proportion of successes. This estimate is very natural: we estimate the unknown probability of success by the observed proportion of successes.

ii) The posterior distribution of π equals

$$p(\pi|X=2) = \frac{p(\pi)L(\pi; X=2)}{\int p(\pi')L(\theta'; X=2)d\theta'} \propto p(\pi)L(\pi; X=2) = 1 \times \binom{3}{2}\pi^2(1-\pi)^1,$$

for $0 \leq \pi \leq 1$ ($p(\pi|X=2) = 0$ for other values on π). The density is hence proportional to $\pi^2(1-\pi)^1$ for $0 \leq \pi \leq 1$. This is recognized as the Beta-distribution with parameters $\alpha = 3$ and $\beta = 2$. The mean of this distribution equals $\hat\pi_{post} = \alpha/(\alpha+\beta) = 3/5 = 0.6$. It is hence a compromise between the prior mean 0.5 (which is a sensible mean if there is no additional information) and the ML-estimate 0.67.

iii) If $n = 30$ and $x = 20$ the ML-estimate is still $\hat\pi = x/n = 2/3$. The posterior is obtained as above, and it is shown to be proportional to $\pi^{20}\pi^{10}$, i.e., the Beta-distribution with parameters $\alpha = 21$ and $\beta = 11$. This distribution has mean $\hat\pi_{post} = \alpha/(\alpha+\beta) = 21/32 \approx 0.66$. The compromise between the prior mean of 0.5 and the ML-estimate 0.67 is now shifted so that it is very close to the ML-estimate. The reason for this is that we now have much more data making the ML-estimate more reliable. This is a general situation for Bayesian analyses: when little data is available, the prior distribution is influential, whereas it is less influential when more data is available. Put differently: the choice of prior distribution plays a major role when data is sparse, but has less, or hardly any effect when much data is available.

Exercise 15.2 i) The prior of the three parameters equals $\lambda\alpha\beta e^{-(\lambda+\alpha+\beta)}$. The likelihood for the complete data $L(\lambda, \alpha, \beta; i_0 = 0, i_1, \ldots, i_5, r_0 = 7, r_1 = 6, \ldots, r_5 = 15)$, is obtained as in Section 5.4 (cf. 5.35 for the log-likelihood when the infectious period is exponentially distributed):

$$L(\lambda, \alpha, \beta) = \prod_{k=1}^{5} \frac{\lambda I(i_k-)S(i_k-)}{10} e^{-\int_0^{r_5} \lambda I(t)S(t)/10 dt} \prod_{k=0}^{5} f_{\alpha,\beta}(r_k - i_k).$$

In the expression, $I(t)$ and $S(t)$ denote the numbers observed at t, quantities that are known given the complete data; $I(i_k-)$ and $S(i_k-)$ denote the same quantities, but now at the time instant immediately prior to time i_k. Finally, $f_{\alpha,\beta}(r_k - i_k)$ denotes the Gamma-density with parameters α and β observed in the point $r_k - i_k$ (the length of the infectious period). The posterior distribution is proportional to the product of the prior and the likelihood.

ii) As a proposal distribution for α, β and λ one can multiply the existing parameter value with some non-negative random variable, for example a uniform number between 0.5 and 1.5. This means that the proposal would be sampled from $q(x|y) = 1/y$ for $0.5/y \leq x \leq 1.5/y$. For the infection times one can update $i_1, \ldots, i_5$ sequentially. At each step one can update i_k to a uniform value between 0 and r_k (this is the range of possible infection times). A candidate point changing the infection time such that some other infection time occurs when no one is infectious will never be accepted since the likelihood for such a point will be 0.

Bibliography

This bibliography lists only books. The aim is to provide the reader with pointers to a selection of textbook sources from mathematics and population dynamics that contain much of the backbone of the material presented in the text. The books are chosen for direct relevance, and are biased by the present authors' preferences. The bibliography starts with a list of books devoted to various aspects of epidemic modeling. In no category do we aim to be exhaustive.

EPIDEMIC MODELS

R.M. Anderson (ed.): *Population Dynamics of Infectious Diseases: Theory and Applications.* Chapman & Hall, London, 1982.

R.M. Anderson & R.M. May: *Population Biology of Infectious Diseases.* Springer-Verlag, Berlin, 1982.

R.M. Anderson & R.M. May: *Infectious Diseases of Humans: Dynamics and Control.* Oxford University Press, Oxford, 1991.

H. Andersson & T. Britton: *Stochastic Epidemic Models and their Statistical Analysis.* Springer-Verlag, Berlin, 2000.

N.J.T. Bailey: *The Mathematical Theory of Infectious Diseases and its Applications.* Griffin, London, 1975.

N.J.T. Bailey: *The Biomathematics of Malaria.* Griffin, London, 1982.

M. Bartlett: *Stochastic Population Models in Ecology and Epidemiology.* Methuen, London, 1960.

N.G. Becker: *Analysis of Infectious Disease Data.* Chapman & Hall, London, 1989.

F. Brauer & C. Castillo-Chavez: *Mathematical Models in Population Biology and Epidemiology.* Springer-Verlag, Berlin, 2001.

F. Brauer, P. Van den Driessche & J. Wu (eds.): *Mathematical Epidemiology.* Springer-Verlag, Berlin, 2008.

S. Busenberg & K. Cooke: *Vertically Transmitted Diseases: Models and Dynamics.* Springer-Verlag, Berlin, 1993.

V. Capasso: *Mathematical Structures of Epidemic Systems.* Springer-Verlag, Berlin, 1993.

C. Castillo-Chavez (ed.): *Mathematical Approaches for Emerging and Reemerging Infectious Diseases.* Springer-Verlag, Berlin, 2002.

G. Chowell, J.M. Hyman, L.M.E. Bettencourt & C. Castillo-Chavez (eds.): *Mathematical and Statistical Estimation Approaches in Epidemiology.* Springer-Verlag, Berlin, 2009.

A.D. Cliff & P. Haggett: *Atlas of Disease Distributions: Analytical Approaches to Epidemiological Data.* Blackwell, London, 1988.

D.J. Daley & J. Gani: *Epidemic Modelling: An Introduction.* Cambridge University Press, Cambridge, 1999.

C.A. Donnelly & N.M. Ferguson: *Statistical Aspects of BSE and vCJD: Models for Epidemics.* Chapman & Hall, London, 1999.

J.C. Frauenthal: *Mathematical Modeling in Epidemiology.* Springer-Verlag, Berlin, 1980.

J.-P. Gabriel, C. Lefévre & P. Picard (eds.): *Stochastic Processes in Epidemic Theory.* Springer-Verlag, Berlin, 1990.

B.T. Grenfell & A.P. Dobson (eds.): *Ecology of Infectious Diseases in Natural Populations.* Cambridge University Press, Cambridge, 1995.

A.B. Gumel & S. Lenhart (eds.): *Modeling Paradigms and Analysis of Disease Transmission Models.* American Mathematical Society, Providence, 2010.

H.W. Hethcote & J.A. Yorke: *Gonorrhea Transmission Dynamics and Control.* Springer-Verlag, Berlin, 1984.

H.W. Hethcote & J.W. Van Ark: *Modelling HIV Transmission and AIDS in the United States.* Springer-Verlag, Berlin, 1992.

F.C. Hoppensteadt: *Mathematical Methods for Analysis of a Complex Disease.* Courant Lecture Notes 22. American Mathematical Society, Providence, 2011.

P.J. Hudson, A. Rizzoli, B.T. Grenfell, J.A.P. Heesterbeek & A.P. Dobson (eds.): *Ecology of Wildlife Diseases.* Oxford University Press, Oxford, 2002.

V. Isham & G. Medley: *Models for Infectious Human Diseases: Their Structure and Relation to Data.* Cambridge University Press, Cambridge, 1996.

M.J. Keeling & P. Rohani: *Modeling Infectious Diseases in Humans and Animals.* Princeton University Press, 2008.

A. Krämer, M. Kretzschmar & K. Krickeberg: *Modern Infectious Disease Epidemiology: Concepts, Methods, Mathematical Models and Public Health.* Springer-Verlag, Berlin, 2010.

J. Kranz (ed.): *Epidemics of Plant Diseases: Mathematical Analysis and Modelling.* Springer-Verlag, Berlin.

H.A. Lauwerier: *Mathematical Models of Epidemics.* Mathematisch Centrum, Amsterdam, 1981.

Z. Ma & J. Li (eds.): *Dynamic Modeling and Analysis of Epidemics.* World Scientific, Singapore, 2009.

Z. Ma, Y. Zhou & J. Wu (eds.): *Modeling and Dynamics of Infectious Diseases.* World Scientific, Singapore, 2009.

C.J. Mode & C.K. Sleeman: *Stochastic Processes in Epidemiology: HIV/AIDS, Other Infections and Computers.* World Scientific, Singapore, 2000.

D. Mollison, G. Scalia-Tomba & J.A. Jacquez (eds.): *Spread of Epidemics: Stochastic Modeling and Data Analysis.* Special issue of *Math. Biosci.*, **107**, pp. 149 – 562, 1991.

D. Mollison (ed.): *Epidemic Models: Their Structure and Relation to Data.* Cambridge University Press, Cambridge, 1995.

I. Nåsell: *Hybrid Models of Tropical Infections.* Springer-Verlag, Berlin, 1985.

J. Radcliffe & L. Rass: *Spatial Deterministic Epidemics.* Mathematical Surveys and Monographs, American Mathematical Society, Providence, 2003.

R. Ross: *The Prevention of Malaria*, 2nd edition. Churchill, London, 1911.

L. Sattenspiel & A. Lloyd: *The Geographic Spread of Infectious Diseases: Models and Applications.* Princeton University Press, 2009.

M.E. Scott & G. Smith: *Parasitic and Infectious Diseases: Epidemiology and Ecology.* Academic Press, San Diego, 1994.

N. Shigesada & K. Kawasaki: *Biological Invasions: Theory and Practice.* Oxford University Press, Oxford, 1997.

J.E. Van der Plank: *Plant Diseases: Epidemics and Control.* Academic Press, New York, 1963.

E. Vynnycky & R. White: *An Introduction to Infectious Disease Modelling.* Oxford University Press, 2010.

POPULATION DYNAMICS

H. Caswell: *Matrix Population Models.* Sinauer, Massachusetts, 1989 (2nd edition, 2000).

J.M. Cushing: *An Introduction to Structured Population Dynamics.* SIAM, Philadelphia, 1998.

L. Edelstein-Keshet: *Mathematical Models in Biology.* Birkhäuser (McGraw-Hill), New York, 1988.

M. Gilpin & I. Hanski (eds.): *Metapopulation Dynamics: Empirical and Theoretical Investigations.* Academic Press, London, 1991.

I. Hanski & M. Gilpin (eds.): *Metapopulation Biology: Ecology, Genetics and Evolution.* Academic Press, San Diego, 1997.

F. Hoppensteadt: *Mathematical Theories of Populations: Demographics, Genetics and Epidemics*. SIAM, Philadelphia, 1975.

S.A. Levin, T.G. Hallam & L.J. Gross (eds.): *Applied Mathematical Ecology*. Springer-Verlag, Berlin, 1989.

J.A.J. Metz & O. Diekmann: *The Dynamics of Physiologically Structured Populations*. Springer-Verlag, Berlin, 1986.

J. McGlade (ed.): *Advanced Theoretical Ecology: Principles and Applications*. Blackwell Science, London, 1999.

R.M. Nisbet & W.S.C. Gurney: *Modelling Fluctuating Populations*. Wiley, New York, 1982.

N. Shigesada & K. Kawasaki: *Biological Invasions: Theory and Practice*. Oxford University Press, Oxford, 1997.

H.R. Thieme: *Mathematics in Population Biology*. Princeton University Press, 2003.

S. Tuljapurkar & H. Caswell (eds.): *Structured Population Models in Marine, Freshwater and Terrestrial Systems*. Chapman & Hall, London, 1997.

NON-NEGATIVE MATRICES AND OPERATORS

A. Berman & R.J. Plemmons: *Nonnegative Matrices in the Mathematical Sciences*. Academic Press, New York, 1979.

M.A. Krasnosel'skij, Je.A. Lifshits & A.V. Sobolev: *Positive Linear Systems: The Method of Positive Operators*. Heldermann Verlag, Berlin, 1989.

H. Minc: *Nonnegative Matrices*. Wiley, New York, 1988.

E. Seneta: *Non-negative Matrices*. George Allen & Unwin, London, 1973.

DYNAMICAL SYSTEMS AND BIFURCATIONS

V.I. Arnold: *Ordinary Differential Equations*. 2nd edition. Springer-Verlag, New York, 2006.

J.K. Hale: *Ordinary Differential Equations*. Wiley, New York, 1969.

J.K. Hale & H. Koçak: *Dynamics and Bifurcations*. Springer-Verlag, New York, 1991.

M. Hirsch & S. Smale: *Differential Equations, Dynamical Systems and Linear Algebra*. Academic Press, New York, 1974.

Y.A. Kuznetsov: *Elements of Applied Bifurcation Theory*. 2nd edition. Springer-Verlag, New York, 1998.

ANALYSIS

R. Courant & D. Hilbert: *Methods of Mathematical Physics*, Vol. II. Interscience, New York, 1962.

J. Dieudonné: *Foundations of Modern Analysis.* Academic Press, New York, 1969.

J.A. Jacquez: *Compartmental Analysis in Biology and Medicine.* 3rd edition. BioMedware, Ann Arbor, 1996.

A.N. Kolmogorov & S.V. Fomin: *Introductory Real Analysis.* Prentice-Hall, Englewood Cliffs, 1970/Dover, New York, 1975.

W. Rudin: *Real and Complex Analysis.* 2nd edition. McGraw-Hill, New York, 1974.

PROBABILITY AND STOCHASTIC PROCESSES

K.L. Chung: *Elementary Probability Theory with Stochastic Processes.* Springer-Verlag, Berlin, 1979.

S.N. Ethier & T.G. Kurtz: *Markov Processes: Characterization and Convergence.* Wiley, New York, 1986.

N.S. Goel & N. Richter-Dyn: *Stochastic Models in Biology.* Academic Press, New York, 1974.

G.R. Grimmett & D.R. Stirzaker: *Probability and Random Processes.* 3rd edition. Oxford University Press, 2001.

A. Gut: *An Intermediate Course in Probability.* 2nd edition. Springer-Verlag, New York, 2009.

P. Haccou, P. Jagers & V.A. Vatutin: *Branching Processes - Variation, Growth, and Extinction of Populations.* Cambridge University Press, 2005.

T.E. Harris: *The Theory of Branching Processes.* Springer-Verlag, Berlin, 1963.

P. Jagers: *Branching Processes with Biological Applications.* Wiley, London, 1975.

I. Karatzas & S.E. Shreve: *Brownian Motion and Stochastic Calculus.* Springer-Verlag, Berlin, 1991.

T.G. Kurtz: *Approximation of Population Processes.* SIAM, Philadelphia, 1981.

T.M. Liggett: *Continuous Time Markov Processes: An Introduction.* American Mathematical Society, Providence, 2010.

D. Ludwig: *Stochastic Population Theories.* Springer-Verlag, Berlin, 1974.

C.J. Mode: *Multitype Branching Processes, Theory and Applications.* Elsevier, New York, 1971.

E. Renshaw: *Stochastic Population Processes: Analysis, Approximations, Simulations.* Oxford University Press, 2011.

S.M. Ross: *Introduction to Probability Models.* 10th edition. Academic Press, 2009.

S.M. Ross: *A First Course in Probability.* 8th edition. Pearson publishing, 2010.

H.M. Taylor & S. Karlin: *An Introduction to Stochastic Modeling.* Academic Press, Orlando, 1984.

STATISTICS

P.K. Andersen, O. Borgan, R.D. Gill & N. Keiding: *Statistical Models Based on Counting Processes.* Springer, New York, 1993.

N. Balakrishnan & V.B. Nevzorov: *A Primer on Statistical Distributions.* Wiley, Hoboken, 2003.

G. Casella & R. L. Berger: *Statistical Inference.* Duxburry, 2002.

W.R. Gilks, S. Richardson & D.J. Spiegelhalter (eds): *Markov chain Monte Carlo in practice.* Chapman and Hall, London, 1996.

J. Lindsey: *Statistical Analysis of Stochastic Processes in Time.* Cambridge University Press, 2004.

Y. Pawitan *In All Likelihood: Statistical Modeling and Inference Using Likelihood.* Oxford University Press, 2001.

J. Rice: *Mathematical Statistics and Data Analysis.* 2nd edition. Thomson Learning, 2006.

RANDOM NETWORKS/GRAPHS

B. Bollobás: *Random graphs.* 2nd edition. Cambridge University Press, 2001.

R. Durrett: *Random Graph Dynamics.* Cambridge University Press, 2006.

Index

www.ingramcontent.com/pod-product-compliance
Ingram Content Group UK Ltd.
Pitfield, Milton Keynes, MK11 3LW, UK
UKHW010827161224
452264UK00001B/19